Understanding Regression Analysis

Understanding Regression Analysis

A Conditional Distribution Approach

Peter H. Westfall

Andrea L. Arias

CRC Press is an imprint of the
Taylor & Francis Group, an **informa** business
A CHAPMAN & HALL BOOK

CRC Press
Taylor & Francis Group
6000 Broken Sound Parkway NW, Suite 300
Boca Raton, FL 33487

and by

CRC Press
2 Park Square, Milton Park, Abingdon, Oxon OX14 4RN

Printed on acid-free paper

International Standard Book Number-13: 978-0-367-45852-2 (Hardback)
978-0-367-49351-6 (Paperback)

Library of Congress Cataloging-in-Publication Data

Names: Westfall, Peter H., 1957- author. | Arias, Andrea L., author.
Title: Understanding regression analysis : a conditional distribution approach / Peter H. Westfall, Andrea L. Arias.
Description: Boca Raton : CRC Press, [2020] | Includes bibliographical references and index. | Summary: "This book unifies diverse regression applications including the classical model, ANOVA models, generalized models including Poisson, Negative binomial, logistic, and survival, neural networks and decision trees under a common umbrella; namely, the conditional distribution model. It explains why the conditional distribution model is the correct model, also explains why the assumptions of the classical regression model are wrong. This one takes a realistic approach from the outset that all models are just approximations. The emphasis is to model Nature's processes realistically, rather than to assume that Nature works in particular, constrained ways"-- Provided by publisher.
Identifiers: LCCN 2020003321 (print) | LCCN 2020003322 (ebook) | ISBN 9780367458522 (hardback) | ISBN 9781003025764 (ebook)
Subjects: LCSH: Regression analysis.
Classification: LCC QA278.2 .W475 2020 (print) | LCC QA278.2 (ebook) | DDC 519.5/36--dc23
LC record available at https://lccn.loc.gov/2020003321
LC ebook record available at https://lccn.loc.gov/2020003322

Visit the Taylor & Francis Web site at
http://www.taylorandfrancis.com

and the CRC Press Web site at
http://www.crcpress.com

Contents

Preface xiii
Authors xvii

1. Introduction to Regression Models 1
1.1 The Regression Model in Terms of Conditional Distributions 2
1.1.1 Randomness of the Measured Area of a Circle as Related to Its Measured Radius 3
1.1.2 Randomness of a Person's Financial Assets as Related to Their Age 4
1.2 Models and Generalization 6
1.3 The "Population" Terminology and Reasons Not to Use It 8
1.4 Data Used in Regression Analysis 10
1.5 Random-*X* Versus Fixed-*X* 12
1.5.1 The Trashcan Experiment: Random-*X* Versus Fixed-*X* 13
1.6 Some Preliminary Regression Data Analyses Using `R` 14
1.6.1 The Production Cost Data and Analysis 14
1.6.2 The Personal Assets Data and Analysis 15
1.6.3 The Grade Point Average Data and Analysis 16
1.7 The Assumptions of the Classical Regression Model 17
1.7.1 Randomness 18
1.7.2 Correct Functional Specification 19
1.7.3 Constant Variance (Homoscedasticity) 20
1.7.4 Uncorrelated Errors (or Conditional Independence) 21
1.7.5 Normality 21
1.7.6 Putting Them All Together: The Classical Regression Model 21
1.8 Understanding the Regression Model by Using Simulation 22
1.8.1 Random-*X* Simulation 27
1.9 The Linear Regression Function, and Why It Is Wrong 28
1.10 LOESS: An Estimate of the True (Curved) Mean Function 32
Appendix A: Conditional Distributions of the Bivariate Normal Distribution, and Origin of the Term "Regression" 36
Reference 40
Exercises 40

2. Estimating Regression Model Parameters 43
2.1 Estimating Regression Models via Maximum Likelihood 43
2.2 Maximum Likelihood in the Classical (Normally Distributed) Regression Model, Which Gives You Ordinary Least Squares 45
2.2.1 Simulation to Illustrate the Concept of "Least Squares Estimates" 47
2.2.2 Summarizing 48
2.3 Maximum Likelihood with Non-normal Distributions Gives Non-OLS Estimates 49
2.4 The Gauss-Markov Model and Theorem 52
Exercises 55

3. The Classical Model and Its Consequences ... 57
3.1 Unbiasedness ... 59
3.2 Unbiasedness of OLS Estimates Assuming the Classical Model: A Simulation Study ... 61
3.3 Biasedness of OLS Estimates When the Classical Model Is Wrong ... 63
3.4 Estimation and Practical Use of σ^2 ... 65
3.5 Standard Errors ... 70
3.6 Exact Inferences: Confidence Intervals for the β's ... 73
3.6.1 Understanding "Exactness" and "Non-exactness" via Simulation ... 75
3.6.2 Practical Interpretation of the Confidence Interval for β_1 ... 76
3.7 Exact Inferences: Confidence Intervals for $E(Y \mid X = x)$ ... 78
3.8 Exact Inferences: Prediction Intervals for $Y \mid X = x$ ... 80
3.9 Hypothesis Testing and p-Values: Is the Observed Effect of X on Y Explainable by Chance Alone? ... 82
3.9.1 Is the Last Digit of a Person's Identification Number Related to Their Height? ... 83
3.9.2 Simulation Study to Understand the Null Distribution of the T Statistic ... 86
3.9.3 The p-Value ... 88
Reference ... 91
Exercises ... 91

4. Evaluating Assumptions ... 93
4.1 Graphical/Descriptive Methods Versus Testing Methods for Checking Assumptions ... 95
4.2 Which Assumptions Should You Evaluate First? ... 96
4.3 Evaluating the Linearity Assumption Using Graphical Methods ... 97
4.3.1 Production Cost Data (x_i, y_i) Plot with LOESS Smooth and $(\hat{y}_i, e_i)$ Plot with LOESS Smooth ... 98
4.3.2 Car Sales Data (x_i, y_i) Plot with LOESS Smooth and $(\hat{y}_i, e_i)$ Plot with LOESS Smooth ... 99
4.4 Evaluating the Linearity Assumption Using Hypothesis Testing Methods ... 100
4.4.1 Testing for Curvature with the Production Cost Data ... 100
4.4.2 Testing for Curvature with the Car Sales Data ... 102
4.5 Practical Versus Statistical Significance ... 103
4.5.1 Simulation Study to Demonstrate Practical vs. Statistical Significance ... 103
4.6 Evaluating the Constant Variance (Homoscedasticity) Assumption Using Graphical Methods ... 104
4.6.1 Production Cost Data $(\hat{y}_i, e_i)$ and $(\hat{y}_i, |e_i|)$ Plots, with LOESS Smooths ... 105
4.6.2 Personal Assets Data $(\hat{y}_i, e_i)$ and $(\hat{y}_i, |e_i|)$ Plots, with LOESS Smooths ... 106
4.7 Evaluating the Constant Variance Assumption Using Hypothesis Testing Methods ... 107
4.7.1 Testing for Heteroscedasticity Using the Production Cost Data ... 108
4.7.2 Testing for Heteroscedasticity Using the Personal Assets Data ... 108
4.8 Evaluating the Uncorrelated Errors Assumption Using Graphical Methods ... 109
4.8.1 The Car Sales Data (t, e_t) and (e_{t-1}, e_t) Plots ... 110

4.9 Evaluating the Uncorrelated Errors Assumption Using Testing Methods 111
4.10 Evaluating the Normality Assumption Using Graphical Methods.................. 111
4.10.1 Evaluating the Normality Assumption Using the Car Sales Data 113
4.11 Evaluating the Normality Assumption Using Testing Methods....................... 114
4.12 A Caution about Using Residuals to Evaluate Normality 114
References 115
Exercises 115

5. Transformations 117
5.1 Transformation of the X Data Only 118
5.1.1 What Should I Use with My Data, $\ln(x)$, or Inverse of X, or No Transform at All? 120
5.1.2 Comparing Transformations of X with the Car Sales Data 121
5.2 Logarithmic Transformation of the Y data 125
5.2.1 Log Transforming Income 127
5.2.2 What Should I Use with My Data, $\ln(Y)$ or No Y Transform at All? ... 129
5.2.3 Comparing Log Likelihoods with the Charity Data Set 132
5.3 The $\ln(Y)$ Transformation and Its Use for Heteroscedastic Processes 132
5.4 An Example Where the Inverse Transformation $1/Y$ Is Needed 134
5.5 The Box-Cox Transformation 136
5.6 Transforming Both Y and X 139
5.7 Elasticity 140
Exercises 143

6. The Multiple Regression Model 145
6.1 Prediction 146
6.1.1 Predicting Loan Repayment 146
6.1.2 Simulation Demonstrating the Law of Total Expectation 147
6.1.3 Simulation Demonstrating the Law of Total Variance 151
6.2 Why Prediction Is Different from Causation? 153
6.2.1 Does Eating Ice Cream Cause You to Drown? 154
6.3 The Classical Multiple Regression Model and Interpretation of Its Parameters 156
Appendix A: Use of Instrumental Variables to Estimate Causal Effect 161
A.1 Foundations 161
A.2 The Causal Model 165
A.3 The Instrumental Variable Method 165
Reference 166
Exercises 167

7. Multiple Regression from the Matrix Point of View 169
7.1 The Least Squares Estimates in Matrix Form 170
7.2 The Regression Model in Matrix Form 172
7.3 Unbiasedness of the OLS Estimator $\hat{\beta}$ Under the Gauss-Markov Model 174
7.3.1 Unbiasedness of the OLS Estimates $\hat{\beta}$ Conditional on the X Data 174
7.3.2 Unbiasedness of the OLS Estimates $\hat{\beta}$, *not* Conditional on the Values of the X Data 175

7.4 Measurement Error ... 176
7.5 Standard Errors of OLS Estimates ... 178
7.6 Application of the Theory: The Graduate Student GPA Data Analysis, Revisited ... 181
Exercises ... 183

8. *R*-Squared, Adjusted *R*-Squared, the *F* Test, and Multicollinearity ... 185
8.1 The *R*-Squared Statistic ... 185
8.2 The Adjusted *R*-Squared Statistic ... 188
8.3 The *F* Test ... 189
8.3.1 Simulation Study to Understand the *F* Statistic ... 190
8.4 Multicollinearity ... 192
8.4.1 The Effects of Multicollinearity on the *T* Statistics ... 195
8.4.2 Possible Actions to Take with Multicollinear *X* Variables ... 198
Exercises ... 199

9. Polynomial Models and Interaction (Moderator) Analysis ... 201
9.1 The Quadratic Model in One *X* Variable ... 202
9.2 The Quadratic Model in Two or More *X* Variables ... 205
9.3 Interaction (or Moderator) Analysis ... 206
9.3.1 Path Diagrams ... 207
9.3.2 Parameter Interpretation in Interaction Models ... 208
9.3.3 Effect of Misanthropy on Support for Animal Rights: The Moderating Effect of Idealism ... 209
9.4 The Variable Inclusion Principle ... 213
9.4.1 Why You Should Always Include the Intercept Term ... 214
9.4.2 Why You Should Include the Linear Term in a Quadratic Model ... 216
9.4.3 Why You Should Include the Linear Terms in an Interaction Model ... 217
References ... 217
Exercises ... 218

10. ANOVA, ANCOVA, and Other Applications of Indicator Variables ... 219
10.1 Using a Single Indicator Variable to Represent a Single Nominal Variable Having Two Levels (Two-Sample Comparison) ... 219
10.1.1 Does It Matter Whether the Indicator Variable Is Coded as 1,0 vs. 0,1? ... 222
10.2 Using Multiple Indicator Variables to Represent a Single Nominal Variable Having Three or More Levels (ANOVA) ... 223
10.3 Using Indicator Variables and "Ordinary" *X* Variables in the Same Model (ANCOVA) ... 229
10.4 Interaction Between Indicator Variables and "Ordinary" *X* Variables (ANCOVA with Interaction) ... 232
10.4.1 Does Location Affect House Price, Controlling for House Size? ... 236
10.5 Full Model versus Restricted Model *F* Tests ... 240
10.5.1 Computing the *F* Statistic to Compare Full and Restricted Models ... 242
10.5.2 Simulation to Understand the Null (Chance-Only) Model ... 243

10.6 Two Nominal Variables (Two-Way ANOVA) 246
10.6.1 Nested Model Sequence, Version 1 255
10.6.2 Nested Model Sequence, Version 2 255
10.7 Additional Applications of Indicator Variables 257
10.7.1 Piecewise Linear Regression; Regime Analysis 257
10.7.2 Relationship Between Commodity Price and Commodity Stockpile 259
10.7.3 Using Indicator Variables to Represent an Ordinal *X* Variable 261
10.7.4 Repeated Measures, Fixed Effects, and Unobserved Confounding Variables 266
10.7.5 The Independence Assumption and Repeated Measurements 268
References 270
Exercises 270

11. Variable Selection 273
11.1 The Effect of Estimating Parameters on Prediction Accuracy 275
11.1.1 Predicting Hans' Graduate GPA: Theory Versus Practice 275
11.2 The Bias-Variance Tradeoff 278
11.2.1 Simulation Study to Demonstrate the Bias-Variance Tradeoff 279
11.3 Variable Selection Based on Penalized Fit 284
11.3.1 Identifying Models with Low BIC for Predicting Crime Rate 285
11.4 Variable Selection Based on Out-of-Sample Prediction Accuracy 287
11.4.1 Example Showing Decrease in SSE but Increase in SSPE 289
Exercises 294

12. Heteroscedasticity and Non-independence 295
12.1 Maximum Likelihood and Weighted Least Squares 296
12.2 The Gauss-Markov Theorem, Revisited 303
12.2.1 Simulation Study to Illustrate That WLS Is More Efficient than OLS 303
12.3 More General Standard Deviation Functions 305
12.4 The Effect of Estimating Parameters in Variance Functions 310
12.5 The Blunt Axe Approach: Heteroscedasticity-Consistent Standard Errors 311
12.5.1 Simulation to Investigate Whether e_i^2 Is a Reasonable Estimate of σ_i^2 313
12.6 Generalized Least Squares for Non-independent Observations 317
12.6.1 Generalized Least Squares Estimates and Standard Errors for the Charitable Contributions Study 322
Appendix A: Likelihood Ratio Tests 324
Appendix B: Wald Standard Errors 326
Reference 327
Exercises 327

13. Models for Binary, Nominal, and Ordinal Response Variables ... 329
13.1 The Logistic Regression Model for Binary *Y* ... 331
13.1.1 Estimating the Probability of Successfully Throwing a Piece of Wadded-up Paper into a Trash Can ... 334
13.2 The Multinomial Regression for Nominal *Y* ... 343
13.2.1 Who Does the Laundry? ... 346
13.3 Models for Ordinal *Y* ... 351
13.3.1 A Note on Comparing Classical, Normally Distributed Models with Ordinal Regression Models ... 358
Exercises ... 358

14. Models for Poisson and Negative Binomial Response ... 361
14.1 The Poisson Regression Model ... 361
14.1.1 Predicting Number of Financial Planners Used by a Person as a Function of Gender and Age ... 365
14.2 Negative Binomial Regression ... 370
14.2.1 Predicting Number of Financial Planners Used by a Person as a Function of Gender and Age, Using Negative Binomial Regression ... 372
14.2.2 A Note on Replicability and Preregistration ... 376
Exercises ... 377

15. Censored Data Models ... 379
15.1 Regression Analysis with Censored Data ... 384
15.1.1 Survival of Marriage as a Function of Education ... 385
15.2 The Proportional Hazards Regression Model ... 392
15.3 The Tobit Model ... 395
15.3.1 Predicting Number of Days Lost to Back Injury ... 398
15.4 Interval Censored Data ... 402
Reference ... 403
Exercises ... 403

16. Outliers: Identification, Problems, and Remedies (Good and Bad) ... 405
16.1 What Is the Problem with Outliers? ... 408
16.2 Why Outliers Are Important ... 409
16.3 Identifying Outliers in Regression Data: Overview ... 411
16.4 Using the "Leverage" Statistic to Identify Outliers in *X* Space ... 413
16.5 Using Standardized Residuals to Identify Outliers in $Y|X$ Space ... 417
16.6 Cook's Distance ... 419
16.6.1 Outlier Analysis Using the Data of the Crime Rate Prediction Model ... 421
16.7 Strategies for Dealing with Outliers ... 425
16.7.1 Analysis of Data with an Extreme Outlier by Using Heavy-Tailed Distributions ... 427
16.8 Quantile Regression ... 428
16.8.1 Simulation Study to Validate the Quantile Regression Estimates ... 433
16.8.2 Quantile Regression Models for Personal Assets ... 435
16.9 Outlier Deletion *en masse* and Winsorization ... 436

Appendix A: R Code to Perform the Simulation Study Given Table 16.3 ... 446
References ... 447
Exercises ... 448

17. Neural Network Regression ... 451
17.1 Universal Approximators ... 452
17.2 Neural Network and Polynomial Approximations of a Known Noiseless Function ... 453
17.3 Neural Network and Polynomial Approximations in a Real Example: Predicting Charitable Contributions ... 461
Exercises ... 469

18. Regression Trees ... 471
18.1 Tree Regression with One *X* Variable and One Split ... 474
18.2 Choosing the Split Value ... 478
18.3 Multiple Splits on a Single *X* Variable ... 479
18.4 Tree Regression with Multiple *X* Variables ... 481
Exercises ... 485

19. Bookend ... 487

Index ... 491

Preface

This book distills years of experience teaching regression analysis to research-oriented students who are not in the field of statistics. For this group, the main barrier to learning regression, and much of statistics, for that matter, is the difficulty to model data probabilistically. Once that barrier is cleared, all the rest becomes much more easily understood.

The probabilistic point of view in regression is specifically embodied in the model for the variability of the Y data given particular, fixed values of the X data. This variability is modeled using conditional distributions; hence, the subtitle: "A conditional distribution approach." The entire subject of regression is couched in terms of conditional distributions; this point of view unifies diverse methods such as classical regression, ANOVA, Poisson regression, logistic regression, heteroscedastic regression, quantile regression, models for nominal Y data, causal models, neural network regression, and tree regression. All are conveniently viewed in terms of models for the conditional distributions of Y given particular values of X.

Unlike regression models (e.g., linear) presented in other books, conditional distributions are the *correct* models for regression data. They tell you that, for a given value of your X variable, there is a distribution of potentially observable Y values. If you happen to know this distribution, then you know everything that you can possibly know about your response variable, Y, as it relates to a given value of your predictor, X. The model explains 100% of the potentially observable Y data, unlike typical regression approaches based on the R^2 statistic, which explain only a fraction of the Y data, and are also *incorrect* in that assumptions are nearly always violated.

Consistent with the conditional distribution framework, which describes models for the real processes that produced the data, regression models are likewise presented as producers of data, rather entities that are produced by data. This simple probabilistic view can be difficult at first for students who have come to understand statistical models (regression models in particular) as entities that are produced by data. Once this hurdle is cleared, however, all of the notoriously "difficult" concepts of statistics become understandable. For one example, to understand standard errors in regression, one must first understand the idea of "potentially observable data sets," which is easy to understand when the regression model is viewed as a producer of potentially observable data.

A related distinguishing feature of this book is that the "population" terminology is almost entirely avoided in favor of the "data-generating process" terminology. In all but the most pristine finite-population sampling studies, the "population" framework is simply wrong. Knowledgeable sources, therefore, place the term "population" within quotation marks ("") to underscore that this terminology is incorrect. But in regression applications, the "population" interpretation of model parameters is even less coherent, hence it is essential to use the "process" terminology. Detailed explanations of the fallacy of the "population" terminology are given in Chapter 1 of this book.

Yet another key difference between this book and other regression books is that the X data are assumed to be random throughout this book. Many regression books assume that the X data are fixed by design, but the fixed-X case is actually rare in real-world applications of regression. Hence, regression books that employ only the fixed-X approach can apply only to a small subset of the actual uses of regression. In many cases, it does not matter that you assume the X data are fixed, even when the X data are random—that is a main point of the conditional distribution approach. On the other hand, it is a bad idea to teach improper probabilistic views of data. Because difficulty with probabilistic modeling is perhaps the single largest barrier to learning statistics, it is essential to learn correct probabilistic views of data from the start. The special case where the X data are fixed is covered under the conditional distribution approach of this book because when the X data are fixed, they are just a special case of random X data, but with degenerate probability distributions.

Further, to understand more complex topics, one needs the random-X approach. For one example, the effect of measurement error is most naturally described in a random-X framework. For another example, in time-series data, it is common to regress the Y variable on its previous value (or its lag). If Y is random, then so must all of its lags be random. Also, the classical R^2 statistic is best understood in the random-X framework using the "Law of Total Variance." The "bias-variance tradeoff," which is essential to understand the logic of the variable and model selection methods that are foundational to data science, also uses the Law of Total Variance. This book takes the random-X approach from the very beginning, allowing a seamless transition to the more complex concepts that appear later, as well as allowing greater realism.

Another way this book differs from others is that it fully incorporates the statements and the suggestions of the American Statistical Association (in 2016, and again in 2019) concerning p-values. This book mostly eschews p-values (and tests of hypotheses in general) for assumption-checking, model specification testing, and data analysis. By learning the conditional distribution framework, along with the fact that a model is a producer of data, readers instead will learn what the assumptions mean, what the consequences of their violations are, and how to assess their degree of violation. In this book, assumptions are evaluated using subject matter considerations, simulation, graphics, and model comparisons, rather than p-value-based methods. On the other hand, p-values and hypothesis testing methods are explained thoroughly and correctly in this book as measures of what is *explainable by chance alone* (as opposed to what is *explained by chance alone*), so that readers can use this book to interpret them with confidence. In particular, the rather extreme limitations of p-value-based methods are emphasized repeatedly throughout this book.

Also, unlike other books, this one states from the beginning that in most cases, all assumptions, including linearity, are simply wrong. Logical arguments for the fact that all assumptions are wrong are given throughout the book, starting with Chapter 1. Thus, rather than focus on proving assumptions to be right (which is as futile as proving that $1 + 1 = 3$), this book instead discusses how to assess deviations from the assumptions to reality, and the book also discusses the consequences of varying degrees of such deviations. Hence, the emphasis is to model Nature's processes realistically, rather than to assume (incorrectly) that Nature works in particular constrained ways. Readers will learn how to properly interpret regression analyses, in light of the fact that the data-generating process they assume when they perform regression analysis is only an approximation to the real data-generating process.

The conditional distribution point of view naturally leads to likelihood-based procedures that are discussed in detail throughout the book. Likelihood-based procedures are the default for logistic, Poisson, multinomial, and other regression methods when the conditional distributions are non-normal; they also provide optimality of ordinary least squares, weighted least squares, generalized least squares, and least absolute deviation estimates. The book, therefore, has a strong likelihood orientation, although non-likelihood-based methods are also discussed.

While Bayesian methods are mentioned only peripherally, the conditional distribution point of view naturally leads to the likelihood function, which is the foundation of Bayesian inference. Bayesian inference with uniform priors is essentially identical to likelihood-based analysis. In addition, understanding the conditional distribution point of view makes understanding Bayesian methods easier, where all distributions are conditional distributions. Thus, this book is an ideal precursor to more advanced studies that use Bayesian methods.

A final distinguishing feature of this book from others is that all probabilistic concepts, such as unbiasedness, 95% coverage, optimal prediction, efficient estimation, etc., are illustrated concretely by using simulation analyses. Rather than expecting readers to take complicated and abstract probabilistic results "on faith," this book instead shows readers how such results can be understood easily in terms of actual numbers.

We assume that readers have a nodding acquaintance with basic calculus, discrete and continuous probability distributions, conditional distributions, likelihood-based analysis, and basic inferential statistics, including tests of hypotheses and confidence intervals. All examples use the `R` software. There are ample resources to get started with `R` on the internet, so this should not pose a barrier to understanding the material. For the `R` purists, we apologize for not using the "<-" assignment operator very often—in the interest of simplicity we have generally opted for the "=" operator instead. All `R` code is available at https://github.com/andrea2719/URA-Rcode. Modifications might be needed for `R` versions 4.0 and higher.

We thank students of ISQS 5349 (regression), as well as various faculty at Texas Tech University, and colleagues worldwide, for "test driving" this material and providing numerous constructive suggestions.

Authors

Peter H. Westfall has a PhD in Statistics from the University of California at Davis, as well as many years of teaching, research, and consulting experience, in a variety of statistics-related disciplines. He has published more than 100 papers on statistical theory, methods, and applications; and he has written several books, spanning academic, practitioner, and textbook genres. He is former editor of *The American Statistician* and a fellow of the American Statistical Association.

Andrea L. Arias is a Senior Operations Research Specialist at BNSF Railway. She has a PhD in Industrial Engineering with a minor in Business Statistics from Texas Tech University, and a Doctoral Degree in Industrial Engineering from Pontificia Universidad Católica de Valparaiso, Chile. Her main areas of expertise include Mathematical Programming, Network Optimization, Statistics and Simulation. She is an active member of the Institute for Operations Research and the Management Sciences (INFORMS).

1

Introduction to Regression Models

Regression models are used to relate a variable, Y, to a single variable X, or to multiple variables, $X_1, X_2, \ldots, X_k$.

Here are some examples of questions that these models help you answer:

- How does a person's choice of toothpaste (Y) relate to the person's age (X_1) and income (X_2)?
- How does a person's cancer remission status (Y) relate to their chemotherapy regimen (X)?
- How does the number of potholes in a road (Y) relate to the material used in surfacing (X_1) and time since installation (X_2)?
- How does a person's ability to repay a loan (Y) relate to the person's income (X_1), assets (X_2), and debt (X_3)?
- How does a person's intent to purchase a technology product (Y) relate to their perceived usefulness (X_1) and perceived ease of use of the product (X_2)?
- How does today's return on the S&P 500 stock index (Y) relate to yesterday's return (X)?
- How does a company's profitability (Y) relate to its investment in quality management (X)?

Understanding such relationships can help you to *predict* what an unknown Y will be for a given fixed value of X, it can help you to make decisions as to what course of action you should choose, and it can help you to understand the subject that you are studying in a scientific way.

Regression models can help you to *forecast* the future as well. Forecasting is a special case of prediction: Forecasting means prediction of the *future*, while prediction includes any type of "what-if" analysis, not only about what might happen in the future, but also about what *might have happened* in the past under different circumstances.

In some subjects, you learn to make predictions using equations such as

$$Y = f(X),$$

where the function f might be a linear, quadratic, exponential, or logarithmic function; or it might not have any "named" function form at all. In all cases, though, this is a *deterministic* relationship: Given a particular value, x, of the variable X, the value of Y is completely determined by $Y = f(x)$.

Notice that there is a distinction between upper-case X and lower-case x. The convention followed in this book regarding lower-case and upper-case Y and X is standard: Upper-case refers to the variable *in general*, which can be many different possible values, while lower-case refers to a *specific value* of the variable. For example, X = Age can be many different values in general, whereas $X = x$ identifies the subset of people having age x, e.g., the subset of people who are 25 years old.

In regression analysis, the variable "Y" goes by many names, including *response variable, target variable, criterion variable, dependent variable, endogenous variable,* and others. The X variable(s) are called, variously, *predictor variable(s), descriptor variable(s), feature variable(s), independent variable(s), exogenous variable(s),* and others. In this book, we typically use the *response/predictor* terms.

For example, let Y = the area of a mathematically perfect circle whose radius is X. The following relationship is exactly true and deterministic:

$$Y = 3.14159265X^2$$

(Note: The trigonometric constant π is not *exactly* equal to 3.14159265 because the decimals continue infinitely. However, it is close enough. In general, throughout this book, rounded versions of numbers will be (technically, incorrectly) called equal to the actual number when the round-off error is very small.)

Thus, if you know that the radius is $X = 10$ meters (meaning that you are considering the subset of possible circles with radius 10 meters), then the model states that the area is 314.159265 meters^2.

However, virtually all real-world phenomena are non-deterministic. Mathematically perfect circles do not exist in the real world, and any measurements you take in the real word are subject to measurement errors. If you saw a big circle, say drawn into a landscape, and measured its radius to be x, its area would not be precisely $3.14159265x^2$, because of imperfections in the border of the circle, and because of imperfections in your measuring instrument. Instead, the area would differ from $3.14159265x^2$ by an unpredictable amount. Hence, the deterministic model is simply wrong in the real world, despite what you learned in your grade school mathematics course about how the area of a circle is precisely equal to $\pi \times r^2$.

On the other hand, the example of the circle is one where the deterministic model provides an excellent approximation, so engineers and physicists use it all the time with little problem. But in many scientific disciplines, including especially the social sciences, relationships are nowhere close to deterministic. For example, suppose we told you that Jane is 27 years old ($X = 27$). Do you now know how many financial assets she has accrued (her Y)? There is *some relationship*—presumably, 35-year-old people tend to have more assets than 27-year-old people. But, unlike the example of the circle's area being approximately a deterministic function of its radius, a person's assets cannot be determined to any reasonable degree of approximation from the person's age.

So, deterministic models are inadequate, and not even approximately correct, in the vast majority of areas of scientific inquiry. Probabilistic models are ideal in these cases because they explain 100% of the outcomes, both deterministic and non-deterministic. The regression model is an example of a probabilistic model, and it also explains 100% of the outcomes.

1.1 The Regression Model in Terms of Conditional Distributions

The regression model assumes that data values Y come from different distributions, where there is a different distribution for every specific value x of the general variable X. This model is written symbolically as follows:

The regression model

$$Y \mid X = x \sim p(y \mid x)$$

The expression $p(y \mid X = x)$ is the distribution of potentially observable Y values when $X = x$, and is called the *conditional distribution of* Y*, given that* $X = x$. Sometimes these conditional distributions will be written as $p(y \mid x)$ for brevity.

The following examples clarify this abstraction.

1.1.1 Randomness of the Measured Area of a Circle as Related to Its Measured Radius

A circle in nature has its radius (X) measured. Suppose the measurement is $X = 10$ meters. Still, there are many potentially observable measurements of its area, Y, due to imperfections in the circle and imperfections in the measuring device. The regression model states that Y is a random observation from the conditional distribution $p(y \mid X = 10)$.

This model is reasonable, because it perfectly matches the reality that there are many potentially observable measurements of the area Y, even when the radius X is measured to be precisely 10 meters. Note that this model does not say anything about the mean of the distribution $p(y \mid x)$: It might be $3.14159265x^2$, but it is more likely not, because of biases in the data-generating process (again, this data-generating process includes imperfections in circles, and also imperfections in the measuring devices).

This model also does not say anything about the nature of the probability distributions $p(y \mid x)$, whether they are discrete, continuous, normal, lognormal, etc., or even whether you have one type of distribution for one x (e.g., normal) and another for a different x (e.g., Poisson). It simply says there is a distribution $p(y \mid x)$ of potential outcomes of Y when $X = x$, and that the number you measure will appear as if produced at random from this distribution, i.e., as if simulated using a random number generator. As such, there is no arguing with the model—the measured data really will look this way (random, variable), hence you may even say that this model is a correct model because it produces data that are random and variable (non-deterministic). Further, because the model is so general, no data can ever contradict, or "reject" it.

Thus, the model $p(y \mid x)$ is *correct*. It is only when you make assumptions about the nature of $p(y \mid x)$, for example, about the specific distributions (e.g. normal), and about how are distributions related to x (e.g., linearly), that you must consider that the model is *wrong* in certain ways.

The model $p(y \mid x)$ does not require that the distribution of Y change for different values of X. If the distributions $p(y \mid x)$ are the same, for all values $X = x$, then, by definition, Y is *independent* of X. In the example above, one may logically assume that the distributions of Y (measured area) will differ greatly for different X (measured radius), and that Y and X are thus strongly *dependent*.

The following `R` code and resulting graph of Figure 1.1 illustrate how the distributions of area (Y) might look for circles whose radius (X) is measured to be 9.0 meters versus circles whose radius is measured to be 10.0 meters. In this example, we assume that $p(y \mid x)$ is a normal distribution with mean πx^2 meters^2 and standard deviation of 1 meter^2.

R code for Figure 1.1

```
area = seq(240, 380,.001)
pdf.9 = dnorm(area, pi*9^2, 1)
pdf.10 = dnorm(area, pi*10^2, 1)
plot(area, pdf.9, type="l", yaxs="i", ylim = c(0, 1.2*max(pdf.9)),
  ylab = "p(y|x)", xlab = "area, y", cex.axis=0.8, cex.lab=0.8)
points(area, pdf.10, type="l", lty=2)
legend("topright", inset=0.05, c("radius=9", "radius=10"),
  lty = c(1,2), cex=0.8)
```

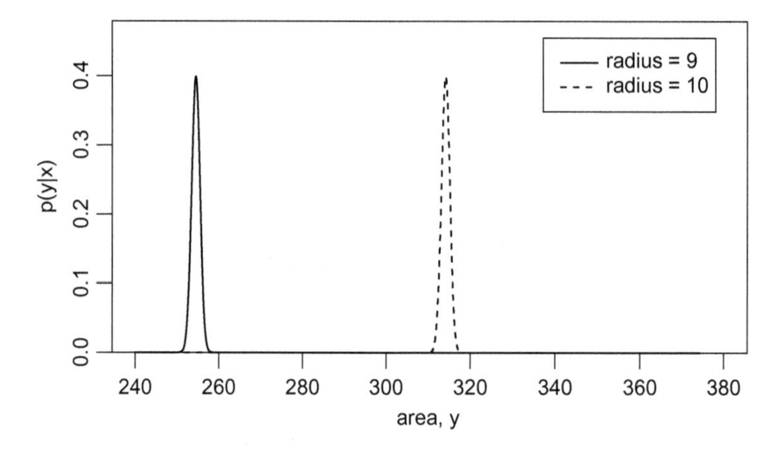

FIGURE 1.1
Possible conditional distributions of areas of circles whose radii are measured to be 9.0 and 10.0 meters. The area is variable because of imperfections in the circles themselves, because of imperfections in the measurement of their radii, and because of imperfections in the measurement of their areas.

1.1.2 Randomness of a Person's Financial Assets as Related to Their Age

Can you predict a person's financial assets from their age? Suppose you know that a person's age is $X = 27$ years, but you know nothing else about that person. The regression model states that the person's assets, Y, comes from a distribution $p(y \mid X = 27)$.

Again, this model also does not say anything about the nature of the distributions $p(y \mid x)$, whether they are discrete, continuous, normal, lognormal, etc. It simply says that, for a particular value of Age (x), there is a distribution of potential assets, and that the assets of a particular person having Age = x can be viewed as a number produced at random from this distribution, as if simulated using a random number generator. As such, there is no arguing with the model—the assets data really do look this way (variable), hence this model is correct. Again, it is only when you make assumptions about the nature of $p(y \mid x)$, for example, what is the specific distribution (e.g., normal), and how are distributions related to x (e.g., linearly), that you must acknowledge that the model is "wrong" in certain ways.

The regression models $p(y \mid x)$ assumed in the following code are "lognormal" probability distributions, chosen this way because assets are non-negative, and because lognormal distributions (among many other distributions) can be used to model non-negative observations. (Normal distributions can also be used to model non-negative observations, provided that the tail area to the left of zero is very small, as in Figure 1.1) But lognormal distributions are used for illustration only. The general regression model $p(y \mid x)$ makes no assumption about the nature of the distributions.

R code for Figure 1.2

```
assets = seq(0, 1200,.1)
pdf.27 = dlnorm(assets, 4.2 , .8)
pdf.35 = dlnorm(assets, 5.7, .8)
plot(assets, pdf.27, type="l", yaxs="i", ylim = c(0, 1.2*max(pdf.27)),
  ylab = "p(y|x)", xlab = "assets, y (thousands)", cex.axis=0.8,
  cex.lab=0.8)
points(assets, pdf.35, type="l", lty=3)
legend("topright", inset=0.05, c("age=27", "age=35"), lty = c(1,2),
  cex = 0.9)
```

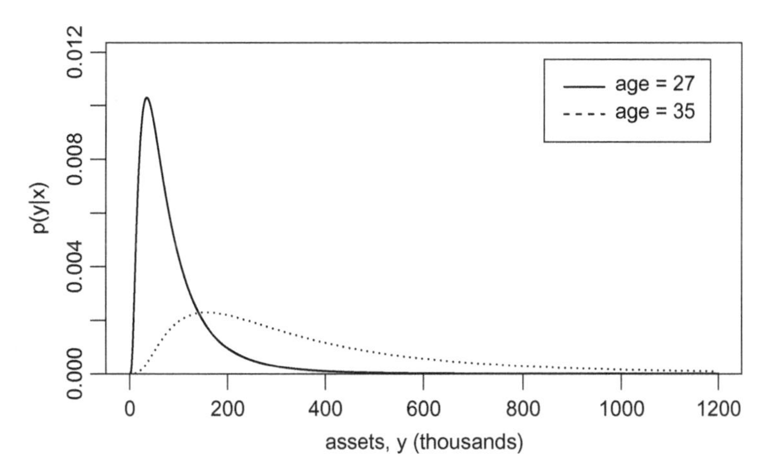

FIGURE 1.2
Possible conditional distributions of assets among people who are 27 years old, and among people who are 35 years old. The area under any density curve must be 1.0, and the variability in possible assets is less for 27-year-olds, which explains why the density curve is higher for 27-year-olds.

You do not have to assume a particular family of the distributions $p(y \mid x)$ such as lognormal, Poisson, normal, etc. However, specific families of distributions $p(y \mid x)$ imply particular types of regression models, as shown in the following box to highlight this crucial point.

When the $p(y\|x)$ are Assumed to be...	Then you Have ...
Poisson	Poisson regression
Negative binomial	Negative binomial regression
Bernoulli	Logistic regression
Normal	Classical regression
Multinomial	Multinomial logistic regression
Beta	Beta regression
Lognormal	Lognormal regression
...	...

This list is not complete. The intent is simply to show that different assumed families of conditional distributions $p(y \mid x)$ give rise to different types of regression models.

1.2 Models and Generalization

The following statement is indisputable:

The data target the processes that produced the data.

The model $p(y \mid x)$ is the model for these processes; therefore, the data specifically target $p(y \mid x)$.

Depending on the context of the study, these data-producing processes may involve biology, psychology, sociology, economics, physics, etc. The processes that produce the data also involve the measurement processes: If the measurement process is faulty, then the data will provide misleading information about the real, natural processes, because, as the note in the box above states, the data target the processes that produced the data. In addition to natural and measurement processes, the process also involves the type of observations sampled, where they are sampled, and when they are sampled. This ensemble of processes that produces the data is called the *data-generating process*, abbreviated DGP.

Consider the (Age, Assets) example introduced in the previous section, for example. Suppose you have such data from a Dallas, Texas-based retirement planning company's clientele, from the year 2003. The processes that produced these data include people's asset accrual habits, socio-economic nature of the clientele, method of measurement (survey or face-to-face interview), extant macroeconomic conditions in the year 2003, and regional effects specific to Dallas, Texas. All of these processes, as well as any others we might have missed, collectively define the *data-generating process* (DGP).

The regression model $Y \mid X = x \sim p(y \mid x)$ is a model for the DGP. Like all models, this model allows *generalization*. Not only does the model explain how the actual data you collected came to be, it also generalizes to an infinity (or near infinity) of other data values that you did not collect. To visualize such "other data," consider the (Age, Assets) example of the preceding paragraph, and imagine being back in the year 1998, well prior to the data collection in 2003. Envision the (Age, Assets) data that *might be* collected in 2003, from your standpoint in 1998. There are nearly infinitely many *potentially observable* data values, do you see? The regression model Assets | Age $= x \sim p(\text{Assets} \mid \text{Age} = x)$ describes not only how the *actual* 2003 data arose, but it also describes all the other *potentially observable data* that *could have arisen*. Thus, the model *generalizes* beyond the observed data to the *potentially observable* data.

However, the model $p(y \mid x)$ generalizes only narrowly to other potentially observable data *from the same data-generating process*. In the (Age, Assets) example described above, the model $p(y \mid x)$ does not refer to the clientele of a different retirement planning company based in Washington, DC in the year 2018. However, one common, and sometimes reasonable, use of regression models $p(y \mid x)$ is to make precisely such "external" generalizations. The validity of such generalizations depends on the similarity of the DGPs: If $p(y \mid x)$ models the DGP where the data were collected (e.g., Dallas, 2003), and $p_{\text{New}}(y \mid x)$ models the DGP to which you wish to generalize (e.g., to Washington DC, 2018), then the validity of the generalization depends on how similar the function $p_{\text{New}}(y \mid x)$ is to $p(y \mid x)$. Such generalization is suspect, because your data target $p(y \mid x)$, not $p_{\text{New}}(y \mid x)$.

Interpolation and extrapolation are related concepts. Interpolation is similar to generalization to other data from the same DGP, whereas extrapolation is similar to generalization to other data from a different DGP. Figure 1.3 illustrates the concepts of *extrapolation* versus *interpolation*. Extrapolation is considerably more suspect than interpolation: With interpolation, there are nearby data values, but with extrapolation, there are no nearby data values.

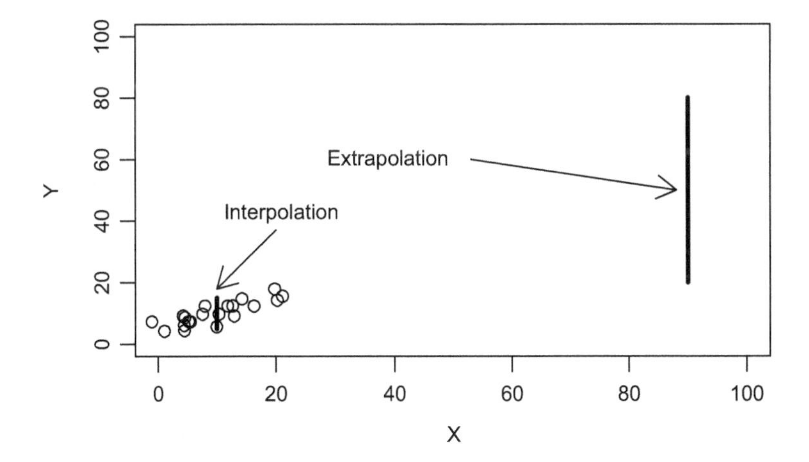

FIGURE 1.3
Illustration of extrapolation versus interpolation. The circles show the observed data. When $X = 10$, the vertical range shows possible values of Y that are consistent with the observed data. When $X = 90$, however, there is much less certainty about the possible values of Y because there are no data nearby. Thus, extrapolation is much less certain than interpolation.

To generalize from your data to a DGP other than what produced your data, you must assume that the two DGPs are similar. You can make an objective argument concerning similarity by first collecting data across a variety of DGPs of the type that you wish to generalize to, and second by comparing the data from these different DGPs. When the data are similar across all DGPs, then generalization to another DGP of the same type is more supportable. For example, if the (Age, Assets) data are similar across a variety of companies across a variety of years and locations, then the generalization to Washington, DC in 2018 is more reasonable, even if you never collected such data.

But even with data from multiple DGPs, extrapolation-style generalization is largely a subjective exercise, requiring subject matter-based arguments about the similarity of the DGPs. For example, you can argue subjectively from biological science that the effect of human DNA on a certain disease condition will be quite stable from the recent past to the near future, even though the statistical analysis, being based on past data, cannot guarantee it. On the other hand, the effect of monetary policy on economic outcomes in the U.S. cannot be assumed to be identical to the effect of monetary policy on economic outcomes in Norway, due to major differences in economic systems. Arguing from economic and political science, generalization is unsafe in this case.

To summarize, you can legitimately generalize from your data to *other potentially observable data from the same DGP*. The models $p(y \mid x)$ discussed throughout this book refer specifically to the DGP that produced your data, thus all generalizations (such as confidence intervals and prediction intervals) discussed in this book refer specifically to $p(y \mid x)$. It is also logical to assume that generalization to DGPs ($p_{\text{New}}(y \mid x)$) that are similar to your DGP ($p(y \mid x)$) is safer than generalization to DGPs that are quite different from your DGP, just like interpolation is safer than extrapolation, as shown in Figure 1.3.

1.3 The "Population" Terminology and Reasons Not to Use It

In the previous section, we emphasized that a regression model is a model for the data-generating process, which is comprised of measurement, scientific, and other processes at the given time and place of data collection. Some sources describe regression (and other statistical) models in terms of "populations" instead of "processes." The "population" framework states that $p(y \mid x)$ is defined in terms of a finite population of values from which Y is randomly sampled when $X = x$. This terminology is flawed in most statistics applications, but is especially flawed in regression; in this section, we explain why.

Suppose you are interested in estimating the mean amount of charitable contributions (Y) that one might claim on a U.S. tax return, as a function of taxpayer income ($X = x$). This mean value is denoted by $\mathrm{E}(Y \mid X = x)$, and is mathematically calculated either by $\mathrm{E}(Y \mid X = x) = \int_{\text{all } y} y \, p(y \mid x) \, dy$ when $p(y \mid x)$ is a continuous distribution, or by $\mathrm{E}(Y \mid X = x) = \sum_{\text{all } y} y \, p(y \mid x)$ when $p(y \mid x)$ is a discrete distribution.

To estimate $\mathrm{E}(Y \mid X = x)$, you obtain a random sample of all taxpayers by (a) identifying the population of all taxpayers (maybe you work at the IRS!), and (b) using a computer random number generator to select a random sample from this population.

Because each taxpayer is randomly sampled, it is correct to infer that the observed Y in your sample for which $X = \$1{,}000{,}000.00$ are a random sample from the subpopulation of

U.S. taxpayers having $X = \$1,000,000.00$. However, in regression analysis, the distribution of this subpopulation of Y values is not what is usually meant by $p(y \mid x)$.

To explain, suppose that, in the entire taxpayer population, there are *only two* taxpayers having income precisely $X = \$1,000,000.00$. Suppose also that one of these taxpayers claimed \$0.00 and the other \$10,000.00 in charitable contributions. Then the "population" definition of the conditional distribution $p(y \mid X = \$1,000,000.00)$, in this case, is given in Table 1.1.

But the model shown in Table 1.1 seems wrong. Do you really think that the potentially observable charitable contributions among taxpayers with \$1,000,000.00 income can only be \$0.00 or \$10,000.00? Further, do you think that the *mean* charitable contributions among people with \$1,000,000.00 income is $E(Y \mid X = \$1,000,000.00) = \$5,000.00$? These conclusions are true in the population framework, but they make no practical sense as regards the processes that give rise to taxpayer behavior.

The only sensible definitions of $p(y \mid X = \$1,000,000.00)$ and $E(Y \mid X = \$1,000,000.00)$ require a conceptualization of a charitable contribution-generating *process*, as opposed to a "population." In this process, there are many *potentially observable* values of Y when $X = \$1,000,000.00$. The distribution of *potentially observable* values of Y is modeled as $p(y \mid X = \$1,000,000.00)$, and is quite different from the population distribution shown in Table 1.1. The mean of these potentially observable Y values is what $E(Y \mid X = \$1,000,000.00)$ refers to. You cannot know what $E(Y \mid X = \$1,000,000.00)$ is, but any rational individual would agree that it is not equal to \$5,000.00. (It is certainly much larger than \$5,000.00).

So, you should use a model that assumes Y is produced from a process rather than a population. In this book, the regression model $Y \mid X = x \sim p(y \mid x)$ will *always* refer to processes, and *never* to populations.

In multiple regression, where there is more than one X variable, the logic for why the "population" model is wrong becomes even more compelling, because cases where there are extremely few (often no) data values in the entire population having a given set of X values is even more common. For example, try to conceptualize how many people in the population of U.S. taxpayers will have the specific combination $X_1 = \$90,100.00$ income, X_2 = Male, X_3 = Retired, $X_4 = 67$ years old, X_5 = Bachelor's degree, X_6 = Lives in New York, X_7 = Married, and X_8 = Four children. Probably there is no person having exactly this profile, hence there is no "population" to speak of at all in this case. Hence, with multiple regression it is even more essential that you adopt the "process" model (i.e., the model that states that your Y data are produced by the conceptual distribution $p(y \mid \mathbf{X} = \mathbf{x})$), rather than the population model. (Note that the boldface font indicates a vector, or list, of possible X variables: $\mathbf{X} = (X_1, X_2, \ldots, X_k)$).

TABLE 1.1

Conditional Distribution of $Y \mid X = \$1,000,000.00$ Using the "Population" Framework

Y	$p(y \mid X = \$1,000,000.00)$
\$0.00	0.5
\$10,000.00	0.5
Total	**1.0**

The randomness in the model that produces data does not come just from the random sampling from the population, it also comes from the production of the population data from the more general data-generating process. The distribution $p(y \mid X = x)$ referred to in the regression model should never be considered a "population distribution"; rather, $p(y \mid X = x)$ should be considered as a "process distribution," one that describes all the potentially observable data, both population (if a population exists at all) and sample.

By adopting the process model, we are not saying that random sampling from a population is bad, or that random sampling from a population is not helpful. On the contrary, random sampling from a population is very helpful to reduce or eliminate bias. Specifically, a non-random sample will target a subset of the population, and will, therefore, target a subset of the process that produced the population data, and hence will be biased.

One way to understand the "process" versus "population" distinction is by the words "infinite" and "finite." There are infinitely many potentially observable streams of data in the process model, even when the sample space is finite. For example, even if the observation is binary, like "Yes/No", the process model describes infinitely many streams such as Yes, Yes, No, No, No, Yes, No, Yes, Yes, Yes, Yes, No, No, …, continuing infinitely. On the other hand, populations are always finite, so there are likewise only a finite number of potential samples from a population.

Finally, if the data really can be considered as, in some sense, a sample from a finite population, such as with the taxpayer example above, the regression models we use will always assume that these population data themselves come from a larger data-generating process. In the taxpayer example, this larger process is the taxpayers' behavioral process.

1.4 Data Used in Regression Analysis

A typical data set used in simple regression analysis looks as shown in Table 1.2.

In Table 1.2, "Obs" refers to "observation number." It is important to recognize what the observations refer to, because the regression model $p(y \mid x)$ describes variation in Y at that level of observation, as discussed in the following examples. For example, with *person-level data*, each "Obs" is a different person, person 1, person 2, etc. The variation in Y modeled by $p(y \mid x)$ refers to variation between people. For example, $p(y \mid X = 27)$ might refer to the potentially observable variation in Assets among people who are 27 years old. With *firm-level data*, each "Obs" is a different firm, e.g. firm 1 is Pfizer, firm 2 is Microsoft, etc., and the variation in Y modeled by $p(y \mid x)$ refers to variation between firms. For example,

TABLE 1.2

The Regression Data Set

Obs	X	Y
1	x_1	y_1
2	x_2	y_2
3	x_3	y_3
…	…	…
n	x_n	y_n

$p(y \mid X = 2{,}000)$ might refer to the potentially observable variation in net worth among firms that have 2,000 employees.

As a reminder, it is essential for understanding this entire book, to always remember the following two points:

The regression model $p(y \mid x)$ does *not* come from the data.
Rather, the regression model $p(y \mid x)$ is assumed to *produce* the data.

To illustrate the crucial concept that the regression model is a *producer* of data, consider the data set, available in `R`, called "`EuStockMarkets`," having Daily Closing Prices of Major European Stock Indices, from 1991 to 1998. These data are time-series data, where the "Obs" are consecutive trading days.

```
> data(EuStockMarkets)
> tail(EuStockMarkets[,1:2])[1:5,]
            DAX    SMI
[1855,] 5598.32 7952.9
[1856,] 5460.43 7721.3
[1857,] 5285.78 7447.9
[1858,] 5386.94 7607.5
[1859,] 5355.03 7552.6
[1860,] 5473.72 ?????
```

In the `EuStockMarkets` data set, DAX represents the closing price of the German stock index for a particular day, and SMI the closing price of the Swiss stock index on the same day. Consider the last observation in the data set (which is not in the `EuStockMarkets` data) for which DAX is 5,473.72 and SMI is indicated by "?????." What is the value of this particular SMI? The regression model $p(y \mid x)$ states that this SMI is produced at random from a distribution $p(y \mid X = 5473.72)$, which is the distribution of potentially observable outcomes of SMI when DAX is fixed at 5,473.72.

When you do not know the SMI value, as in the case where it is given as "?????" above, it is hopefully easy for you to view it as random. The harder mental exercise, but one that is absolutely essential to understanding regression, is to view *all the observed SMI* values *in the same way*. In other words, according to the regression model, SMI = 7,952.9 was produced at random from a distribution $p(y \mid \text{DAX} = 5598.32)$, SMI = 7,721.3 was produced at random from a distribution $p(y \mid \text{DAX} = 5460.43)$, and so on.

To facilitate this mental exercise, simply imagine that *all* of the values in the SMI column have "?????" instead of actual data values. Then imagine selecting each value at random from the appropriate $p(\text{SMI} \mid \text{DAX} = x)$ distribution. Then imagine that the actual data values that you see are the results of such random selections. Did you do it? Did it make sense? If so, good! You now understand the regression model. Do not read any further until you have performed this mental exercise and understand it, because none of the rest of the book will make any sense if you do not understand this point of view.

Self-study question: Did you perform the mental exercise suggested in the preceding paragraph?

TABLE 1.3

Representation of a Regression Data Set, with Reference to the Regression Model

Obs	X	Y	
1	x_1	y_1	$\leftarrow p(y \mid X = x_1)$
2	x_2	y_2	$\leftarrow p(y \mid X = x_2)$
3	x_3	y_3	$\leftarrow p(y \mid X = x_3)$
...	...	...	...
n	x_n	y_n	$\leftarrow p(y \mid X = x_n)$

TABLE 1.4

The Regression Model for the European Stock Market Data. See the Bottom Row: Note That the Model Structure is the Same, Regardless of Whether the Y Value Is Actually Observed

Obs	DAX	SMI	
[1855,]	5,598.32	7,952.9	$\leftarrow p(\text{SMI} \mid \text{DAX} = 5598.32)$
[1856,]	5,460.43	7,721.3	$\leftarrow p(\text{SMI} \mid \text{DAX} = 5460.43)$
[1857,]	5,285.78	7,447.9	$\leftarrow p(\text{SMI} \mid \text{DAX} = 5285.78)$
[1858,]	5,386.94	7,607.5	$\leftarrow p(\text{SMI} \mid \text{DAX} = 5386.94)$
[1859,]	5,355.03	7,552.6	$\leftarrow p(\text{SMI} \mid \text{DAX} = 5355.03)$
[1860,]	5,473.72	?????	$\leftarrow p(\text{SMI} \mid \text{DAX} = 5473.72)$

Table 1.3 represents the regression data set in a way that clarifies the foundational assumption of the regression model; namely, that the Y data are produced by distributions $p(y \mid x)$.

In the regression model, *every single* Y observation is produced by a *different distribution*, depending on the value x of the X variable. So again, the distributions $p(y \mid x)$ are not from the data. Rather, they are assumed to *produce* the data.

For the European stock market data, the regression model applies as shown in Table 1.4.

1.5 Random-X Versus Fixed-X

So far, we have skipped over the issue of whether X is random or fixed, and how to model the X data probabilistically if X is random. First, let's make it very clear what "random-X" and "fixed-X" mean. If the X data are *known in advance of the data collection*, then X is fixed, otherwise, X is random. Another way to identify that the X values are fixed is to imagine doing another study like the one you just did: Would the X's be the same? If so, then the X's are fixed; otherwise, they are random.

1.5.1 The Trashcan Experiment: Random-*X* Versus Fixed-*X*

Here is something you can do (or at least imagine doing) with a group of people. You need a crumpled piece of paper (call it a "ball"), a tape measure, and a clean trashcan. Let each person attempt to throw the ball into the trashcan. The goal of the study is to identify the relationship between success at throwing the ball into the trash can (*Y*), and distance from the trashcan (*X*).

In a fixed-X version of the experiment, place markers 5 feet, 10 feet, 15 feet and 20 feet from the trashcan. Have *all* people attempt to throw the ball into the trashcan from *all* those distances. Here the *X*'s are fixed because they are known in advance. If you imagine doing another experiment just like this one (say in a different class), then the *X*'s would be the same: 5, 10, 15 and 20.

In a random-X version of the same experiment, you give a person the ball, then *tell the person to pick a spot* where he or she thinks the probability of making the shot might be around 50%. Have the person attempt to throw the ball into the trashcan multiple times from that distance that he or she selected. Repeat for all people, letting each person pick where they want to stand. Here the *X*'s are random because they are not known in advance. If you imagine doing another experiment just like this one (say in a different class), then the *X*'s would be different because different people will choose different places to stand.

The fixed-X version gives rise to *experimental data*. In experiments, the experimenter first sets the X and then observes the Y. The random-X version gives rise to *observational data*, where the *X*'s are simply observed, and not controlled by the researcher.

Experimental data are the gold standard, because with observational data the observed effect of *X* on *Y* may not be a causal effect. With experimental data, the observed effect of *X* on *Y* can be more easily interpreted as a causal effect. Issues of causality in more detail in later chapters.

Self-study question: In what way will the "trashball" random-X experiment above be biased relative to the fixed-X experiment?

Table 1.5 modifies Table 1.3 to show how to view random *X* data in your data set.

The random-X model is actually a two-stage model. In the "trashcan" experiment, there is a distribution, $p(x)$, of distances that people choose from the basket. The model states

TABLE 1.5

Representation of a Regression Data Set, with Reference to the Assumed Regression Model, Allowing for Random-X Data Generation

Obs		*X*	*Y*	
1	$p(x)\rightarrow$	x_1	y_1	$\leftarrow p(y \mid X = x_1)$
2	$p(x)\rightarrow$	x_2	y_2	$\leftarrow p(y \mid X = x_2)$
3	$p(x)\rightarrow$	x_3	y_3	$\leftarrow p(y \mid X = x_3)$
…		…	…	…
n	$p(x)\rightarrow$	x_n	y_n	$\leftarrow p(y \mid X = x_n)$

that the first person's chosen distance, x_1, is produced at random from that distribution. Next, that person's success outcome, Y, is assumed to be produced by the distribution $p(y \mid X = x_1)$; and so on for the remaining observations 2, 3, …, n. (Since there are only two possible outcomes for Y, these distributions are Bernoulli distributions.)

1.6 Some Preliminary Regression Data Analyses Using R

In this section, we introduce some basic regression data analyses of some data sets that we will use throughout the book.

1.6.1 The Production Cost Data and Analysis

A company produces items, classically called "Widgets," in batches. Here Y = production cost for a given job (a batch), and X = the number of "Widgets" produced during that job. It stands to reason that it will cost more to produce more widgets. Regression is used to clarify the nature of this relationship by identifying the additional cost per widget produced (called "variable cost"), and the set-up cost for any job, regardless of the number of widgets produced (called "fixed cost").

The following R code (i) reads the data, (ii) draws a scatterplot of the (Widgets, Cost) data, (iii) adds the best-fitting straight line to the data, and (iv) summarizes the results of the linear regression analysis. Do not worry if you do not understand everything at this time; it will all be explained in detail later.

R code for Figure 1.4

```
ProdC = read.table("https://raw.githubusercontent.com/andrea2719/
URA-DataSets/master/ProdC.txt")
attach(ProdC)
plot(Widgets, Cost)
abline(lsfit(Widgets, Cost), lwd=2)
```

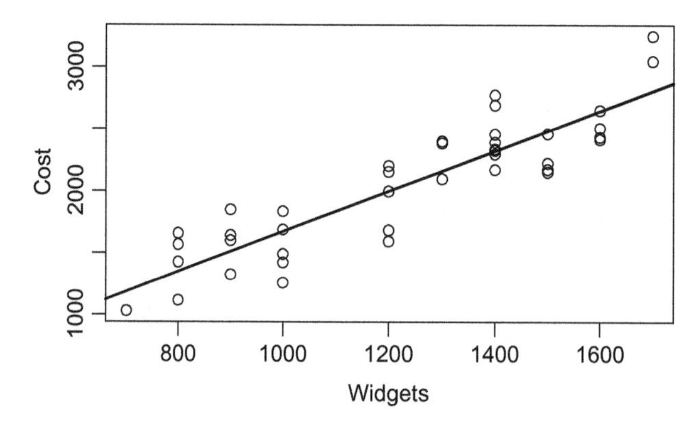

FIGURE 1.4
Scatterplot of the Production Cost data, with best-fitting line superimposed.

Results of the analysis that gives the line in Figure 1.4 are obtained as follows:

```
> fit = lm(Cost ~ Widgets)
> summary(fit)
Call:
lm(formula = Cost ~ Widgets)

Residuals:
    Min      1Q  Median      3Q     Max
-415.31 -193.30    4.76  172.20  451.75

Coefficients:
            Estimate Std. Error t value Pr(>|t|)
(Intercept)  55.4592   168.0455    0.33    0.743
Widgets       1.6199     0.1326   12.22 9.88e-15 ***
---
Signif. codes:  0 '***' 0.001 '**' 0.01 '*' 0.05 '.' 0.1 ' ' 1

Residual standard error: 239.1 on 38 degrees of freedom
Multiple R-squared:  0.7971,    Adjusted R-squared:  0.7917
F-statistic: 149.3 on 1 and 38 DF,  p-value: 9.877e-15
```

Notice the "`Estimate`" values under "`Coefficients`": The numbers 55.4592 and 1.6199 are the intercept and slope of the straight-line function $55.4592 + 1.6199 \times \text{Widgets}$ that is graphed in Figure 1.4. The remainder of the output will be explained later in this book.

Self-study question: What numbers in the analysis above correspond to the "fixed cost" and to the "variable cost"?

1.6.2 The Personal Assets Data and Analysis

Figure 1.2 displays graphs of distributions of accumulated assets for people of different ages. Such distributions can be estimated from data such as the Personal Assets data set, which contains personal assets and age measurements on $n = 53$ people.

`R` code for Figure 1.5

```
Worth = read.table("https://raw.githubusercontent.com/andrea2719/
URA-DataSets/master/Pass.txt")
attach(Worth)
plot(Age, P.assets, ylab = "Accumulated Personal Assets",
  cex.lab = 0.8)
abline(lsfit(Age, P.assets), lwd=2)
```

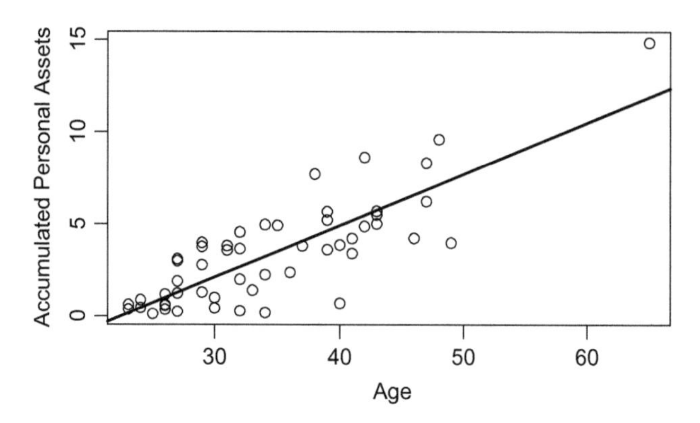

FIGURE 1.5
Scatterplot of (Age, Assets), from the Personal Assets data set, with best-fitting straight line superimposed.

Notice in Figure 1.5 that the variation in Personal Assets, which you can visualize as the "typical" vertical deviation from the line, increases as age increases. Such changes in variation are called *heteroscedasticity*, which is discussed in much more detail later in the book.

1.6.3 The Grade Point Average Data and Analysis

School administrators wonder whether college entrance exams (like the SAT, GRE, GMAT, MCAT, etc.) can predict success in college. Data sets such as the Grade Point Average data set can help to answer this question. The Grade Point Average data set contains information on grade point averages, GMAT scores (an entrance exam for business students), among other variables. The following R code analyzes the Grade Point Average data separately for Masters and Ph. D. students. The slope of the best-fitting straight line is a measure of how well the GMAT predicts grade point averages.

R code for Figure 1.6

```
G.P.A. = read.table("https://raw.githubusercontent.com/andrea2719/
URA-DataSets/master/gpa_gmat.txt")
attach(G.P.A.)

## Data for Ph.D. students.
gpa.phd = gpa[degree=='P']
gmat.phd = gmat[degree=='P']

## Data for Masters students.
gpa.mas = gpa[degree=='M']
gmat.mas = gmat[degree=='M']

par(mfrow=c(1,2))
plot(gmat.phd, gpa.phd, ylim = c(2.4,4.0), xlim = c(400,800),
  ylab= 'Grade Point Average ', xlab = 'GMAT Score ',
  main = 'Ph.D. Students ')
abline(lsfit(gmat.phd, gpa.phd), lwd=2)
```

(Continued)

```
plot(gmat.mas, gpa.mas, ylim = c(2.4,4.0), xlim = c(400,800),
  ylab= 'Grade Point Average ', xlab = 'GMAT Score ',
  main = 'Masters Students ')
abline(lsfit(gmat.mas, gpa.mas), lwd=2)
```

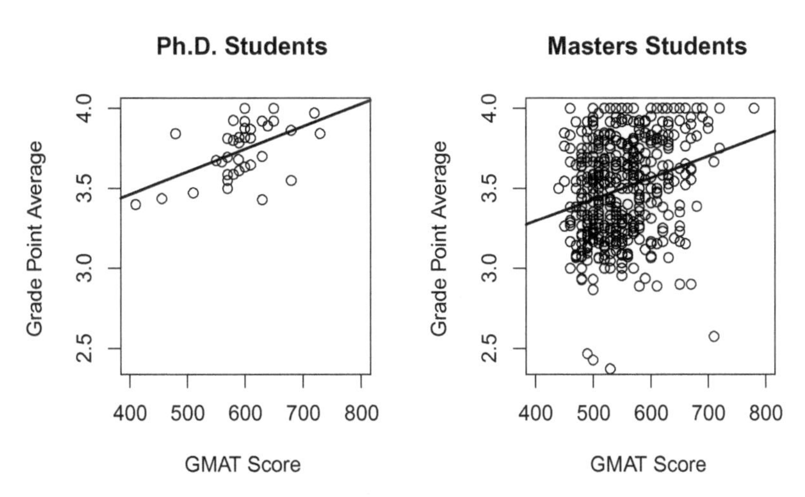

FIGURE 1.6
Scatterplots of (GMAT, GPA), with best-fitting lines, for PhD and Masters students.

As shown in Figure 1.6, it appears that there is a relationship between GMAT scores and grade point averages. It is not a strong relationship, but it appears to be a stronger relationship for the PhD students than for the Masters students.

1.7 The Assumptions of the Classical Regression Model

A definitive source for the definition of a statistical model is the article "What is a Statistical Model?" by Peter McCullagh (2002), who defines a statistical model as "... a set of probability distributions on the sample space ..." Translating, a statistical model is simply an assumption that your sample data are produced randomly by a particular probabilistic process that lies in a prescribed set of possible probabilistic processes. This "prescribed set of possible probabilistic processes" is what is meant by a "statistical model."

For a simple example, one often refers to the "normal model" in statistics. This model does not prescribe a particular normal distribution as the model for the DGP; instead, it states that the data are randomly generated from a particular normal distribution within the general class of $N(\mu,\sigma^2)$ distributions.

There are several assumptions that you make when you analyze the data using regression models. The first and most important assumption is that the data are produced probabilistically, which is specifically stated as $Y|X = x \sim p(y|x)$. Different types of regression models then make further assumptions regarding the prescribed sets of distributions, and regarding the prescribed way that these distributions are related to x. The assumptions are important because they determine the adequacy of the model.

Adequacy of the regression model refers to the closeness of the approximation of the model, as a producer of data, to the real data-generating process.

1.7.1 Randomness

Statistical models, including regression models, are statements about how the potentially observable data are produced, in general. They are quantifications of your subject matter theory. If you are writing a research paper in a scientific discipline, you will typically explain all this theory in words that state how and why such generalities occur. Your statistical model is simply a concise, mathematical and probabilistic summary of all that general theory. Your research hypotheses, which are also statements about how your data will appear (or might have appeared), are also defined in terms of your statistical model for your data-generating process.

Usually, you do not see any "randomness" assumption explicitly stated in research articles or other texts. Instead, the assumption is implicit, which you will often see stated in a model form such as

$$Y = \beta_0 + \beta_1 X + \varepsilon$$

Implicit in that model formulation is that ε is random. This assumption is necessary because the data Y are not a deterministic function of X. If your relationships are in fact deterministic, then stop reading this book immediately! You should read a book on differential equations instead.

Anticipating multiple regression, where there is one or more X variables, we introduce the boldface term $\boldsymbol{X}$ to denote a set of possible X variables: $\boldsymbol{X} = (X_1, X_2, \ldots, X_k)$.

Randomness assumption

The value of Y is variable, coming randomly from a distribution $p(y \mid \boldsymbol{x})$ of potentially observable Y values that is specific to the particular values ($\boldsymbol{X} = \boldsymbol{x}$) of the $\boldsymbol{X}$ variables.

It may be the case that Y is independent of $\boldsymbol{X}$, in which case $Y \mid \boldsymbol{X} = \boldsymbol{x} \sim p(y)$, a distribution that is the same, no matter what is $\boldsymbol{x}$. This violates no assumption of the regression model. In fact, many research hypotheses involve a question as to whether Y is related to $\boldsymbol{X}$ at all; these hypotheses are evaluated by estimating both the unrestricted model $Y \mid \boldsymbol{X} = \boldsymbol{x} \sim p(y \mid \boldsymbol{x})$ and the restricted model $Y \mid \boldsymbol{X} = \boldsymbol{x} \sim p(y)$, and then by comparing the results.

Note that the randomness assumption by itself makes no assumption about distributions (Poisson, normal, lognormal, or otherwise); and it makes no assumptions about the functional relationships between Y and $\boldsymbol{X}$ (linear, quadratic, logarithmic, or otherwise). As such, the model is a valid and correct model: Statistical data really do look as if generated from distributions, simply because they exhibit variability.

Also, notice that there is no assumption here concerning how the $\boldsymbol{X}$ data are generated, or even that they are random at all. We will generally assume the $\boldsymbol{X}$ data are random, and as mentioned above, the randomness of $\boldsymbol{X}$ is sometimes important, as is discussed in later chapters.

The remaining assumptions make more specific statements about distributions and functional forms of relationships. They define the classical regression model. When you use the usual output from any standard regression software, you are making *all* of the assumptions given in the following sections. However, these assumptions are very restrictive, and much of this book concerns alternative models that are more realistic. This book also explains when you can use the more restrictive models, despite the fact that such models are not precisely correct in the way that they mimic the true data-generating process. As George Box is credited with saying, "All models are wrong but some are useful."

1.7.2 Correct Functional Specification

The *conditional mean function* is $f(x) = \mathrm{E}(Y \mid X = x)$, the collection of means of the conditional distributions $p(y \mid x)$ (a different mean for every x), viewed as a function of x. The conditional mean function $f(x)$ is the deterministic portion of the more general regression model $Y \mid X = x \sim p(y \mid x)$.

Definition of the true *conditional mean function*

The true conditional mean function is given by $f(x) = \mathrm{E}(Y \mid X = x)$.

Note that the true conditional mean function is different from the true *regression model*, which was already given in Section 1.1, but is repeated here to make the distinction clear.

Definition of the true *regression model*

The true regression model is given by $Y \mid X = x \sim p(y \mid x)$.

When the distributions $p(y \mid x)$ are continuous, you can obtain the true conditional mean function from the true regression model via $\mathrm{E}(Y \mid X = x) = \int y\, p(y \mid x) dy$. However, you cannot obtain the true regression model from the true conditional mean function, for the simple reason that you cannot tell anything about a distribution from its mean. For example, even if you know that the mean of Y is 10.0 (for any $X = x$), you still do not know anything about the distribution of Y (normal, lognormal, Poisson, etc.), or even its variance.

Whether you realize it or not, whenever you instruct the computer to analyze your regression data, you are making an assumption about the mean function. The correct functional specification assumption is simply the assumption that the mean function that you assume correctly specifies the true mean function of the data-generating process.

Correct functional specification assumption

The collection of true conditional means $f(x) = \mathrm{E}(Y \mid X = x)$ fall exactly on a function that is in the family of functions $f(x; \beta)$ that *you assume when you analyze your data*, for some vector β of fixed, unknown parameters.

Some examples of functions you might assume for the collection of conditional means are given as follows:

- Linear Regression: Here, you assume $f(\boldsymbol{x};\boldsymbol{\beta}) = \beta_0 + \beta_1 x$. Flat lines, where $\beta_1 = 0$, are in this family and hence do not violate the assumption.
- Quadratic Regression: Here, you assume $f(\boldsymbol{x};\boldsymbol{\beta}) = \beta_0 + \beta_1 x + \beta_2 x^2$. Straight lines, where $\beta_2 = 0$, and flat lines, where $\beta_1 = \beta_2 = 0$, are in this family and hence do not violate the assumption.
- Inverse function regression: Here, you assume $f(\boldsymbol{x};\boldsymbol{\beta}) = \beta_0 + \beta_1(1/x)$.
- Exponential regression function: Here, you assume $f(\boldsymbol{x};\boldsymbol{\beta}) = \exp(\beta_0 + \beta_1 x)$.
- Multiple regression with interaction: Here, you assume $f(\boldsymbol{x};\boldsymbol{\beta}) = \beta_0 + \beta_1 x_1 + \beta_2 x_2 + \beta_3 x_1 x_2$.

Generally, whatever form you assume (and instruct the computer to use) for $f(\boldsymbol{x};\boldsymbol{\beta})$ defines the correct functional specification assumption. If the true conditional mean function $\mathrm{E}(Y \mid \boldsymbol{X} = \boldsymbol{x})$ for the data-generating process is exactly equal to the $f(\boldsymbol{x};\boldsymbol{\beta})$ that you specify, for some values of the parameters $\boldsymbol{\beta}$ (some of which can be 0), then the correct functional specification assumption is valid.

In more advanced applications such as logistic regression (Chapter 13), the correct functional specification assumption refers to the relationship of a transformation of the mean of Y (called a *link function*), to the $\boldsymbol{X}$ vector. In other applications such as quantile regression (discussed in Chapter 16), the correct functional specification assumption refers to the relationship of a *quantile* of the distribution of Y to $\boldsymbol{X} = \boldsymbol{x}$, rather than the mean. For example, the assumption might be that the *median* of the conditional distribution of $Y \mid \boldsymbol{X} = \boldsymbol{x}$ is a linear function of $\boldsymbol{x}$.

In the classical regression model, the correct functional specification assumption is that the model is *linear in the β parameters*. All but the exponential regression function example above are linear in the β parameters. Non-standard regression methods that are discussed later, such as logistic regression and neural network regression, are models that are not linear in the parameters and require special methods for their analysis.

1.7.3 Constant Variance (Homoscedasticity)

The correct functional specification assumption refers to the means of the conditional distributions $p(y \mid \boldsymbol{x})$, as a function of $\boldsymbol{x}$. The constant variance assumption refers to the variances of the conditional distributions $p(y \mid \boldsymbol{x})$, as a function of $\boldsymbol{x}$. Letting $\mu_x = \mathrm{E}(Y \mid X = x)$, these conditional variances are calculated as $\sigma_x^2 = \mathrm{Var}(Y \mid X = x) = \int_{\text{all } y} (y - \mu_x)^2 p(y \mid x)\,dy$.

Like the conditional mean function, the conditional variance function can have any function form in reality, linear, exponential, or any generic form whatsoever, provided that the function is non-negative. However, in the classical regression model, this function is assumed to have a very restrictive form: It is assumed to be a flat function that gives the same function value, regardless of the value of $\boldsymbol{x}$.

The constant variance (homoscedasticity) assumption

The variances of the conditional distributions $p(y \mid \boldsymbol{x})$ are constant (i.e., they are all the same number, σ^2) for all specific values $\boldsymbol{X} = \boldsymbol{x}$.

1.7.4 Uncorrelated Errors (or Conditional Independence)

Unlike the previous assumptions, which do not refer to any specific data set, this assumption refers to the data that you can collect, with observations $i = 1,2,\ldots,n$.

Uncorrelated errors (or conditional independence) assumption

The potentially observable "error term," $\varepsilon_i = Y_i - f(x_i;\beta)$, is uncorrelated with the potentially observable error $\varepsilon_j = Y_j - f(x_j;\beta)$, for all sample pairs (i,j), $1 \le i, j \le n$.

Alternatively, you can assume that the Y_i are independent, given all the X data. This alternative form is used in the construction of likelihood functions for maximum likelihood estimation and Bayesian analysis.

1.7.5 Normality

Normality is important for various reasons, as will be discussed in great detail throughout the book.

Normality assumption

The conditional probability distribution functions $p(y\,|\,x)$ are normal distributions (as opposed to Bernoulli, Poisson, or other) for every $\mathbf{X} = \mathbf{x}$.

Specifically, the normality assumption states that for a given $\mathbf{X} = \mathbf{x}$, Y is produced by a model having the function form of the normal distribution:

$$p(y\,|\,x) = \frac{1}{\sqrt{2\pi}\sigma_x} \exp\left[-\frac{(y-\mu_x)^2}{2\sigma_x^2}\right]$$

Normality does not imply any of the other assumptions. A model can stipulate that the distributions are all normal, *and that* μ_x is a curved function of x, *and that* σ_x is a nonconstant (or *heteroscedastic*) function of x. Heteroscedastic models, where the conditional variance function is not a flat function, are discussed and estimated in Chapter 12.

1.7.6 Putting Them All Together: The Classical Regression Model

None of the assumptions make any sense without the first assumption of a random data-generating process. But given random generation, the remaining assumptions can be all true, all false, or some true and others false. They are not connected. But when they are all true, and when $f(\mathbf{x};\beta) = \beta_0 + \beta_1 x_1 + \beta_2 x_2 + \ldots + \beta_k x_k$, you have the classical (multiple) regression model.

The classical regression model

$$Y_i \mid X_i = x_i \sim_{\text{independent}} \mathrm{N}(\beta_0 + \beta_1 x_{i1} + \beta_2 x_{i2} + \ldots + \beta_k x_{ik}, \sigma^2), \text{ for } i = 1,2,\ldots,n.$$

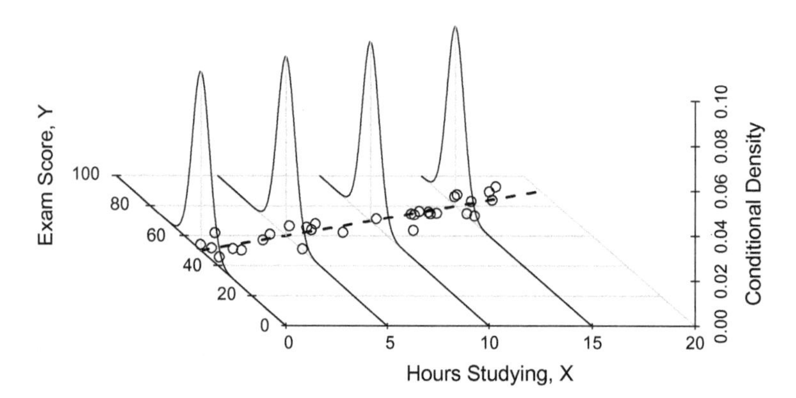

FIGURE 1.7
Graphs of four conditional distributions $p(y \mid x)$. For given $X = x$, you assume that the response variable data on the "floor" (indicated by o's) are produced by such distributions when you perform regression analysis.

Figure 1.7 illustrates how these distributions look in the case where there is one explanatory (X) variable, in a setting where Y = exam score and X = hours spent studying for the exam.

1.8 Understanding the Regression Model by Using Simulation

Simulation is an essential tool to understand all statistical models, particularly the more advanced ones. Simulation allows you to understand the regression model as a *producer* of data, just like the real process you are studying, which also produces data. Simulation also makes it easy to understand the meaning and importance of the regression assumptions. In particular, simulation clarifies the often confusing, but actually quite simple notion that the output from regression software provides *estimates* of true parameter values, rather than the true values themselves: With simulation, you know the true targets of the estimates (the true values) because you specify them yourself in your simulation code.

All statistical models, including regression models, are *recipes* for how the data are produced. You should be able to carry out the instructions of these recipes using simulation. If it is not clear how to simulate data using a model that someone has presented to you, then they have not specified the model correctly. When you analyze regression data, you assume that your data have been produced at random by such a model.

For example, consider the Production Cost data. The random generation model is reasonable if the original data are similar to randomly produced data. In particular, the original data scatterplot should look like the scatterplots of data simulated from the model. The scatterplot of the original data shown in Figure 1.4 was obtained as follows:

```
ProdC = read.table("https://raw.githubusercontent.com/andrea2719/
URA-DataSets/master/ProdC.txt")
attach(ProdC)
plot(Widgets, Cost)
```

Now, look at scatterplots of data produced by the classical regression model. The parameters $\beta_0 = 55$, $\beta_1 = 1.5$, $\sigma = 250$ shown in the simulation code are unknown in practice; here we pick these values to produce data that are similar to our observed data.

```
## Plausible true parameters of your simulation model.
beta0 = 55    # A plausible value of the intercept
beta1 = 1.5   # A plausible value of the slope
sigma = 250   # A plausible value of the conditional standard deviation
n = length(Cost); par(mfrow=c(2,2))
sim.Cost = beta0 + beta1*Widgets + rnorm(n,0,sigma)
plot(Widgets, sim.Cost)
```

Figure 1.8 shows four such simulations. Comparing these scatterplots with the actual data scatterplot shown in Figure 1.4, it is not difficult to believe that the model is reasonable.

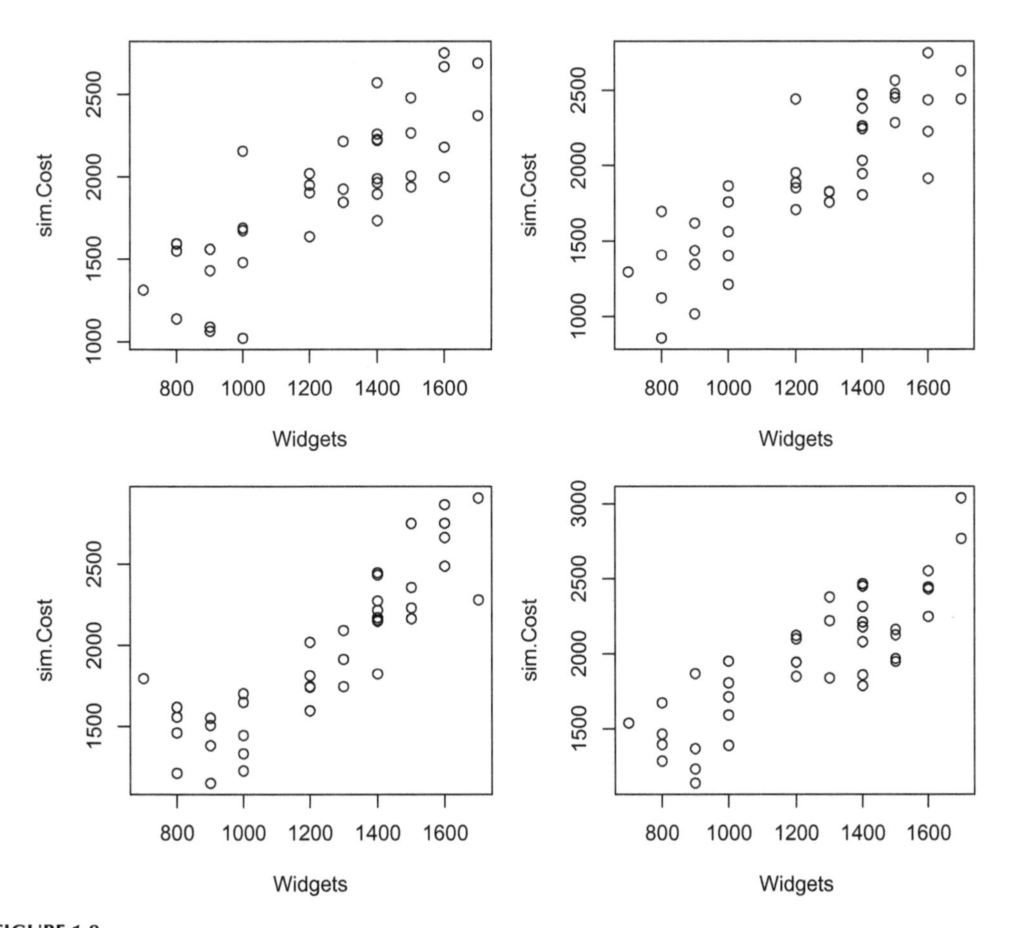

FIGURE 1.8
Four data sets simulated according to the classical regression model in the Production Cost setting. The real data in Figure 1.4 appear similarly, thus the model is reasonable.

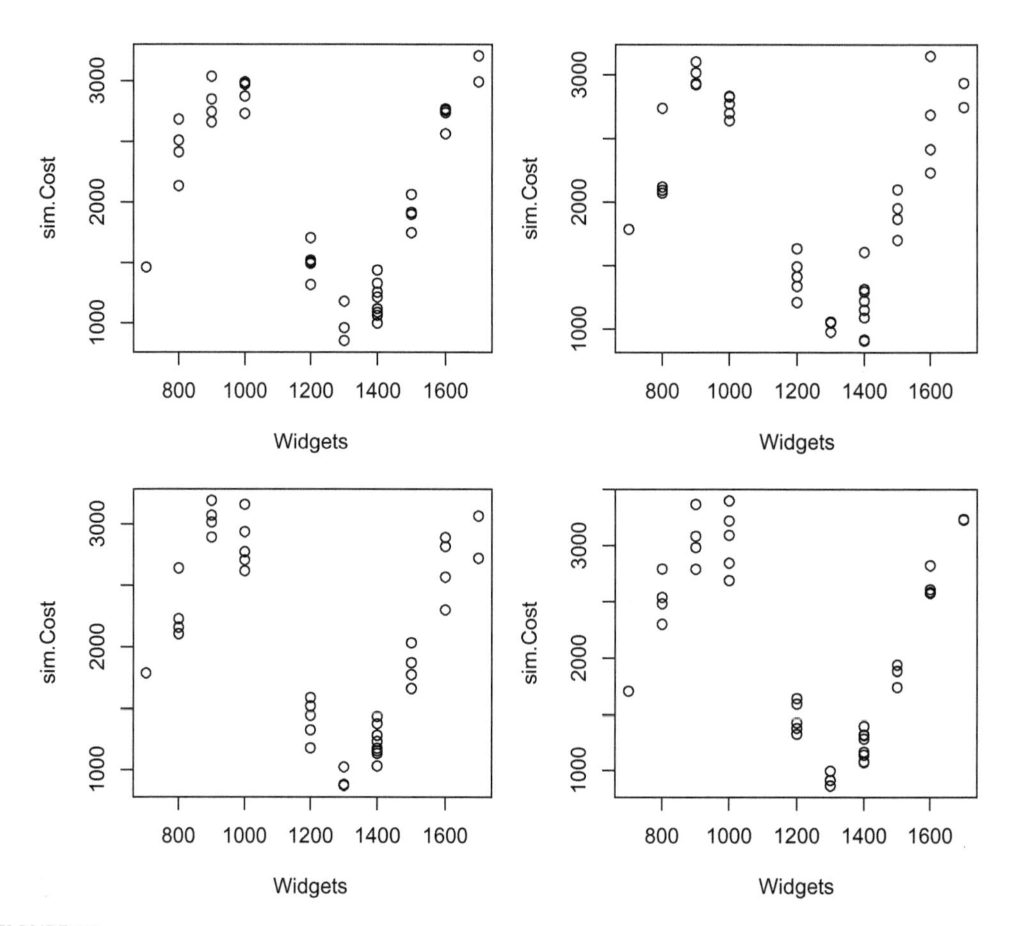

FIGURE 1.9
Four data sets simulated according to a regression model with nonlinear conditional mean function in the Production Cost setting. The real data in Figure 1.4 do not appear similarly.

To understand how violations of assumptions appear in the data, you can look at scatterplots of simulated data where the assumptions are violated. The following code generates and plots data that come from conditional distributions with a nonlinear (curved) conditional mean function. Four such random generations are shown in Figure 1.9. This model is bad because it does not produce data that look like the actual data shown in Figure 1.4.

```
sim.Cost = 1000*(sin(Widgets/120) + rnorm(n,0,.2)) + 2000
plot(Widgets, sim.Cost)
```

The other assumptions, when violated, also affect how the data appear in the scatterplot. You can also look at scatterplots of data produced by conditional distributions having non-constant variance. Results are graphed in Figure 1.10.

```
sim.Cost = beta0 + beta1*Widgets + rnorm(n,0,1)*(Widgets - 600)
plot(Widgets, sim.Cost)
```

Now, look at the regression estimates of the parameters. They are not the same as the true values of β_0 and β_1 that you pick in the simulation model. This is a lesson you must

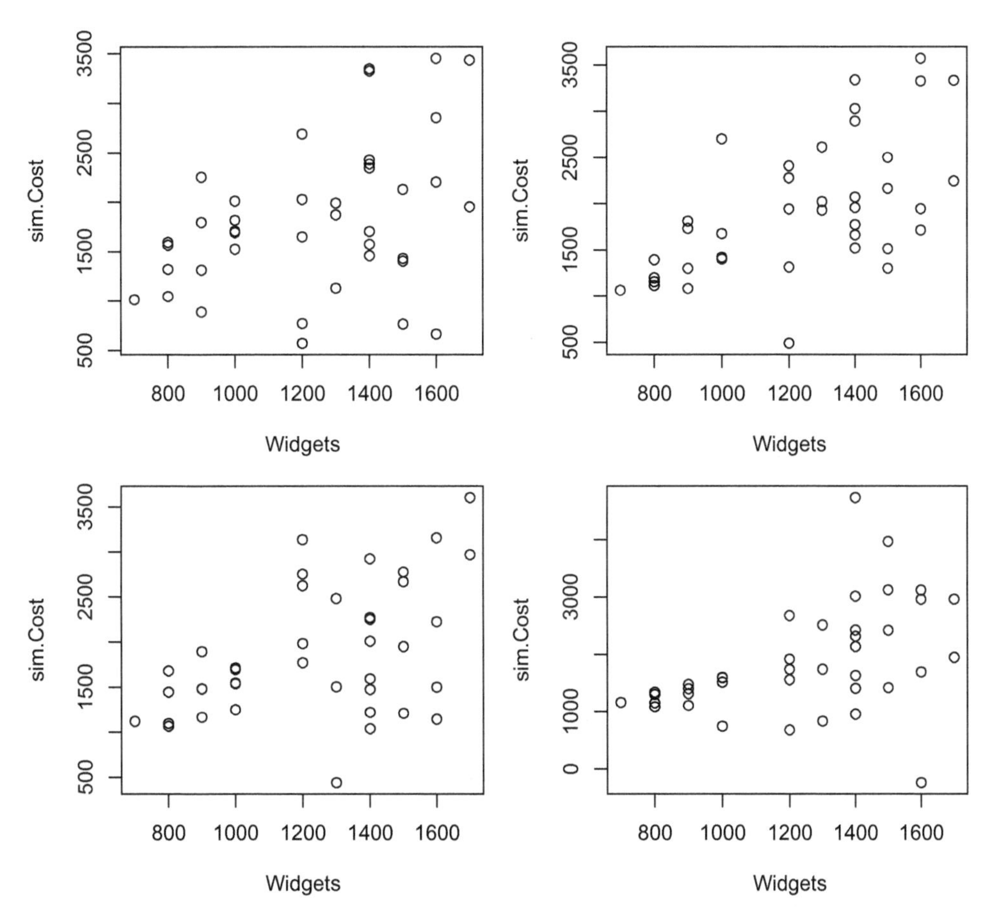

FIGURE 1.10
Four data sets simulated according to a regression model with non-constant conditional variance function in the Production Cost setting. The real data in Figure 1.4 do not appear similarly in that the vertical range (variability) in the original data does not increase as shown here.

understand for any analysis of real data: Your estimates are not the true values there either. A major benefit of simulation is that you can understand precisely how the estimates differ from the true values because you know the true values.

For one simulated set of data, the `R` code `summary(lm(sim.Cost ~ Widgets))` gives the following output:

```
Coefficients:
              Estimate Std. Error t value Pr(>|t|)
(Intercept) -559.4506   561.1705  -0.997    0.325
Widgets        1.9890     0.4427   4.492 6.39e-05 ***
---
Signif. codes:  0 '***' 0.001 '**' 0.01 '*' 0.05 '.' 0.1 ' ' 1

Residual standard error: 798.3 on 38 degrees of freedom
Multiple R-squared:  0.3469,    Adjusted R-squared:  0.3297
F-statistic: 20.18 on 1 and 38 DF,  p-value: 6.388e-05
```

While the true values of the parameters are $\beta_0 = 55,\ \beta_1 = 1.5$ the estimates shown above are $\hat{\beta}_0 = -559.4506$ and $\hat{\beta}_1 = 1.9890$. A different run of the code would give different estimates. An important lesson of this simulation is that you must view the estimates as random quantities, even though the true parameters are fixed.

The normality assumption, when violated, also has an effect on the appearance of the scatterplots. There are many non-normal probability models, the model used for the simulated data shown in Figure 1.11 is just one example, given by the code:

```
sim.Cost = beta0 + beta1*Widgets + sigma *(rt(n,4)^2 -2)
plot(Widgets, sim.Cost)
```

Notice in the simulation, that we picked the parameters $(\beta_0, \beta_1, \sigma) = (55, 1.5, 250)$. Why did we pick those values? They seemed to come from thin air. They certainly are not the values as estimated using the software.

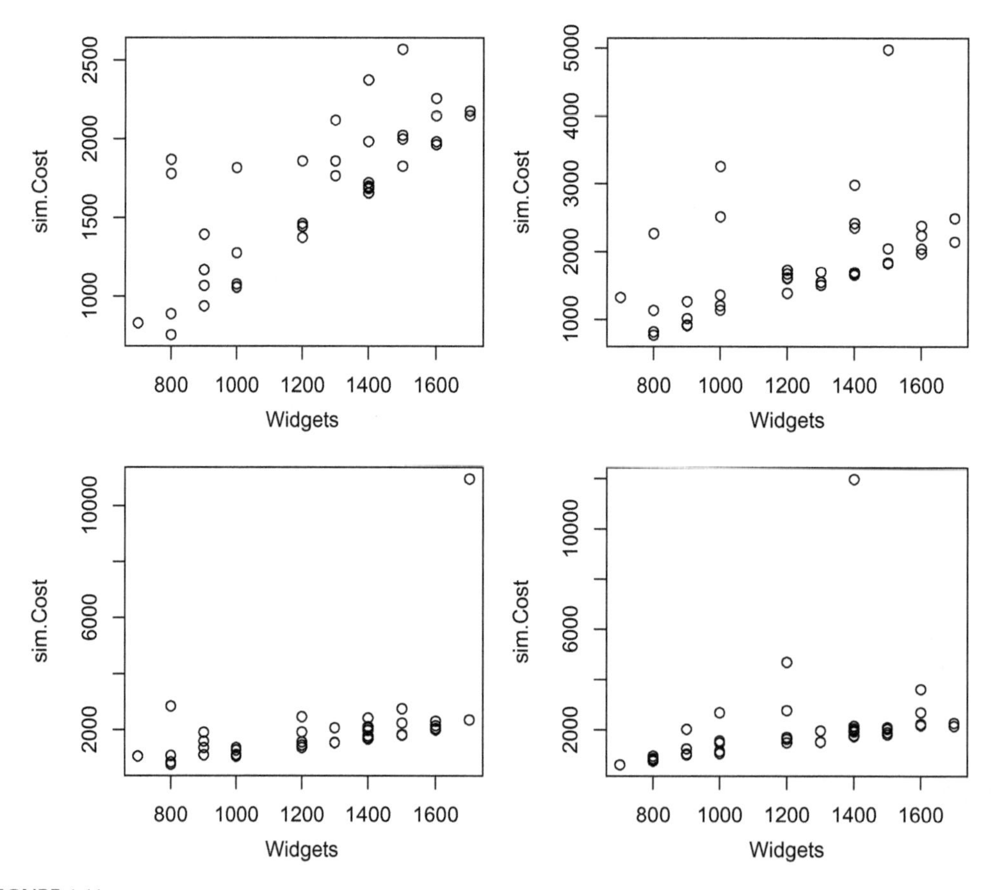

FIGURE 1.11
Four data sets simulated according to a regression model with non-normal conditional distributions in the Production Cost setting. The real data in Figure 1.4 do not appear similarly, in that there are no outliers or evidence of positive skewness in the original data.

The following analogy explains the choice of simulation parameters:

Analogy that explains simulation models

Data produced by the simulation model	*is to*	The parameters of the simulation model
	as	
Data produced by the true data-generating process	*is to*	The parameters of the true data-generating process

According to the analogy that explains simulation models, it does not matter too much which parameters you pick for your simulation model. The point is to understand how the data produced by the simulation relate to those parameters. As the analogy implies, you can view the difference between your simulated data and your simulation model in exactly the same way as you view the difference between your real data and the natural process that produced your real data. Since you know the parameters of your simulation model, and you know how your simulated data look, you know via direct observation how the simulated data relate to your choice of the parameters for your simulation model. This knowledge, in turn, informs you how to view your real data, especially in regard to how they target the processes you are studying.

In particular, from one of the simulations, possible estimates for (β_0, β_1) were $(-559.4506, 1.9890)$, while true values for (β_0, β_1) were $(55, 1.5)$. In the real world, you also will have estimates, but you will not know the true values. Simulation makes it easy to understand this concept, as well as the "standard errors" that you see in the column to the right of the "Estimate" column; these standard errors measure how far the estimates are from the true (usually unknown, except in simulation) values. Standard errors will be discussed in much more detail later in this book.

When you understand how the simulation data relates to the simulation model, you will understand how your real data relates to the parameters of the natural processes that you are studying in your research. It is a perfect analogy. Please become comfortable with this analogy, quickly, as we will use simulation throughout this book to make the more difficult concepts in regression concrete and understandable.

1.8.1 Random-*X* Simulation

The simulations above assume that the Widgets data are always the same, from one set of 40 jobs to another. It is likely that this is not true in the real world of Widget production. Instead, the actual Widgets will vary from one collection of 40 jobs to another, and the simulation model given above is therefore unrealistic.

The following code modifies the codes used above to generate Widgets data randomly (and puts it into a variable called `Widgets.r`), but with data rounded to the nearest 100 to mimic the observed data.

```
Widgets.r = round(rnorm(40, 1200, 300), -2)
Cost.r = beta0 + beta1*Widgets.r + rnorm(n,0,sigma)
plot(Widgets.r, Cost.r)
```

You should run that code a few times and compare with the fixed-x simulation so you can see the differences from the fixed-x simulations, whose code was as follows:

```
sim.Cost = beta0 + beta1*Widgets + rnorm(n,0,sigma)
plot(Widgets, sim.Cost)
```

Fortunately, the lessons learned from the fixed-x simulations do not differ dramatically when random-X simulation is used, at least in this case.

1.9 The Linear Regression Function, and Why It Is Wrong

Usually, when people learn regression, they learn to understand the relationship between Y and X as a linear function. Specifically, the linearity assumption states that the means of the conditional distributions $p(y \mid x)$ fall precisely on a straight line of the form $\beta_0 + \beta_1 x$, i.e., that $\mu_x = \mathrm{E}(Y \mid X = x) = \beta_0 + \beta_1 x$.

See Figure 1.7 above for a graphic illustration of what this assumption tells you about the means of the conditional distributions: In that graph, four conditional distributions are shown, corresponding to four distinct values $X = x$. The linearity assumption states that the means of those four distributions, as well as the means for all other conditional distributions that are not shown in Figure 1.7, fall precisely on a straight line $\beta_0 + \beta_1 x$, for some values of the parameters β_0 and β_1. The linearity assumption does not require that you know the numerical values of β_0 and β_1; rather, it simply states that the conditional means fall on some line $\beta_0 + \beta_1 x$, for some (usually unknown) numerical values of the parameters β_0 and β_1.

The parameter β_0 is called the *intercept* of the line. When $\mathrm{E}(Y \mid X = x) = \beta_0 + \beta_1 x$, it follows that $\mathrm{E}(Y \mid X = 0) = \beta_0 + \beta_1(0) = \beta_0$. In words, if the linearity assumption is true, then the mean of the distribution of Y when $X = 0$ is equal to β_0. Often, the range of X does not include 0, in which case that interpretation is not particularly useful. In such cases, you can vaguely interpret β_0 as a parameter related to the unconditional mean of Y: If the mean of Y is larger, then β_0 will be larger to reflect the vertical height, or distance from zero, of the regression function.

The parameter β_1 tells you something about the relationship between Y and X. If the linearity assumption is true, then this parameter is the difference between the conditional means of the distributions of Y where the X variable differs by 1.0, which can be demonstrated as follows:

$$
\begin{aligned}
\mathrm{E}(Y \mid X = x+1) - \mathrm{E}(Y \mid X = x) &= \{\beta_0 + \beta_1(x+1)\} - (\beta_0 + \beta_1 x) \\
&= \{\beta_0 + \beta_1 x + \beta_1\} - \beta_0 - \beta_1 x \\
&= \beta_0 + \beta_1 x + \beta_1 - \beta_0 - \beta_1 x \\
&= \beta_1
\end{aligned}
$$

No matter what is x, the mean of the distribution of Y when $X = x+1$ is exactly β_1 higher than the mean of the distribution of Y when $X = x$, if the linearity assumption is true.

Interpretations of β_0 and β_1 are shown in Figure 1.12.

R code for Figure 1.12

```
library(shape)
x = seq(0,4,1); y = 2+3*x
plot(x,y, type="l", ylim = c(0,15), xlim = c(-2,4),
  ylab="E(Y|X=x)" , cex.axis=0.8, cex.lab=0.8)
x1 = seq(-1,0,1); y1 = 2+3*x1
points(x1,y1,type="l", lty=2); abline(v=0)
text(-1.2,2, bquote(paste(beta[0])), adj=c(.5,.5), cex=1.2)
Arrows(-1. ,2., 0, 2, arr.type="triangle",arr.adj=1)
points(c(1,2), c(5,5), type="l",lty=3)
points(c(2,2), c(5,8), type="l",lty=3)
text(3.2,6.5, bquote(paste(beta[1])), adj=c(.5,.5), cex=1.2)
Arrows(3.0 ,6.5, 2, 6.5, arr.type="triangle",arr.adj=1)
```

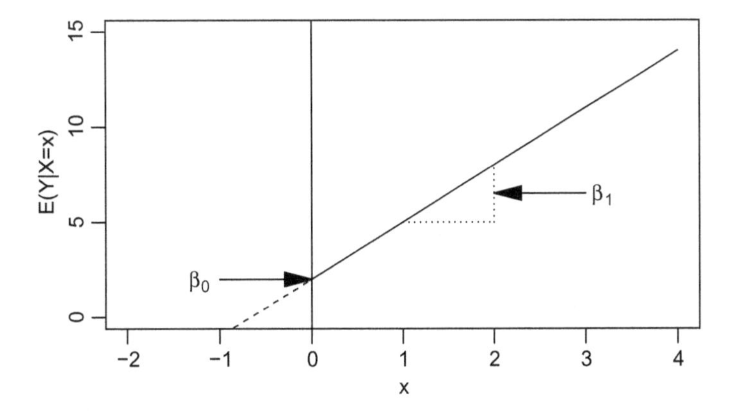

FIGURE 1.12
Graph of $E(Y \mid X = x)$ when the linearity assumption is true, in which case $E(Y \mid X = x) = \beta_0 + \beta_1 x$. The intercept, β_0, is the mean of the distribution of Y when $X = 0$. The slope, β_1, is the difference between the means of two conditional distributions of Y where the X variable differs by one unit. In this example, the two conditional distributions are $p(y \mid X = 2)$ and $p(y \mid X = 1)$.

When the linearity assumption is not true, then the interpretations given above for β_0 and β_1 are not precisely true. At best they are only approximately true, depending on how close the true regression function $E(Y \mid X = x)$ is to a straight line. (Figure 1.13 illustrates the approximation of the linear function to the true regression function.)

And further, *linearity is nearly always false in practice.* The following paragraphs offer a logical argument (essentially, a mathematical proof) to support this (jarring) statement.

Consider Figure 1.7 again. The particular four X values are 0, 5, 10 and 15. Denote the means of the corresponding conditional distributions $p(y \mid x)$ shown in Figure 1.7 by μ_0, μ_5, μ_{10}, and μ_{15}.

If the linearity assumption is true, then the conditional means μ_0 and μ_5 must fall precisely on the straight line as follows:

$$\mu_0 = \beta_0 + \beta_1(0)$$
$$\mu_5 = \beta_0 + \beta_1(5)$$

So far so good: A straight line passes exactly through any two points. Solving for the β's, these equations are both true when $\beta_0 = \mu_0$ and $\beta_1 = (\mu_5 - \mu_0)/5$. This illustrates one case where the linearity assumption is precisely true: It is precisely true when there are only two possible values of the X variable since a line passes perfectly through any two points having different X values.

But with one additional possible value of the X variable (i.e., with three or more possible values of X), it is usually impossible for the linearity assumption to be precisely true. Suppose, in the case above, that $\mu_0 = 51$ and $\mu_5 = 59$ (it will not matter which values you pick here, the argument will remain the same). Then $\beta_0 = 51.0$ and $\beta_1 = (59-51)/5 = 1.6$. If the linearity assumption is true, then the mean of the distribution of Y when $X = 10$ *must precisely equal* $\mu_{10} = 51.0 + 1.6(10) = 67.0000000000000000000000000000...$

The zeros go on forever because that is what the linearity assumption states. So if the linearity assumption is true, then the mean of the distribution of Y when $X = 10$ is determined, to the infinite decimal, by the means of the two distributions when $X = 0$ and $X = 5$, to be precisely $\mu_{10} = 67.00000000000....$

How likely is this to be true? The true mean μ_{10} is an unknown parameter that can take any value in an interval range of possible values. Even if that interval range is small, and centered around 67.000…, such as from 66.8 to 67.2, the *a priori* chance that μ_{10} can be any particular value in that interval, like 67.0000…, is 0.0. The reason this is true is that there are infinitely many possible values of μ_{10} in any interval range, and the value $\mu_{10} = 67.0000...$ is just one of the infinitely many possible values of μ_{10}. Hence, the probability that μ_{10} is precisely equal to 67.0000... is 1/infinity, or zero. (This argument can be made mathematically rigorous by assuming that μ_{10} has a continuous Bayesian prior distribution.)

To put it another way, why would Nature select the number 67.000… for the mean μ_{10}? Why would she (or he or it) not select 67.000271902? Or 66.997110? There is no natural constraint that forces Nature to behave in such a precise fashion. Nature will do whatever Nature chooses to do. She (or he or it) does not particularly care that we humans like simple linear models.

In summary, with just three possible X values, it is usually impossible for the linearity assumption to be precisely true. Obviously, with more than three possible X values, it is also impossible; the argument given above simply becomes more compelling, because additional mean values would also have to lie precisely on that same line, all having the same 0 probability of being true.

An important but rare exception is the case where Y is independent of X. In this case, the linearity assumption is true, since $\mathrm{E}(Y \mid X = x) = \mu$, where μ is a constant that does not depend on x, so that $\mathrm{E}(Y \mid X = x) = \beta_0 + \beta_1 x$, where $\beta_0 = \mu$ and $\beta_1 = 0$, no matter how many possible values X may take.

This same type of argument extends to more complex models such as quadratics, cubics, exponentials, etc. For example, when the X variable can take on three and only three distinct values, then the conditional mean values fall precisely on a quadratic function. But with four or more possible X values, it is ordinarily impossible for the quadratic model to be precisely true.

Stating that the fundamental assumption of regression is nearly always wrong may seem like a bad way to start a book on regression! But it is actually a very good way to start because it sets a tone of realism that we will maintain throughout. There is no sense in learning things wrong, especially at the beginning!

Like linearity, the constant variance and normality specifications are also usually wrong, *a priori*, as a model for the true data-generating process. But it is not necessarily a problem that any or all of the assumptions of the model are wrong. Oh wait, let's set that statement off clearly:

It is not necessarily a problem that any or all of the assumptions of the model are wrong.

It is ok to make assumptions about the model that are obviously incorrect, *provided that the assumptions are not too badly violated.* That is, you *can* assume linearity, provided the curvature in the true regression function $f(x) = \mathrm{E}(Y \mid X = x)$ is not so extreme that a linear approximation will give bad results. Figure 1.13 shows a case where the linear approximation to $f(x)$ is reasonable (despite being incorrect), and another case where it is not reasonable.

R code for Figure 1.13

```
par(mfrow=c(1,2))
x = seq(0, 40, 0.01)
mu1 = 30 - 3*x +.1*x^2
plot(x,mu1, type="l", cex.axis=0.8, cex.lab=0.8,
 ylab = bquote(paste(mu[1])))
abline(lsfit(x,mu1), lty=2)
mu2 = 30 - .3*x +.002*x^2
plot(x,mu2, type="l", cex.axis=0.8, cex.lab=0.8,
 ylab = bquote(paste(mu[2])))
abline(lsfit(x,mu2), lty=2)
```

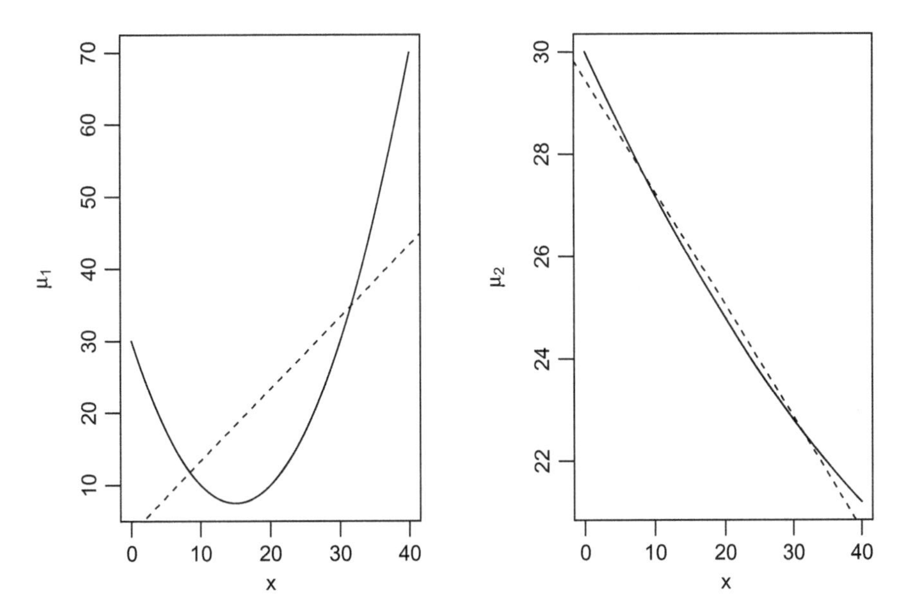

FIGURE 1.13
A case where the linear approximation (dashed line) to the true mean function (solid curve) is not reasonable is shown in the left panel, and a case where it is reasonable is shown in the right panel.

The same logic that is demonstrated by Figure 1.13 as regards the linearity assumption also applies to other assumptions such as normality and constant variance. These assumptions are nearly always false, but it is reasonable to make these model assumptions, provided that the data-generating process of the model is reasonably close to the true data-generating process.

Of course, the trick for real data analysis is to answer the question "What is 'reasonably close'?" The answer is "It depends on the assumption, the kinds of analysis that is being performed, the true data-generating process, and the sample size." There can be no single, "one size fits all" answer, but you can use simulation to help answer the question for your particular context.

1.10 LOESS: An Estimate of the True (Curved) Mean Function

So, the linearity assumption $\mathrm{E}(Y \mid X = x) = \beta_0 + \beta_1 x$ is wrong. What is right? What is right is that $\mathrm{E}(Y \mid X = x) = f(x)$, which is some function $f(x)$ that you do not know. However, data allow you to estimate such unknown quantities.

If your data set had lots of repeats on particular x values, you could use the average of the Y data values where $X = x$ to estimate the function $f(x)$. For example, consider the data in Table 1.6 below obtained from a survey of students in a class. The Y variable is "rating of the instructor," on a discrete 1 to 5 scale (where 5 means "best"), and the X variable is "expected grade in course," where 0 = "F", 1 = "D", 2 = "C", 3 = "B", and 4 = "A."

Using the data shown in Table 1.6, an obvious estimate of $\mathrm{E}(Y \mid X = 2)$ is $\hat{f}(2) = (2+3)/2 = 2.5$ (the hat ("^") signifies that this is just an estimate, not the true expected value). Similar, intuitively obvious estimates are $\hat{f}(3) = (5+2+4+4)/4 = 3.75$, and $\hat{f}(4) = (5+4+4+5)/4 = 4.5$.

TABLE 1.6

Data Where There Are Repeats of the Value of Y for Particular Values $X = x$

Obs	X	Y
1	2	2
2	2	3
3	3	5
4	3	2
5	3	4
6	3	4
7	4	5
8	4	4
9	4	4
10	4	5

The data and the estimated mean function are shown in Figure 1.14. Notice that the function $\hat{f}(x)$ is not perfectly linear, as is expected since there are three distinct X values.

R code for Figure 1.14

```
x = c(2,2,3,3,3,3,4,4,4,4)
y = c(2,3,5,2,4,4,5,4,4,5)
x1 = c(2,3,4)
f.hat = c(2.5,3.75,4.5)
plot(x, jitter(y,.5), ylab="Rating of Instructor (jittered)",
 xlab="Expected Grade" , cex.axis=0.8, cex.lab=0.8)
points(x1, f.hat, pch = "X")
points(x1,f.hat, type="l", lty=2)
```

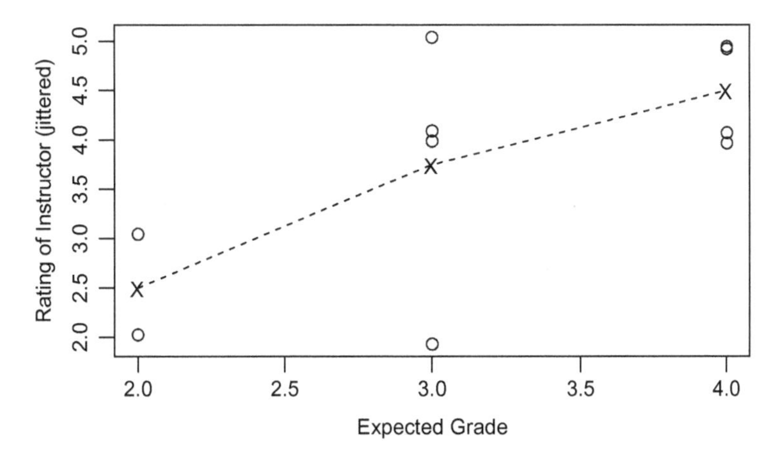

FIGURE 1.14
Estimated mean function (dashed line) using within-group averages (X's). The Y variable is "jittered" to show that each of the combinations (3,4), (4,4), and (4,5) actually refer to two distinct person's responses, rather than a single person's response.

While averaging as shown in Figure 1.14 works when there are repeats on the X values, what if there are few repeats, or none at all? In that case, you can estimate the mean $\text{E}(Y \mid X = x) = f(x)$ by averaging all the Y values for X in a *neighborhood* of $X = x$.

For example, if you are trying to estimate average household monthly expenditures (Y) for people having annual income (X) equal to \$60,000.00, it is likely that you will have no one in your data set with precisely \$60,000.00 income, so you cannot use the averaging trick described above. However, you can "borrow strength" from nearby observations, and consider people whose income is in a *neighborhood* of $X = \$60,000.00$, such as all people having income between \$55,000.00 and \$65,000.00. There will likely be several people in the data set in this range, and the average monthly expenditures of people in this group provides a reasonable estimate of $\text{E}(Y \mid X = \$60,000) = f(60,000)$. You can similarly estimate $f(x)$ for values other than $X = \$60,000.00$ by averaging Y data within similar ±\$5,000 neighborhoods of x, e.g., $\hat{f}(70,000)$ = (average monthly household expense among people whose income lies between \$65,000 and \$75,000).

The LOESS method does something similar, although there is much more to it that will not be mentioned here. The important point to remember is that the "LO" in LOESS refers to "local," as in local neighborhoods. You can view these methods as more elaborate (and better) local averaging methods than what we just described, but the essential idea is the same: If you want to estimate the mean of Y where $X = x$, you should use the data in a neighborhood x, and use an average of the Y values to estimate this mean. The resulting mean function estimate will not be a straight line, just like Figure 1.14 does not show a straight line.

In some versions of LOESS, the extreme data points are down-weighted, producing a measure of center that is not a mean value, but one that is more like a median value. The "`gaussian`" (normal distribution) default setting for the "family" parameter inside the `add.loess` function defined below makes the fitted classical regression line and the fitted LOESS curves comparable in the sense that both estimate mean values.

Throughout this book, the LOESS curves will be accessed using the `add.loess` function.

`R` code "`add.loess`" for accessing LOESS, as used in this book

```
add.loess <- function(X,Y, span = .75,col="black", lty=1, lwd=1){
   lines(spline(X,predict(loess(Y~X, span=span, family="gaussian"))),
col=col,lty=lty, lwd=lwd) }
```

The size of the "local" neighborhoods over which the averaging occurs is governed by the "`span`" parameter, with smaller values indicating narrower neighborhoods, and larger indicating wider neighborhoods. To illustrate how the span parameter affects the result, we analyze two data sets, one real and one simulated. The first is the Dow Jones Industrial Average (DJIA) from 2/13/2017 to 2/13/2018, and the other is simulated (X, Y) data where the true regression function is a line. We overlay LOESS fits using span parameter 0.75 (the default) and 0.10 (a much smaller window). You can see in Figure 1.15 that the smaller window results in a much more "wiggly" function, which appears desirable in the case of the DJIA data, but not the simulated linear data.

`R` code for Figure 1.15

```
djia = read.csv("https://raw.githubusercontent.com/andrea2719/
URA-DataSets/master/DJIA.csv", header=T)
par(mfrow=c(1,2))
attach(djia)
Trading.Day = 1:length(DJIA)
plot(Trading.Day, DJIA, xlab="2/13/2017 - 2/13/2018 Trading Day")
add.loess(Trading.Day, DJIA, lty=2, lwd=2)
add.loess(Trading.Day, DJIA, span=.1, lwd=2)
```

(*Continued*)

```
set.seed(12346)
X = rnorm(253)
Y = X + rnorm(253)
plot(X,Y)
abline(0,1, lwd=3)
add.loess(X,Y, lty=2, lwd=2)
add.loess(X,Y, span=.1, lwd=2)
```

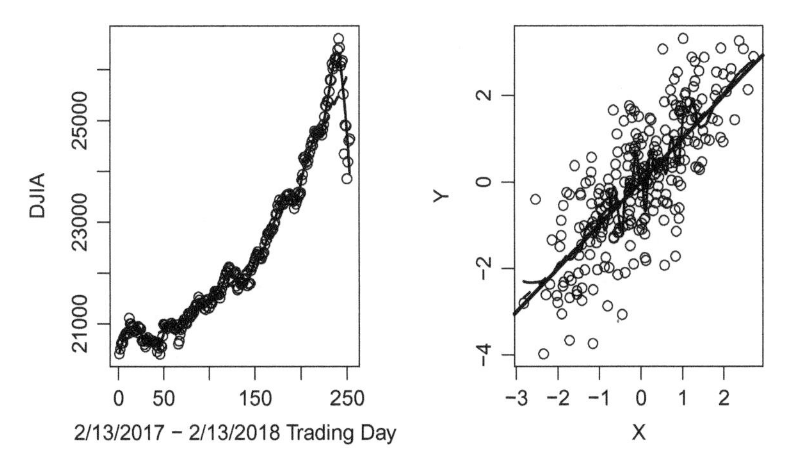

FIGURE 1.15
Data sets illustrating LOESS fits with default span 0.75 (dashed curve in both plots) and with span 0.10 (solid wiggly curve in both plots). In the right plot, the solid straight line is the true regression function.

Figure 1.16 shows a scatterplot of data where it is not clear what is the relationship, whether linear, curved or no relationship at all. The LOESS curve shows clearly that the relationship is curved. The curvature makes sense from a subject matter standpoint, as indicated in the figure's legend.

R code for Figure 1.16

```
cp = read.table("https://raw.githubusercontent.com/andrea2719/
URA-DataSets/master/complex.txt")
attach(cp)
plot(Complex, jitter(Pref), xlab="Complexity",
  ylab="Preference (jittered)")
abline(lsfit(Complex, Pref))
add.loess(Complex, Pref, lty=2)
```

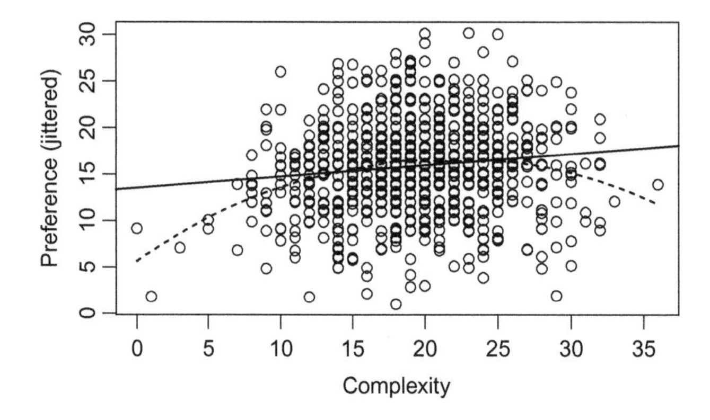

FIGURE 1.16
LOESS (dashed) and linear (solid) fits to the product complexity data. The LOESS fit shows that as Complexity increases, preference for the product increases to a point, then declines as the product becomes too complex. The linear fit, on the other hand, shows that preference keeps increasing as product complexity increases.

If the linearity assumption is always wrong, then why bother with linear functions at all? Why not always use LOESS, which almost always gives a curved function? One answer lies in the right panel of Figure 1.15, which shows that the LOESS curve might be "too wiggly" in the case where the true function is smooth.

Another answer lies in the theory of estimators and their efficiency, a theme that will recur throughout this book. Even incorrect models can provide better (more efficient) estimates than correct models, provided, of course, that the incorrect model is not "too incorrect." You will see much more on this topic in Chapter 11.

Appendix A: Conditional Distributions of the Bivariate Normal Distribution, and Origin of the Term "Regression"

The term "regression" appeared in the late 19th century in Sir Francis Galton's writings. Using statistical analysis of data on Y = son's adult height and X = father's adult height, Galton noticed that, among fathers of above-average height, their sons tended to be shorter than their fathers, but still taller than the general average male. Conversely, among fathers of below-average height, their sons tended to be taller than their fathers, but still shorter than the general average male. Galton coined the phrase "regression to the mean" to describe this phenomenon, because the sons' heights tended away from their fathers' heights, toward the overall mean height.

Galton's observations can be illustrated in terms of the *bivariate normal distribution*, a joint distribution of (X, Y) pairs that (i) requires that both X and Y have marginal normal distributions, and (ii) allows that X and Y are correlated. Thus, the bivariate normal distribution has five parameters, namely μ_x, σ_x, μ_y, σ_y, and ρ, where ρ is the correlation parameter.

The joint probability density function of a bivariate normal (X, Y) pair gives the relative likelihood of the various (x, y) combinations, and its mathematical form is

$$p(x,y) = \frac{1}{2\pi\sigma_x\sigma_y\sqrt{1-\rho^2}} \exp\left\{-\frac{1}{2(1-\rho^2)}\left(z_x^2 + z_y^2 - 2\rho z_x z_y\right)\right\}$$

where $z_x = (x-\mu_x)/\sigma_x$, and $z_y = (y-\mu_y)/\sigma_y$.

To illustrate the meaning of this distribution using some realistic numbers, suppose fathers and sons' heights both have the same mean, $\mu_x = \mu_y = 175$ cm, as well as the same standard deviation $\sigma_x = \sigma_y = 8$ cm. The correlation parameter ρ governs similarity between fathers' and sons' heights, with $\rho = 1$ meaning sons have exactly the same height as their fathers, and $\rho = 0$ meaning sons' heights are completely unrelated to fathers' heights. Clearly, there is *some* relationship, but not a perfect one, so $0 < \rho < 1$ in this case. Figure A.1 displays the bivariate normal density where $\rho = 0.4$, indicating a moderately weak relationship between X and Y.

R code for Figure A.1

```
mux = 175; muy=mux; sx = 8; sy=sx; r = .4;
sxy = r*sx*sy; x = seq(150,200,length=41); y=x;

pxy = function(x,y) {
  zx = (x-mux)/sx; zy = (y-muy)/sy;
  volume = 2*pi*sx*sy*sqrt(1-r^2)
  kernel = exp(-.5*(zx^2+zy^2-2*r*zx*zy)/(1-r^2))
  kernel/volume }
density = outer(x,y,pxy)

persp(x, y, density, theta=10, phi=20, r=50, d=0.1, expand=0.5,
  ltheta=90, lphi=180, shade=0.75, ticktype="detailed", nticks=5,
  xlab="\nX=Father's Height (cm)", ylab= "\nY=Son's Height (cm)",
  zlab="\n\np(x,y)", cex.axis=0.8, cex.lab=0.9)
```

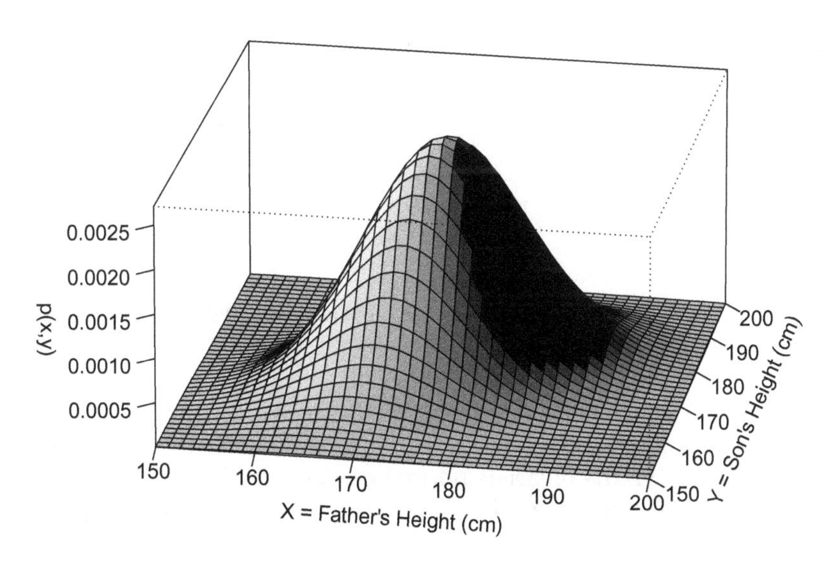

FIGURE A.1
Bivariate normal density model for (X, Y) = (Father's Height, Son's Height), where both means are assumed to be 175 (cm), both standard deviations are assumed to be 8 (cm), and the correlation is assumed to be 0.4. The density value $p(x, y)$ indicates relative likelihood of individual (x, y) combinations. The most likely combination is where the two variables are at their mean, (175,175).

The meaning of $p(x,y)$ as "density" becomes more clear when you look at a scatterplot of (X, Y) data from the $p(x,y)$ distribution: Places where the data are more dense have higher density! Figure A.2 shows a graph of $n = 10{,}000$ data points (x_i, y_i) sampled randomly from the distribution shown in Figure A.1, with contours showing points of equal density superimposed.

R code for Figure A.2

```
library(MASS)
Mu = c(175, 175)
Sigma = matrix(c( 8^2,     .4*8*8,
                .4*8*8,  8^2) , nrow = 2)
XY.sim = mvrnorm(n = 10000, Mu, Sigma)
plot(XY.sim, pch=".", xlab="Father's Height (cm)",
 ylab = "Son's Height (cm)")
contour(x = x, y = y, z = density,nlevels=5, lwd=1.6,
 add = TRUE, drawlabels=FALSE)
```

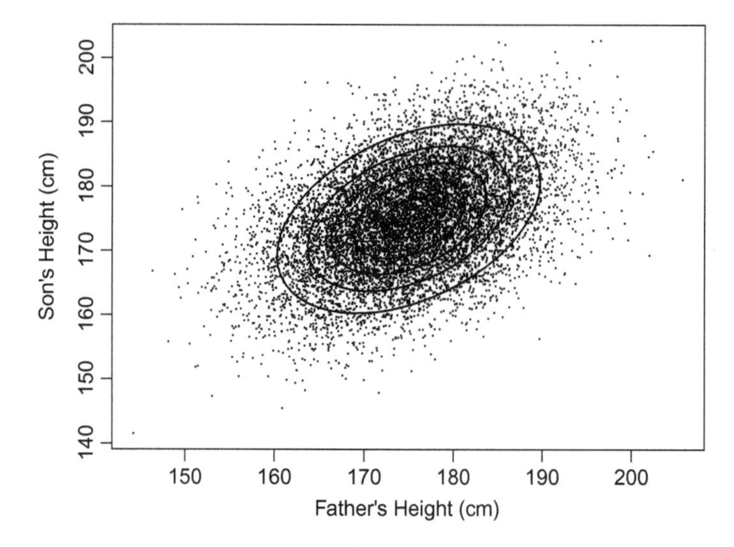

FIGURE A.2
Scatterplot of 10,000 pairs (x_i, y_i) from the joint distribution of Figure A.1, with contours of equal height of the density $p(x,y)$ superimposed. Points (x, y) closer to the center have higher density.

The bivariate normal distribution is especially useful to illustrate the conditional distribution form of the regression model. A mathematical theorem states as follows: *If* the (X, Y) variables come from the bivariate normal model, *then* the conditional distributions $p(y \mid X = x)$ obey the assumptions of the classical regression model, having the following form:

$$Y \mid X = x \sim \mathrm{N}(\beta_0 + \beta_1 x, \sigma^2)$$

Specifically, the parameters of the conditional distribution are given by $\beta_1 = \rho\sigma_y/\sigma_x$, $\beta_0 = \mu_y - \beta_1\mu_x$, and $\sigma^2 = (1-\rho^2)\sigma_y^2$.

Using the parameter settings of the (Father's Height, Son's Height) example, we therefore have:

$$\text{Son's Height} \mid \text{Father's Height} = x \sim \mathrm{N}\left(105 + 0.4x, 7.332^2\right).$$

Self-study question: How did we calculate the numbers 105, 0.4, and 7.332?

The conditional distribution helps explain Galton's expression "regression to the mean." In Figure A.3, the conditional mean function is graphed, along with the "naïve guess line," where a person naively guesses that a son will have height equal to his father's height. It is clear that the height of sons is "regressed toward the mean" compared to the naïve guess.

R code for Figure A.3

```
Father = 150:200
Naive.Guess = Father
Cond.Mean = 105 + .4*Father
plot(Father, Naive.Guess, type="l", xlab = "Father's Height",
  ylab = "Son's Height")
points(Father, Cond.Mean, type="l", lty=2)
abline(h = 175, lty=3)
arrows(190,189.5, 190, 182)
arrows(160,160.5, 160, 168)
```

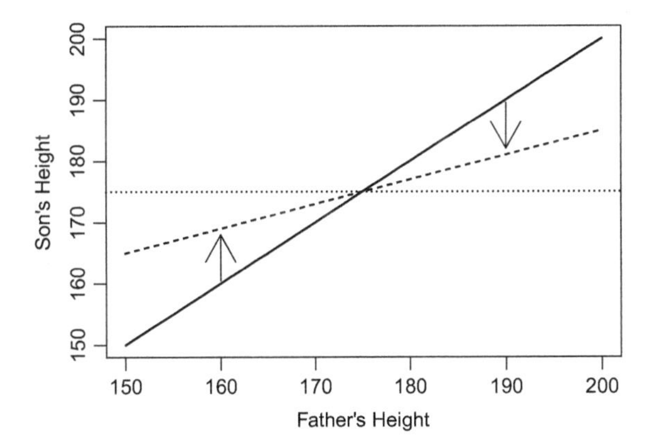

FIGURE A.3
Naïve guess (solid line) of son's height, given father's height, along with the conditional mean height of sons, given father's height (dashed line). The conditional mean is "regressed" toward the overall mean height of sons, as shown by arrows that move the naïve guess toward the overall mean (dotted line).

There is no magic to the "regression toward the mean" concept; rather, it is just common sense when you think about it. In cases where there is no correlation between X and Y, the conditional mean line is flat, so there is complete regression toward the mean. This makes sense because when there is no correlation, the value of X does not help you predict Y at all, hence the naïve ($y = x$) prediction line is obviously wrong. And in cases where the correlation is perfect, then the naïve and conditional mean lines coincide, which also makes sense. Thus, regression toward the mean is a function of correlation: Less correlation implies greater regression toward the mean.

The theoretical conditional distribution in the father/son example was found to be $Y \mid X = x \sim N(105 + 0.4x, 7.332^2)$; this distribution can be empirically verified by using simulated data. Using the simulated data plotted in Figure A.2, there are 161 Y values where the X value is in the range 190 ± 1; the mean of these 161 Y values is 180.6, and the standard deviation is 7.4. These statistics nicely corroborate the theoretical values $105 + 0.4(190) = 181$ and 7.332. These statistics also demonstrate the "regression to the mean" concept, because the mean height of the simulated "sons" whose fathers' heights were around 190 cm "regressed" to 180.6 cm, toward the mean of 175.0 cm, similar to the phenomenon Galton demonstrated with real data.

A final comment regarding the bivariate normal distribution: While the bivariate normal distribution is a specific model where the regression assumptions are automatically satisfied, the regression assumptions do not require that X be normally distributed. But the bivariate normal model is still useful, in that it provides an example where you can easily identify the true regression model $p(y \mid x)$, and also in that it provides a way to explain Galton's phrase "regression to the mean."

Reference

McCullagh, P. (2002). What is a statistical model? *Annals of Statistics, 30*, 1225–1267.

Exercises

1. Your health club charges you $100 for initial membership, and $5 per visit thereafter.
   a. After $X = x$ visits, how much money (Y) have you spent? Explain why the model is deterministic.
   b. Give $p(y|X = 3)$ for problem 1. (Note: It is a probability distribution that puts 100% probability on a single number; this is called a degenerate probability distribution.)
   c. How could you change the problem statement so that the amount of money you spend spent after $X = 3$ visits has a non-degenerate probability distribution? Keep the problem real, about the health club, the money Y. Do not discuss simulation or other fake data.
   d. Sketch a graph (by hand) of $p(y \mid X = 3)$ for your scenario of 1.c.
2. Suppose the classical regression model holds, with $\beta_0 = 10$, $\beta_1 = 1$ and $\sigma = 5$.
   a. Graph the distribution of $Y \mid X = 4$, and the distribution of $Y \mid X = 8$ on the same axes, with different plot line types. Include labels.

b. Find $\Pr(10 < Y < 20 \mid X = 4)$ and $\Pr(10 < Y < 20 \mid X = 8)$. Relate these numbers to the graphs in 2.a.

c. Simulate $n = 10$ Y data values from the model by first generating $n = 10$ X values from the Poisson distribution with $\lambda = 12$, then by generating the Y values specific to those X values. Is this a fixed-X or a random-X model?

d. Using the simulated data in 2.c., draw the scatterplot, and overlay (i) the true regression function (ii) the least-squares estimate of the true regression function, and (iii) the LOESS estimate of the true regression function.

e. Repeat 2.d with $n = 100$.

f. Repeat 2.d with $n = 1{,}000$.

g. Comparing 2.d, 2.e, and 2.f, what benefit do you see of having a larger sample size? (Hint: The true regression function is the target of all estimated regression functions.)

3. Use the "Product Complexity" data set discussed in Chapter 1,

```
cp = read.table("https://raw.githubusercontent.com/andrea2719/
URA-DataSets/master/complex.txt")
```

Let Y = `Pref` and X = `Complex`.

a. Draw the scatterplot, and overlay (i) the least-squares estimate of the true regression function, and (ii) the LOESS estimate of the true regression function.

b. Using the definition in Chapter 1, explain the meaning of the "true conditional mean function" for this case.

c. Which estimate do you think is closer to the true conditional mean function here, the least-squares line or the LOESS fit?

d. Experiment with different span values for the LOESS fit until you find one that is "too large, but barely," and one that is "too small, but barely."

2

Estimating Regression Model Parameters

So, LOESS is good but not necessarily best. Sometimes linear models are good even though they are almost always wrong. How can you know which estimates to use?

Besides simulation, another guiding principle we will use throughout this book is *likelihood*. Methods based on likelihood are usually excellent. While not infallible, they can be considered as a "gold standard" of statistical methods: If your data come from particular models $p(y \mid x)$, and if you analyze your data using maximum likelihood that assumes those same particular models, then your analysis will be nearly ideal.

Least squares estimation, the most common method for analyzing regression data, is itself motivated by likelihood, since the least squares estimates are in fact the maximum likelihood estimates that you get when you assume $p(y \mid x)$ is a normal distribution. This fact can be viewed as a lucky coincidence: If the normal distribution did not have a squared term in its exponent, then you would not use least squares. Instead, you would use least absolute deviations, or some other method, as the default for regression analysis.

In addition, likelihood-based methods are an essential first step toward Bayesian methods, which are rapidly becoming an essential statistical tool for all scientists. Finally, standard methods for regression with non-normal distributions, such as logistic regression and Poisson regression use likelihood-based analyses by default, so you need to understand likelihood in order to read the computer output.

We begin this chapter by reviewing likelihood-based methods, with special attention to their use in regression.

2.1 Estimating Regression Models via Maximum Likelihood

Table 1.3 of Chapter 1 is reproduced here as Table 2.1. This table shows how you should view regression data.

Table 2.1 also shows you how to get the likelihood function and the associated maximum likelihood estimates: Just plug the observed y_i values into the conditional distributions and multiply them. Table 2.2 shows how you can obtain the likelihood function.

Now, under particular regression assumptions, each distribution $p(y \mid X = x_i)$ is a function of parameters $\boldsymbol{\theta}$, where $\boldsymbol{\theta}$ is a *vector* (or list) containing all the β's and other parameters such as σ. Displaying this dependence explicitly, Table 2.2 becomes Table 2.3.

TABLE 2.1

Representation of a Regression Data Set, with Reference to the Regression Model

Obs	X	Y	
1	x_1	y_1	$\leftarrow p(y \mid X = x_1)$
2	x_2	y_2	$\leftarrow p(y \mid X = x_2)$
3	x_3	y_3	$\leftarrow p(y \mid X = x_3)$
…	…	…	…
n	x_n	y_n	$\leftarrow p(y \mid X = x_n)$

TABLE 2.2

Contributions to the Likelihood Function

Obs	X	Y	Contribution to the Likelihood Function
1	x_1	y_1	$p(y_1 \mid X = x_1)$
2	x_2	y_2	$p(y_2 \mid X = x_2)$
3	x_3	y_3	$p(y_3 \mid X = x_3)$
…	…	…	…
n	x_n	y_n	$p(y_n \mid X = x_n)$

TABLE 2.3

Contributions to the Likelihood Function in Terms of the Parameters

Obs	X	Y	Contribution to the Likelihood Function
1	x_1	y_1	$p(y_1 \mid X = x_1, \theta)$
2	x_2	y_2	$p(y_2 \mid X = x_2, \theta)$
3	x_3	y_3	$p(y_3 \mid X = x_3, \theta)$
…	…	…	…
n	x_n	y_n	$p(y_n \mid X = x_n, \theta)$

By assuming conditional independence of the potentially observable $Y_i \mid X_i = x_i$ observations, you then can multiply the individual likelihoods to get the joint likelihood as follows:

$$L(\theta \mid \text{data}) = p(y_1 \mid X = x_1, \theta) \times p(y_2 \mid X = x_2, \theta) \times p(y_3 \mid X = x_3, \theta) \times \ldots \times p(y_n \mid X = x_n, \theta)$$

To estimate the parameters $\boldsymbol{\theta}$ via maximum likelihood, you must identify the specific $\hat{\theta}$ that maximizes $L(\theta \mid \text{data})$. The resulting values of the vector $\hat{\theta}$ are called the *maximum likelihood estimates.*

Thus, to get maximum likelihood estimates, you need to specify three things: (1) probability distributions $p(y \mid X = x, \boldsymbol{\theta})$, such as normal, Poisson, etc., (2) dependence structure of the collection of n observations $Y_i \mid X_i = x_i$ (conditional independence is the easiest), and (3) linkage of the distributions to the specific $X = x$ values. The computer software will make default choices here (e.g., (1) normality, (2) independence, and (3) linearity), but the default choices are nearly always wrong, so you need to know how to specify different models. If you specify a more realistic model, then you will get more realistic estimates.

See Figure 1.13 again for examples of reasonable and unreasonable models for the mean function. Realize that there are many other aspects of the model that may or may not be reasonable, including assumed forms of distributions, assumed variance function, and assumed dependence structure between observations. You can handle all such issues easily using likelihood-based methods using available software.

Examples of distributions $p(y \mid X = x, \boldsymbol{\theta})$ that you can use include Poisson, negative binomial, Bernoulli, normal, beta, normal mixture, and multinomial, but there are infinitely many other possible distributions. The multinomial distribution is particularly noteworthy because it allows you to model nominal data, such as choice of brand to purchase, using regression techniques. Thus, regression models do not require numeric Y data.

When the Y data are numeric, the distributions $p(y \mid X = x, \boldsymbol{\theta})$ are typically tied to specific $X = x$ values through the conditional means of the distributions of $Y \mid X = x$, which are denoted by $\mathrm{E}(Y \mid X = x) = f(x)$. Commonly used examples include (i) the linear function, where $f(x) = \beta_0 + \beta_1 x$, (ii) the quadratic function, where $f(x) = \beta_0 + \beta_1 x + \beta_2 x^2$, (iii) the exponential function, where $f(x) = \exp(\beta_0 + \beta_1 x)$, and (iv) the logistic function, where $f(x) = \exp(\beta_0 + \beta_1 x) / \{1 + \exp(\beta_0 + \beta_1 x)\}$. Like choices of the probability distribution $p(y \mid X = x, \boldsymbol{\theta})$, there are infinitely many possible models for the mean function $f(x)$.

2.2 Maximum Likelihood in the Classical (Normally Distributed) Regression Model, Which Gives You Ordinary Least Squares

When you assume the classical regression model (see Section 1.7), where the distributions are all normal, homoscedastic, and linked to x linearly, the probability distributions $p(y \mid X = x)$ that you assume to produce your $Y \mid X = x$ data are the $\mathrm{N}(\beta_0 + \beta_1 x, \sigma^2)$ distributions. These distributions have mathematical form given by:

$$p(y \mid X = x, \boldsymbol{\theta}) = p(y \mid X = x, \beta_0, \beta_1, \sigma) = \frac{1}{\sqrt{2\pi}\sigma} \exp\left[-\frac{\{y - (\beta_0 + \beta_1 x)\}^2}{2\sigma^2} \right]$$

You need this mathematical form to construct the likelihood function.

See Figure 1.7 (again!) for a graphic illustration of these models. Notice that the parameter vector is $\boldsymbol{\theta} = \{\beta_0, \beta_1, \sigma\}$; i.e., there are three unknown parameters of this model that you must estimate using maximum likelihood. Assuming conditional independence, the likelihood function is

$$L(\theta \mid \text{data}) = p(y_1 \mid X = x_1, \theta) \times p(y_2 \mid X = x_2, \theta) \times \ldots \times p(y_n \mid X = x_n, \theta)$$

$$= \frac{1}{\sqrt{2\pi}\sigma} \exp\left[-\frac{\{y_1 - (\beta_0 + \beta_1 x_1)\}^2}{2\sigma^2} \right] \times \frac{1}{\sqrt{2\pi}\sigma} \exp\left[-\frac{\{y_2 - (\beta_0 + \beta_1 x_2)\}^2}{2\sigma^2} \right]$$

$$\times \cdots \times \frac{1}{\sqrt{2\pi}\sigma} \exp\left[-\frac{\{y_n - (\beta_0 + \beta_1 x_n)\}^2}{2\sigma^2} \right]$$

$$= (2\pi)^{-n/2} (\sigma^2)^{-n/2} \exp\left[-\frac{\sum_{i=1}^{n} \{y_i - (\beta_0 + \beta_1 x_i)\}^2}{2\sigma^2} \right]$$

A technical note: In the random-X case, the likelihood should also contain the multiplicative terms $p(x_1) \times p(x_2) \times \ldots \times p(x_n)$. But as long as $p(x)$ does not depend on the unknown parameters $(\beta_0, \beta_1, \sigma)$, these extra terms have no effect and are thus usually left off of the likelihood function.

Taking the logarithm of the likelihood function and simplifying, you get the log-likelihood function.

The log-likelihood function for the classical regression model

$$LL(\theta \mid \text{data}) = -\frac{n}{2}\ln(2\pi) - \frac{n}{2}\ln(\sigma^2) - \frac{1}{2\sigma^2}\sum_{i=1}^{n} \{y_i - (\beta_0 + \beta_1 x_i)\}^2$$

Notice that there is a "–" sign in front of the "Σ" term in the log-likelihood function. Because of this "–" sign, the values of β_0 and β_1 that *maximize* the log likelihood function must *minimize* the summation $\Sigma_{i=1}^{n} \{y_i - (\beta_0 + \beta_1 x_i)\}^2$. The resulting values $\hat{\beta}_0$ and $\hat{\beta}_1$ give the line $\hat{f}(x) = \hat{\beta}_0 + \hat{\beta}_1 x$ such that the *sum of squared vertical deviations from points* y_i *to the line* $\hat{\beta}_0 + \hat{\beta}_1 x_i$ *is the minimum*. Thus, in addition to being maximum likelihood (ML) estimates, these numbers $\hat{\beta}_0$ and $\hat{\beta}_1$ are also called "ordinary least squares" (OLS) estimates. The "least squares" phrase refers to the fact that these estimates minimize a sum of squares, and the "ordinary" term signals that there are different types of least squares estimates, specifically *weighted* least squares (WLS) and *generalized* least squares (GLS), as we shall see.

The vertical deviation between a point y_i and the corresponding value on the prediction line $\beta_0 + \beta_1 x_i$ is often called an "error," although this is somewhat of a misnomer, since there is nothing "wrong" with the y_i value. Nevertheless, the sum of squared vertical deviations is sometimes called "sum of squared errors," and abbreviated as "SSE." Specifically,

$$\text{SSE}(\beta_0, \beta_1) = \sum_{i=1}^{n} \{y_i - (\beta_0 + \beta_1 x_i)\}^2.$$

To understand the concept of ordinary least squares, you can run the following `R` code and change the input values to the "`sseplot`" function as indicated in the comments.

2.2.1 Simulation to Illustrate the Concept of "Least Squares Estimates"

```
x = rnorm(10,5,1)
#True beta0=2.0; true beta1=0.7; true sigma=1
y = 2.0 + 0.7*x + rnorm(10,0,1)
plot(x,y)
sseplot = function(a,b) {
   abline(a,b, col="gray")
   sse = sum ((y - (a+b*x))^2)
   sse
}
```

Here are some results of running the `sseplot` function a few times. Notice that the first argument of the `sseplot` function is the intercept of the line and the second argument is the slope.

```
> sseplot(1.5,.8)    # intercept = 1.5, slope = .8
[1] 4.184854
> sseplot(1.6,.5)
[1] 22.79862
> sseplot(3.0,.2)
[1] 31.31375
> sseplot(5.0,.2)
[1] 14.62507
> sseplot(1.2,.5)
[1] 34.57111
> sseplot(2.0,.7)
[1] 4.844624
```

Figure 2.1 shows the graph of the original data and the six lines represented by the six "`sseplot`" function evaluations.

The better the line fits through the middle of the data, the smaller the sum of squared vertical deviations (SSE). Note also that, in this example, the true line ($2.0+0.7x$) is not the one with the smallest SSE: `sseplot(1.5,.8)` gave SSE = 4.184854, while `sseplot(2.0,.7)` gave SSE = 4.844624.

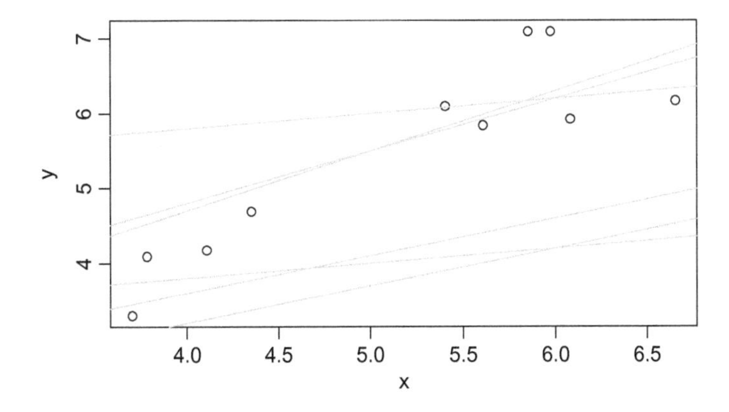

FIGURE 2.1
Results of running the "`sseplot`" function six times. Each line has a different sum of squared vertical deviations from the points (o's) to the line.

The OLS estimates that minimize $\text{SSE}(\beta_0, \beta_1)$ are obtained using the following R code:

```
b0.hat = lsfit(x,y)$coefficients[1]
b1.hat = lsfit(x,y)$coefficients[2]
```

The line with the minimum SSE, or the OLS line, is graphed as `sseplot(b0.hat, b1.hat);` in the simulation above (your results will vary by randomness), the minimum possible SSE occurs for the line with intercept −0.197 and slope 1.095. The SSE for that line was 2.984, smaller than any other line you tried or might have tried, and that is the meaning of "least squares."

2.2.2 Summarizing

- The OLS line is closer to the data than all the other lines in the sense that it gives a smaller SSE than all the other lines.
- The OLS line is not the same as the true line but is an estimate of the true line.
- The OLS line is random—every time you run the code `x = rnorm(10,5,1); y = 2.0 + 0.7*x + rnorm(10,0,1)` you get different OLS lines, even though the true conditional mean line is always the same, $\beta_0 + \beta_1 x = 2.0 + 0.7x$.

The estimates $\hat{\beta}_0$ and $\hat{\beta}_1$ (called `b0.hat` and `b1.hat` in the R code above) are maximum likelihood estimates. They are obtained by taking derivatives of the log-likelihood function, setting the derivatives to zero, and solving for $\hat{\beta}_0$ and $\hat{\beta}_1$. The resulting solutions are given as follows in the simple regression case:

$$\hat{\beta}_1 = \frac{\hat{\sigma}_{xy}}{\hat{\sigma}_x^2}, \quad \hat{\beta}_0 = \hat{\mu}_y - \hat{\beta}_1 \hat{\mu}_x$$

In this formula, $\hat{\mu}_y = \sum_{i=1}^n y_i / n$, $\hat{\mu}_x = \sum_{i=1}^n x_i / n$, $\hat{\sigma}_x^2 = \sum_{i=1}^n (x_i - \hat{\mu}_x)^2 / n$, and $\hat{\sigma}_{xy} = \sum_{i=1}^n (x_i - \hat{\mu}_x)(y_i - \hat{\mu}_y) / n$.

These formulas for $\hat{\beta}_0$ and $\hat{\beta}_1$ are valid *only* when there is one X variable. In multiple regression, where there is more than one X variable, the formulas involve matrices as given in Chapter 6.

In R, these formulas are illustrated as follows:

```
b1.ols = cov(x,y)/var(x); b1.ols
b0.ols = mean(y) - b1.ols*mean(x); b0.ols
```

You will notice that `b1.ols` and `b0.ols` are exactly the same as `b1.hat` and `b0.hat`.

The maximum likelihood estimate of the conditional variance parameter, σ^2, is

$$\hat{\sigma}^2 = \frac{1}{n} \sum_{i=1}^{n} \left\{ y_i - (\hat{\beta}_0 + \hat{\beta}_1 x_i) \right\}^2.$$

Regression software usually displays a slight modification of this conditional variance estimate to reduce its bias, even though the bias is negligible. These issues are discussed in the next chapter.

2.3 Maximum Likelihood with Non-normal Distributions Gives Non-OLS Estimates

The ordinary least squares (OLS) estimates are maximum likelihood estimates from the classical, normally distributed model. But just as linearity is never precisely true, normality is never precisely true either. There are always asymmetries, levels of discreteness, levels of outlier potential, and boundedness characteristics that make all real data-generating processes non-normal. Can you still use OLS, then? The answer is yes—as with any statistical procedure based on the assumption of normality, you can still use it with non-normal distributions. The procedure will be reasonably good if the distributions that produced the data are reasonably close to normal distributions. But, if the distributions are far from normal, other methods may be better.

An interesting alternative to the normal distribution is the Laplace distribution, for which

$$p(y) = \frac{1}{\sqrt{2}\sigma} \exp\left[-\sqrt{2}\frac{|y-\mu|}{\sigma} \right].$$

The mathematical form of the Laplace distribution looks similar to that of the normal distribution, but since the values in the exponent are *absolute deviations* from the mean rather than *squared deviations*, the Laplace distribution allows much higher probability that an observation can be far from the mean. In other words, the Laplace distribution allows a higher probability of an extreme observation, commonly called an *outlier*. The excess kurtosis of the Laplace distribution is 3 (that of the normal distribution is 0), which also implies that the Laplace distribution is more outlier-prone than the normal distribution.

Figure 2.2 compares the normal distribution with $\mu = 0$, $\sigma = 1$ with the corresponding Laplace distribution. Notice that the Laplace distribution extends farther into the tails, despite the fact both distributions have the same standard deviation.

R code for Figure 2.2

```
par(mfrow=c(1,2))
z = seq(-5,5,.01)
p.normal = dnorm(z)
p.laplace = (1/sqrt(2))*exp(-sqrt(2)*abs(z))
plot(z, p.laplace, type="l", lty=2, yaxs="i",
     ylim=c(0,1.1*max(p.laplace)), ylab="density")
points(z, p.normal,type="l")
plot(z, p.laplace, type="l", lty=2, yaxs="i", ylim=c(0,.03),
     xlim = c(2.5,5), ylab="density")
points(z, p.normal,type="l")
```

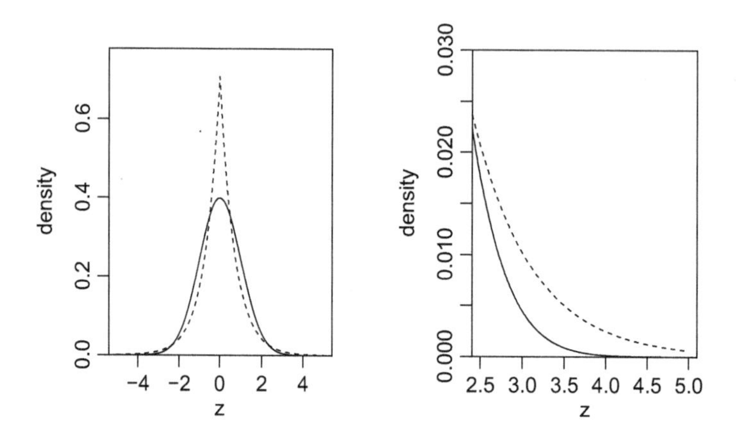

FIGURE 2.2
Normal (solid line) and Laplace (dashed line) distributions. Both distributions have mean 0.0 and standard deviation 1.0. Right-hand panel shows the distributions at 2.5 standard deviations and higher, indicating that it is more likely to see a Laplace random variable that is 3, 4, and 5 standard deviations from the mean than it is to see a normal random variable that far from the mean.

Suppose you think the Laplace model is a good one, or at least that it is better than the normal model, because you know that your data-generating process produces outliers (e.g., values that are further than 3 standard deviations from the mean) more often than does the normal distribution. Suppose also that you are willing to assume linearity, so you specify $\mu = \beta_0 + \beta_1 x$. Then your likelihood function is given by

$$L(\theta \mid \text{data}) = p(y_1 \mid X = x_1, \theta) \times p(y_2 \mid X = x_2, \theta) \times \ldots \times p(y_n \mid X = x_n, \theta)$$

$$= \frac{1}{\sqrt{2}\sigma} \exp\left[-\sqrt{2}\frac{|y_1 - (\beta_0 + \beta_1 x_1)|}{\sigma}\right] \times \frac{1}{\sqrt{2}\sigma} \exp\left[-\sqrt{2}\frac{|y_2 - (\beta_0 + \beta_1 x_2)|}{\sigma}\right]$$

$$\times \cdots \times \frac{1}{\sqrt{2}\sigma} \exp\left[-\sqrt{2}\frac{|y_n - (\beta_0 + \beta_1 x_n)|}{\sigma}\right]$$

$$= 2^{-n/2}(\sigma^2)^{-n/2} \exp\left[-\frac{\sqrt{2}}{\sigma}\sum_{i=1}^{n} |y_i - (\beta_0 + \beta_1 x_i)|\right]$$

Taking the logarithm, you get the log-likelihood function for the Laplace regression model.

The log-likelihood function for the Laplace regression model

$$LL(\theta \mid \text{data}) = -\frac{n}{2}\ln(2) - \frac{n}{2}\ln(\sigma^2) - \frac{\sqrt{2}}{\sigma}\sum_{i=1}^{n} |y_i - (\beta_0 + \beta_1 x_i)|$$

Notice the similarity between the log-likelihood functions for the classical model (Section 2.2) and the Laplace model. Specifically, notice again that there is a "–" sign in front of the "Σ" term. Thus, the values of β_0 and β_1 that *maximize* the log-likelihood function for the Laplace model must *minimize* the sum of *absolute* vertical deviations:

$$\text{SAE}(\beta_0, \beta_1) = \sum_{i=1}^{n} |y_i - (\beta_0 + \beta_1 x_i)|.$$

As mentioned in the discussion of SSE in the previous section, the deviations from y_i to the line $\beta_0 + \beta_1 x_i$ are often called "errors," hence the "SAE" acronym stands for *sum of absolute errors*.

The estimates that minimize the SAE are different from the OLS estimates (which minimize SSE) because they are not so sensitive to outlying values: If there is a particular y_i that is far from the middle of the data, then its *squared residual* $\{y_i - (\hat{\beta}_0 + \hat{\beta}_1 x_i)\}^2$ will be very large, and consequently make SSE large. So the least-squares criterion will force the OLS line toward this y_i to minimize the SSE.

On the other hand, the *absolute residual* $|\, y_i - (\hat{\beta}_0 + \hat{\beta}_1 x_i)\,|$ will not be so extreme (relative to the other absolute residuals), thus the resulting line will not be so greatly influenced by an outlier in the case of minimizing SAE. The estimates that minimize the SAE are often called the "LAD" or "Least Absolute Deviations" estimates.

You can find the LAD estimates by using the "`rq`" function in the "`quantreg`" library, which is also used for quantile regression, discussed in Chapter 16. Figure 2.3 compares OLS with LAD estimated lines when there is an outlier.

R code for Figure 2.3

```
set.seed(123) # To fix the random numbers.
x = rnorm(10,5,1)
y = 2.0 + 0.7*x + rnorm(10,0,1)
x[11]=7; y[11]=-1.0  # An outlier
plot(x,y); points(x[11], y[11], pch=20)
abline(lsfit(x,y), lty=2)
library(quantreg)
abline(rq(y~x)$coefficients)
```

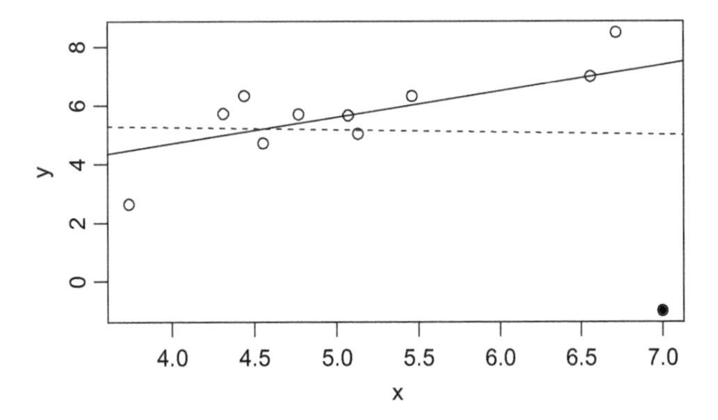

FIGURE 2.3
The ordinary least squares (OLS) line (dashed line) minimizes the sum of squared vertical deviations from the line and is the ML-estimated mean function assuming normal distributions. The least absolute deviation (LAD) line (solid line) minimizes the sum of absolute vertical deviations from the line and is the ML-estimated mean function assuming Laplace distributions. The solid circle in the lower right-hand corner is an outlier. The OLS (normality-assuming) line has a negative slope because it is strongly influenced by the outlier.

The ordinary least squares (OLS) estimates are best when the data come from normal distributions, and the LAD estimates are best when the data come from Laplace distributions. Other distributions will give still different estimates that will be best for those distributions.

On the other hand, you can use OLS estimates with non-normal distributions, and you can use LAD estimates with non-Laplace distributions. In general, you can answer the question, "Which estimate is best?" by using simulation studies.

2.4 The Gauss-Markov Model and Theorem

While the OLS estimates are best under the assumptions of the classical regression model, including, in particular, the assumption of normality, the OLS estimates still have a good mathematical property when you drop the normality assumption. The Gauss-Markov theorem states that *if* the data come from the classical regression model, minus the normality assumption, *then* these estimates are still "good," in a certain sense.

To be more precise, recall the classical regression model, stated here in terms of simple regression:

$$Y_i \mid X_i = x_i \sim_{\text{independent}} \text{N}(\beta_0 + \beta_1 x_i, \sigma^2), \text{ for } i = 1,2,\ldots,n$$

It is common to write the observable $Y_i \mid X_i = x_i$ as follows:

$$Y_i = \beta_0 + \beta_1 x_i + \{Y_i - (\beta_0 + \beta_1 x_i)\}$$

or as

$$Y_i = \beta_0 + \beta_1 x_i + \varepsilon_i,$$

where $\varepsilon_i = Y_i - (\beta_0 + \beta_1 x_i)$ is the deviation from the Y value to the conditional mean for observation i. These ε_i terms are called "errors," as noted above, or more specifically as "true errors," because they involve vertical deviations from the true regression line. Note that the true errors are not observable in practice, because you do not know the true β's.

The classical regression model

$$Y_i \mid X_i = x_i \sim_{\text{independent}} \text{N}(\beta_0 + \beta_1 x_i, \sigma^2), \text{ for } i = 1,2,\ldots,n$$

is equivalent to the model

$$Y_i = \beta_0 + \beta_1 X_i + \varepsilon_i$$

under the assumptions that

$$\text{(i) E}(Y_i \mid X_i = x_i) = \beta_0 + \beta_1 x_i, \text{ and (ii) } \varepsilon_i \sim_{\text{iid}} \text{N}(0, \sigma^2)$$

The premise of the Gauss-Markov theorem is that the data are produced by the Gauss-Markov model.

The Gauss-Markov model

$$Y_i = \beta_0 + \beta_1 X_i + \varepsilon_i$$

where

1. $E(Y_i \mid X_i = x_i) = \beta_0 + \beta_1 x_i$, and
2. $\varepsilon_i \sim_{\text{iid}} p(\varepsilon)$, with $\text{Var}(\varepsilon_i) = \sigma^2 < \infty$

Notice that the Gauss-Markov model simply removes the normality assumption from the classical model. The remaining assumptions (linearity, constant variance, independence) are still there. An example of a model obeying the assumptions of the Gauss-Markov model is the Laplace regression model described in Section 2.3. In that model, the error distribution $p(\varepsilon)$ is the Laplace distribution.

To visualize the Gauss-Markov model, look at Figure 1.7 again (again!). That graph shows four normal distributions of Y at four different X values. The Gauss-Markov model simply allows that those four distributions are non-normal distributions (perhaps Laplace, perhaps discrete, perhaps skewed), but it still assumes that they all have the same variance, and it still assumes that the means of the distributions lie precisely on a straight line $\beta_0 + \beta_1 x$.

The statement and the discussion of the Gauss-Markov (G-M) theorem that follows will be given in terms of $\hat{\beta}_1$, the OLS estimate of the slope parameter, but it also applies to the estimated intercept $\hat{\beta}_0$, as well as to the estimate of $E(Y \mid X = x)$, which is $\hat{\beta}_0 + \hat{\beta}_1 x$.

Note that the statement of the G-M theorem introduces the terms "estimator" instead of "estimate" to emphasize the random nature of the estimate.

The Gauss-Markov theorem

If the data are produced by the Gauss-Markov model, *then* (i) $\hat{\beta}_1$ is an unbiased estimator of β_1, and (ii) $\hat{\beta}_1$ has minimum (conditional on the X data) variance in the class of all other estimators of β_1 that are unbiased and linear functions of the data $\{Y_i\}$.

In short, the G-M theorem says two good things about the OLS estimators with non-normality: (i) They are unbiased, and (ii) they are better (have smaller variance) than any other unbiased estimator that is a linear function of the data. Thus, OLS estimators are often called best linear unbiased estimators, or BLUEs.

As it turns out, the unbiasedness of OLS estimators is guaranteed under even more general conditions, including non-constant variance and correlated errors. However, unbiasedness by itself does not say much: An estimator can be very, very far from the estimand, for every possible sample, and still be an unbiased estimator.

So the most interesting conclusion of the G-M theorem is that the OLS estimator $\hat{\beta}_1$ is "best," in the sense of having minimum variance: If you have two unbiased estimators of

β_1, and one of them has smaller variance, then the estimator with smaller variance is the better estimator because it tends to be closer to β_1.

Self-study question: How can an estimator be very far from the estimand and still be unbiased?

However, notice that the G-M theorem states that the OLS estimator is best only in the class of unbiased estimators that are *linear functions of the data*. Specifically, the G-M theorem says that the OLS estimator $\hat{\beta}_1$ is better than any other unbiased estimator $\hat{\beta}_{1*}$ that is given by the linear function

$$\hat{\beta}_{1*} = a_1Y_1 + a_2Y_2 + \ldots + a_nY_n$$

An example of such an "other estimator" $\hat{\beta}_{1*}$ is given as follows. Suppose you design an experiment to see the effect on production output of increasing the humidity in the room in the range 20%, 30%, 40%. You set the humidity at 20, observe Y_1; then set the humidity to 30, observe Y_2; then set the humidity to 40, and observe Y_3. A linear function of the Y_i values is then $a_1Y_1 + a_2Y_2 + a_3Y_3$.

Assuming the usual linear regression model $Y_i = \beta_0 + \beta_1X_i + \varepsilon_i$, along with the Gauss-Markov assumptions (i.e., all but normality) the estimator

$$\hat{\beta}_{1*} = -0.1Y_1 + 0.1Y_2 + 0.0Y_3$$

is an unbiased estimator of β_1, which you can prove as follows:

$$\begin{aligned}
E(\hat{\beta}_{1*}) &= E(-0.1Y_1 + 0.1Y_2 + 0.0Y_3) \\
&= E[-0.1\{\beta_0 + \beta_1(20) + \varepsilon_1\} + 0.1\{\beta_0 + \beta_1(30) + \varepsilon_2\} + 0.0\{\beta_0 + \beta_1(40) + \varepsilon_3\}] \\
&= E\left[\beta_1(-0.1\times 20 + 0.1\times 30 + 0.0\times 40) + (-0.1\times\varepsilon_1 + 0.1\times\varepsilon_2 + 0.0\times\varepsilon_3)\right] \\
&= \beta_1 + E\left[(-0.1\times\varepsilon_1 + 0.1\times\varepsilon_2 + 0.0\times\varepsilon_3)\right] \\
&= \beta_1
\end{aligned}$$

The estimator $\hat{\beta}_{1*} = -0.1Y_1 + 0.1Y_2 + 0.0Y_3$ is also seen to be a linear function of the Y's by letting $a_1 = -0.1$, $a_2 = 0.1$, and $a_3 = 0.0$.

Should you use this estimator, $\hat{\beta}_{1*} = -0.1Y_1 + 0.1Y_2 + 0.0Y_3$, to estimate β_1? It is unbiased, so does that mean it is "good"? The Gauss-Markov theorem says no, do not use it, because the OLS estimator is also unbiased and has a smaller variance. Thus, the OLS estimator of β_1, $\hat{\beta}_1$, will tend to be closer to β_1 than this other linear estimator, $\hat{\beta}_{1*} = -0.1Y_1 + 0.1Y_2 + 0.0Y_3$.

The "other estimator" seems silly because it discards the Y_3 data value. Often, estimators that are linear functions of the data, like this "other estimator," are similarly contrived.

On the other hand, maximum likelihood estimators typically involve logarithms, products, exponentials, etc., and hence they are not simple linear functions having the form $a_1Y_1 + a_2Y_2 + \cdots + a_nY_n$. Further, maximum likelihood estimators are the gold standard, arguably "best."

Thus, in essence, the Gauss-Markov theorem states that the OLS estimator is "a big fish in a small pond." OLS is best (smallest variance, the big fish), but only when compared to a very limited class of estimators (those that are linear functions of the data, the small pond). The Gauss-Markov theorem certainly does *not* say that the OLS estimators are better than maximum likelihood or other estimators; in fact, maximum likelihood estimators are often much better.

Not convinced? Simulation to the rescue! All the abstract details of unbiasedness, minimum variance, best estimation, non-normality, etc., become concrete in simulation studies, as the exercises show.

Exercises

1. Perform a simulation study to verify that LAD estimates of β_1 tend to be closer to the OLS estimates when the conditional distributions are Laplace distributions. Here is an R function to simulate data from the Laplace distribution:

```
rLaplace <- function(n, m, s) {
   U = runif(n)
   Laplace <- m + ifelse(U<.5, s*log(2*U)/sqrt(2),
   -s*log(2-2*U)/sqrt(2))
   }
```

2. Perform a simulation study to verify that the estimator $\hat{\beta}_{1*} = -0.1Y_1 + 0.1Y_2 + 0.0Y_3$ described in the final section is (i) unbiased and (ii) has a larger variance than the OLS estimator, as the G-M theorem predicts.

3. A simulation study available in https://raw.githubusercontent.com/andrea2719/URA-Rcode/master/Compare_OLS_with_LAD.R demonstrates that, when the conditional distributions are Laplace, then the ML estimates using the Laplace distribution (which turn out to be LAD estimates) are better than the OLS estimates. The study also shows that, when the conditional distributions are normal, then the ML estimates using the normal distribution (which turn out to be OLS estimates) are better than the LAD estimates.

   However, in the real world, distributions are neither Laplace nor normal. What happens then? Well, you can use another simulation: Simulate data from a distribution that is neither normal nor Laplace, and then compare LAD and OLS. Which distribution should you use? Use one that is similar to the distribution of your real data-generating process.

Use the code provided above to compare the OLS and LAD estimates when the conditional distributions of Y are (i) Uniform and (ii) $T_{2.5}$ (the T distribution with 2.5 degrees of freedom). To understand when these distributions may be appropriate in reality, note that the uniform distribution does not produce outliers, while the $T_{2.5}$ distribution does produce outliers.

To simulate from these distributions, change the two lines in the code from

```
Y.normal  = beta0 + beta1*X +      rnorm(n, 0 , sigma)
Y.Laplace = beta0 + beta1*X + rLaplace(n, 0 , sigma)
```

to

```
Y.uniform  = beta0 + beta1*X + sigma*runif(n, -1 , 1)/sqrt(1/3)
Y.T2p5 = beta0 + beta1*X + sigma*rt(n,2.5)/sqrt(5)
```

Also make the appropriate (numerous) labeling changes elsewhere in the code.

Use the resulting graphs and summary statistics to answer the questions: (i) Which is better, OLS or LAD, when the distributions are uniform?, and (ii) Which is better, OLS or LAD, when the distributions are $T_{2.5}$? Also, offer an explanation as to why your conclusions make sense, based on outliers.

4. How long is a person willing to wait on the phone for customer service? Suppose Y = length of time (in minutes) a person is willing to wait on the phone for customer service, and X = the urgency of their need for customer service, as measured using a 1, 2, 3, 4, 5 scale, with 5 = most urgent need.

Here is a small data set of (x_i, y_i) values for a survey of $n = 4$ people: (2, 1), (5, 20), (5, 10), (1,0.5).

A commonly used probability model for time to event data is the exponential distribution, given as

$$p(y \mid \lambda) = \lambda e^{-\lambda y}$$

The parameter λ is the inverse of the mean; that is $E(Y) = 1/\lambda$.

Suppose that $E(Y \mid X = x) = x/\beta$, so that $\lambda = \beta/x$. (The slope of the assumed regression line is $1/\beta$ and the intercept is 0).

a. Write down the likelihood function of β.
b. Graph the likelihood function in 4.a.
c. Find the value $\hat{\beta}$ that maximizes the likelihood function in 4.a by setting the derivative of the log likelihood function to 0 and solving.
d. Use the estimated model to graph the estimated distributions $p(y \mid X = x)$ when (i) $X = 2$ and (ii) when $X = 4$, on the same axes, with different plot symbols.
e. Use the estimates in 4.d to estimate $P(Y > 10 \mid X = x)$ when (i) $X = 2$ and (ii) when $X = 4$.

3

The Classical Model and Its Consequences

The classical regression model assumes normality, independence, constant variance, and linearity of the conditional mean function, and is (once again) stated as follows:

$$Y_i \mid X_i = x_i \quad \sim_{\text{independent}} \quad \mathrm{N}(\beta_0 + \beta_1 x_i, \sigma^2), \;\; \text{for } i = 1,2,\ldots,n.$$

Whether you like it or not, this model is also what your computer assumes when you ask it to analyze your data via standard regression methods. The parameter estimates you get from the computer are best under this model, and the inferences (p-values and confidence intervals) are exactly correct under this model. If the assumptions of the model are not true, then the estimates are not best, and the inferences are incorrect. You might think we are saying that assumptions must be true in order to use statistical methods that make such assumptions, but we are not. As we noted in Chapter 1, it is not necessarily a problem that any or all of the assumptions of the model are wrong, *depending on how badly violated is the assumption*. And the easiest way to understand whether an assumption is violated "too badly" is to use simulation.

We have found that students in statistics classes often resist learning simulation. After all, the data that researchers use is usually real, and not simulated, so the students wonder, what is the point of using simulation? Here are some answers:

- Simulation shows you, clearly and concretely, how to interpret the regression analysis of your real (not simulated) data.
- Simulation helps you to understand how a regression model can be useful even when the model is wrong.
- Simulation models help you to understand the meaning of the regression model parameters.
- Simulation models help you to understand the meaning of the regression model assumptions.
- Simulation models help you to understand the meaning of a "research hypothesis."
- Simulation helps you to understand how to interpret your data in the presence of chance effects.
- Simulation helps you to understand all the commonly misunderstood concepts in statistics, like "unbiasedness," "standard error," "p-value," and "confidence interval."
- Simulation methods are commonly used in the analysis of real data; examples include the bootstrap and Markov Chain Monte Carlo.

An alternative to using simulation is to use advanced mathematics, typically involving multidimensional calculus. But this is much, much harder than simulation. Further, simulation can solve many problems very easily that cannot be solved mathematically, even by the most brilliant mathematicians on the planet!

Many of the results stated in this chapter are actually the results of mathematical theorems. Mathematical theorems are always statements having the following form:

General form of a mathematical theorem

If "condition A" is true, *then it follows logically that* "condition B" is also true.

In mathematical statistics, the "condition A" of a theorem typically refers to the model assumptions: *If* the model assumptions are true, *then* …. The "condition B" refers to some desirable property that follows from the validity of the model assumptions, such as, "*then* the confidence level is exactly 95%." Such theorems are proven in sources on mathematical statistics, step by step, showing with perfect logic that "B" is a precise consequence of "A." The reason that statistical assumptions (the premise "A") are so important is that, when they are not true (that is, when the premise "A" is false), then the desirable statistical consequences (the conclusion "B") cannot be guaranteed.

A simple example of a mathematical theorem: *If* $x = -2$, *then* $x^2 = 4.0$. This theorem is simple to prove using what you learned in high school algebra. The theorem provides a good analogy to explain why statistical assumptions are important: Just like the conclusion "$x^2 = 4.0$" (the "condition B" of the theorem) cannot be guaranteed when the premise "$x = -2$" (the "condition A") is false, the good consequences of the statistical analysis (the "condition B") cannot be guaranteed when the assumptions of the model (the "condition A") are false. It also highlights an important point about a theorem of the form *if* "A", *then* "B": The condition "A" need not be true to arrive at the truth of "B." You can have $x^2 = 4.0$ when $x \neq -2$ (only when $x = 2$); and by the same token, you can have the desirable statistical consequences, even when the model assumptions are violated.

Thus, truths of the desirable statistical consequences are guaranteed when the regression model assumptions are true. The desirable consequences might also occur when the assumptions are false; to address this issue you need to use higher mathematics or simulation studies. Some of the simpler mathematical theorems will be proven in this book, and you can find proofs of the more advanced theorems in mathematical statistics books. But anyone can understand the results of these more advanced theorems via simulation, regardless of their mathematical training.

Simulation is the key to understanding everything in this book.

We do not expect you to understand advanced mathematics beyond what we present in this book. But we insist that you understand simulations because they will help you to understand all of the more advanced mathematical results in a natural, visceral way. Every time a simulation result is presented in this book, you should (i) run the `R` code, (ii) understand the lesson of the output of the `R` code, and most importantly, (iii) understand how to relate the results of the simulation to the real (not simulated) data that you may encounter in your life.

3.1 Unbiasedness

The Gauss-Markov (G-M) theorem states that, under certain model assumptions (the premise, "A" of the theorem), the OLS estimator has minimum variance among linear unbiased estimators (that is the consequence, the "condition B" of the theorem). To understand the G-M theorem, you first need to understand what "unbiasedness" means. Recall the view of regression data shown in Chapter 2, shown again in Table 3.1.

To be specific, please consider the Production Cost data set from Chapter 1. The actual data are shown in Table 3.2, along with the random data-generation assumption of the regression model.

In particular, the value 2,224 is assumed to be produced at random from a distribution of potentially observable Cost values among jobs having 1,500 widgets, the value 1,660 is assumed to be produced at random from a distribution of potentially observable Cost values among jobs having 800 widgets, and so on. If you are having trouble visualizing these different distributions, just have a look at Figure 1.7 again, and put yourself in the position of the job manager at this company: In two different jobs where the number of widgets is the same, will the costs also be the same? Of course not; see the first and third observations in the data set, for example. There is an entire distribution of potentially observable Cost values when Widgets = 1500, and this is what is meant by $p(y \mid X = 1500)$.

TABLE 3.1

Representation of a Regression Data Set; Repeat of Table 2.1

Obs	X	Y	
1	x_1	y_1	$\leftarrow p(y \mid X = x_1)$
2	x_2	y_2	$\leftarrow p(y \mid X = x_2)$
3	x_3	y_3	$\leftarrow p(y \mid X = x_3)$
…	…	…	…
n	x_n	y_n	$\leftarrow p(y \mid X = x_n)$

TABLE 3.2

How Table 3.1 Looks for the Production Cost Data Set

Obs	Widgets, X	Cost, Y	
1	1,500	2,224	$\leftarrow p(y \mid X = 1500)$
2	800	1,660	$\leftarrow p(y \mid X = 800)$
3	1,500	2,152	$\leftarrow p(y \mid X = 1500)$
…	…	…	…
40	1,000	1,260	$\leftarrow p(y \mid X = 1000)$

TABLE 3.3

How Table 3.2 Looks for Another Collection of 40 jobs with Unobserved Costs, When You Assume the Classical Regression Model

Obs	Widgets, X	Cost, Y	
1	1,500	Y_1	$\leftarrow N(\beta_0+\beta_1(1500),\sigma^2)$
2	800	Y_2	$\leftarrow N(\beta_0+\beta_1(800),\sigma^2)$
3	1,500	Y_3	$\leftarrow N(\beta_0+\beta_1(1500),\sigma^2)$
…	…	…	…
40	1,000	Y_{40}	$\leftarrow N(\beta_0+\beta_1(1000),\sigma^2)$

Now, use your imagination. Imagine another collection of 40 jobs, from the same process that produced the data above, with the widgets data exactly as observed, but with specific costs not observed. Further, imagine that the classical model is true so that the distribution $p(y\,|\,X=x)$ is the $N(\beta_0+\beta_1 x,\sigma^2)$ distribution. The specific costs are not observed, but the potentially observable data will appear as shown in Table 3.3.

In Table 3.3, the Y_i are random variables, coming from the same distributions that produced the original data. Again, use your imagination: There are infinitely many potentially observable data sets as shown in Table 3.3, because there are infinitely many sequences of *potentially observable* values for Y_1; infinitely many sequences of *potentially observable* values for Y_2,...; and there are infinitely many sequences of *potentially observable* values for Y_{40}. Again, if you are having a hard time visualizing this, just look at Figure 1.7 again: There are an infinity of possible values under each of the normal curves shown there. The $n=40$ Y_i values in Table 3.3 are one set of random selections from such distributions.

For each of these potentially observable data sets of $n=40$ jobs, you will get different parameter estimates. Because there are infinitely many potentially observable data sets, and because each data set gives different parameter estimates, there are also infinitely many different potentially observable values of $\hat{\beta}_0$ and $\hat{\beta}_1$.

Thus, your task is to use your imagination and view the one data set you *actually* observed (the one above with 40 observations, for example) as one of infinitely many potentially observable data sets that you *could have* observed from the same data-generating process. As such, you must also view the particular parameter estimates you *actually* observed, for example, the OLS estimates $\hat{\beta}_0=55.5$, $\hat{\beta}_1=1.62$, as one of the infinitely many pairs of parameter estimates that you *could have* observed.

In this context, unbiasedness of a parameter estimate $\hat{\theta}$ is defined as follows. The Greek letter "θ" refers to any generic parameter or function of parameters, for example θ might refer to β_1 (where $\theta=\beta_1$) or to σ (where $\theta=\sigma$) or to a conditional mean such as $\beta_0+\beta_1(15)$ (where $\theta=\beta_0+\beta_1(15)$), etc.

When is $\hat{\theta}$ an unbiased estimator of θ?

There are infinitely many potentially observable parameter estimates $\hat{\theta}$, one for each potentially observable sample of (x_i, y_i) data from the same data-generating process. These estimates $\hat{\theta}$ have a distribution $p(\hat{\theta})$ of potentially observable estimates. If the mean of this distribution is precisely equal to θ, then $\hat{\theta}$ is an unbiased estimator of θ.

It is a mathematical theorem that, *if* the assumptions of the classical model are true, *then* the OLS estimates are unbiased. But if the assumptions of the classical model are wrong in certain ways, then OLS estimates can be biased. Thus, unbiased estimates are a consequence of whether or not the assumptions of the model are valid. Bias is one reason in a long list of reasons that you should be concerned about the validity of the assumptions of the regression model that you plan to use.

We will not prove unbiasedness right now—strangely enough, it is easier to prove in the more complicated case of multiple regression with matrix formulations, so we will prove it later in Chapter 7. But for now, we want you to understand what unbiasedness means by using simulation, as shown in the following section. Be sure you can apply the lessons learned from the simulations to a real-life example of your own choosing.

3.2 Unbiasedness of OLS Estimates Assuming the Classical Model: A Simulation Study

To start a simulation study, you must specify the model and its parameter values, which in the case of the classical model will be the $N(\beta_0 + \beta_1 x, \sigma^2)$ probability distribution, along with the three parameters $(\beta_0, \beta_1, \sigma)$. These parameters are unknown, so just pick any values that make sense. No matter what values you pick for those parameters, the estimates you get are (i) random, and (ii) when unbiased, neither systematically above nor below those parameter values, in an average sense.

In reality, Nature picks the actual values of the parameters $(\beta_0, \beta_1, \sigma)$, and you do not know their values. In simulation studies, you pick the values $(\beta_0, \beta_1, \sigma)$. The estimates ($\hat{\beta}_0$, $\hat{\beta}_1$, and $\hat{\sigma}$) target those particular values, but *with error that you know precisely because you know both the estimates and the true values.* In the real world, with your real (not simulated) data, your estimates $\hat{\beta}_0$, $\hat{\beta}_1$, and $\hat{\sigma}$ also target the true values β_0, β_1, and σ, but since you do not know the true values for your real data, you also do not know the error. Simulation allows you to understand this error, so you can better understand how your estimates $\hat{\beta}_0$, $\hat{\beta}_1$, and $\hat{\sigma}$ relate to Nature's true values β_0, β_1, and σ.

In the Production Cost example, the values $\beta_0 = 55$, $\beta_1 = 1.5$, $\sigma^2 = 250^2$ produce data that look reasonably similar to the actual data, as shown in Chapter 1. So let's pick those values for the simulation. No matter which values you pick for your simulation parameters β_0, β_1, and σ, the statistical estimates $\hat{\beta}_0$, $\hat{\beta}_1$, and $\hat{\sigma}$ "target" those values.

To make the abstractions concrete and understandable, run the following simulation code, which produces data exactly as indicated in Table 3.3.

```
beta0 = 55; beta1 = 1.5; sigma = 250  # Nature's parameters
Widgets = c(1500,800,1500,1400,900,800,1400,1400,1300,1400,700,
   1000,1200,1200,900,1200,1700,1600,1200,1400,1400,1000,
   1200,800,1000,1400,1400,1500,1500,1600,1700,900,800,1300,
   1000,1600,900,1300,1600,1000)
n = length(Widgets)
# Examples of potentially observable data sets:
Sim.Cost = beta0 + beta1*Widgets + rnorm(n, 0, sigma)
head(cbind(Widgets, Sim.Cost))
```

The first six observations of one randomly produced data set are as follows:

```
     Widgets  Sim.Cost
[1,]    1500  2657.253
[2,]     800  1191.036
[3,]    1500  2505.507
[4,]    1400  2270.826
[5,]     900  1685.990
[6,]     800  1414.286
```

Notice that the X data (the widget values) are the same as in the original sample, but the Y data (the cost values) are randomly produced from the various $N(\beta_0 + \beta_1 x, \sigma^2)$ distributions. Note also that the simulated cost data do not look quite like the original cost data, because the simulated data have decimals, whereas the original observed cost values are integers. This is one way (among many ways) that you know for sure that the normality assumption is incorrect as a model for the real, potentially observable values of the Cost variable, but don't worry about that for now.

The least-squares estimate $\hat{\beta}_1$ for one randomly generated data set is given by entering the command `lm(Sim.Cost ~ Widgets)$coefficients[2]`. We got the estimate $\hat{\beta}_1 = 1.4185$ for one simulation, which is close to the true value $\beta_1 = 1.5$. But "closeness" of an estimate tells you nothing about unbiasedness of an estimate.

Rather, to understand unbiasedness, you have to simulate many times, rather than just once. If, in all those simulations, the estimate $\hat{\beta}_1$ is most often *below* the target 1.5, then $\hat{\beta}_1$ is a biased (low) estimator. Conversely, if the estimate $\hat{\beta}_1$ is most often *above* the target 1.5, then $\hat{\beta}_1$ is a biased (high) estimator. And finally, if, in all those simulations, the estimate $\hat{\beta}_1$ is sometimes above, sometimes below the target 1.5, but on average (over infinitely many simulations) equal to the target 1.5, then $\hat{\beta}_1$ is an unbiased estimator.

In another simulation, we got the estimate $\hat{\beta}_1 = 1.5373$, this time above the target $\beta_1 = 1.5$. But two simulations are not enough to demonstrate unbiasedness. Figure 3.1 shows the histogram of 100,000 such estimates. Notice that the estimates are neither systematically larger nor systematically smaller than the target $\beta_1 = 1.5$, as expected for an unbiased estimator.

R code for Figure 3.1

```
Nsim = 100000; b1.ols = numeric(Nsim)
for (i in 1:Nsim) {Sim.Cost = beta0 + beta1*Widgets + sigma*rnorm(n)
  b1.ols[i] = lm(Sim.Cost ~ Widgets)$coefficients[2]  }
hist(b1.ols, freq=F, breaks=100, main="",
  xlab = expression ("OLS estimate of" ~beta[1]))
abline(v=1.5, lwd=2.5)
```

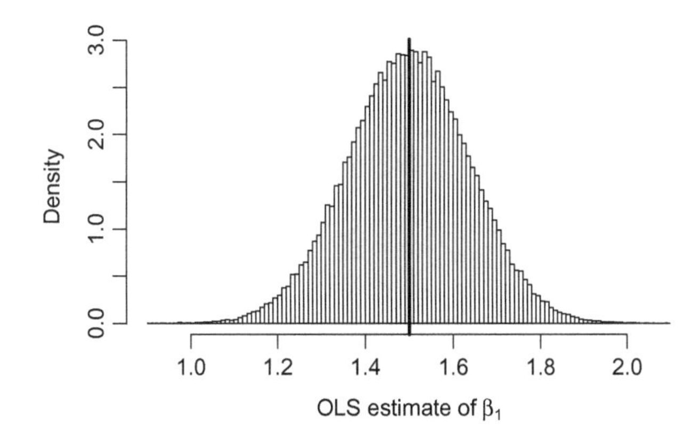

FIGURE 3.1
Histogram of 100,000 estimates $\hat{\beta}_1$, each calculated from a sample of 40 observations simulated from the $N(55+1.5x, 250^2)$ model. The vertical line is at $\beta_1 = 1.5$. As expected for an unbiased estimator, the estimates $\hat{\beta}_1$ are neither systematically larger nor systematically smaller than $\beta_1 = 1.5$.

In practice, you do not repeat your sampling 100,000 times, you just take one sample (e.g., one sample of $n = 40$ observations). But look at Figure 3.1: The estimate $\hat{\beta}_1$ you get from your one sample is a number on the horizontal axis. So put your finger somewhere on the horizontal axis in Figure 3.1, but *not at 1.5*. That location where your finger is, is a possible $\hat{\beta}_1$ that you might get in your one data sample.

Did you put your finger on the graph as we asked? If not, then read the previous paragraph again and do it!

3.3 Biasedness of OLS Estimates When the Classical Model Is Wrong

Unbiasedness of the estimates $\hat{\beta}_0$ and $\hat{\beta}_1$ also implies unbiasedness of the OLS-estimated conditional mean value, $\hat{\mu}_x = \hat{\beta}_0 + \hat{\beta}_1 x$, when the classical model is valid. But when the classical model does not correspond to Nature's model, the OLS-estimated conditional mean value $\hat{\mu}_x = \hat{\beta}_0 + \hat{\beta}_1 x$ can be biased. Motivated by the Product Complexity example of Figure 1.16, where Y = Preference and X = Complexity, suppose that Nature's mean function is not linear, but instead a curved function $E(Y \mid X = x) = f(x)$. But you do not know Nature's ways, so you assume the classical model $Y \mid X = x \sim N(\beta_0 + \beta_1 x, \sigma^2)$. Then your OLS-estimated conditional mean value, $\hat{\mu}_x = \hat{\beta}_0 + \hat{\beta}_1 x$, is biased.

Suppose in particular that Nature's data-generating process is $Y \mid X = x \sim \mathrm{N}(7 + x - 0.03x^2, 10^2)$, where $X \sim \mathrm{N}(15, 5^2)$. A scatterplot of $n = 1{,}000$ data values from this process is shown in the left panel of Figure 3.2, with OLS line and LOESS fit superimposed. Notice that the LOESS fit looks more like the true, quadratic function than the incorrect linear function.

The right panel of Figure 3.2 shows that the OLS estimates $\hat{\mu}_{15} = \hat{\beta}_0 + \hat{\beta}_1(15)$, based on samples of size $n = 1{,}000$, are biased (low) estimates of the true mean.

R code for Figure 3.2

```
par(mfrow=c(1,2))
n = 1000 ;X = rnorm(n, 15,5); Xsq = X^2
Y = 7 + X -.03*Xsq + rnorm(n,0,10)
plot (X,Y, pch="."); abline(lsfit(X,Y)); add.loess(X,Y, lty=2)
Nsim = 10000; mu.hat = numeric(Nsim)
for (i in 1:Nsim) {
  X = rnorm(n, 15,5); Xsq = X^2
  Y = 7 + X -.03*Xsq + rnorm(n,0,10)
  fit = lm(Y ~ X)
  mu.hat[i] = fit$coefficients[1] + fit$coefficients[2] * 15
 }

hist(mu.hat, breaks=20, main="", cex=0.8,
  xlab = "Estimated Mean of Y | X = 15")
abline(v = 7 + 15 -.03*15^2, lwd=2)
```

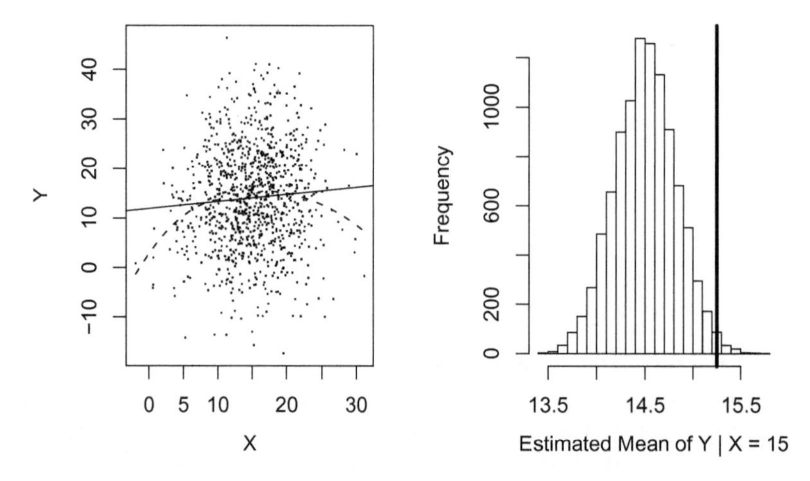

FIGURE 3.2
Left panel: Scatterplot of 1,000 data points (X, Y) having non-linear (quadratic in this example) conditional mean function, with linear and LOESS fits superimposed. Right panel: Histogram of 10,000 estimates of $E(Y \mid X = 15)$ using the (incorrect) linear fits to each data set having $n = 1{,}000$ observations. The vertical line is the true conditional mean $E(Y \mid X = 15) = 15.25$, showing that the OLS estimated mean is biased low when Nature is modeled incorrectly.

Self-study question: Connect the information in Section 1.9 with the information in Section 3.3 above to answer the question, why OLS estimates are *nearly always* biased, in practice?

3.4 Estimation and Practical Use of σ^2

The parameter σ^2 is perhaps the most important parameter of a regression model because it measures prediction accuracy. As shown previously, another way to write the model is $Y = \beta_0 + \beta_1 x + \varepsilon$, where $\varepsilon \sim N(0, \sigma^2)$. Thus the prediction error terms are the ε values, and these differ from zero with a variance of σ^2.

If the β's were known (as is true in simulations but not in reality), you could calculate errors $\varepsilon_i = \{Y_i - (\beta_0 + \beta_1 x_i)\}$ and obtain an unbiased estimate of σ^2 as:

$$(\text{An unbiased estimator of } \sigma^2) = \frac{1}{n}\sum_{i=1}^{n} \varepsilon_i^2$$

This estimator is unbiased because *each individual* ε_i^2 *is an unbiased estimator of* σ^2, which you can see as follows:

$$E(\varepsilon_i^2) = E\{(Y_i - \beta_0 - \beta_1 x_i)^2\} = \text{Var}(Y_i \mid X = x_i) = \sigma^2.$$

Self-study question: Using the fact that each ε_i^2 is an unbiased estimator of σ^2, along with the linearity and additivity properties of expectation, demonstrate mathematically that estimator $\frac{1}{n}\sum_{i=1}^{n} \varepsilon_i^2$ is an unbiased estimator of σ^2.

However, in practice, you cannot use this estimator because the β's are unknown; thus the ε's are unknown (or unobservable) as well. But you can use a similar estimator based on the residuals $e_i = \{Y_i - (\hat{\beta}_0 + \hat{\beta}_1 x_i)\}$, which are observable:

$$(\text{Another estimator of } \sigma^2) = \frac{1}{n}\sum_{i=1}^{n} e_i^2$$

This is, in fact, the maximum likelihood estimator, as given in Chapter 2. However, this estimator is biased: Recall that the values $\hat{\beta}_0$, $\hat{\beta}_1$ are chosen to minimize SSE; that is, $\text{SSE} = \sum_{i=1}^{n} e_i^2$ is a minimum. In particular, $\text{SSE} = \sum_{i=1}^{n} e_i^2 \le \sum_{i=1}^{n} \varepsilon_i^2$, which means that the estimator $\frac{1}{n}\sum_{i=1}^{n} e_i^2 = \text{SSE}/n$ is *biased low* since $\frac{1}{n}\sum_{i=1}^{n} \varepsilon_i^2$ is unbiased.

In basic statistics, you learned that the variance estimator uses "n – 1" in the denominator instead of "n" to remove similar bias; the quantity "n – 1" is sometimes called *degrees of freedom*. You may have also heard that you lose a degree of freedom for every parameter you estimate. In regression, these parameters refer to the β's, so in simple regression, you lose two degrees of freedom. This leads to the following estimator of σ^2.

The standard, unbiased estimator of conditional variance in the classical regression model

$$\hat{\sigma}^2 = \frac{1}{n-(\#\text{ of }\beta\text{'s})}\sum_{i=1}^{n} e_i^2 = \frac{1}{n-(\#\text{ of }\beta\text{'s})}\text{SSE}$$

One theorem from mathematical statistics states that *if* the assumptions of the classical regression model are all true, *then* $\hat{\sigma}^2$ is an unbiased estimator of σ^2. But if the assumptions are not true, then $\hat{\sigma}^2$ can be badly biased.

We will not prove the unbiasedness of $\hat{\sigma}^2$ in this book, because the math is harder than usual. Simulation provides a better way to understand the concept because it makes the concept transparently clear. Simulation also makes it easy to show how the estimator is biased when the assumptions are not valid. So, in the case of the estimates of σ^2, we most want you to understand what bias means from the standpoint of simulation, rather than from the standpoint of mathematical proof.

In the simulation study of Section 3.2 above, $\sigma^2 = 250^2 = 62{,}500$. Figure 3.3 shows the estimated distributions of the biased estimator (SSE/n) and of the unbiased estimator $\left(\hat{\sigma}^2 = \text{SSE}/(n-2)\right)$, based on 100,000 simulated data sets.

R code for Figure 3.3

```
n=40
Nsim = 100000; sig2.hat = numeric(Nsim);
sig2.biased = numeric(Nsim)
for (i in 1:Nsim) {Sim.
Cost = beta0 + beta1*Widgets + rnorm(n,0,sigma)
     SSE = sum(lm(Sim.Cost~Widgets)$residuals^2)
     sig2.hat[i] = SSE/(n-2); sig2.biased[i] = SSE/n }

par(mfrow=c(1,2))
hist(sig2.hat, freq=F, breaks=100, xlab="Unbiased estimates of
sig.squared", main="")
abline(v=250^2, lwd=2, col="gray")
abline(v=mean(sig2.hat), lwd=2, lty=2)
hist(sig2.biased, freq=F, breaks=100, xlab="Biased estimates of
sig.squared", main = "")
abline(v=250^2, lwd=2, col="gray")
abline(v=mean(sig2.biased), lwd=2, lty=2)
```

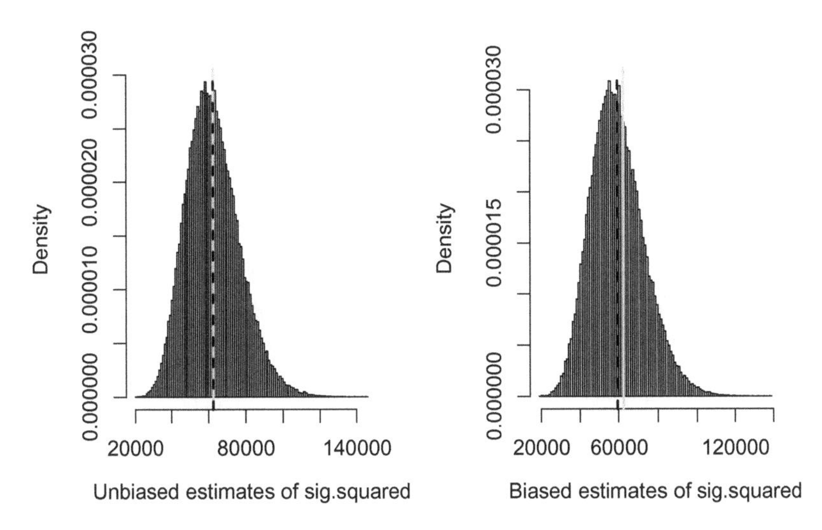

FIGURE 3.3
Histograms of unbiased estimates of σ^2, SSE/(n – 2) (left panel), and biased estimates SSE/n (right panel). The true variance is $250^2 = 62{,}500$, shown as vertical solid lines. The dashed line indicates the average of the 100,000 simulated estimates. No bias is noted in the left panel, while slight downward bias is seen in the right panel.

More interesting than the estimated variance $\hat{\sigma}^2$ is the estimated standard deviation $\hat{\sigma}$, because $\hat{\sigma}$ measures how far the individual Y values are from their conditional means. In particular, under the classical model, 95% of the potentially observable $Y \mid X = x$ data are in the range $(\beta_0 + \beta_1 x) \pm 1.96\sigma$. The estimate of σ has a special name:

Estimated conditional standard deviation of *Y* in the classical regression model

$$\hat{\sigma} = \sqrt{\frac{1}{n-(\#\text{ of }\beta\text{'s})}\sum_{i=1}^{n} e_i^2}$$

This statistic goes by various names, including "Root mean squared error," and "Estimated conditional standard deviation." The `R` software calls it "Residual standard error." But "standard error" also refers to the accuracy of the estimated β parameters, thus we will give $\hat{\sigma}$ the more precise label, "Estimated conditional standard deviation," because it is an estimate of the standard deviation of the conditional distribution $p(y \mid x)$.

For all the attention given to making $\hat{\sigma}^2$ unbiased, it is ironic that the more interesting statistic, $\hat{\sigma} = \sqrt{\hat{\sigma}^2}$, is biased low because of Jensen's inequality. But, similar to what Figure 3.3 shows for the biased "n in the denominator" estimator of variance, the bias of the estimator $\hat{\sigma} = \sqrt{\hat{\sigma}^2}$ is negligible. So we will use that estimator, despite its bias.

Self-study question: How can you use Jensen's inequality to show that $\hat{\sigma} = \sqrt{\hat{\sigma}^2}$ is biased low?

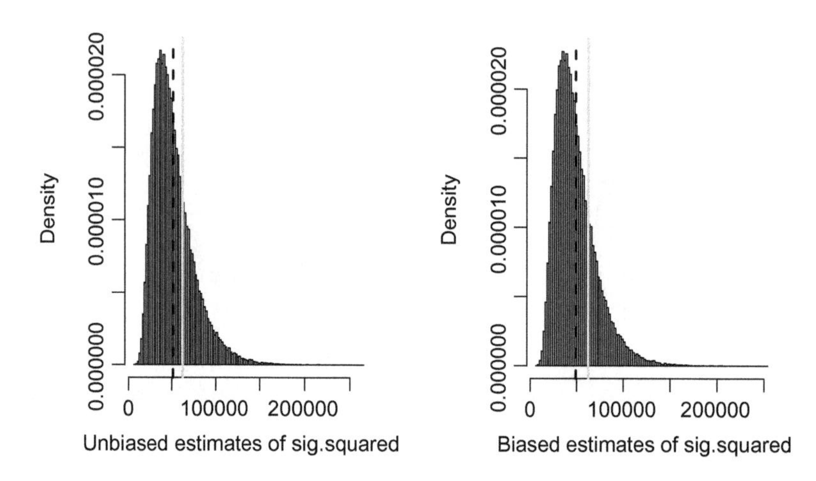

FIGURE 3.4
Histograms of (supposedly) unbiased estimates of σ^2, SSE/$(n-2)$ (left panel), and biased estimates SSE/n (right panel) when the error terms are not independent. The true variance is $250^2 = 62{,}500$, shown as vertical solid lines. The dashed line indicates the average of the 100,000 simulated estimates. Downward bias is noted for both estimates, caused by violation of the independence assumption.

A source of much greater bias is failure of model assumptions. When the error terms ε_i are not independent, as occurs, for example, with time series and spatial data, then the estimated variance can be noticeably biased. If you replace "`rnorm(n,0,sigma)`" with "`arima.sim(n = n, list(ar = c(0.8)),sd = sigma*sqrt(1-.8^2))`" in the code for Figure 3.3, the errors follow an *autoregressive model*. This model states that the residual ε_i is closely related to the previous residual ε_{i-1} with autocorrelation 0.8 (more details on these kinds of models appear in Chapter 12). Figure 3.4 is identical to Figure 3.3, except that the data come from a process where the independence assumption of the classical regression model is violated.

The message from Figure 3.4 is clear: Violations of model assumptions can have dramatically negative consequences. Here, the consequence is that failure of model assumptions caused the supposedly unbiased estimate of σ^2 to be biased low.

We now discuss the abstract unbiasedness concepts, and the use of the estimate $\hat{\sigma}$, in the context of actual data. Here are the results of the data analysis of the original (not simulated) Production Cost data set using R:

```
> summary(lm(Cost~Widgets, data = ProdC))

Call:
lm(formula = Cost ~ Widgets, data = ProdC)

Residuals:
    Min      1Q  Median      3Q     Max
-415.31 -193.30    4.76  172.20  451.75

Coefficients:
            Estimate Std. Error t value Pr(>|t|)
(Intercept)  55.4592   168.0455    0.33    0.743
Widgets       1.6199     0.1326   12.22 9.88e-15 ***
---
Signif. codes:  0 '***' 0.001 '**' 0.01 '*' 0.05 '.' 0.1 ' ' 1

Residual standard error: 239.1 on 38 degrees of freedom
Multiple R-squared:  0.7971,    Adjusted R-squared:  0.7917
F-statistic: 149.3 on 1 and 38 DF,  p-value: 9.877e-15
```

The estimates are $\hat{\beta}_0 = 55.4592$, $\hat{\beta}_1 = 1.6199$, and $\hat{\sigma} = 239.1$. You can understand what biased and unbiasedness means in this study as follows:

What unbiasedness means with real data

The estimate $\hat{\beta}_1 = 1.6199$ is one of infinitely possible potentially observable slope estimates because different data sets produced by the same data-generating process give different estimates $\hat{\beta}_1$. If the classical regression model is a perfect representation of the data-generating process, then the average of all other potentially observable values of $\hat{\beta}_1$ is equal to β_1, and that is what it means for $\hat{\beta}_1$ to be unbiased.

Notes:

- A similar interpretation holds for $\hat{\beta}_0$.
- A similar interpretation does not hold for $\hat{\sigma}$, since the average of all similarly potentially observable $\hat{\sigma}$ values is *smaller* than the true unknown value of σ, even when the classical model is valid. So the result $\hat{\sigma} = 239.1$ is from a process where such results tend to be too small; however, the specific estimate $\hat{\sigma} = 239.1$ is not necessarily smaller than σ, because there is variability in the estimate as well as bias.
- Unbiasedness of an estimate tells you that there is no *systematic* tendency for the estimates to be lower, or higher, on average, than their true values.
- Unbiasedness does *not* imply that the estimate is *close* to the true value.

To further apply the theory to the real world, suppose the "next job" will have 1,000 widgets. Under the classical model, 95% of the potentially observable values of the Cost variable when Widgets = 1,000 will be in the range $\beta_0 + \beta_1(1000) \pm 1.96\sigma$. The parameters β_0, β_1 and σ are all unknown, but are *approximated* by $\hat{\beta}_0 = 55.4592$, $\hat{\beta}_1 = 1.6199$, and $\hat{\sigma} = 239.1$. Hence, when there are 1,000 widgets, *approximately* 95% of the potentially observable values of the Cost variable will be in the range $(55.4592 + 1.6199(1000)) \pm 1.96(239.1)$, or in the range from 1,206.72 to 2,143.99. This interval is called an *approximate prediction interval* and is shown in Figure 3.5. Notice in Figure 3.5 that the interval range appears reasonable for capturing 95% of potentially observable Cost values when Widgets = 1,000.

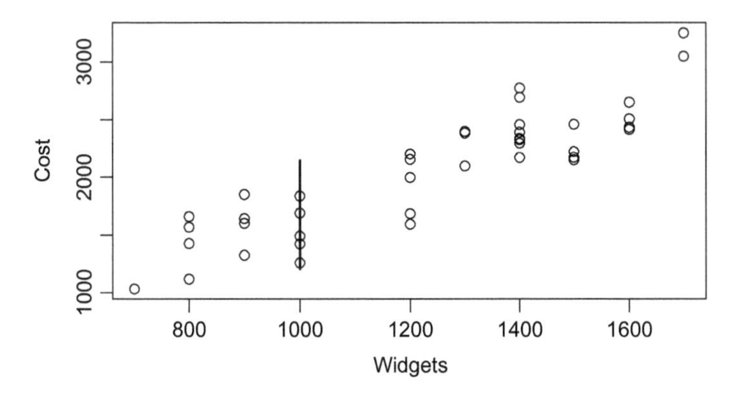

FIGURE 3.5
Scatterplot of the Production Cost data, with approximate 95% prediction interval (solid line), showing a range of the potentially observable values of Cost when Widgets = 1,000.

3.5 Standard Errors

The Gauss-Markov theorem states that the OLS estimator has minimum variance among linear unbiased estimators. What does "variance" of the OLS estimator refer to? Please look at Figure 3.1 again: You can see that there is variability in the possible values of $\hat{\beta}_1$ ranging from 1.0 to 2.0. Variance of the estimator $\hat{\beta}_1$, denoted symbolically by $\text{Var}(\hat{\beta}_1)$, refers to the variance of the distribution $p(\hat{\beta}_1)$ that is shown in Figure 3.1.

If the assumptions of the Gauss-Markov model are true, *then* the following formula gives the exact variance of the OLS estimator $\hat{\beta}_1$.

Variance of the OLS estimator $\hat{\beta}_1$

$$\text{Var}(\hat{\beta}_1) = \frac{\sigma^2}{(n-1)s_x^2}$$

In the formula for $\text{Var}(\beta_1)$, note that $s_x^2 = \sum(x_i - \hat{\mu}_x)^2 / (n-1)$ is the usual estimate of the variance of X. Note that the $\text{Var}(\hat{\beta}_1)$ formula is conditional on the observed values of the X data; this is apparent because s_x^2 is specifically a function of the observed X data.

When coupled with unbiasedness of $\hat{\beta}_1$, smaller $\text{Var}(\hat{\beta}_1)$ implies a more accurate estimate, i.e., an estimate that tends to be closer to β_1. Hence, we have the following interesting conclusions regarding the accuracy of the OLS estimate $\hat{\beta}_1$:

The OLS estimate $\hat{\beta}_1$ of β_1 is more accurate when:

- n is larger, and/or
- s_x^2 is larger, and/or
- σ^2 is smaller.

As mentioned above, the formula given for $\text{Var}(\hat{\beta}_1)$ can be mathematically derived from the assumptions of the Gauss-Markov model. Violation of assumptions renders the formula incorrect. In particular, violation of the homoscedasticity assumption is the rationale for using *heteroscedasticity-consistent standard errors*, which are covered in Chapter 12.

Strangely enough, the mathematics needed to prove the variance formula is easier in the multiple regression model, so we will prove it later in Chapter 7. But for now, you should understand the assumptions that imply the result (e.g., the classical model) and the result itself (the formula for $\text{Var}(\hat{\beta}_1)$) by using simulation: If you simulate many thousands of data sets from the same model, with the same sample size, and with the same X data, then the sample variance estimate of the resulting thousands of $\hat{\beta}_1$ estimates will be (within simulation error) equal to $\sigma^2 / \{(n-1)s_x^2\}$. The simulation also clarifies the "conditional on observed values of the X data" interpretation because the X data are the same for every simulated data set.

Here is the R code to verify the mathematical variance formula:

```
b1.ols = numeric(100000)
for (i in 1:100000) {Sim.Cost = beta0 + beta1*Widgets + sigma*rnorm(n)
 b1.ols[i] = lm(Sim.Cost ~ Widgets)$coefficients[2]  }
var(b1.ols)                       # empirical
sigma^2/((n-1)*var(Widgets))      # theoretical
```

The theoretical (fixed) variance is $\sigma^2/\{(n-1)s_x^2\} = 250^2/\{(40-1)83358.97\} = 0.01922485$, while the variance of our 100,000 $\hat{\beta}_1$ estimates was 0.01911871, which differs from the theoretical value 0.01922485 by random simulation error only.

The square root of the variance is more interesting: The true standard deviation of the $\hat{\beta}_1$ estimates in the simulation is $250/\{(40-1)83358.97\}^{1/2} = 0.1387$. Recall, from the discussion of unbiasedness above, that the mean of the distribution of $\hat{\beta}_1$ is $\beta_1 = 1.5$ in this simulation study. Thus, 95% of the estimates $\hat{\beta}_1$ will be in the range $1.5 \pm 1.96(0.1387)$, or between 1.228 and 1.772. (Yet another mathematical theorem is that, *if* the classical model is true, *then* $p(\hat{\beta}_1)$ is exactly a normal distribution.)

Figure 3.6 shows the true distribution of $\hat{\beta}_1$ for this simulation, indicating the 95% range.

R code for Figure 3.6

```
par(mar=c(5,4.5,4,2)+0.1)
b1.list = seq(1.0, 2.0,.001)
pdf.b1 = dnorm(b1.list, 1.5, 0.1387)
plot(b1.list, pdf.b1, type="l", yaxs="i",
  ylim=c(0,1.2*max(pdf.b1)), xlab=bquote(hat(beta)[1]),
  ylab = bquote("density of " ~ hat(beta)[1]))
points(c(1.228,1.228),c(0, dnorm(1.228,1.5,.1387)),
  type="l", col="gray")
points(c(1.772,1.772),c(0, dnorm(1.772,1.5,.1387)),
  type="l", col="gray")
```

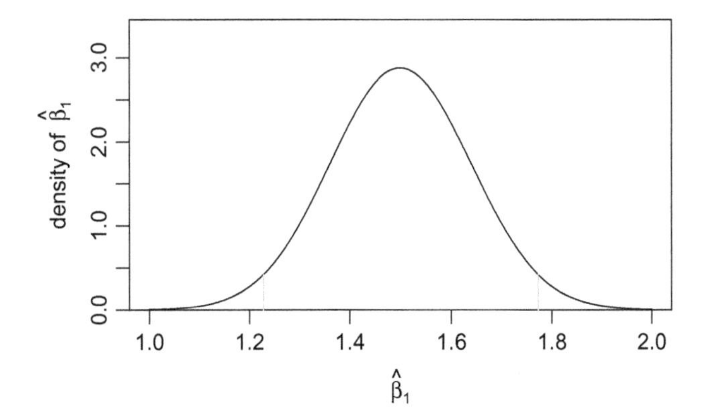

FIGURE 3.6
True distribution of the estimates $\hat{\beta}_1$ in the simulation study, indicating exact 95% bounds. The histogram shown in Figure 3.1 is a simulation-based approximation to this true distribution.

Under the classical model, the random estimate $\hat{\beta}_1$ is within $\pm 1.96\sqrt{\text{Var}(\hat{\beta}_1)}$ of the fixed β_1 for 95% of the random samples from the same process. Equivalently, the fixed β_1 is within $\pm 1.96\sqrt{\text{Var}(\hat{\beta}_1)}$ of the random $\hat{\beta}_1$ for 95% of the random samples from the same process. Since $\text{Var}(\hat{\beta}_1) = \sigma^2 / \{(n-1)s_x^2\}$ in the simple, classical regression model, it follows that $\sqrt{\text{Var}(\hat{\beta}_1)} = \sigma/(s_x\sqrt{n-1})$, and that the fixed β_1 is within $\pm 1.96\sigma/(s_x\sqrt{n-1})$ of the random $\hat{\beta}_1$ for 95% of the random samples from the same process. This fact gives the exact 95% confidence interval for the fixed, unknown β_1:

Exact (when the classical model is true) 95% confidence interval for β_1

$$\hat{\beta}_1 \pm 1.96\sigma/(s_x\sqrt{n-1})$$

The exact confidence interval is fine in theory, but it includes the unknown parameter σ, so you cannot use it. But you can estimate σ as discussed in the previous section via $\hat{\sigma}$, called "Residual standard error" by the `lm` function in `R`. Since $\hat{\sigma} \neq \sigma$, you cannot just substitute one for the other and expect everything is ok. What changes is that the interval is no longer exact, it is only approximate.

Approximate 95% confidence interval for β_1; Standard error

$$\hat{\beta}_1 \pm 1.96\,s.e.(\hat{\beta}_1),$$

where $s.e.(\hat{\beta}_1) = \hat{\sigma}/(s_x\sqrt{n-1})$ is called the *standard error of* $\hat{\beta}_1$.

Standard errors are usually displayed next to the parameter estimates by regression software: From the `R` analysis of the Production Cost data, we see that $s.e.(\hat{\beta}_1) = 0.1326$. This value is calculated by the software as follows:

$$s.e.(\hat{\beta}_1) = \hat{\sigma}/(s_x\sqrt{n-1}) = 239.1 / \left\{83358.97^{1/2}(40-1)^{1/2}\right\} = 0.1326$$

Standard errors are important in regression analysis, so it is important to have a good definition handy in your mind so you can understand them:

Definition of standard error

The standard error of an estimate is the *estimated standard deviation* of the estimator that corresponds to the estimate.

There are lots of "estimate" terms in the standard error definition! To explain, consider the Production Cost example above, where 0.1326 is the estimated standard deviation of the distribution of the random estimator $\hat{\beta}_1$. Recall (again!) that $\hat{\beta}_1$ is a random variable because it takes a different value for every one of the infinitely many potentially observable samples, all from the same data-generating process. The estimated standard deviation of the distribution of possible $\hat{\beta}_1$ values is 0.1326, and is calculated using data from just one of these infinitely many samples. The *true standard deviation* of the estimator $\hat{\beta}_1$ is $\sigma / (s_x\sqrt{n-1})$; the *estimated standard deviation* is $\hat{\sigma}/(s_x\sqrt{n-1})$; this is called an *estimated standard deviation* because $\hat{\sigma}$ is an *estimate* of σ.

Think about the above explanation for a second: Isn't it remarkable that you can estimate the standard deviation of *infinitely many* potentially observable $\hat{\beta}_1$ values just using your *one data set*, which gives you *just one* $\hat{\beta}_1$? It is not magic. Rather, it is a result of a mathematical theorem: *If* the data are produced as the G-M model stipulates, *then* the formula for $\text{Var}(\hat{\beta}_1)$ given above is correct. The standard error comes from that variance formula, and all terms in the formula can be estimated using a single sample of data.

Assumptions are important: If the assumptions of the model are grossly violated (particularly independence and constant variance), then the standard error reported by the software will be a very poor estimate of the standard deviation of those infinitely many potentially observable $\hat{\beta}_1$ values produced by the true data-generating process.

There is also a standard error for $\hat{\beta}_0$, shown in the `lm` output above as $s.e.(\hat{\beta}_0) = 168.0455$, and is interpreted and used the same way: 168.0455 is the estimated standard deviation of the potentially observable values of $\hat{\beta}_0$, and an approximate 95% confidence interval for the true, fixed unknown β_0 of the data-generating process is $55.4592 \pm 1.96(168.0455)$.

3.6 Exact Inferences: Confidence Intervals for the β's

To interpret the estimate and its standard error, you should have a mental conversation with yourself, saying something like this:

How to think about the estimate and its standard error

Hmmm, the estimated slope is shown in the output as 1.6199, and the standard error is shown in the output as 0.1326. So the actual slope is most likely in the range $1.6199 \pm 2(0.1316)$, or roughly between 1.6 ± 0.26. AHA! The true slope is most likely a positive number! So the X variable has a positive relation to Y!

We used 2.0 rather than 1.96 as a multiplier of the standard error because the result is only approximate anyway, so why not? We might as well simplify things by using another approximation, 2.0 instead of 1.96. It just makes life easier. And it works well in practice, so we generally recommend that you follow the advice given by the above mental conversation.

But there are precise, *mathematically exact* results that you can use in the case where the data are produced by the classical model. The theory is mathematically deep, but you probably have seen it before, to one degree or another. It involves "Student's T distribution," which is ubiquitous in statistics. In a nutshell, the issue revolves around how to deal with the estimate $\hat{\sigma}$ of σ in the standard error formula. After all, as shown above, the first interval formula involving 1.96 and σ is exact; the only reason for calling the second interval formula "approximate" is because of the substitution of $\hat{\sigma}$ for σ. The effect of using $\hat{\sigma}$ rather than σ can be precisely, exactly, quantified. A mathematical theorem states that *if* the classical regression model produces the real data, *then* the additional variability incurred when you use $\hat{\sigma}$ rather than σ is precisely accounted for by using the T (Student's T) distribution rather than the Z (standard normal) distribution.

Specifically, the critical value 1.96 is from the Z (standard normal) distribution, the number that puts 95% probability between –1.96 and 1.96. It is, therefore, the 0.975 quantile of the standard normal distribution. In `R` it is `qnorm(.975)`, which returns the even more precise value 1.959964.

To account for the error in using the estimate $\hat{\sigma}$ of σ in the standard error formula, you need to use the T distribution rather than the Z distribution. The T distribution involves a "degrees of freedom" parameter, which in essence measures the accuracy of $\hat{\sigma}$ as an estimator of σ. This degrees of freedom quantity is mathematically identical to the divisor used to make the estimated variance an unbiased estimate:

$$dfe = n - (\#\text{ of } \beta\text{'s})$$

The "*e*" on "*df*" refers to "error": Recall that, σ, the conditional standard deviation of $Y \mid X = x$, is also the standard deviation of the error term ε. You can think of *dfe* as the "effective sample size" that is used to estimate the error standard deviation.

There is also a "model degrees of freedom" that we will discuss later, using the symbol *dfm*. The model degrees of freedom means something completely different: It refers to the flexibility (freedom) of the regression model; essentially the number of free parameters (β's) in the model, excluding the intercept.

To get exact intervals for regression coefficients, you use the quantiles of the T_{dfe} distribution, rather than the quantiles of the Z distribution. The mathematics is precise but will not be proved here: It states that, *if* the data are produced by the classical regression model, *then* you have the following result.

Exact (assuming the classical model) $100(1-\alpha)\%$ confidence interval for a β

The exact confidence interval for any β in the classical regression model is given by

$$\hat{\beta} - c \times s.e.(\hat{\beta}) < \beta < \hat{\beta} + c \times s.e.(\hat{\beta}).$$

The critical value is $c = T_{dfe,1-\alpha/2}$, the $1-\alpha/2$ quantile of the T_{dfe} distribution.

When $\alpha = 0.05$, this interval is very close to the approximate "plus or minus 2 standard errors" interval described above, but unlike that one, which is approximate, the one above is exact in the sense that the confidence level is precisely $100(1-\alpha)\%$, not approximately $100(1-\alpha)\%$.

Consider the `lm` output for the Production Cost data. The parameter estimate is $\hat{\beta}_1 = 1.6199$, and its standard error is $s.e.(\hat{\beta}) = 0.1326$. The degrees of freedom for error are shown in the output as 38 (n = 40, and there are two β's, so $dfe = 40 - 2 = 38$). The critical value is $c = T_{38,0.975} = 2.024394$ (available using `R` as `qt(.975,38)`), so the exact 95% confidence interval for β_1 is $1.6199 \pm 2.024394 \times 0.1326$, or $1.351 < \beta_1 < 1.888$.

You can get these exact intervals for the β's easily by using the command `confint(lm(Cost~Widgets, data=ProdC))`in `R`, which gives the following output:

```
                 2.5 %     97.5 %
(Intercept) -284.731009 395.64950
Widgets        1.351451   1.88825
```

So, you can conclude with exactly 95% confidence, provided the true data-generating process is the classical regression model, that $1.351 < \beta_1 < 1.888$. Further, you should understand that the interval can only be considered as an exact 95% interval when the assumptions of the classical regression model are all true.

3.6.1 Understanding "Exactness" and "Non-exactness" via Simulation

What does "exact" mean in these discussions? It means that the true confidence level is exactly 95% when you use a 95% confidence interval. Non-exactness means that the true confidence level is not equal to 95%—it may be higher or lower than 95%. Further, "true confidence level" refers to the true probability that the parameter lies within the prescribed confidence limits.

Here is a simple simulation to illustrate "exactness." The data are simulated according to the classical model, the 95% interval for β_1 is calculated, and we check whether the true β_1 lies within the interval. Then we repeat that process 100,000 times, finding the proportion of the 100,000 intervals that contain the true β_1. This proportion should be close to 95% and will be exactly 95% with infinitely many (rather than 100,000) simulations.

On the other hand, when data are simulated from a model where the assumptions are violated, the proportion will be different from 95%, even with infinitely many simulations. The simulation code that follows simulates data from the classical model, and also from the model with non-normal conditional distributions used to obtain Figure 1.11.

```
Nsim = 100000; Exact.LCL = numeric(Nsim); Exact.UCL = numeric(Nsim)
Nonexact.LCL = numeric(Nsim); Nonexact.UCL = numeric(Nsim)
for (i in 1:Nsim) {Sim.Cost.1 = 55 + 1.5*Widgets + 250*rnorm(40)
Sim.Cost.2 = 55 + 1.5*Widgets + 250*(rt(40,4)^2 -2)  #non-normal case
 CL.1 = confint(lm(Sim.Cost.1 ~ Widgets))
 CL.2 = confint(lm(Sim.Cost.2 ~ Widgets))
 Exact.LCL[i] = CL.1[2,1]; Exact.UCL[i] = CL.1[2,2]
 Nonexact.LCL[i] = CL.2[2,1]; Nonexact.UCL[i] = CL.2[2,2]    }
# Exact case
mean((Exact.LCL < 1.5) & (Exact.UCL > 1.5))
#non-exact case
mean((Nonexact.LCL < 1.5) & (Nonexact.UCL > 1.5))
```

Results are as follows; yours will vary slightly by randomness:

```
> mean((Exact.LCL < 1.5) & (Exact.UCL > 1.5))
[1] 0.94907
> #non-exact case
> mean((Nonexact.LCL < 1.5) & (Nonexact.UCL > 1.5))
[1] 0.96058
```

Thus, in the case where the classical model is true, 94.907% of the 100,000 samples gave a confidence interval that contained the true $\beta_1 = 1.5$. According to the mathematical theory, this percentage will be exactly 95% with infinitely many simulated data sets.

On the other hand, in the simulation where the conditional distributions are non-normal as illustrated in Figure 1.11, 96.058% of the 100,000 samples gave a confidence interval that contained the true $\beta_1 = 1.5$. The mathematical theory does *not* state that this percentage will be exactly 95% with infinitely many simulated data sets. In fact, the true percentage with infinitely many data sets will be more than 95% in this case.

The non-exactness of the confidence interval is not a huge problem for the given simulation study, because the actual confidence level is close to 95% in the non-normal case. This study provides an example of our common refrain: You can best understand why and whether violations of assumptions are problematic via simulation.

Violations of assumptions other than normality can cause bigger problems. Figure 3.2 shows a case where the estimates are biased, and in such cases the intervals will systematically miss the target on the low side, leading to coverage rates close to 0% in extreme cases. Similarly, heteroscedasticity (non-constant variance) can cause the standard errors to be too small, also leading to coverage rates much lower than 95%, which you can verify by using simulation.

As it turns out, violation of the normality assumption is not usually a major concern for the validity of confidence intervals for the β parameters: Even with non-normal conditional distributions $p(y \mid x)$, the Central Limit Theorem dictates that the distribution of the parameter estimates will be approximately normal. Other inferences are not so robust to non-normality: The prediction interval discussed in Section 3.8 below will behave quite poorly with non-normal processes. Inferences for variance parameters are similarly non-robust. Further, even when OLS-based inferences are robust in the sense of having confidence levels near 95% under non-normality, the OLS estimates themselves can be quite inaccurate relative to ML estimates under non-normality.

3.6.2 Practical Interpretation of the Confidence Interval for β_1

We now discuss the practical interpretation of the confidence interval for the slope parameter. As with everything in regression, these interpretations involve conditional distributions.

If the linearity assumption is true, *then* the parameter β_1 is the difference between the means of the conditional distributions of Y for cases where the X variable differs by one unit. Specifically:

$$\mathrm{E}(Y \mid x+1) - \mathrm{E}(Y \mid x) = \{\beta_0 + \beta_1(x+1)\} - (\beta_0 + \beta_1 x) = \beta_0 + \beta_1 x + \beta_1 - \beta_0 - \beta_1 x = \beta_1$$

Thus, the mean of the distribution of potentially observable Y when $X = x+1$ is precisely β_1 higher than the mean of the distribution of potentially observable Y when $X = x$. In particular, the mean of the distribution of Cost when Widgets = 1,001 is exactly β_1 higher than the mean of the distribution of Cost when Widgets = 1,000. And it does not matter which two values $(x+1, x)$ that you compare: The mean of the distribution of Cost when Widgets = 1,601 is exactly β_1 higher than the mean of the distribution of Cost when Widgets = 1,600.

Here and throughout the book, we will refer to β_1 as a measure of the *effect* of X on Y. In general, the word *effect* has the following meaning:

The meaning of the phrase "X has an effect on Y"

When the conditional distribution $p(y \mid X = x_1)$ differs from $p(y \mid X = x_2)$, for some specific values x_1 and x_2 of the variable X, then X has an effect on Y.

Often, the effect occurs through the mean function; i.e., if $\beta_1 \neq 0$, then the mean of $p(y \mid X = x_1)$ is $\beta_0 + \beta_1 x_1$, which differs from the mean of $p(y \mid X = x_2)$, which is $\beta_0 + \beta_1 x_2$. When the means of two distributions are different, then the distributions must be different, hence a non-zero β_1 implies that there is an effect of X on Y. (But note that it is possible that the effect of X on Y occurs in the variance function, not the mean; therefore, even if $\beta_1 = 0$, it is still possible that X can have an effect on Y.)

The confidence interval $1.351 < \beta_1 < 1.888$ tells you that the difference between the means of any two conditional distributions where the Widgets differ by one, is between 1.351 and 1.888 Cost units, with 95% confidence. Thus, an additional Widget is associated with an increase in *average* Cost that is somewhere between 1.351 and 1.888 units. Since the interval includes only positive values, the conditional distribution of Cost when Widgets is higher by one unit is shifted to the right, and you can, therefore, state that Widgets has a positive effect on Cost.

The word "average" is crucial; otherwise, the interpretation is wrong. Just because one job has more widgets than another job does not mean that the *actual* cost will be higher. Figure 3.5 shows many cases where the *actual* cost is lower when the number of widgets is higher.

Self-study question: In Figure 3.5, consider all jobs having Widgets = 1,200 or Widgets = 1,400. How many jobs when Widgets = 1,400 have a lower cost than a job where Widgets = 1,200?

You can modify this interval for cases where the X variable differs by more than 1 unit: Consider two types of jobs, one with Widgets = 1,100 and the other with Widgets = 1,000. Then

$$\begin{aligned} E(Y \mid X = 1100) - E(Y \mid X = 1000) &= \{\beta_0 + \beta_1(1100)\} - \{\beta_0 + \beta_1(1000)\} \\ &= \beta_0 + \beta_1(1100) - \beta_0 - \beta_1(1000) \\ &= 100\beta_1 \end{aligned}$$

Since $1.351 < \beta_1 < 1.888$ with 95% confidence, it follows that $(100)1.351 < 100\beta_1 < (100)1.888$ with 95% confidence, or that $135.1 < 100\beta_1 < 188.8$. Thus, the *average* of the potentially observable Cost is somewhere between 135.1 and 188.8 units more for jobs having 1,100 widgets than for jobs having 1,000 widgets.

Note that the confidence interval $1.351 < \beta_1 < 1.888$ shows that there is an *increase* in the average Cost associated with jobs having more widgets. Had the interval included 0, we would not be so sure: Suppose the interval was $-1.351 < \beta_1 < 1.888$. In this case, we would say that jobs with 1,001 widgets might cost less, on average, by up to 1.351 less, than jobs having 1,000 widgets. So in the case where the interval for β_1 *includes* 0, we are not sure whether the conditional distribution of Y when $X = x + 1$ is *shifted to the left* of the distribution of Y when $X = x$ (corresponding to $\beta_1 < 0$), or whether it is *shifted to the right* (corresponding to $\beta_1 > 0$).

When the confidence interval for β_1 *excludes* zero, it has been stated historically that X is a *statistically significant* predictor of Y, because the *direction* of the effect is ascertained. When the confidence interval for β_1 includes zero, X has been called a *statistically insignificant* predictor of Y, because the *direction* of the effect cannot be ascertained. This way of determining statistical significance using confidence intervals is equivalent to the (equally) historical way of determining statistical significance using the "p–value < 0.05 " criterion, to be discussed in Section 3.9.

3.7 Exact Inferences: Confidence Intervals for $E(Y \mid X = x)$

Recall that $E(Y \mid X = x)$ is the mean of the (infinitely many) potentially observable values of Y when $X = x$. If you had many observations (say, m, for "many") on Y when $X = x$, you would not need to use regression at all to find the confidence interval for $E(Y \mid X = x)$. Instead, you could use the ordinary confidence interval that you learned in elementary statistics, given by

$$\bar{y}_x \pm t_{m-1,1-\alpha/2}\hat{\sigma}_{y|x} / \sqrt{m},$$

where $\hat{\sigma}^2_{y|x} = \sum_{i=1}^{m}(y_{i,x} - \bar{y}_x)^2 / (m-1)$ is the usual estimate of the variance based on m observations. The "x" subscript tells you that these values are all observed for a particular $X = x$.

However, in regression analysis, there are usually very few observations on Y at any particular $X = x$. (See Table 1.6 for example.) Instead, you can use regression analysis to "borrow strength" from nearby data. In the Production Cost case there are no data values where $X = 850$, yet that does not stop you from constructing an interval for $E(Y \mid X = 850)$: Under the linearity assumption of the regression model, $E(Y \mid X = 850) = \beta_0 + \beta_1(850)$, which you estimate by using $\hat{\beta}_0 + \hat{\beta}_1(850)$.

To construct a confidence interval, you need a standard error. But before you can get the standard error, you need the standard deviation. And before you get the standard deviation, you need the variance.

So consider $\text{Var}(\hat{\beta}_0 + \hat{\beta}_1 x)$ for some fixed $X = x$ (like X = 850). Without giving a proof, we will just give the formula for the (conditional on the X data) variance. (It turns out that the proof is easier in the multiple regression case using matrix algebra.) The variance formula, like so much else in regression, is actually the result of yet another mathematical theorem that states, *if* the assumptions are true, *then* the variance, standard deviation, and standard error are given as follows.

Variance, standard deviation, and standard error of $\hat{\beta}_0 + \hat{\beta}_1 x$

$$\text{Var}(\hat{\beta}_0 + \hat{\beta}_1 x) = \sigma^2 \left\{ \frac{1}{n} + \frac{(x - \hat{\mu}_x)^2}{(n-1)s_x^2} \right\}$$

$$s.d.(\hat{\beta}_0 + \hat{\beta}_1 x) = \sigma \sqrt{\frac{1}{n} + \frac{(x - \hat{\mu}_x)^2}{(n-1)s_x^2}}$$

$$s.e.(\hat{\beta}_0 + \hat{\beta}_1 x) = \hat{\sigma} \sqrt{\frac{1}{n} + \frac{(x - \hat{\mu}_x)^2}{(n-1)s_x^2}}$$

These formulas lead to the confidence interval, which is exact when the true data-generating process is the classical regression model.

The exact $100(1-\alpha)\%$ confidence interval for $\text{E}(Y \mid X = x)$

$$\hat{\beta}_0 + \hat{\beta}_1 x \pm t_{dfe,1-\alpha/2} s.e.(\hat{\beta}_0 + \hat{\beta}_1 x)$$

The law of large numbers (LLN) provides insight into this interval: It states that as the sample size gets larger, the mean of the observed data becomes closer to the mean of the potentially observable data. This increasing accuracy is reflected in the fact that the "elementary statistics" confidence interval for the mean, $\bar{y}_x \pm t_{m-1,1-\alpha/2}\hat{\sigma}_{y|x}/\sqrt{m}$, has width which shrinks to 0 width as $m \to \infty$. The shrinking of the interval width to zero also occurs for the regression model-based confidence interval for $\text{E}(Y \mid X = x)$ as $n \to \infty$, as you can see because there are n's in the denominator of the standard error formula.

You can get these intervals easily by using R:

```
fit = lm(Cost ~ Widgets, data=ProdC)
x = data.frame(c(850))
colnames(x) = c("Widgets")
predict(fit,x, interval="confidence")
```

The results are as follows:

```
       fit      lwr      upr
1 1432.332 1303.753 1560.912
```

So you can say, with exactly 95% confidence (recall that exactness requires all model assumptions to be precisely true) that the *mean* of all potentially observable costs, for jobs with 850 widgets, is some number between 1,303.753 and 1,560.912 Cost units.

3.8 Exact Inferences: Prediction Intervals for $Y \mid X = x$

Notice that the range of the confidence interval, 1,303.753 and 1,560.912, is around 250 Cost units, much narrower than the range shown by the vertical bar in Figure 3.5, which is around 1,000 Cost units. Why? Is this yet another approximation? No. The reason for the big difference is that the *mean of all potentially observable values of the Cost variable* for jobs with any fixed number of widgets, is very much different than a *single potentially observable value of the Cost variable.*

It makes sense, intuitively and by the Law of Large Numbers, that you can estimate a *mean value* more and more precisely, with a margin of error tending towards 0, when you increase your sample size. However, you cannot estimate a *single value* with such precision, no matter how large is your sample size.

To understand the distinction between a *single value* of Y and the *mean value* of the potentially observable Y's, suppose you have a model for how stock returns behave:

$$\left(\text{Return on Company A's stock}\right) = \beta_0 + \beta_1(\text{Return on the } S\&P \text{ 500 index}) + \varepsilon.$$

Suppose also that you have lots of historical data (maybe your n is in the tens of thousands) with which you can estimate β_0 and β_1. Suppose you have used these data to estimate this model as

$$\left(\text{Return on Company A's stock}\right) = 0.0032 + 0.67(\text{Return on the } S\&P \text{ 500 index}) + e$$

Now, suppose we tell you that the return on the *S&P 500* index was 0.005 (0.5%) yesterday. What can you tell us about the return on Company A's stock? Was it exactly $0.0032 + 0.67(0.005) = 0.00655$ (0.655%)? We hope you can see that the answer is no, it was not 0.655%. In all likelihood, it was not even very close to 0.655%. It might even have been a negative number—it is quite common that individual stock prices move in a direction counter to the market. You cannot predict the actual value of company A's stock return with precision using this model, because company A's stock return is not a deterministic function of the *S&P 500* return. There are unique features of Company A that make it differ from the market, sometimes substantially, on any given day.

Having a large sample size here means that the estimate of the *mean of all potentially observable returns* of Company A's stock, given the *S&P 500* market return is 0.005, is very close to $0.0032 + 0.67(0.005) = 0.00655$. However, the *mean of all potentially observable returns* is very different from a *single potentially observable return*: The individual return differs from the conditional mean (which is approximately 0.00655), by the random error term e. This error term can be quite large, leading possibly to a negative Company A return when the *S&P 500* return is positive.

The error term e plays a crucial role here. To construct the prediction interval, you need to know the standard error of e. Let Y denote the value you are trying to predict, for a given $X = x$, e.g., the companies return (Y) when the *S&P 500* return (X) is 0.005 (i.e., when $X = 0.005$). Here

$$e = Y - (\hat{\beta}_0 + \hat{\beta}_1 x).$$

Since the data Y is not yet observed, it is independent of the data used to get the estimates, assuming the classical model, which makes this independence assumption. Using the linearity and additivity properties of (conditional) variance,

$$\begin{aligned} \mathrm{Var}(e) &= \mathrm{Var}(Y) + \mathrm{Var}(\hat{\beta}_0 + \hat{\beta}_1 x) \\ &= \sigma^2 + \sigma^2 \left\{ \frac{1}{n} + \frac{(x - \hat{\mu}_x)^2}{(n-1)s_x^2} \right\} \end{aligned}$$

The second summand of this formula shows that the variance of the error is somewhat reduced with larger n because of a more accurate estimate of the regression function: As discussed above, $\mathrm{Var}(\hat{\beta}_0 + \hat{\beta}_1 x)$ tends to zero for large n. However, no matter how large is n (even if n is infinite), you cannot reduce the variance of the error term e below σ^2, which is the conditional variance of the distribution of $Y \mid X = x$.

Plugging in the estimate $\hat{\sigma}$ for σ and taking the square root gives the standard error for e:

$$s.e.(e) = \hat{\sigma} \sqrt{1 + \frac{1}{n} + \frac{(x - \hat{\mu}_x)^2}{(n-1)s_x^2}}$$

This gives the exact prediction interval, which again, is exact only when the true data-generating process is the classical regression model.

Exact prediction interval for $Y \mid X = x$

$$\hat{\beta}_0 + \hat{\beta}_1 x \pm t_{dfe,1-\alpha/2} s.e.(e)$$

Compared to the confidence interval formula for the mean $E(Y \mid X = x)$, the only difference is the extra "1" under the square root of the standard error formula. This may seem like a trivial difference, but actually, it makes all the difference in the world: While the *mean* of all potentially observable Y can be estimated more and more accurately with larger n because the interval width converges to 0, a *single* potentially observable Y cannot be estimated ever more accurately with larger n. Regardless of n, the standard error of e is always greater than $\hat{\sigma}$, as you can see from the formula for *s.e.(e)* given just above.

You can obtain the prediction interval easily in `R` by using the command `predict(fit, x, interval="prediction")`. This gives you the result:

```
        fit      lwr      upr
1 1432.332 931.6039 1933.061
```

So you can say, with exactly 95% confidence, that a single potentially observable (perhaps for next week's job) value of Cost, given Widgets = 850, will be between 931.6039 and 1,933.061 Cost units. Symbolically,

$$931.6039 \text{ Cost units } < \text{Cost} \mid \text{Widgets} = 850 \ < 1933.061 \text{ Cost units}$$

As discussed above, this interval is much wider than the confidence interval for the mean of the conditional distribution, which is symbolically written as follows:

$$1303.753 \text{ Cost units } < \text{E}(\text{Cost} \mid \text{Widgets} = 850) \ < 1560.912 \text{ Cost units}$$

See Figure 3.5 again: In the interval, you see an *approximate* 95% prediction interval, obtained as $\hat{\beta}_0 + \hat{\beta}_1 x \pm 1.96\hat{\sigma}$ which is the estimated mean of the conditional distribution of $Y \mid X = x$, plus or minus 1.96 estimated standard deviations of the conditional distribution of $Y \mid X = x$. The exact interval is slightly wider because it accounts for randomness in the estimated parameters.

Because estimates become more accurate with larger n, the difference between the exact and approximate 95% intervals will diminish as n increases. However, the difference between the prediction intervals and the confidence interval for the mean becomes greater: As n increases, the prediction intervals converge to $\beta_0 + \beta_1 x \pm 1.96\sigma$, while the confidence interval for the mean converges to $\beta_0 + \beta_1 x \pm 0$.

As a final note, the prediction interval is *very sensitive* to any violation of the normality assumption, since it is based on individual observations and not averages. Intervals based on averages are not as sensitive to non-normality because averages have approximately normal distributions (by the Central Limit Theorem), even when the data distributions are non-normal.

3.9 Hypothesis Testing and *p*-Values: Is the Observed Effect of *X* on *Y* Explainable by Chance Alone?

Some researchers will do nearly anything to get a publication. The incentives are great: Fame, tenure, promotion, annual salary, raises, prime class assignments, and clout in one's department are a function of quality and quantity of publications.

Historically, statistical results were required to be "statistically significant" to be publishable. In terms of confidence intervals, this means that the interval for the effect (e.g., the β) in question must exclude 0 so that you can confidently state the direction of the effect (positive or negative) of the given X variable on Y.

Researchers used the p-values that are reported routinely by regression software to determine "statistical significance." But p-values are easily manipulated, and unscrupulous researchers can analyze data "creatively" to get nearly any p-value they would like to see. This has led to an unfortunate practice known as *p-hacking*, where researchers try analyses many different ways until they get a p-value that is statistically significant, and then try to publish the results. Because of their potential for misuse, there is a strong movement in the scientific community away from use of p-values, as well as the phrase "statistical significance," in favor of other statistics and characterizations.

When interpreted correctly and not misused, the p-value does provide interesting and somewhat useful information. Thus, we insist that you understand p-values very well, so that you can use them correctly and effectively, and so that you will not become a "p-hacker."

To interpret the p-value correctly, you must consider the question, "Is the estimate of the effect of X on Y *explainable* by chance alone?" But to answer that question, you must first understand what it means for an estimated effect to be *explained* by chance alone. The following example explains this concept.

3.9.1 Is the Last Digit of a Person's Identification Number Related to Their Height?

On the surface of it, this is a silly question. But it provides a great example to help you to understand what it means for a phenomenon to be "*explained* by chance alone," which is the first thing you need to know before you can ascertain whether a phenomenon is "*explainable* by chance alone."

Suppose you have a data set containing heights of 100 adult males in the United States, along with the last digit of their social security number (SSN). Since adult male heights are approximately normally distributed with mean 70 inches and standard deviation 4 inches, and since the last digit of the SSN is uniformly distributed on the numbers 0,1,2,...,9, the following code simulates a quite realistic example of how such a data set would look.

```
## Simulation of data relating Height to SSN
n = 100
set.seed(12345) # so that the results will replicate perfectly
height = round(rnorm(100, 70, 4))
ssn = sample(0:9, 100, replace=T)
ssn.data = data.frame(ssn, height)
head(ssn.data)
```

This code gives you the following (hypothetical but realistic) data on last digit of social security number (SSN) and Height:

```
  ssn height
1   5     72
2   8     73
3   1     70
4   5     68
5   6     72
6   7     63
```

Now, perform a regression analysis using the model Height | SSN = $x \sim N(\beta_0 + \beta_1 x, \sigma^2)$:

```
> summary(lm(height~ssn))

Call:
lm(formula = height ~ ssn)
Residuals:
     Min       1Q   Median       3Q      Max 
-10.1190  -3.1035   0.7017   2.5069   8.9499 

Coefficients:
            Estimate Std. Error t value Pr(>|t|)    
(Intercept)  71.9741     0.8123  88.608   <2e-16 ***
ssn          -0.2138     0.1465  -1.459    0.148    
---
Signif. codes:  0 '***' 0.001 '**' 0.01 '*' 0.05 '.' 0.1 ' ' 1

Residual standard error: 4.422 on 98 degrees of freedom
Multiple R-squared:  0.02126,   Adjusted R-squared:  0.01127 
F-statistic: 2.129 on 1 and 98 DF,  p-value: 0.1478
```

Based on these data, $\hat{\beta}_0 = 71.9741$, $\hat{\beta}_1 = -0.2138$, and $\hat{\sigma} = 4.422$ (inches). Particularly intriguing is the estimate $\hat{\beta}_1 = -0.2138$, which suggests that, on average, a one unit *increase* in the last digit of SSN corresponds to a 0.2138 inch *reduction* in average Height. Further, the results suggest that people with last SSN digit = 9 have average Height that is $9(0.2138) = 1.9242$ inches *less than* people whose last SSN digit is 0!

These conclusions are, of course, silly. We hope you agree that the difference between $\hat{\beta}_1 = -0.2138$ and 0 is completely *explained* by chance alone. In other words, in this idiosyncratic collection of 100 people, the data lined up, by chance alone, in such a way that the people with lower SSN tended to be somewhat taller than those with higher SSN. In another sample, it would come out differently. It is hard to argue that the true β_1 is anything other than zero here.

Even though the data above are simulated, we hope you can see that the simulation matches the real world perfectly. Had these data been real, you would also get a non-zero estimated slope $\hat{\beta}_1$, and the difference between that value and 0.0 would also be explained by chance alone.

Have doubts? Just ask a bunch of your friends and colleagues their height and last digit of SSN (or other ID), then do the regression on your own data. You will see that the estimated slope $\hat{\beta}_1$ is not 0.0, and you should agree with us that this difference is *explained by chance alone*, regardless of whether the result is "significant" or "insignificant."

Definition of "*Explained by chance alone*"

When the only reason for a difference between statistical estimates is chance alone, and not any systematic effect, then that difference is said to be *explained by chance alone.*

The example above, where SSN seems to affect Height, is arguably such an example where the difference between $\hat{\beta}_1 = -0.2138$ and 0.0 would seem to have no other explanation than chance.

However, in the real world, it is very difficult to state with 100% certainty that chance is the only explanation for an observed effect. Someone might argue that there really is a systematic, scientific relationship between SSN and Height. It is going to be a tortured argument! Please read the following argument carefully, as it explains the difference between "explained by chance alone" and "not explained by chance alone." It also provides an additional example to understand the concept of a "data-generating process." Are you ready for this? Take a deep breath. Here we go:

People find out their SSN at an early age before they are fully grown. Those people with low last digits of their SSN feel like they need to be "more." Who wants a zero or a one for a last digit of SSN? That's for losers! So, to compensate for their feelings of inadequacy imparted by this low last digit of their SSN, they decide to adopt a lifestyle (at an early age!) conducive to good health and fitness, eating well and exercising well. All this good eating and good exercise have a modest effect on their eventual adult height, leading to slightly taller people having low last digits of their SSN.

If the "theory" given by the previous paragraph is really true, then the negative coefficient $\hat{\beta}_1 = -0.2138$ would not be explained by chance alone, it would also be explained by the systematic effect of low SSN on personal habits.

See, we told you it was going to be a stretch! Nevertheless, the silly theory cannot be completely ruled out. Who knows, maybe there is some truth to it? In the real world, it is very difficult to argue, with 100% certainty, that certain effects are *explained* by chance alone. There is always some doubt about the nature of true data-generating process.

The data used in the SSN/Height example was simulated data, not real data. When you simulate successive random numbers using the computer, they are in fact independent, because that is the way random number generators work. So the difference between −0.2138 and 0 is in fact *explained by chance* with the simulated data because the true β_1 is really 0 in the simulation model.

Simulation is the safest way to obtain data where you know for certain that statistical differences are "explained by chance alone." When data are simulated from a model where there are no systematic differences *in the model itself*, any corresponding differences in the resulting analysis of the simulated data are *explained by chance alone.*

Now that you understand what it means for a difference to be *explained* by chance alone, you can understand what it means for a difference to be *explainable* by chance alone.

Definition of "*Explainable* by chance alone"

When a difference between statistical estimates is *within a typical range* of differences that are *explained by chance alone*, then that difference is said to be *explainable by chance alone.*

As mentioned above, it seems logical that the supposed SSN effect on Height is *explained* by chance alone, but someone can always argue to the contrary. Thus, the most convincing way to be sure that you are looking at results that are *explained* by chance alone is to use simulation. Simulation also allows you to see the "typical range of differences that are explained by chance alone," because you can simulate many times, and calculate the typical range from the simulated data.

In regression, a difference from a parameter estimate to 0 is usually measured in terms of number of standard errors from 0. Recall that the standard error is the estimated standard deviation. Thus, if the true β is zero, it will be very unlikely to observe a $\hat{\beta}$ that is three standard errors away from 0.

So, the distance from $\hat{\beta}$ to zero is measured in terms of standard errors, given by $\hat{\beta}/s.e.(\hat{\beta})$. If $\hat{\beta}\,/\,s.e.(\hat{\beta}) = 3.2$, for example, then $\hat{\beta}$ is 3.2 standard errors from 0, and it would be very unlikely that β could possibly be zero in this case.

This measure of distance from a $\hat{\beta}$ to 0 in terms of standard errors is famous—it is called the *T* statistic.

The *T* statistic

The *T* statistic measures how many standard errors $\hat{\beta}$ is from 0, and is given by

$$T = \hat{\beta}\,/\,s.e.(\hat{\beta})$$

In the SSN/Height example, $\hat{\beta}_1 = -0.2138$, and $T = -0.2138\,/\,0.1465 = -1.459$. So the estimated coefficient is only 1.459 standard errors below 0.

A model where any non-zero $\hat{\beta}_1$ is *explained* by chance alone is the model where $\beta_1 = 0$. Hence a *T* statistic of –1.459 is *explainable* by chance alone, since a data value that is 1.459 standard deviations from the mean is within a "typical range" of *T* statistics that you will observe when the data come from a data-generating process having $\beta_1 = 0$.

Yet another mathematical theorem: *If* the true data-generating process is the classical regression model, and *if* that model has $\beta = 0$ (for some particular β), *then* the *T* statistic corresponding to that β has an exact, known distribution.

Distribution of the *T* statistic under the chance-only (null) model

If the data are produced by the classical regression model where the particular β of interest is equal to 0, then

$$T = \hat{\beta}\,/\,s.e.(\hat{\beta}) \sim T_{dfe}$$

Not coincidentally, the T_{dfe} distribution is the same distribution that is used in the confidence interval for that β.

3.9.2 Simulation Study to Understand the Null Distribution of the *T* Statistic

The following `R` code generates the data under the null model. It is the same model used in the SSN/Height example above, but re-written in a way to make the constraint $\beta_1 = 0$ explicit. Because we did not use `set.seed` as in the previous code, the output is random. Your results will differ, but that is good because the point here is to understand randomness.

```
n = 100
beta0 = 70; beta1 = 0    # The null model; true beta1 = 0
ssn = sample(0:9, 100, replace=T)
height = beta0 + beta1*ssn + rnorm(100,0,4)
ssn.data = data.frame(ssn, height)
fit.1 = lm(height~ssn, data=ssn.data)
summary(fit.1)
```

This code gives the following output (yours will vary by randomness):

```
Call:
lm(formula = height ~ ssn, data = ssn.data)

Residuals:
    Min      1Q  Median      3Q     Max
-9.4952 -2.8261 -0.3936  2.2521 11.6764

Coefficients:
            Estimate Std. Error t value Pr(>|t|)
(Intercept) 69.77372    0.71865  97.089   <2e-16 ***
ssn          0.01915    0.14336   0.134    0.894
---
Signif. codes:  0 '***' 0.001 '**' 0.01 '*' 0.05 '.' 0.1 ' ' 1

Residual standard error: 4.155 on 98 degrees of freedom
Multiple R-squared:  0.000182,  Adjusted R-squared:  -0.01002
F-statistic: 0.01784 on 1 and 98 DF,  p-value: 0.894
```

In our simulation, the estimate $\hat{\beta}_1 = 0.01915$ is $T = 0.134$ standard errors from zero, and you know that this difference is *explained* by chance alone *because* the data are *simulated from the null model* where $\beta_1 = 0$.

But one simulation is not enough. To see the range of T values that are *explained* by chance alone, you must repeat the above simulation over and over. The following code repeats the process 100,000 times.

```
n = 100; beta0 = 70; beta1 = 0    # The null model; true beta1 = 0
nsim = 10000; t.stat = numeric(nsim)
for (i in 1:nsim) {ssn = sample(0:9, 100, replace=T)
 height = beta0 + beta1*ssn + rnorm(100,0,4)
 t.stat[i] = summary(lm(height~ssn))$coefficients[2,3] }

hist(t.stat, freq=F, breaks=50)
curve(dt(x,98), from=-3, to=3, add=T)
abline(v=c(qt(.025,98), qt(.975,98)), col="gray")
```

The result of the simulation is shown in Figure 3.7.

The range of T values in Figure 3.7 that are inside the 95% critical values comprise 95% of the potentially observable values of T under the null model. It is standard practice to say that an observed T statistic inside such a 95% range indicates that the difference between $\hat{\beta}_1$ and zero is *explainable by chance alone,* since such a difference is within the central 95% range of T statistics that are *explained by chance alone.*

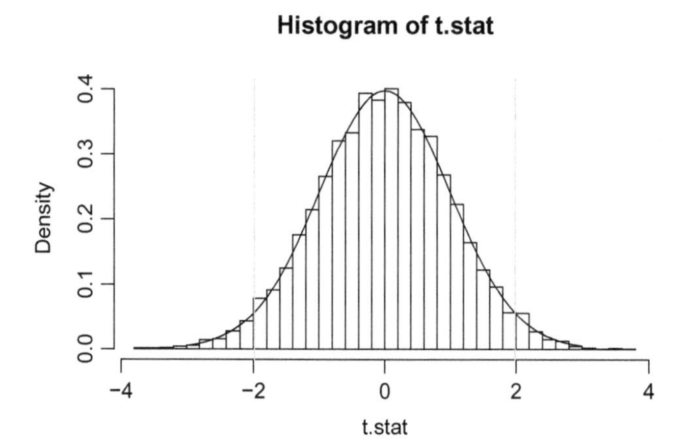

FIGURE 3.7
Distribution of distances from $\hat{\beta}_1$ to zero (measured in terms of standard errors, i.e., in terms of T statistics), that are explained by chance alone. The histogram displays 100,000 simulated T statistics, estimating the true distribution of the T statistics. The overlaid curve is the true distribution of the T statistics, the T_{98} distribution. The 95% critical values, −1.9845 and +1.9845, are shown as vertical solid lines.

So, in the farcical example above relating SSN to Height, the estimated effect of SSN on Height is $\hat{\beta}_1 = -0.2138$, which is −1.459 standard errors from 0 ($T = -1.459$). This value falls well within the 95% range of T statistics that are *explained by chance alone*, as shown in Figure 3.7, so you can say that the estimated non-zero effect of SSN on Height is *explainable by chance alone*.

3.9.3 The *p*-Value

In the example above, the thresholds to determine which real T values are *explainable by chance alone* are the numbers that put 95% of the T values that are *explained by chance alone* between them; these are −1.9845 and +1.9845 in the case of the T_{98} distribution. If the observed T statistic falls *outside* that range, then we can say that the difference between $\hat{\beta}_1$ and 0 is *not easily explained by chance alone*.

See Figure 3.7 again. Notice that there is 5% total probability *outside* the ±1.9845 range, simply because there is 95% probability *inside* the range. Now, if the T statistic falls *inside* the 95% range, then there has to be *more than* 5% total probability outside the $\pm T$ range. See Figure 3.7 again, and suppose $T = 1.7$, which is inside the range. Then there has to be *more than* 5% probability outside the ±1.7 range, right? See Figure 3.7 again, and locate ±1.7 on the graph. Make sure you understand this; it is not hard at all. Do not just read the words, because then you will not understand. Instead, look at Figure 3.7, put your finger on the graph at 1.7, and think about the area outside the ±1.7 range. It is more than 0.05, do you see?

Now, suppose $T = 2.5$, and look at Figure 3.7 again. Then there has to be *less than* 5% probability outside the ±2.5 range, right? See Figure 3.7 again, and locate ±2.5 on the graph. Make sure you understand this; it is not hard at all. Look at the graph! Do not just read the words! Instead, put your finger on the graph at 2.5 and think about the area outside the ±2.5 range. It is less than 0.05, do you see?

Self-study question: Did you put your finger on the graph as we asked in the last two paragraphs?

For the T statistics reported in the regression output, the p-value is simply the probability outside the $\pm T$ range of the T_{dfe} distribution. Since the p-value will be less than 0.05 whenever the T statistic is outside the central 95% range (and vice versa), it follows that, when the data are produced by the null (chance only) model, *and when the classical model assumptions are all true,* then the p-value will be less than 0.05 exactly 5% of the time. When the null model is true but the classical assumptions are not, then the p-value will be less than 0.05 perhaps more than 5% of the time, perhaps less. Thus, "exactness" of the p-value refers to how often the p-value can be less than 0.05 when the null model is true.

The threshold 0.05 that is commonly used for p-values is called the *significance level.* Just like the true confidence level of a 95% confidence interval is exactly 95% when the assumptions are all true, the true significance level of a $p < 0.05$ test is exactly 5% when the assumptions are all true. When the assumptions are violated, the true confidence level and significance level differ from their targets, 95% and 5%, respectively. Simulation is the best tool to answer the question, "How much different?"

Using `R`, you can calculate the p-value by summing the area of the two tails. If $T = -1.459$, as in the farcical SSN/Height example, then the p-value is computed as follows:

```
> Lower.tail = pt(-1.459, 98) ;  Upper.tail = 1 - pt(1.459, 98)
> Lower.tail + Upper.tail
[1] 0.1477629
```

Notice that this value is printed automatically in the `R` output of the `lm` function under the column heading "`Pr(>|t|)`."

In some cases, such as with the `F` test in regression analysis (see multiple regression analysis), the p-value is the probability in one tail of the distribution. The most general definition of a p-value, which includes one-tailed and two-tailed cases, is given as follows:

Definition of p-value, in general

The p-value is the probability of observing a test statistic as extreme as what was actually observed, assuming the data were produced by the null model.

If the p-value is less than 0.05, then the test statistic is *outside* the 95% range of test statistics that are *explained* by chance alone, hence you conclude that the results are *not easily explained by chance alone.* If the p-value is more than 0.05, then the test statistic is *inside* the 95% range of statistics that are explained by chance alone, hence you can conclude that the results are *explainable by chance alone.* The smaller the p-value, the more difficult it is to explain the results by chance alone. Hence, with small p-values, the results are better explained as being the result of systematic effects, in addition to the ubiquitous chance effects.

Sometimes people interpret p-values very badly. The worst is the interpretation "if $p > 0.05$, then the null model is true." That interpretation badly conflates the concept *explained by chance alone* with the concept *explainable by chance alone*. Unfortunately, that misinterpretation is commonly made by ignorant researchers who test for the validity of model assumptions where the null model states that the assumption is true, and if $p > 0.05$, they claim that the assumption is valid. This is a horribly misguided practice! The next chapter discusses the evaluation of model assumptions in more detail and shows how to interpret p-values correctly in that context.

There are several other bad interpretations of p-values. The American Statistical Association (ASA) described many of these in the document "The ASA's Statement on p-Values: Context, Process, and Purpose," (Wasserstein and Lazar 2016). Perhaps the second-worst misinterpretation of a p-value is that it is the probability that the null model is true, or equivalently that it is the probability that the results are explained by chance alone. This misinterpretation is so common that we need to set it off:

A grossly incorrect interpretation of p-value

The p-value is *not* the probability that the null model (or null hypothesis) is true.
In other words, the p-value is *not* the probability that the results are explained by chance alone.

The grossly incorrect interpretation seems plausibly correct, on its surface. After all, small p-values allow you to rule out the null model. However, large p-values *do not* imply that the null model is more likely. For example, suppose you relate a son's height (Y) to his father's height (X), using a sample of $n = 12$ father/son pairs. Suppose you get an estimate $\hat{\beta}_1 = 0.40$, and a standard error $s.e.(\hat{\beta}_1) = 0.80$, giving $T = 0.50$. The p-value is then given by `2*(1-pt(.50, 10))`, giving $p = 0.628$. Is there a 62.8% chance that father's and son's heights are unrelated? Certainly not. We know, *a priori* from genetics that there is 100% probability that father's and son's heights *are* related. Further, the data better support models where $\beta_1 \neq 0$ than they support the model where $\beta_1 = 0$. In particular, the model where $\beta_1 = 0.48$ is better supported by the data than the model where $\beta_1 = 0$: The estimate $\hat{\beta}_1 = 0.40$ is 0.5 standard errors from 0.0, but only 0.10 standard errors from 0.48. So if the model where $\beta_1 = 0$ is considered 62.8% likely, then the model where $\beta_1 = 0.48$ must be more than 62.8% likely. We are already at greater than 100% probability, and we have not even considered all the other possible values of β_1! This argument shows why you cannot interpret the p-value as the probability that the null model is true.

The correct interpretation is given above: The p-value is the probability of observing a more extreme result, assuming that the null model is true. Thus, the p-value states that, under the (strictly) hypothetical assumption that father's and son's heights are unrelated, you will see an estimate $\hat{\beta}_1$ that is 0.50 or more standard errors from 0.0 in 62.8% of samples of $n = 12$ father/son pairs. Thus, the observed value $\hat{\beta}_1 = 0.40$ is well within the range of $\hat{\beta}_1$ values that you will observe when $\beta_1 = 0$. But this does not prove that $\beta_1 = 0$, because $\hat{\beta}_1 = 0.40$ is also well within the range of $\hat{\beta}_1$ values that you will observe when $\beta_1 = 0.48$.

As the discussion above shows, the p-value provides only indirect evidence concerning the null model. This indirectness is one reason why the p-value is not particularly useful. It is also the reason for the interpretation that a large p-value means the results are *explainable* by the null model. The term "explainable" correctly allows that there are other, more plausible explanations (i.e., the non-null model), regardless of how large is the p-value.

You should not use p-values at all if you are going to interpret them incorrectly. They only provide an objective measure of what is *explainable* by chance alone. You should use p-values *only* if you understand their precise meanings and their limitations; and even then, you should only use them with extreme caution.

Reference

Wasserstein, R. L., & Lazar, N. A. (2016). The ASA's statement on p-values: Context, process, and purpose. *The American Statistician, 70*(2), 129–133.

Exercises

1. The following study investigates the relationship between salary (X) and acreage on which a home is built (Y). Run and report the results of the following code, then answer the questions.

```
inc = read.csv("https://raw.githubusercontent.com/andrea2719/
URA-DataSets/master/t11_1.csv")
head(inc)
plot(acreage~salary, data=inc)
fit1 = lm(acreage~salary, data=inc)
summary(fit1)
```

   a. Write down the classical model for how the *potentially observable data* are produced. Use the "i" subscript, and indicate the sample size (n) in your model specification. Do not use hats (^'s) on the β's.
   b. Give the OLS estimates of the β's of your model in 1.a. Use the hats (^'s) now.
   c. Give the standard error of the OLS estimate $\hat{\beta}_1$.
   d. Explain what the standard error means in terms of the model in 1.a. Specifically, the model gives a recipe for different data sets other than this one that were potentially observable; explain how the standard error refers to these other data sets. Since the standard error is a "fixed X" statistic, make your interpretation similarly "fixed X."
   e. Give the approximate 95% confidence interval $\hat{\beta}_1 \pm 2s.e.(\hat{\beta}_1)$ for β_1.

f. Can you tell from the interval in 1.e. whether people with more salary tend to choose larger lots for their homes? Why?

g. Use your common sense, and not any data. Do you think that the average acreage chosen by people with an annual salary \$300,000 will be more or less than the average acreage chosen by people with \$50,000 annual salary? Why do you think so?

h. In what way are your answers to 1.f. and 1.g. contradictory? What is the resolution to this contradiction?

2. See Figure 3.1. Change the regression model to $N(110 + 3.0x,\ 500^2)$, but use the same X (Widgets) data. Redraw the graph, and compare it to Figure 3.1. There are two major differences that you should notice. Explain these two differences in terms of (i) Unbiasedness (be specific) and (ii) the formula given for the standard deviation (not standard error) of the OLS estimator $\hat{\beta}_1$ (be specific).

4

Evaluating Assumptions

As shown in Chapter 1, the linearity assumption in regression is almost always false, to one degree or another, when there are more than two levels of the X variable. The constant variance assumption and normality assumptions are also almost always false, to one degree or another. Independence is often violated as well. When violated, these assumptions can make the results of the analyses dubious: The model does not represent reality very well, estimates are biased, and confidence intervals, prediction intervals, and p-values do not behave as expected.

Nevertheless, the goal of evaluating model assumptions in statistics is *not* to prove that the assumptions of the model are true. You may as well try to prove that $1 + 1 = 3$! Rather, the goal is to assess *degree of violation of the assumptions.* If the assumptions of your chosen model are not too badly violated, then your model represents reality reasonably well, it is usable for making predictions, and the inferential techniques behave approximately as expected (e.g., ≈95% confidence, ≈5% significance level). If the assumptions of your chosen model are badly violated, then you should choose a different model. You can most easily answer the question, "how badly violated must the assumptions be before I must choose a different model?" by using simulation, where you simulate data with violated assumptions, and see how well the estimates and inferences work when you use different models.

To answer the question, "How badly violated are the assumptions for my study?", you should first and foremost consider the assumptions *without using your data set.* You can do this by using your subject matter knowledge about the *data-generating process.* Only then can you analyze graphs and summary statistics of your data appropriately. The data values cannot tell you everything about the violations of the assumptions—they can incorrectly suggest that assumptions are badly violated when the assumptions are reasonable, and they can incorrectly suggest that the assumptions are reasonable when they are badly violated. The assumptions do not refer to the data; they refer to the processes that generated your data. Thus, your diagnosis of the assumptions should always involve subject matter, in addition to data analysis.

Since researchers often think that assumptions refer to the data, let us be clear and emphatic:

> ***ALL* OF THE ASSUMPTIONS REFER TO THE *DATA-GENERATING PROCESS.***
> ***NONE* OF THE ASSUMPTIONS REFER TO THE ACTUAL OBSERVED DATA.**

The data shed light on the validity of the assumptions only insofar as the data reflect the underlying data-generating process. In particular, data sets with small samples sizes do not represent the entirety of the underlying data-generating process very well, so it is difficult to assess the degree of violations of assumptions when you have a small sample size, n.

The data analysis tools described in this chapter involve graphical/descriptive methods and testing methods. The graphical and descriptive methods are informal, and require some judgment, as well as care in producing clear graphs, to make proper interpretation. Testing methods, which are always based on test statistics and their associated p-values, are much more problematic because they only answer the question, "Is the apparent violation of the assumption that is seen in the data explainable by chance alone?", and not the more relevant question, "Is the assumption so badly violated that an alternative model or method is needed?"

Testing methods, while problematic, have been widely used in history, and you still see them to this day. But realize that p-value based testing of model assumptions is a dead-end road. As given by the American Statistical Association (ASA) statement on p-values, "By itself, a p-value does not provide a good measure of evidence regarding a model or hypothesis." In particular, the ASA statement implies that you should not rely only on p-values to address model assumptions, e.g., as often recommended by econometrics sources that recommend using "specification tests" to evaluate model assumptions. As the ASA statement explains, they should use alternative techniques.

One reason that tests of assumptions are not very useful is that nearly all assumptions are false *a priori*. So if the test "passes," in the sense that the assumption is not rejected (i.e., $p > 0.05$; see examples below in this chapter), are you somehow supposed to conclude that the assumption is valid? Of course not. The assumption was likely violated in reality, to one degree or another, so any test of an assumption that "passes" was likely to be a Type II error.

Self-study question: If a test of model assumption "passes" ($p > 0.05$), why is it likely to be a Type II error?

Another reason that tests are not very useful is that, if the test does not pass (i.e., $p < 0.05$; see below in this chapter for examples), it does not necessarily imply that the assumption is so badly violated that an alternative method is needed. The test does not address the question, "Is the assumption *badly* violated?" It only addresses the question "*Is* the assumption violated?" and it does not even provide a particular good answer to that question.

Tests can be very misleading because of their extreme dependence on the sample size. All tests have extremely high power with large sample sizes; thus, a rejection of the assumption can easily occur even when the assumption is only slightly violated. If the assumption is only slightly violated, it is still a reasonable assumption to make, regardless of the result of the test. Review Figure 1.13: The case where linearity is reasonable, although technically wrong, would be rejected by a p-value based test with large enough sample size. The test would give a correct conclusion because linearity is indeed false. But it is not a useful conclusion, because the linear model is a reasonable model to use, even though the true regression function is slightly curved.

Because the testing methods are so sensitive to sample size, scientists have moved toward reporting effect sizes instead of p-values. An "effect size" is a measure of the "size" of the deviation, rather than its "significance" (i.e., its p-value). The concern over the effect of sample size is also the reason that there are many alternative fit statistics other than the p-value in "structural equation models," which are models that are commonly used to analyze survey data.

Paradigms in research methods have been shifting away from testing methods for decades now, driven by improved understanding of statistical theory and by better software. Thankfully, p-value based testing methods are becoming less popular relative to improved (often simulation-based) methods for assumption-checking that have been developed in recent years.

On the other hand, chance variation is real and has an effect. If you see something in the graphs or data summaries that suggests a deviation from the assumption, the difference might be explainable by chance alone, so you need to interpret your data carefully. The p-value from the test of the assumption answers the question, "Is the observed effect explainable by chance alone?" and that question alone. So the testing methods and their associated p-values are useful for one purpose only: To attempt to rule out chance as a possible explanation for what you see in your data.

To reiterate, the test of the assumption and its associated p-value do not answer the question, "Is the assumption satisfied?", nor does it answer the question, "Is the assumption violated to the extent that alternative methods are needed?" That is why you should rely mainly on graphical and descriptive methods, supplemented by subject matter knowledge and simulation, to evaluate assumptions, rather than tests of hypotheses and their associated p-values.

4.1 Graphical/Descriptive Methods Versus Testing Methods for Checking Assumptions

One benefit of using graphical/descriptive methods to check assumptions, rather than hypothesis testing (p-value based) methods, is *transparency*: The graphs show the data, as they are. The p-values of the statistical tests give information that is distorted by the sample size. Another benefit is that you can determine the *practical significance* of a result using graphical methods and descriptive statistics, but not by statistical tests and their p-values. Tests can tell you whether a result is *statistically significant* (again, historically, $p < 0.05$), but statistically significant results can be practically unimportant, and vice versa, because of the sample size distortion. Unlike statistical tests of assumptions, larger sample sizes always point you closer to the best answer when you use well-chosen graphs and descriptive statistics.

But, care is needed in interpreting and constructing graphs. Interpreting graphs requires practice, judgment, and some knowledge of statistics. In addition, producing good graphs requires skill, practice, and in some cases, an artistic eye. A classic and very helpful text on the use and construction of statistical graphics is *The Visual Display of Quantitative Information*, by Edward Tufte (Tufte 2001).

The only good thing about tests is that they answer the question, "Is the apparent deviation from the assumption that is seen in the data explainable by chance alone?" The question of whether a result is explainable by chance alone is indeed important because researchers are prone to over-interpret idiosyncratic (chance) aspects of their data. Hypothesis testing provides a reality check to guard against such over-interpretation. But other methods, simulation in particular, are better for assessing the effects of chance deviation. Hence, p-value based hypothesis testing methods are not even needed for their one use, which is to assess the effect of chance variation.

Tests of model assumptions have been used for much of statistical history and are still used today in some quarters. Perhaps the main reason for their historical persistence is *simplicity*. Researchers have routinely applied the rule, "p-value greater than

0.05 → assumption is satisfied; p-value less than 0.05 → assumption is not satisfied," because it is simple, despite it being a horribly misguided practice. We have already mentioned many concerns with tests, but here they are, in set-off form, so that you can easily refer to them.

Drawbacks of using hypothesis tests to check assumptions

- In nearly all cases, the regression assumptions are not precisely true. Attempting to prove that they are true is like trying to prove that $1 + 1 = 3$.
- With small sample sizes, all tests have low power. Thus you will likely fail to reject the null, and thus may be inclined to report, incorrectly, that "the assumption is valid" when your sample size, n, is small.
- With large sample sizes (e.g., in "big data" applications), even small but practically unimportant deviations will be flagged as "statistically significant." The statistical conclusion will be correct, i.e., the assumption really is violated, but the extent of the violation might not be large enough to require remedial action.

Despite the problems with using tests (and p-values) for assumption-checking, you should understand how to use them and interpret them. One reason that you should understand them is that they appear in the historical literature. Another is that it is possible that a reviewer will insist that you use them in your own research. Finally, you should always be concerned with the effect of chance alone on how your data appear. Even if hypothesis testing is not the best methodology to use, at least it brings the important question, "Is this result explainable by chance alone?" to the fore.

4.2 Which Assumptions Should You Evaluate First?

We suggest (only mildly; this is not a hard-and-fast rule) that you evaluate the linearity and constant variance assumptions first. The reason is that, for checking the assumptions of independence and normality, you often will use the residuals $e_i = y_i - \hat{y}_i$, where the predicted values $\hat{y}_i = \hat{\beta}_0 + \hat{\beta}_1 x_i$ are based on the linear fit. If the assumption of linearity is badly violated, then these estimated residuals will be badly biased. In such a case you should evaluate the normality and independence assumptions by first fitting a more appropriate (non-linear) model, and then by using that model to calculate the predicted values and associated residuals.

Furthermore, if the linearity assumption is reasonably valid but the homoscedasticity (constant variance) assumption is violated, then the residuals e_i will automatically look non-normal, even when the conditional distributions $p(y \mid x)$ are normal because some residuals will come from distributions with larger variance and some will come from distributions with smaller variance, lending a heavy-tailed appearance to the pooled $\{e_i\}$ data.

For these reasons, we mildly suggest that you evaluate the assumptions in the order (1) linearity, (2) constant variance, (3) independence, and (4) normality. But there are cases where this sequence is logically flawed, so please just treat it as one of those "ugly rules of thumb."

4.3 Evaluating the Linearity Assumption Using Graphical Methods

While we are not big fans of data analysis "recipes," in regression or elsewhere, which instruct you to perform step 1, step 2, step 3, etc. for the analysis of your data, we are happy to recommend the following first step for the analysis of regression data.

Step 1 of any analysis of regression data

Plot the ordinary (x_i, y_i) scatterplot, or scatterplots if there are multiple X variables.

The simple (x_i, y_i) scatterplot gives you immediate insight into the viability of the linearity, constant variance, and normality assumptions (see Section 1.8 for examples of such scatterplots). It will also alert you to the presence of outliers.

To evaluate linearity using the (x_i, y_i) scatterplot, simply look for evidence of curvature. You can overlay the LOESS fit to better estimate the form of the curvature. Recall, though, that all assumptions refer to the data-generating process. Thus, if you are going to claim there is curvature, such curvature should make sense in the context of the subject matter. For one example, boundary constraints can force curvature: If the minimum Y is zero, then the curve *must* flatten for X values where Y is close to zero. For another example, in the case of the product preference vs. product complexity shown in Figure 1.16, there is a subject matter rationale for the curvature: People prefer more complexity up to a point, after which more complexity is less desirable. Ideally, you should be able to justify curvature in terms of the processes that produced your data.

A refinement of the (x_i, y_i) scatterplot is the residual (x_i, e_i) scatterplot. This scatterplot is an alternative, "magnified" view of the (x_i, y_i) scatterplot, where the $e = 0$ horizontal line in the (x_i, e_i) scatterplot corresponds to the least-squares line in the (x_i, y_i) scatterplot. Look for upward or downward "U" shape to suggest curvature; overlay the LOESS fit to the (x_i, e_i) data to help see these patterns.

You can also use the $(\hat{y}_i, e_i)$ scatterplot to check the linearity assumption. In simple regression (i.e., one X variable), the $(\hat{y}_i, e_i)$ scatterplot is identical to the (x_i, e_i) scatterplot, with the exception that the horizontal scale is linearly transformed via $\hat{y}_i = \hat{\beta}_0 + \hat{\beta}_1 x_i$. When the estimated slope is negative, the horizontal axis is "reflected"—large values of x map to small values of $\hat{y}_i$ and vice versa. You can use this plot just like the (x_i, e_i) scatterplot. In simple regression, the $(\hat{y}_i, e_i)$ scatterplot offers no advantage over the (x_i, e_i) scatterplot. However, in multiple regression, the $(\hat{y}_i, e_i)$ scatterplot is invaluable as a quick look at the overall model, since there is just one $(\hat{y}_i, e_i)$ plot to look at, instead of several (x_{ij}, e_i) plots (one for each X_j variable). This $(\hat{y}_i, e_i)$ scatterplot, which you can call a "predicted/residual scatterplot," is automatically provided by `R` when you plot a fitted `lm` object.

In all these plots, LOESS is very helpful. But don't over-interpret curvature that is almost always shown by LOESS. Remember the LOESS plot is from the data, not from the model. Remember also that LOESS uses "local averaging"; hence, where the X data are sparse, there is less data to "average," leading to unreliable LOESS estimates.

4.3.1 Production Cost Data (x_i, y_i) Plot with LOESS Smooth and $(\hat{y}_i, e_i)$ Plot with LOESS Smooth

```
ProdC = read.table("https://raw.githubusercontent.com/andrea2719/
URA-DataSets/master/ProdC.txt")
attach(ProdC)
par(mfrow=c(1,2))
plot(Widgets, Cost); abline(lsfit(Widgets, Cost))
add.loess(Widgets, Cost, col = "gray", lty=2)
fit = lm(Cost ~ Widgets)
y.hat = fit$fitted.values; resid = fit$residuals
plot(y.hat, resid); add.loess(y.hat, resid, col = "gray", lty=2)
abline(h=0)
```

The results are shown in Figure 4.1. The horizontal $e = 0$ line in the $(\hat{y}_i, e_i)$ plot corresponds to the least-squares line in the (x_i, y_i) plot. Pronounced curvature in the LOESS fits would suggest curvature in the true regression function, but the curvature seems slight here, perhaps explainable by chance alone. Also, from a subject matter standpoint, economies of scale suggest that, if there is curvature, the slope of the Cost function should decrease as Widgets increases. Thus, based on the data and on subject matter considerations, the linear model appears reasonable here.

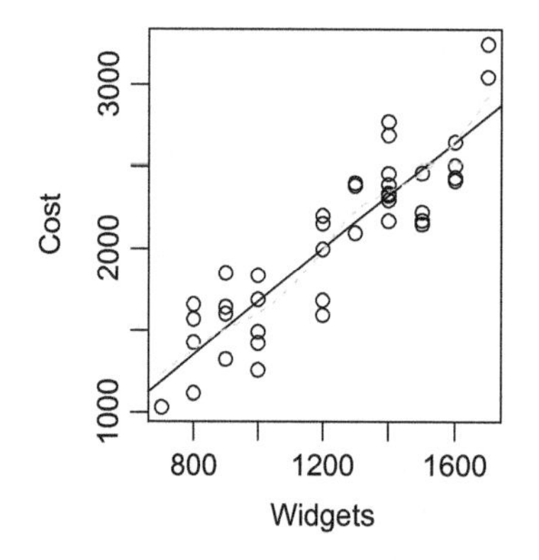

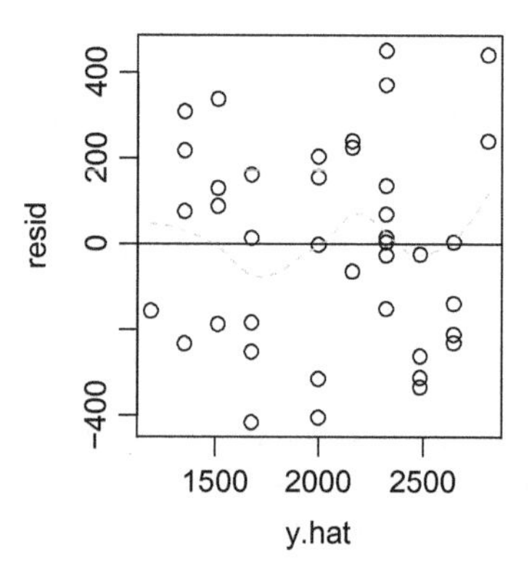

FIGURE 4.1
Scatterplots of the Production Cost data, with LOESS fits (dashed lines). Left panel: (x_i, y_i). Right panel: $(\hat{y}_i, e_i)$. The solid line in the (x_i, y_i) plot is the OLS line, which corresponds to the flat line at $e = 0$ in the $(\hat{y}_i, e_i)$ plot. Notice that any possible curvature, as estimated by the LOESS fit (dashed line) is magnified in the $(\hat{y}_i, e_i)$ plot.

Notice the phrasing in the previous sentence: The model "appears reasonable," rather than "is correct." From Chapter 1, you know, with 100% certainty, that precise linearity is, in fact, untrue for this example.

4.3.2 Car Sales Data (x_i, y_i) Plot with LOESS Smooth and $(\hat{y}_i, e_i)$ Plot with LOESS Smooth

The "Car Sales" data set relates the monthly U.S. Federal interest rate in the 1980's to corresponding U.S. total car sales. Presumably, Sales will tend to be lower when Interest Rate is higher because fewer people will be able to afford the interest payments.

```
CarS = read.table("https://raw.githubusercontent.com/andrea2719/
URA-DataSets/master/CarS.txt")
attach(CarS)
par(mfrow=c(1,2))
plot(INTRATE, NSOLD); abline(lsfit(INTRATE, NSOLD))
add.loess(INTRATE, NSOLD, col = "gray", lty=2)
fit = lm(NSOLD~INTRATE)
y.hat = fit$fitted.values; resid = fit$residuals
plot(y.hat, resid); add.loess(y.hat, resid, col = "gray", lty=2)
abline(h=0)
```

The results are shown in Figure 4.2. Unlike the Production Cost example, there is clear curvature in this case, shown in both plots. The $(\hat{y}_i, e_i)$ plot offers a magnification of the (x_i, y_i) plot and shows the curvature even more clearly. Also, you should note that the convex curvature that is suggested in these plots *does* make sense from subject matter—as Interest Rate increases, the Sales mean function should flatten, because a certain proportion of car sales are cash purchases, and these sales are not directly affected by interest rates.

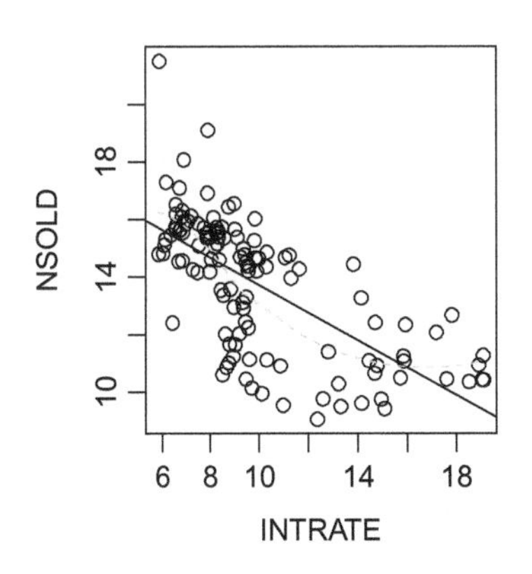

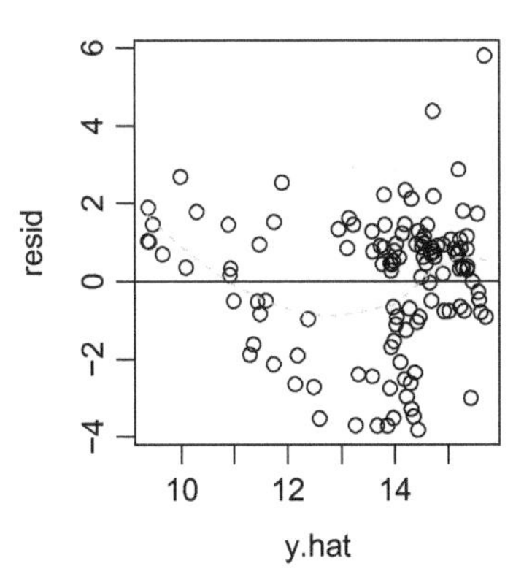

FIGURE 4.2
Scatterplots of the Car Sales data with LOESS fits (dashed lines). Left panel: (x_i, y_i). Right panel: $(\hat{y}_i, e_i)$.

4.4 Evaluating the Linearity Assumption Using Hypothesis Testing Methods

Here, we will get slightly ahead of the flow of the book, because multiple regression is covered in the next chapter. A simple, powerful way to test for curvature is to use a multiple regression model that includes a quadratic term. The quadratic regression model is given by:

$$Y = \beta_0 + \beta_1 X + \beta_2 X^2 + \varepsilon$$

This model assumes that, if there is curvature, then it takes a quadratic form. Logic for making this assumption is given by "Taylor's Theorem," which states that many types of curved functions are well approximated by quadratic functions.

Testing methods require restricted (null) and unrestricted (alternative) models. Here, the null model enforces the restriction that $\beta_2 = 0$; thus the null model states that the mean response is a linear (not curved) function of x. So-called "insignificance" (determined historically by $p > 0.05$) of the estimate of β_2 means that the evidence of curvature in the observed data, as indicated by a non-zero estimate of β_2 or by a curved LOESS fit, is explainable by chance alone under the linear model. "Significance" (determined historically by $p < 0.05$) means that such evidence of curvature is not easily explained by chance alone under the linear model.

But you should not take the result of this p-value based test as a "recipe" for model construction. If "significant," you should not automatically assume a curved model. Instead, you should ask, "Is the curvature dramatic enough to warrant the additional modeling complexity?" and "Do the predictions differ much, whether you use a model for curvature or the ordinary linear model?" If the answers to those questions are "No," then you should use the linear model anyway, even if it was "rejected" by the p-value based test.

In addition, models employing curvature (particularly quadratics) are notoriously poor at the extremes of the x-range(s). So again, you can easily prefer the linear model, even if the curvature is "significant" ($p < 0.05$).

Conversely, if the quadratic term is "insignificant," it does not mean that the function is linear. Recall from Chapter 1 that the linearity is usually false, *a priori*; hence, "insignificance" means that you have failed to detect curvature. If the test for the quadratic term is "insignificant," it is most likely a Type II error.

Even when the curvature does not have a perfectly quadratic form, the quadratic test is usually very powerful; rare exceptions include cases where the curvature is somewhat exotic. If the quadratic model is grossly wrong for modeling curvature in your application, then you should use a test based on a model other than the quadratic model.

4.4.1 Testing for Curvature with the Production Cost Data

The following `R` code illustrates the method.

```
ProdC = read.table("https://raw.githubusercontent.com/andrea2719/
URA-DataSets/master/ProdC.txt")
attach(ProdC)
plot(Widgets, Cost); abline(lsfit(Widgets, Cost))
Widgets.squared = Widgets^2
```

```
fit.quad = lm(Cost ~ Widgets + Widgets.squared); summary(fit.quad)
lines(spline(Widgets, predict(fit.quad)), col = "gray", lty=2)
```

Figure 4.3 shows both the linear and quadratic (curved) fit to the data. Since the linear and quadratic fits are so similar, it (again) appears that there is no need to model the curvature explicitly in this example.

Relevant lines from the summary of fit are shown as follows:

```
Coefficients:
                 Estimate     Std. Error   t value   Pr(>|t|)
(Intercept)      4.564e+02    7.493e+02     0.609     0.546
Widgets          9.149e-01    1.290e+00     0.709     0.483
Widgets.squared  2.923e-04    5.322e-04     0.549     0.586

Residual standard error: 241.3 on 37 degrees of freedom
Multiple R-squared:  0.7987,    Adjusted R-squared:  0.7878
F-statistic: 73.42 on 2 and 37 DF,  p-value: 1.318e-13
```

Notice the p-value for testing the $\beta_2 = 0$ restriction: Since the p-value is 0.586, the difference between the coefficient 0.0002923 (`2.923e-04`) and 0.0 is explainable by chance alone. That is, even if the process were truly linear (i.e., even if $\beta_2 = 0$), you would often see quadratic coefficient estimates ($\hat{\beta}_2$) as large as 0.0002923 when you fit a quadratic model to similar data. If this is confusing to you, just run a simulation from a similar linear process (where $\beta_2 = 0$), and fit a quadratic model. You will see a non-zero $\hat{\beta}_2$ in every simulated data set, and most will be within 2 standard errors of 0.0 (the $\hat{\beta}_2$ above is $T = 0.549$ standard errors from 0.0).

While curvature shown in Figure 4.3 is *explainable* by chance alone, you still cannot conclude linearity. As shown in Chapter 1, linearity cannot be precisely true in this example because (i) there are more than two possible values of X, and (ii) Y is related to X. Further, from a subject matter standpoint, *concave* curvature makes more sense in this example because of economies of scale. Curvature exhibited by the data is *explainable* by chance alone, but it is not *explained* by chance alone.

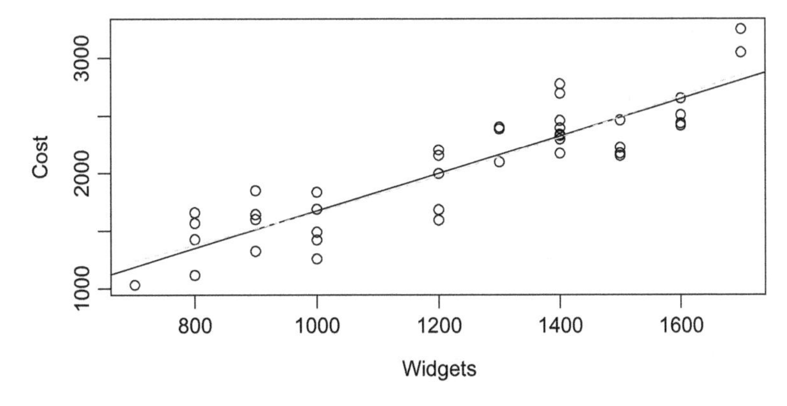

FIGURE 4.3
Linear (solid) and quadratic (dashed) fits to the Production Cost data. It appears that there is no need to model curvature explicitly.

4.4.2 Testing for Curvature with the Car Sales Data

The same method works for the Car Sales data, as shown in the following code.

```
sales = read.table("https://raw.githubusercontent.com/andrea2719/
URA-DataSets/master/CarS.txt")
attach(sales)
plot(INTRATE, NSOLD); abline(lsfit(INTRATE, NSOLD))
INTRATE.squared = INTRATE^2
fit.quad = lm(NSOLD ~ INTRATE + INTRATE.squared); summary(fit.quad)
lines(spline(INTRATE, predict(fit.quad)), col = "gray", lty=2)
```

Figure 4.4 shows both the linear and quadratic (curved) fit to the data. Here, the linear and quadratic fits are not similar, so it appears that there *is* a need to model the curvature explicitly.

Relevant summary of the fit for the Car Sales quadratic model is shown as follows:

```
Coefficients:
                 Estimate Std. Error t value Pr(>|t|)
(Intercept)      24.68608    1.69502  14.564  < 2e-16 ***
pINTRATE         -1.63209    0.30686  -5.319 5.09e-07 ***
INTRATE.squared   0.04809    0.01264   3.804 0.000228 ***
---
Signif. codes:  0 '***' 0.001 '**' 0.01 '*' 0.05 '.' 0.1 ' ' 1

Residual standard error: 1.678 on 117 degrees of freedom
Multiple R-squared:  0.5154,    Adjusted R-squared:  0.5071
F-statistic: 62.22 on 2 and 117 DF,  p-value: < 2.2e-16
```

Since the p-value for testing that $\beta_2 = 0$ is 0.000228, the difference between the coefficient 0.04809 and 0.0 is not easily explained by chance alone. That is, if the process were truly

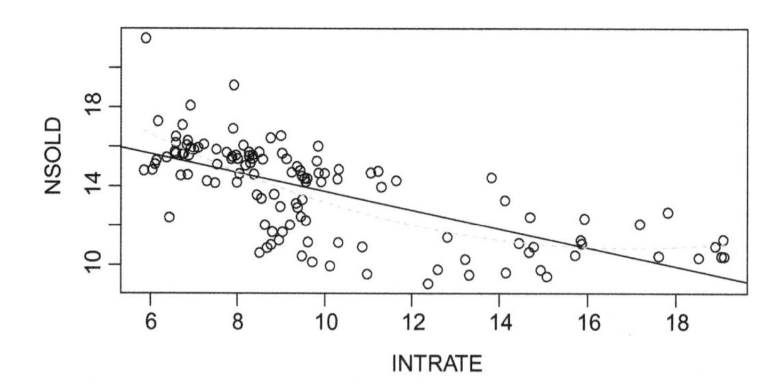

FIGURE 4.4
Linear (solid) and quadratic (dashed) fits to the Car Sales data. It appears that there *is* a need to model curvature explicitly.

linear (i.e., if $\beta_2 = 0$), you would only rarely see quadratic coefficient estimates as large as 0.04809 in similar studies. If this is confusing to you, just run a simulation from a similar, but truly linear process, and fit a quadratic model. You will see a non-zero $\hat{\beta}_2$ in every simulated data set, but most will be within 2 standard errors of 0.0, unlike the actual $\hat{\beta}_2$ from the real data, which is $T = 3.804$ standard errors from 0.0.

Hence, the curvature in the quadratic (dashed) function shown in Figure 4.4 is not easily explained by chance alone. It is, therefore, reasonable to conclude that there is curvature in the function that relates average sales to interest rates. And, as mentioned above, from a subject matter theory standpoint, there *should* be convex curvature because the Sales mean function *should* flatten for large values of the Interest Rate variable, reflecting cash purchases that are not directly affected by interest rates.

On the other hand, the quadratic analysis does not imply that the curvature has a quadratic form. Like linearity, precise quadratic curvature is also usually false. Indeed, comparing Figure 4.4 with the LOESS fit in Figure 4.2, the quadratic model appears to "accelerate" too quickly in the upper left portion of the data.

4.5 Practical Versus Statistical Significance

While the Car Sales example above is a case where statistically significant ($p < 0.05$) curvature corresponds to practically significant (which means practically important) curvature, it is not always the case that statistical significance corresponds to practical significance. This can easily happen in data sets where n is large (e.g., with "big data"), because with large data sets you have the ability to estimate even slight curvature very precisely, with small standard error.

The following simulation illustrates this situation: There is *statistically significant* ($p = 0.000166$) curvature, as shown by the hypothesis test for the quadratic term. However, the curvature is *practically insignificant,* as can be seen by graphing the linear and quadratic fitted functions. The sample size in this simulation is large, n = 1,000,000, but not unusually large for "big data" applications.

4.5.1 Simulation Study to Demonstrate Practical vs. Statistical Significance

```
set.seed(54321)  # For perfect replicability of the random simulation.
x   = 10 + 2*rnorm(1000000);   xsq = x^2
y   = 2 + .6*x + .003*xsq + 4*rnorm(1000000) #beta2=.003 does not equal 0!
fit.quad = lm(y ~ x + xsq)
summary(fit.quad)  # Significant curvature: p-value = 0.000166

## A .1% random sample from the data set is selected to make the scatterplot
## more legible. Otherwise, the points are too dense to view.
select = runif(1000000); x1 = x[select<.001] ; y1 = y[select<.001]
plot(x1, y1, main = "Scatterplot of a 0.1% Subsample" )
abline(lsfit(x, y), col="gray")
lines(spline(x, predict(fit.quad)), lty=3)
```

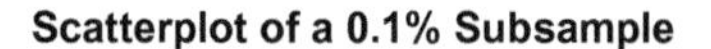

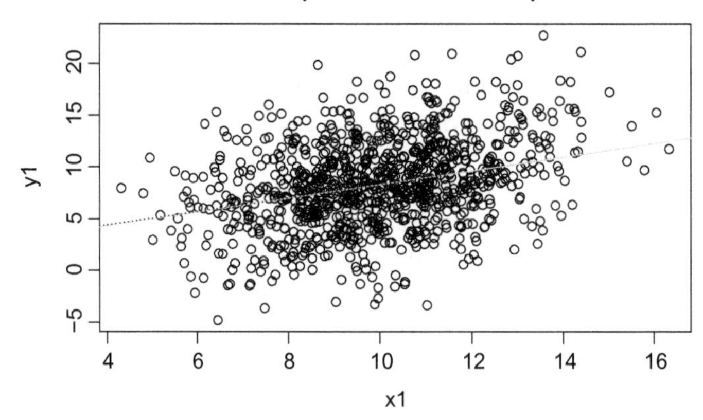

FIGURE 4.5
Comparison of incorrect linear fit (solid line) with correct curved fit (dotted curve). The lack of fit of the linear function is statistically significant ($p < 0.05$), but not practically important. The reason this happened is that the sample size is very large, with n = 1,000,000 observations. (To make the graph legible, only 0.1% of the n = 1,000,000 observations are shown.)

Figure 4.5 shows that the linear fit (solid line) is adequate, even though the true model is quadratic, and not linear.

4.6 Evaluating the Constant Variance (Homoscedasticity) Assumption Using Graphical Methods

The first graph you should use to evaluate the constant variance assumption is the $(\hat{y}_i, e_i)$ scatterplot. Look for changes in the pattern of vertical variability of the e_i for different $\hat{y}_i$. The most common indications of constant variance assumption violation are shapes that indicate either increasing variability of Y for larger $\text{E}(Y \mid X = x)$, or shapes that indicate decreasing variability of Y for larger $\text{E}(Y \mid X = x)$. Increasing variability of Y for larger $\text{E}(Y \mid X = x)$ is indicated by greater variability in the vertical ranges of the e_i when $\hat{y}_i$ is larger.

Recall again that the constant variance assumption (like all assumptions) refers to the data-generating process, not the data. The statement "the data are homoscedastic" makes no sense. By the same logic, the statements "the data are linear" and "the data are normally distributed" also are nonsense. Thus, whichever pattern of variability that you decide to claim based on the $(\hat{y}_i, e_i)$ scatterplot, you should try to make sense of it in the context of the subject matter that determines the data-generating process. As one example, physical boundaries on data force smaller variance when the data are closer to the boundary. As another, when income increases, people have more choice as to whether or not they choose to purchase an item. Thus, there should be more variability in expenditures among people with more money than among people with less money. Whatever pattern you see in the $(\hat{y}_i, e_i)$ scatterplot should make sense to you from a subject matter standpoint.

While the LOESS smooth to the $(\hat{y}_i, e_i)$ scatterplot is useful for checking the linearity assumption, it is not useful for checking the constant variance assumption. Instead, you should use the LOESS smooth over the plot of $(\hat{y}_i, |e_i|)$. When the variability in the residuals is larger, they will tend to be farther from zero, giving larger mean absolute residuals $|e_i|$. An increasing trend in the $(\hat{y}_i, |e_i|)$ plot suggests larger variability in Y for larger $E(Y \mid X = x)$, and a flat trend line for the $(\hat{y}_i, |e_i|)$ plot suggests that the variability in Y is nearly unrelated to $E(Y \mid X = x)$. However, as always, do not over-interpret. Data are idiosyncratic (random), so even if homoscedasticity is true in reality, the LOESS fit to the $(\hat{y}_i, |e_i|)$ graph will not be a perfectly flat line, due to chance alone. To understand "chance alone" in this case you can simulate data from a homoscedastic model, construct the $(\hat{y}_i, |e_i|)$ graph, and add the LOESS smooth. You will see that the LOESS smooth is not a perfect flat line, and you will know that such deviations are *explained* by chance alone.

The hypothesis test for homoscedasticity will help you to decide whether the observed deviation from a flat line is explainable by chance alone, but recall that the test does not answer the real question of interest, which is "Is the heteroscedasticity so bad that we cannot use the homoscedastic model?" (That question is best answered by simulating data sets having the type of heteroscedasticity you expect with your real data, then by performing the types of analyses you plan to perform on your real data, then by evaluating the performance of those analyses.)

4.6.1 Production Cost Data $(\hat{y}_i, e_i)$ and $(\hat{y}_i, |e_i|)$ Plots, with LOESS Smooths

The following code shows how to construct graphs to evaluate the homoscedasticity assumption.

```
ProdC = read.table("https://raw.githubusercontent.com/andrea2719/
URA-DataSets/master/ProdC.txt ")
attach(ProdC); par(mfrow=c(1,2))
fit = lm(Cost ~ Widgets)
y.hat = fit$fitted.values; resid = fit$residuals
plot(y.hat, resid); abline(h=0)
abs.resid = abs(resid)
plot(y.hat, abs.resid); add.loess(y.hat, abs.resid, col = "gray", lty=2)
```

The graphs are shown in Figure 4.6. The left panel shows the $(\hat{y}_i, e_i)$ plot. Heteroscedasticity would be seen if the variability in the residuals (and hence in Y) increases for larger $\hat{y}_i$ or decreases for larger $\hat{y}_i$. There is no clear evidence of either, so the left panel's $(\hat{y}_i, e_i)$ plot shows no evidence of heteroscedasticity.

The right panel offers a refined view by using the absolute values of the residuals. If the variability in the residuals e_i is larger with larger $\hat{y}_i$, then there should be an increasing trend in the $(\hat{y}_i, |e_i|)$ plot; and if the variability in the residuals e_i is smaller with larger $\hat{y}_i$, then there should be a decreasing trend in the $(\hat{y}_i, |e_i|)$ plot. Either case suggests heteroscedasticity. In the right panel of Figure 4.6, the curve is not perfectly flat as expected when the process is homoscedastic, but the deviation from perfect flatness may be explained by chance alone (as is confirmed by using the p-value based test below).

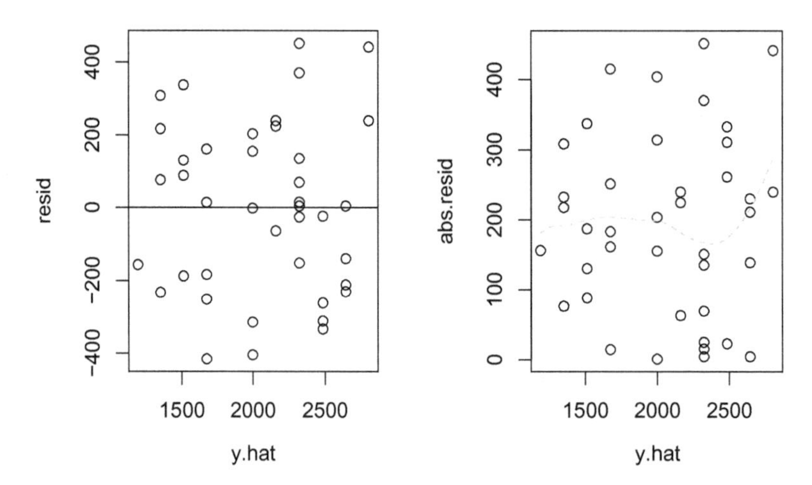

FIGURE 4.6
Plots of ($\hat{y}_i$, e_i) (left panel) and ($\hat{y}_i$, $|e_i|$) with LOESS smooth (right panel) for the Production data. There is no clear evidence of heteroscedasticity in either view. In the right panel, the upward trend would suggest larger variability for large predicted values, but this trend may be explained by chance alone.

4.6.2 Personal Assets Data $(\hat{y}_i, e_i)$ and $(\hat{y}_i, |e_i|)$ Plots, with LOESS Smooths

The Personal Assets data introduced in Chapter 1 reflect heteroscedasticity, as the following analysis shows.

```
Worth = read.table("https://raw.githubusercontent.com/andrea2719/
URA-DataSets/master/Pass.txt")
attach(Worth); par(mfrow=c(1,2))
fit = lm(P.assets ~ Age)
y.hat = fit$fitted.values; resid = fit$residuals
plot(y.hat, resid); abline(h=0)
abs.resid = abs(resid); plot(y.hat, abs.resid)
add.loess(y.hat, abs.resid, col = "gray", lty=2)
```

The graphs are shown in Figure 4.7. The left panel shows the ($\hat{y}_i$, e_i) plot; evidence of heteroscedasticity is seen in that the variability in the residuals (and hence in Y) increases for larger $\hat{y}_i$.

The right panel of Figure 4.7 offers a refined view by using the absolute values of the residuals. The increasing variability in the residuals e_i for larger $\hat{y}_i$ seen in the left panel of Figure 4.7 leads to the increasing trend in the ($\hat{y}_i$, $|e_i|$) plot of the right panel. There is odd "up-then-down" appearance in the LOESS smooth but such a pattern is not particularly logical from a subject matter perspective and is likely due to idiosyncrasies of this particular data set (i.e., due to chance alone).

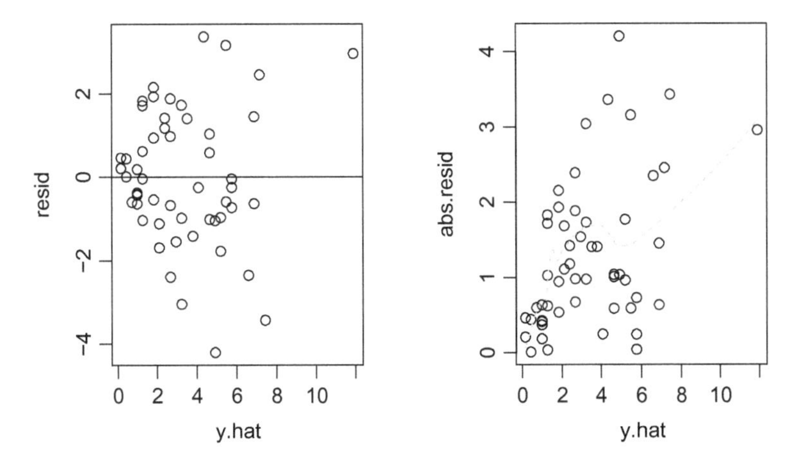

FIGURE 4.7
Plots of $(\hat{y}_i, e_i)$ (left panel) and $(\hat{y}_i, |e_i|)$ with LOESS smooth (right panel) for the Personal Assets data. There is clear evidence of heteroscedasticity in both views.

4.7 Evaluating the Constant Variance Assumption Using Hypothesis Testing Methods

Consider the $(\hat{y}_i, |e_i|)$ scatterplot in the right-hand panel of Figure 4.7. In that plot, there is an increasing trend that suggests heteroscedasticity. You can test for trend in the $(\hat{y}_i, |e_i|)$ scatterplot by fitting an ordinary regression line to those data, and then testing for significance of the slope coefficient. Significance ($p < 0.05$) means that the observed trend is not easily explained by chance alone under the homoscedastic model; insignificance ($p > 0.05$) means that the observed trend is explainable by chance alone under the homoscedastic model. This test is called the Glejser test (Glejser 1969).

There are many tests for heteroscedasticity other than the Glejser test, including the "Breusch-Pagan test" and "White's test." These tests use absolute and/or squared values of the residuals. Because absolute and squared residuals are non-negative, the assumption of normality of the absolute and squared residuals is obviously violated. Hence these tests are only approximately valid.

Another approach to testing heteroscedasticity is to model the variance function $\text{Var}(Y \mid X = x) = g(x, \theta)$ explicitly within a model that uses a reasonable (perhaps non-normal) distribution for $Y \mid X = x$, then to estimate the model using maximum likelihood, and then to test for constant variance in the context of that model using the likelihood ratio test. This approach is better because it identifies the nature of the heteroscedasticity explicitly, which may be an end unto itself in your research. This approach is also better because you can use the resulting heteroscedastic variance function $g(x, \theta)$ to obtain weighted least-squares (WLS) estimates of the β's that are better than the ordinary least-squares (OLS) estimates. Chapter 12 discusses these issues further.

4.7.1 Testing for Heteroscedasticity Using the Production Cost Data

The method is fairly simple as the following code shows.

```
ProdC = read.table("https://raw.githubusercontent.com/andrea2719/
URA-DataSets/master/ProdC.txt")
fit = lm(Cost ~ Widgets, data=ProdC)
y.hat = fit$fitted.values; resid = fit$residuals; abs.resid = abs(resid)
fit.glejser = lm(abs.resid ~ y.hat)
summary(fit.glejser)
```

The relevant output is as follows:

```
Coefficients:
            Estimate Std. Error t value Pr(>|t|)
(Intercept) 1.800e+02  9.464e+01   1.902   0.0648
y.hat       7.196e-03  4.491e-02   0.160   0.8736
```

This output is related to the right panel of Figure 4.6. The positive coefficient `7.196e-03` of `y.hat` suggests that, if there is a linear trend in the graph, it is upward. However, such an upward trend is explainable by chance alone ($p = 0.8736$), so, again, there is no compelling evidence of heteroscedasticity in the process that produced these data.

But the test does not prove homoscedasticity. You know, *a priori*, that it is impossible for the conditional distributions of Cost to all have precisely the same variances, to the infinite decimal. Further, for jobs with more Widgets, there is simply more room for variability in Cost (since the Cost values are farther from 0); thus, subject matter considerations suggest that variability in Cost indeed *does increase* for jobs where number of Widgets is larger. But such heteroscedasticity was not proven using the data, because the positive trend in Cost variance for larger Widgets is *explainable by chance alone.*

4.7.2 Testing for Heteroscedasticity Using the Personal Assets Data

The code looks just like the code used for the Production Cost analysis.

```
Pass = read.table("https://raw.githubusercontent.com/andrea2719/
URA-DataSets/master/Pass.txt")
attach(Pass) ; fit = lm(P.assets ~ Age)
y.hat = fit$fitted.values; resid = fit$residuals; abs.resid = abs(resid)
fit.glejser = lm(abs.resid~y.hat)
summary(fit.glejser)
```

The relevant output is as follows:

```
Coefficients:
            Estimate Std. Error t value Pr(>|t|)
(Intercept)  0.66740    0.21463   3.110 0.003064 **
y.hat        0.18500    0.05233   3.535 0.000877 ***
---
Signif. codes:  0 '***' 0.001 '**' 0.01 '*' 0.05 '.' 0.1 ' ' 1
```

This output is related to the right panel of Figure 4.7. The positive coefficient `0.18500` of `y.hat` suggests what you can plainly see; namely, that there is an upward trend in that graph. Such an upward trend is not easily explained by chance alone ($p = 0.000877$).

As always, the results should make sense from a subject matter standpoint. Here, the increase in variability is expected, because as people get older, they generally have more money and have had more time to accrue personal assets. But most importantly, this ability to accrue more, coupled with people's *personal choice* as to whether they wish to accrue more, is what causes increased variability in personal assets with increasing age. People may have much money but simply *choose* to live more simply, and have few personal assets. Others with the same amount of money *choose* to accrue many items, leading to a large range of variability in the high-income group. On the other hand, people with little money have less *choice* as to whether they may accrue many or few assets, leading to smaller variability in that group. Thus, *choice* and *opportunity* cause the heteroscedasticity that is apparent in the data of this example.

4.8 Evaluating the Uncorrelated Errors Assumption Using Graphical Methods

To evaluate the uncorrelated errors assumption, you first have to consider the type of data set you have, whether it is pure time-series, cross-sectional time-series, spatial, repeated measures, or multilevel (grouped) data. With pure time-series data, it is common to let t denote the observation indicator rather than i, and it is common to let T denote the number of time points in the data set rather than n, so the set of observations is indexed by $t = 1,2,...,T$, rather than by $i = 1,2,\ldots,n$.

The uncorrelated errors assumption is often badly violated with pure time-series processes, because, e.g., today is similar to yesterday, but not so similar to five years ago. Thus, the potentially observable values of today's error term, ε_t, are often highly correlated with potentially observable values of yesterday's error term, ε_{t-1}, implying a violation of the uncorrelated errors assumption.

To diagnose correlated errors with pure time-series data, you should first examine the time-series residual graph, or (t, e_t). Look for systematic, non-random patterns, such as trends or sinusoidal-type functional patterns to suggest failure of this assumption. A completely random appearance of this graph is consistent with uncorrelated errors.

The most common type of residual correlation is the correlation of the current error ε_t with the previous error ε_{t-1}, which is called the "lagged" error term. Such correlation is called *auto*correlation because it refers to the correlation of a variable with itself. Thus, the second graph you can view is the lag scatterplot, or (e_{t-1}, e_t), upon which you can superimpose the OLS or LOESS fit to see the trend. A trend in this plot suggests dependence between the current residual and the immediately preceding residual, a violation of the uncorrelated errors assumption. A random scatter with no trend is consistent with uncorrelated errors.

A third kind of plot is the autocorrelation function of the residuals, which displays lag 1, lag 2, lag 3, and more autocorrelations, thus you can use this plot to examine autocorrelations for lags greater than 1.

For data other than pure time-series data, different methods are needed. For spatial data (points in "space," e.g., data with geographic coordinates), you can use a variogram to check for error correlation, in this case called "spatial autocorrelation." With multilevel (grouped) data, you can examine scatterplots where data are labeled by group to diagnose correlation structure; Chapter 10 touches upon this issue. For now, we will discuss only pure time-series data.

4.8.1 The Car Sales Data (t,e_t) and (e_{t-1},e_t) Plots

The Car Sales data are pure time-series since the data are collected in 120 consecutive months. The following code shows the relevant plots to check for uncorrelated (specifically, non-*auto*correlated) errors.

```
CarS = read.table("https://raw.githubusercontent.com/andrea2719/
URA-DataSets/master/CarS.txt")
attach(CarS); n = nrow(CarS)
fit = lm(NSOLD ~ INTRATE)
resid = fit$residuals
par(mfrow=c(1,2))
plot(1:n, resid, xlab="month", ylab="residual")
points(1:n, resid, type="l"); abline(h=0)
lag.resid = c(NA, resid[1:n-1])
plot(lag.resid, resid, xlab="lagged residual", ylab = "residual")
abline(lsfit(lag.resid, resid))
```

The results are shown in Figure 4.8. There is overwhelming evidence of autocorrelation shown by both plots.

What are the consequences of such an extreme violation of assumptions? According to the mathematical theorems summarized in Chapter 3, *if* the data-generating process is truly given by the regression model, *then* the confidence intervals and *p*-values behave *precisely* as advertised, with *precisely* 95% confidence, and *precisely* 5% significance levels. When the independence assumption is grossly violated as seen here, the true confidence levels may be *far* from 95% and the true significance levels may be *far* from 5%. How far? You guessed it: You can find out by using simulation.

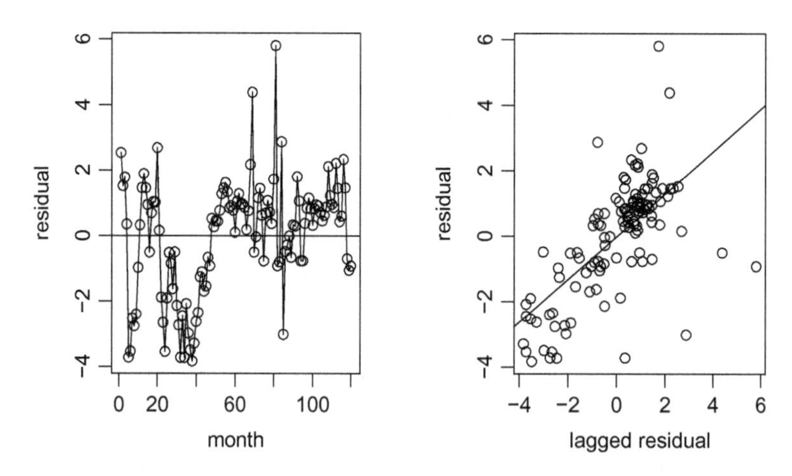

FIGURE 4.8
Plot of (t,e_t) (left panel) and (e_{t-1},e_t) (right panel) for the Car Sales data set. The left plot is clearly non-random, with long runs of consecutive values above or below 0. The right plot shows that the current residual is closely related to the previous residual.

4.9 Evaluating the Uncorrelated Errors Assumption Using Testing Methods

Historically, researchers commonly used a test known as the "Durbin-Watson test" to test for the first-order autocorrelation in pure time-series data. Currently, many tests other than the Durbin-Watson test, such as likelihood ratio tests, are supplied routinely by software that can analyze time-series data. An even simpler (but still useful) test to discover whether the trend in the (e_{t-1}, e_t) plot (the right panel of Figures 4.8) is explainable by chance alone is given by the `cor.test` function in `R`. Simply specify `cor.test(resid, lag.resid)` to get the result. For the Car Sales data you get the following output:

```
                Pearson's product-moment correlation

data:  resid and lag.resid
t = 9.4162, df = 117, p-value = 5.175e-16
alternative hypothesis: true correlation is not equal to 0
95 percent confidence interval:
 0.5404718 0.7481668
sample estimates:
      cor
0.6565924
```

Notice that the estimated correlation is 0.6565924, a positive number, corroborating the positive linear trend shown in the right-hand panel of Figure 4.8. The p-value is 5.175×10^{-16}, indicating that the difference between 0.6565924 and 0.0 is nearly impossible to explain by chance alone. In other words, you will never (for all intents and purposes) see a trend as extreme as in Figure 4.8 in similarly-sized data sets ($n = 120$) that are produced by a model where the errors are in fact uncorrelated.

As always, you should identify a subject matter rationale for any claimed violation of an assumption. In this case, persistent macroeconomic conditions explain that the Sales residual (deviation from Sales prediction based on the interest rate) in a given month is quite similar to that of a previous month, but not so similar to five years ago. In other words, if conditions in the U.S. economy cause higher sales than expected (given interest rates) this month, such conditions are likely to persist through next month, also causing higher sales.

4.10 Evaluating the Normality Assumption Using Graphical Methods

The normality assumption states that each *conditional distribution* $p(y \mid x)$ (one for each x) is a normal distribution. The normality assumption does *not* state that the *marginal distribution* $p(y)$ is normal. It is possible that the distributions $p(y \mid x)$ are all normal yet the distribution $p(y)$ is non-normal; this happens when the distribution of X is non-normal. Thus, you do not assess the normality assumption using the y_i data alone. You have to consider the data Y within specific $X = x$ values.

It is a common error of statistical practice to check normality using the y_i data, so we wish to clarify this issue first, before showing proper methods. Consider a regression process where $Y \mid X = x$ is normal for all $X = x$, with mean $10 + 2x$ and variance 1.0, but where X has the exponential distribution with mean 1.0. The following code generates 200 observations from such a process and shows the *q-q* plot of the y_i data values. Despite all of the assumptions being satisfied, including normality, the *q-q* plot in Figure 4.9 incorrectly suggests that the normality assumption is violated.

R code for Figure 4.9

```
# Normal conditional distributions,
# but non-normal marginal distribution
X = rexp(200,1)
# Y is conditionally normal with mean=10+2x and variance=1
Y = 10 + 2*X + rnorm(200)
qqnorm(Y); qqline(Y)
```

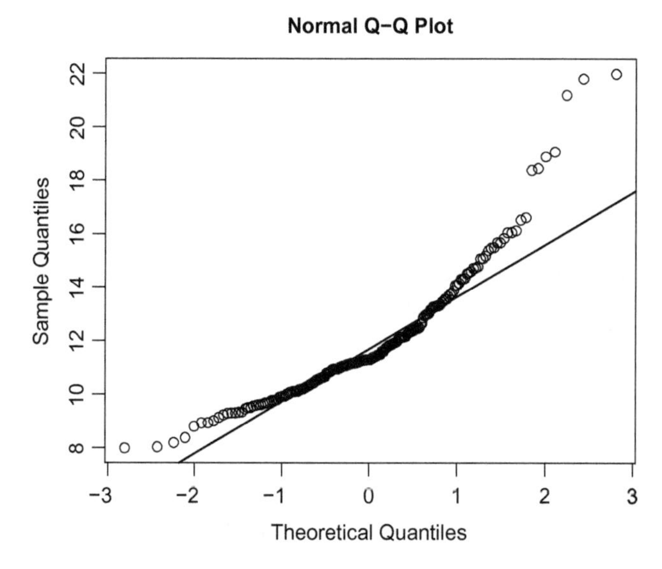

FIGURE 4.9
Normal *q-q* plot of Y data that are conditionally normally distributed, with all regression assumptions valid. The unconditional distribution of Y is clearly non-normal (right skewed), incorrectly suggesting that the normality assumption of regression is violated. The visible right skew is caused by the right skew of the distribution of X.

A related point is that the normality assumption does not refer to the distribution $p(x)$ that produces the X data. There is no assumption in regression that requires the X distribution, $p(x)$, to be a normal or any other distribution. For example, the model used to generate Figure 4.9 produced X from an exponential distribution, yet this violates no assumption. On the other hand, extreme outliers in the X data indicate non-normality of $p(x)$, and while X outliers do not violate any assumption, they do cause problems as discussed in Chapter 16.

The methods for evaluating normality require that the (x, y) data be considered *simultaneously*, either through scatterplots or by using residuals, since the normality assumption refers to the conditional distributions $p(y \mid x)$, which involve both Y *and* X. Your first step should be to look at the (x_i, y_i) scatterplot. The assumption of normality refers to the conditional distributions $p(y\,|\,x)$, which you visualize by considering vertical strips above particular $X = x$ values in the $(x_i,\ y_i)$ scatterplot. In the vertical strips, are there outliers that indicate non-normality? Is there evidence of skewness, in that the outliers tend to be on the high side of the vertical strip rather than the low side? Is their strong evidence of discreteness as shown by the data in the vertical strips only taking a few possible values like 1, 2, 3, 4, or 5? All of these appearances indicate particular types of non-normality of the distributions $p(y\,|\,x)$.

As a second step, you can look at the histogram and normal quantile-quantile (q-q) plots of the residuals $\{e_1, e_2, \ldots, e_n\}$. If the distributions $p(y\,|\,x)$ are all normal *and homoscedastic*, then the histogram and q-q plots of the residuals should have the expected bell-shaped and straight-line appearances, respectively. But caution is needed here: If the distributions are heteroscedastic but normal, then the residuals come from different distributions, some with higher variance, giving the pooled residuals a heavy-tailed appearance, incorrectly suggesting non-normality. For this reason, you should investigate the normality assumption using the residuals e_i *only* when you are reasonably comfortable with the constant variance assumption.

But as always, do not over-interpret chance artifacts in the histogram or q-q plot. Randomness is ubiquitous and always has an effect.

4.10.1 Evaluating the Normality Assumption Using the Car Sales Data

The following code produces both the histogram and the normal q-q plot of the residuals.

```
CarS = read.table("https://raw.githubusercontent.com/andrea2719/
URA-DataSets/master/CarS.txt")
attach(CarS)
fit = lm(NSOLD ~ INTRATE)
Residuals = fit$residuals
par(mfrow=c(1,2))
hist(Residuals)
qqnorm(Residuals); qqline(Residuals)
```

The results are shown in Figure 4.10. Both plots indicate non-normal characteristics of the data-generating process.

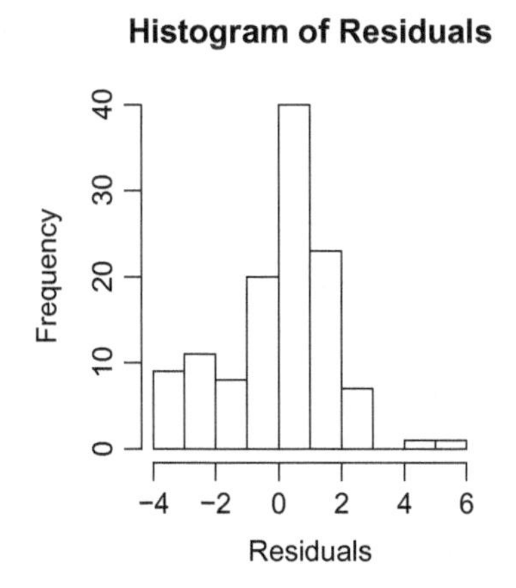

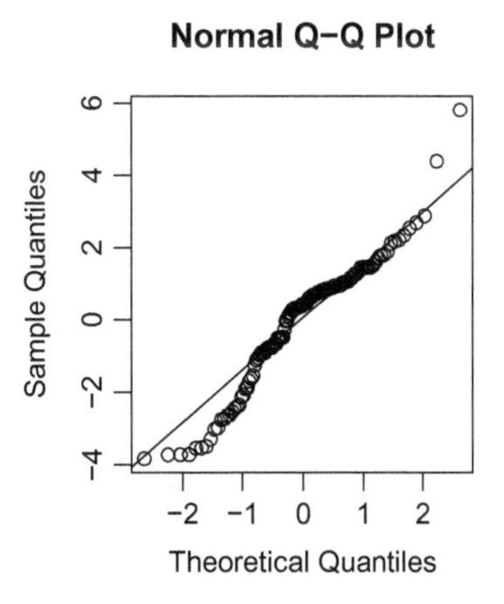

FIGURE 4.10
Histogram (left panel) and normal quantile-quantile plot (right) panel of residuals from the Car Sales regression. The histogram looks somewhat asymmetric rather than bell-shaped. The normal q-q plot also shows a marked deviation from the expected (under normality) straight line.

4.11 Evaluating the Normality Assumption Using Testing Methods

You can use the Shapiro-Wilk test with the residuals e_i (*not* the y_i values!) to test the normality assumption. This test answers the question, "Is the deviation from a straight line seen in the normal q-q plot of the residuals explainable by chance alone?" A "$p < 0.05$" result means that the deviations from the line are not easily explained by chance alone, while a "$p > 0.05$" result means that the deviations from the line are explainable by chance alone. In other words, a $p > 0.05$ result means that the deviations from the line in the q-q plot are similar to deviations that will be seen when data (of the same sample size) are produced by normal distributions.

The R command is simply `shapiro.test(Residuals)`. Applying this command to the Car Sales data gives the following output:

```
        Shapiro-Wilk normality test

data:  Residuals
W = 0.95055, p-value = 0.0002373
```

Since the p-value for the test is $0.0002373 < 0.05$, the deviations from the line in the right panel of Figure 4.10 are not easily explained by chance alone. In other words, in samples of size $n = 120$ from a similar process where the data come from the classical regression model, deviations from the straight line as extreme as what is seen in the q-q plot of Figure 4.10 are seen very rarely, in approximately 2 out of 10,000 samples.

4.12 A Caution about Using Residuals to Evaluate Normality

Methods for assessing normality in regression should involve residuals e_i rather than the raw y_i values. But there is also a concern about using residuals: They can hide discreteness and other non-normal characteristics in the conditional Y distributions.

Here is an example that illustrates the concept where the conditional distributions of Y are highly discrete, and hence clearly non-normal. Yet the normal q-q plot of the residuals and the test for normality using the residuals show no "significant" deviation from normality.

```
par(mfrow=c(1,2)); set.seed(12345)
X = rnorm(1000,3,1); Y = round(X + rnorm(1000,0,1))
table(Y)  # Y is highly discrete and obviously non-normal
model = lm(Y ~ X)
## But the diagnostic plots and tests suggest normality is reasonable
qqnorm(model$residuals); qqline(model$residuals)
shapiro.test(model$residuals); plot(X,Y)
```

The results are shown in Figure 4.11. The q-q plot and the Shapiro test (p-value = 0.1228) of the residuals show no significant deviation from normality, but the non-normal discreteness of the distributions $p(y \mid x)$ is obvious in the simple (x, y) scatterplot.

Thus, you should not rely solely on the residuals to check for normality. You should first examine the ordinary (x_i, y_i) scatterplot. In the example above, the ordinary (x_i, y_i)

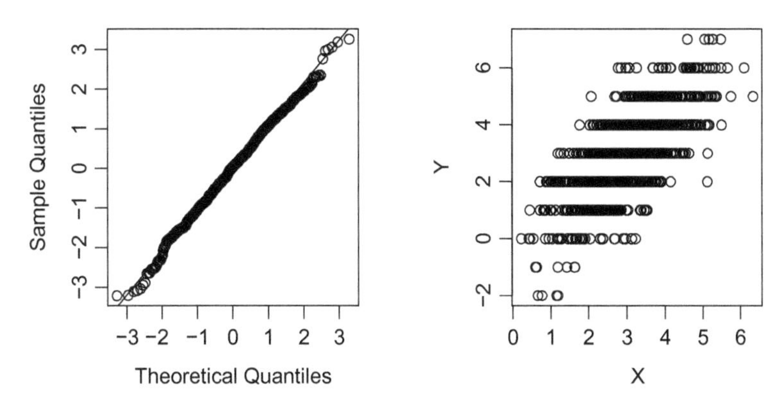

FIGURE 4.11
Normal *q*-*q* plot of residuals (left panel) and (x_i, y_i) scatterplot (right panel) of (*X*, *Y*) data where *Y* is discrete.

scatterplot clearly shows non-normality (in the form of pronounced discreteness) of the $p(y \mid x)$ distributions, but the residuals tell the wrong story by suggesting that the normal model is reasonable.

Remember step 1 of the analysis of any regression data set: Look at the (x_i, y_i) scatterplot.

References

Tufte, E. R. (2001). *The visual display of quantitative information* (Vol. 2). Cheshire, CT: Graphics Press.
Glejser, H. (1969). A new test for heteroskedasticity. *Journal of the American Statistical Association, 64*(325), 316–323.

Exercises

1. Use the "charity" data set,

```
charity = read.csv("https://raw.githubusercontent.com/andrea2719/
URA-DataSets/master/charitytax.csv")
```

Consider the classic regression model for the distribution of charitable contributions (`Charitable.Cont`, the *Y* variable) in terms of income (`Income.AGI`, the *X* variable).

Check the model assumptions (i) linearity, (ii) constant variance, and (iv) normality, using both graphical and testing methods for each of those three assumptions. Give brief written interpretations of the results of each graph or test, so there are six total written interpretations. Make your interpretations specific to the (Charitable contribution vs. Income) data-generating process.

5

Transformations

In Chapter 4, you learned ways to diagnose violations of model assumptions. What should you do if some assumptions are badly violated? The simplest possible remedy is to transform the Y and/or X variables, using logarithmic, inverse, square root, or other simple functions, and then to use the transformed variable(s) in the OLS regression. Among the infinity of possible transformations, the logarithmic transformation is the most common and useful of them all.

You might consider using transformations (on Y and/or on X) when any of the following are true:

- Your assumptions are clearly violated with the untransformed data, and simple transformations make the violations less severe.
- The transformed variables are more interesting than the untransformed variables. For example, per capita variables might be more interesting than totals. Also, log-transformed variables might be more interesting because you can estimate *elasticity* when you log both Y and X (see Section 5.7).
- Visualizations (plots) are enhanced by transformation. For example, in a time-series plot of the Dow Jones Industrial Average (DJIA), the Great Depression from 1929 to 1941 is nearly invisible in the graph of the untransformed DJIA, but it is clearly visible when you log-transform the DJIA.
- Outliers in your raw data have undue influence on the estimated model, and simple transformations (such as log) can reduce the influence of the outliers.
- Your model gives more accurate predictions with appropriately selected transformations.

It is strange that researchers often resist the notion of transformations. As discussed in Chapter 1, the linear function form is nearly always wrong, and for that reason alone, one should at least entertain the notion of alternative function forms.

You can transform Y, X, both Y and X and any combination thereof. But note that transforming Y affects (i) the distribution of Y, as well as (ii) the variance of Y, as well as (iii) the functional form of the relationship between the mean of Y and X. On the other hand, transforming X (or any of the X variables) affects *only* the functional form of the relationship between the mean of Y and X. Transforming X does not affect the distribution of $Y \mid X = x$, so you will not transform X to remedy non-normality or non-constant variance.

When considering transformations, keep in mind the KISS (Keep It Sufficiently Simple) principle. Don't use unusual transformations like $f(v) = \exp\{\sin(v^2/2)\}$. In most cases, you should only consider simple transformations such as log, inverse, and square root.

5.1 Transformation of the *X* Data Only

If you transform *X* to $f(X)$, then your new model is

$$Y = \beta_0 + \beta_1 f(X) + \varepsilon,$$

having mean function

$$E(Y \mid X = x) = \beta_0 + \beta_1 f(x)$$

Your goal is therefore to choose an $f(x)$ function that gives a good model for the conditional mean function, $E(Y \mid X = x)$. Recall that LOESS curves are estimates of the conditional mean function $E(Y \mid X = x)$; therefore, if you transform *X*, you want $\beta_0 + \beta_1 f(x)$ to look somewhat like the LOESS curve, for some values of the parameters β_0 and β_1.

Consider the log-transform $f(x) = \ln(x)$. Figure 5.1 shows different graphs of $\beta_0 + \beta_1 \ln(x)$, as a function of *x*, for different β_0 and β_1.

R code for Figure 5.1

```
x = seq(0.2, 10, .1)
EY1 = 0 + 1*log(x); EY2 = 2 - .5*log(x); EY3 = 1 + 2*log(x)
plot(x, EY1, type="l", ylim = c(-2,5), lty=1, ylab = "E(Y | X=x)")
points(x, EY2, type="l", lty=2); points(x, EY3, type="l", lty=3)
abline(h=0, col="gray"); abline(v=1, lty=1, col="gray")
legend("bottomright", c("b0= 0,  b1=  1", "b0= 2,  b1=-.5",
"b0= 1,  b1=  2"), lty = c(1,2,3), cex=0.8)
```

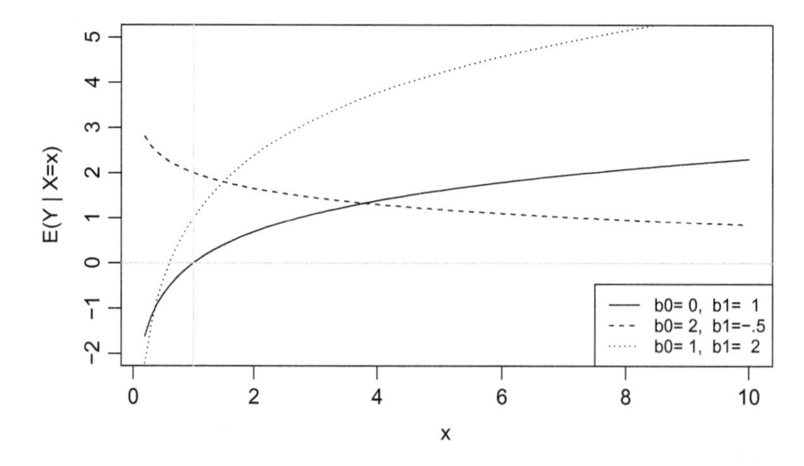

FIGURE 5.1
Graphs of $\beta_0 + \beta_1 \ln(x)$, as a function of *x*, for different β_0 and β_1. Curves intercept the vertical line where $x = 1$ at the value β_0 for that curve.

In Figure 5.1, note the following:

- The graphs are meant to model $E(Y \mid X = x)$, in their original (not transformed) units of measurement. So if a LOESS smooth over the (x_i, y_i) scatterplot (again, in the original, not transformed units) looks something like what you see in Figure 5.1, then the $\ln(x)$ transformation might be suitable.
- The graphs all have $x > 0$. This brings up an important point: You cannot log-transform data that are 0 or negative.
- As $x \to 0$, the function tends to either $+\infty$ or $-\infty$. This behavior may or may not be desirable, from a subject matter standpoint, and is something you should consider when deciding whether to use the $\log(X)$ transformation. If your data have zeros or values close to zero, consider using the transformation $\ln(x+1)$ rather than $\ln(x)$.
- Notice that the parameters β_0 and β_1 in the transformed model have different meanings than they do in the untransformed model: In the transformed model, the parameter β_0 is the place where the function intercepts the $x = 1$ axis (the red line), rather than the $x = 0$ axis. Also, the parameter β_1 is no longer the slope, but rather a multiplier of $\ln(x)$. But you still can say that $\beta_1 > 0$ corresponds to an increasing function and $\beta_1 < 0$ corresponds to a decreasing function.

Now consider the inverse transform $f(x) = x^{-1}$. Figure 5.2 shows different graphs of $E(Y \mid X = x) = \beta_0 + \beta_1 x^{-1}$, as a function of x, for different β_0 and β_1.

R code for Figure 5.2

```
x = seq(0.2, 10, .1)
EY1 = 0 + 1*x^-1
EY2 = 2 - .5*x^-1
EY3 = 1 + 2*x^-1
plot(x, EY1, type="l", lty=1, ylab = "E(Y | X=x)", ylim = c(-1,6))
points(x, EY2, type="l", lty=2)
points(x, EY3, type="l", lty=3)
legend("topright", c("b0= 0,  b1=  1", "b0= 2,  b1=-.5",
"b0= 1,  b1=  2"), lty = c(1,2,3), cex = 0.8)
```

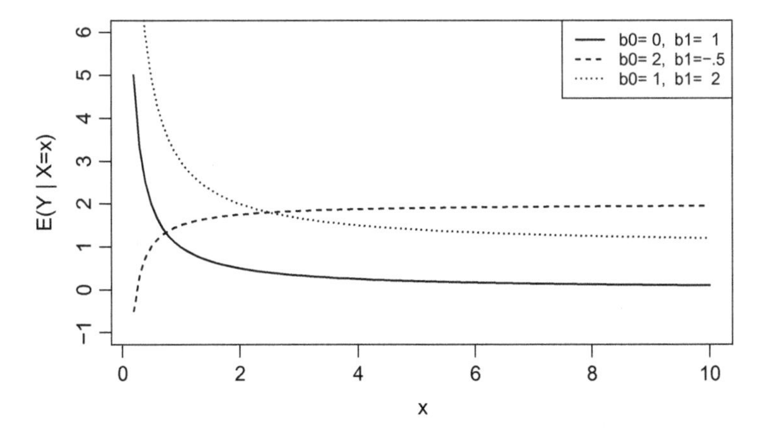

FIGURE 5.2
Graphs of $\beta_0 + \beta_1 x^{-1}$, as a function of x, for different β_0 and β_1.

In Figure 5.2, note the following:

- The graphs are meant to model $E(Y \mid X = x)$, in their original (not transformed) units of measurement. So if a LOESS smooth over the (x_i, y_i) scatterplot (again, in the original, not transformed units) looks something like what you see in Figure 5.2, then the x^{-1} transformation might be suitable.
- Compared to the $\ln(x)$ transformation, the x^{-1} transformation has stronger curvature.
- The graphs have $x > 0$. This brings up an important point—you cannot inverse-transform data that are 0. You can inverse-transform data that are negative, but if the data have both positive and negative numbers, the graph will look strange.

Self-study question: How do the graphs of Figure 5.2 look when the range is $-10 < x < 10$, with $x = 0$ excluded?

- As $x \to 0$, the $E(Y \mid X = x)$ function tends to either $+\infty$ or $-\infty$, like with the log transform. And, like with the log transform, this may be very undesirable behavior if your X variable can take values close to zero.
- The parameter β_0 is the limiting value of $E(Y \mid X = x)$ as x tends to $+\infty$. That's quite a bit different than the value where $x = 0$, as is the case for the intercept in the untransformed model!
- The parameter β_1 is no longer the slope, but rather a multiplier of x^{-1}. Because x^{-1} is inversely related to x, a positive slope parameter ($\beta_1 > 0$) corresponds to a *decreasing* function, while $\beta_1 < 0$ corresponds to an *increasing* function.

5.1.1 What Should I Use with My Data, ln(*x*), or Inverse of *X*, or No Transform at All?

The maximized log likelihood is a great statistic for comparing models. Higher log likelihood implies a better fit of the model to the data. Recall from Chapter 2 that the log likelihood for the classical model is

$$LL(\theta \mid \text{data}) = -\frac{n}{2}\ln(2\pi) - \frac{n}{2}\ln(\sigma^2) - \frac{1}{2\sigma^2}\sum_{i=1}^{n}\{y_i - (\beta_0 + \beta_1 x_i)\}^2$$

Plugging in the MLEs $\hat{\beta}_0$, $\hat{\beta}_1$, and $\hat{\sigma}^2 = \frac{1}{n}\sum_{i=1}^{n}\{y_i - (\hat{\beta}_0 + \hat{\beta}_1 x_i)\}^2$, you get the maximized log likelihood $-\frac{n}{2}\ln(2\pi) - \frac{n}{2}\ln(\hat{\sigma}^2) - \frac{1}{2\hat{\sigma}^2}\sum_{i=1}^{n}\{y_i - (\hat{\beta}_0 + \hat{\beta}_1 x_i)\}^2$. Simplifying this expression gives you the maximized log likelihood.

The maximized log likelihood in the classical regression model

$$LL = -\frac{n}{2}\ln(2\pi) - \frac{n}{2}\ln(\hat{\sigma}^2) - \frac{n}{2}$$

Notice that larger maximized log likelihood corresponds directly to smaller $\hat{\sigma}^2$, which is intuitively logical: Models with a smaller sum of squared deviations from the fitted mean function to the data, fit the data better.

With the X variable expressed as either X, X^{-1}, or $\ln(X)$, you will get three different maximized log likelihoods. The largest of these log likelihoods will correspond to the model giving the minimum sum of squared deviations from the fitted function (i.e., the minimum $\hat{\sigma}^2$). You can get the maximized log likelihood statistic from `R` using the "`logLik(fit)`" command, where "`fit`" is the fitted `lm` object.

5.1.2 Comparing Transformations of *X* with the Car Sales Data

Consider the Car Sales data discussed in Chapter 4. The test for curvature using the quadratic function showed "significant" curvature, corroborating both the curved LOESS smooth, and the subject matter theory, which suggests that as interest rates increase, the mean sales function should flatten because cash sales are not affected by interest rates.

So, first of all, why bother transforming X = Interest Rate? We already showed two different estimates of curvature, one using LOESS, and the other using the quadratic function. Why not use either the LOESS fit or the quadratic model? You could. But there are compelling reasons *not* to use either LOESS or quadratic models when you have curvature. These reasons are also compelling reasons for why you might wish to use a transformation to model curvature.

Problems with LOESS fit

The LOESS function cannot be written in a simple function such as linear, quadratic, exponential, etc. Having a simple function form such as $\mathrm{E}(Y \mid X = x) = \beta_0 + \beta_1 \ln(x)$ makes the model easier to interpret and use.

Problems with quadratic models

Quadratics and higher-order polynomial functions are notoriously bad at the extreme low and high values of the X data. In addition, quadratic models have an extra parameter (β_2) that must be estimated, which can cause loss of accuracy.

So, rather than use the LOESS or the quadratic to model curvature, you might rather fit a function of the form $\mathrm{E}(Y \mid X = x) = \beta_0 + \beta_1 f(x)$. But this is not a hard-and-fast rule. In some cases, the LOESS or quadratic functions might be perfectly adequate for use.

Self-study question: Can you think of an (X, Y) case where the "up-then-down" or "down-then-up" behavior of a quadratic model of $E(Y \mid X = x)$ makes sense?

Here is the `R` code to compare the log likelihood statistics for the model with non-transformed x, the model with logarithmically transformed x (i.e., $f(x) = \ln(x)$) and the model with inverse-transformed x (i.e., $f(x) = x^{-1}$) in the case of the Car Sales data.

```
CarS = read.table("https://raw.githubusercontent.com/andrea2719/
URA-DataSets/master/CarS.txt")
attach(CarS)
```

```
fit.linear.x = lm(NSOLD ~ INTRATE)
ln.INTRATE = log(INTRATE) ; fit.ln.x = lm(NSOLD ~ ln.INTRATE)
inv.INTRATE = INTRATE^-1 ; fit.inv.x = lm(NSOLD ~ inv.INTRATE)
logLik(fit.linear.x); logLik(fit.ln.x); logLik(fit.inv.x)
```

The output is as follows:

```
'log Lik.' -237.8328 (df=3)
'log Lik.' -233.2447 (df=3)
'log Lik.' -232.0109 (df=3)
```

Of the three models, the inverse transform fits best, *according to the log likelihood criterion*. The phrase "according to the log likelihood criterion" reflects an important caveat: Different criteria can favor different models. There is no single statistic, log likelihood or otherwise, that can tell you the correct model. In fact, you already know that all three of these models are wrong.

Self-study question: How do you know that all of these three models, using X, $\log(X)$ and X^{-1} as linear predictors of Y, are wrong as models for $\mathrm{E}(Y \mid X = x)$? (Hint: The answer lies in Chapter 1.)

Notice that the degrees of freedom (`df`) shown in the result of the `logLik` function are not the error degrees of freedom, but rather the number of parameters in each model. There are three parameters in each of the three models, β_0, β_1, and σ.

To understand the inverse-x model better, look at the summarized fit, obtained using the command `summary(fit.inv.x)`:

```
Coefficients:
              Estimate  Std. Error  t value    Pr(>|t|)
(Intercept)     7.6047      0.5792    13.13      <2e-16    ***
inv.INTRATE    55.7597      5.0737    10.99      <2e-16    ***
```

The estimated conditional mean function is thus given by

$$\hat{\mathrm{E}}(\text{Sales} \mid \text{Interest Rate} = x) = 7.6047 + 55.7597x^{-1}$$

If this function is a good fit to the data, then the graph of this function should match the original (non-transformed) (x_i, y_i) data well. Here is `R` code to superimpose the fitted function over the scatterplot of the original, untransformed data:

```
b0 = fit.inv.x$coefficients[1]; b1 = fit.inv.x$coefficients[2]
x = seq(5, 20, .01); y.hat = b0 + b1*x^-1
plot(INTRATE, NSOLD); points(x, y.hat, type="l")
```

The result is shown in Figure 5.3. Comparing this graph to the LOESS fit in Figure 4.2, you can see that the inverse function nicely models the curvature in the relationship between Sales and Interest Rates.

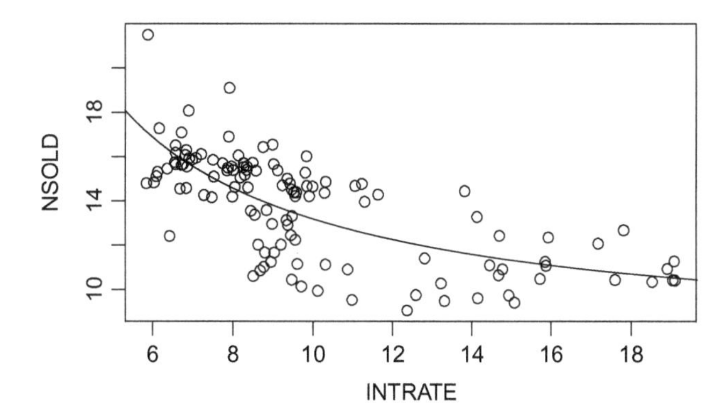

FIGURE 5.3
Scatterplot of (Interest Rates, Sales), with the function $\hat{E}(\text{Sales} \mid \text{Interest Rate} = x) = 7.6047 + 55.7597x^{-1}$ superimposed.

Just because a model has a higher maximized likelihood (L) than another model does not validate the model assumptions. In the case of the classical model, higher L means that the Y data have a smaller sum of squared deviations from the fitted function, but not that the assumptions of the model are valid. You still need to evaluate the assumptions of the transformed model, even when the model has a relatively high likelihood.

Checking the assumptions of the transformed model

To check assumptions of the model when you transform X, simply apply the techniques you learned in the previous chapter with the transformed X variable. In other words, let $U = f(X)$ and check the assumptions of the (U, Y) data in the same way that you check the assumptions with the (X, Y) data.

You can perform all the diagnostic tricks you learned in Chapter 4 to assess the reasonableness of the model that predicts Y as a function of the variable U. For example, if the model using the transformed variable $U = f(X)$ is good, then there should be an approximately linear relationship between Y and U. In Chapter 4, you learned that you can superimpose a LOESS fit, as well as a linear fit, over the scatterplot to assess the reasonableness of the linear fit. The following `R` code shows how.

`R` code for Figure 5.4

```
## Graphically assessing the inverse interest rate transformed model
par(mfrow=c(1,2))
plot(inv.INTRATE, NSOLD); abline(lsfit(inv.INTRATE, NSOLD))
add.loess(inv.INTRATE, NSOLD, col = "gray", lty=2)
fit = lm(NSOLD~inv.INTRATE)
y.hat = fit$fitted.values; resid = fit$residuals
plot(y.hat, resid); add.loess(y.hat, resid, col = "gray", lty=2)
abline(h=0)
```

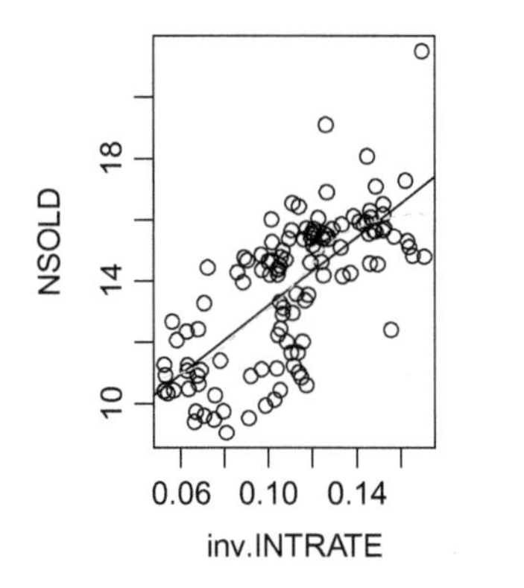

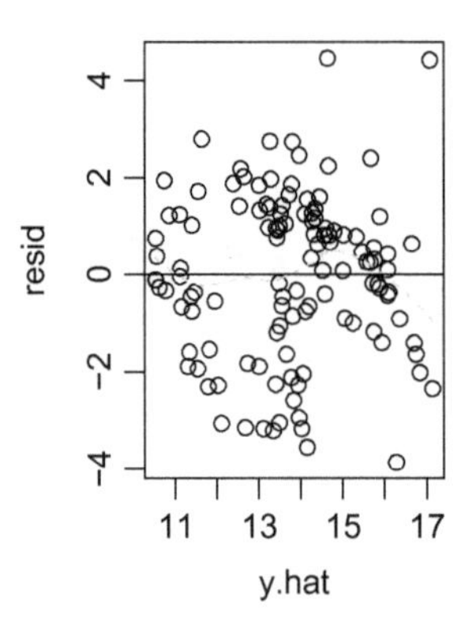

FIGURE 5.4
Replication of Figure 4.2 for the inverse-transformed Interest Rate (X) data in the Car Sales analysis. Scatterplots of (x_i^{-1}, y_i) (left panel) and $(\hat{y}_i, e_i)$ (right panel) are shown with LOESS and linear fits superimposed. The curvature is not as pronounced in the case of the inverse-transformed X data as it was in the case of the non-transformed data.

To see whether the curvature in the left panel of Figure 5.4 is explainable by chance alone, you can use the quadratic test using the inverse-transformed variable $U = X^{-1}$ as follows:

```
## Testing curvature in the transformed model
inv.INT.squared = inv.INTRATE^2
fit.quad = lm(NSOLD~inv.INTRATE+inv.INT.squared)
summary(fit.quad)
```

The p-value for the coefficient of U^2 is given in the output of the R code above is $p = 0.581$, so there is not "significant" evidence of (quadratic) curvature in the model where X is transformed to $U = X^{-1}$.

Interpretation of coefficients β_0 and β_1 in the transformed model $Y = \beta_0 + \beta_1 U + \varepsilon$ is the same as before as well, except that all are in terms of the transformed variable U. In particular, the intercept β_0 is the mean of the distribution of Y when $U = 0$ (implying that $X = \infty$). The estimate $\hat{\beta}_0 = 7.6047$ is an estimate of average car sales when interest rates are infinity. Since no one will buy a car on credit with infinite interest rates, 7.6047 is therefore an estimate of average sales involving cash purchases. But you usually do not have to worry much about the intercept when the 0 value for the explanatory variable is well outside the range of the data because in this case the intercept is an extrapolation. Here, the explanatory variable is $U = 1/(\text{Interest Rate})$ and $U = 0$ means that interest rate equals infinity, obviously well outside the range of the data.

The slope estimate $\hat{\beta}_1 = 55.7597$ is simple to interpret in terms of $U = 1/(\text{Interest Rate})$: The estimated average of the Sales variable is 55.7597 higher in months where $U = u + 1$ than in months where $U = u$. But look at Figure 5.4 above: The range of $U = 1/(\text{Interest Rate})$ is roughly 0.06 to 0.16, so a one unit (1.0) difference in possible values of U is not observable. Instead, consider a 0.01 difference, and compare months where $U = 0.11$ versus months where $U = 0.10$. These correspond to months where actual interest rates are $1/0.11 = 9.09\%$ and $1/0.10 = 10\%$, respectively. The fitted model estimates that the average sales will be 0.557597 higher in the former case (where the interest rate is lower) than in the latter case.

5.2 Logarithmic Transformation of the *Y* data

If you transform the *Y* variable to $f(Y)$ but not the *X* variable, then you think the model

$$f(Y) = \beta_0 + \beta_1 X + \varepsilon$$

is better than the model $Y = \beta_0 + \beta_1 X + \varepsilon$. As with transformation of *X*, in order to use this model successfully, you must understand what this model states in the original (untransformed) (*X*, *Y*) data. Here,

$$Y = f^{-1}\{\beta_0 + \beta_1 X + \varepsilon\},$$

where f^{-1} is the *inverse function* (not the inverse *of* the function). You find the inverse function simply by solving the model equation ($f(Y) = \beta_0 + \beta_1 X + \varepsilon$) for *Y*.

For example, if $f(Y) = \ln(Y)$, then $Y = f^{-1}\{f(Y)\} = \exp\{f(Y)\}$, and the model in terms of the original units is then

$$Y = \exp(\beta_0 + \beta_1 X + \varepsilon),$$

or equivalently,

$$Y = \exp(\beta_0) \times \exp(\beta_1 X) \times \exp(\varepsilon)$$

Notice now that the error term is *multiplicative*, rather than *additive*. Along with Jensen's inequality, the multiplicative error implies that the function $\exp(\beta_0) \times \exp(\beta_1 X)$ is not the conditional mean. To see why not, note that

$$\begin{aligned} E(Y \mid X = x) &= \exp(\beta_0) \times \exp(\beta_1 x) \times E\{\exp(\varepsilon \mid X = x)\} \\ &= \exp(\beta_0) \times \exp(\beta_1 x) \times E\{\exp(\varepsilon)\} \end{aligned}$$

But, since $\exp(\cdot)$ is a convex function, $E\{\exp(\varepsilon)\} > \exp\{E(\varepsilon)\} = \exp(0) = 1$, so that $E(Y \mid X = x) > \exp(\beta_0) \times \exp(\beta_1 x)$. Thus, the back-transformed function, $\exp(\beta_0) \times \exp(\beta_1 x)$, is no longer the mean function of the untransformed data.

On the other hand, if the distribution of ε is symmetric, as is the case where the errors are normally distributed, then the *median* of the distribution of $\exp(\varepsilon)$ is $\exp(0) = 1$, in turn implying that back-transformed function $\exp(\beta_0) \times \exp(\beta_1 x)$ is the *median* of the distribution of untransformed $Y \mid X = x$. We will discuss the estimation of the median (and other quantiles of the distribution of $Y \mid X = x$) again when we discuss *quantile regression* in Chapter 16.

Figure 5.5 displays some of these (median) functions $\exp(\beta_0) \times \exp(\beta_1 x)$ for different combinations of (β_0, β_1).

R code for Figure 5.5

```
x = seq(-1, 4, .1)
E.lnY1 =  0+.5*x; MY1  = exp(E.lnY1)
E.lnY2 = 2 - .5*x; MY2  = exp(E.lnY2)
E.lnY3 = -1 + .9*x; MY3 = exp(E.lnY3)
plot(x, MY1, type="l", lty=1, ylab = "Median of Y|X=x", ylim=c(0,15))
points(x, MY2, type="l", lty=2)
points(x, MY3, type="l", lty=3)
legend(1,14, c("b0=  0,  b1=   .5", "b0=  2,  b1=  -.5", "b0= -1,
   b1=  0.9"), lty = c(1,2,3))
```

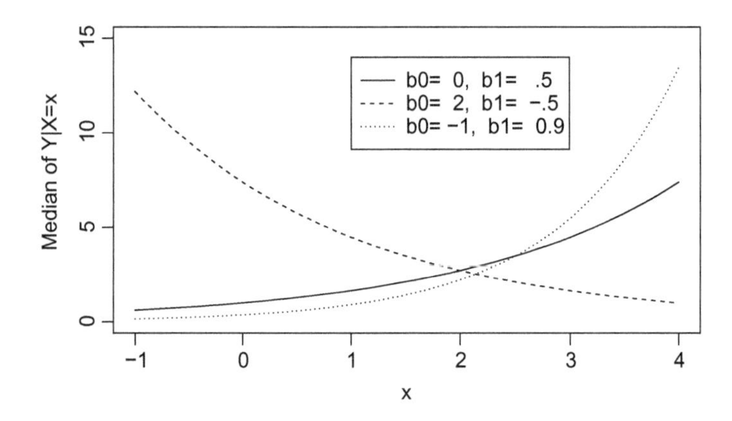

FIGURE 5.5
Graphs of $\exp(\beta_0)\times\exp(\beta_1 x)$, as a function of x, for different β_0 and β_1.

Thus, a first clue that you might want to transform Y logarithmically is if the LOESS smooth of the original (untransformed) (X, Y) data looks something like one of the graphs in Figure 5.5.

As far as parameter interpretation goes, you can interpret $\exp(\beta_0)$ as the *median* of the distribution of Y when $X = 0$, and you can interpret β_1 using the usual "add +1 to x and see what happens" trick. Consider two possible cases:

Case 1: $X = x$. Here, the median of the distribution of Y is $\exp(\beta_0)\times\exp(\beta_1 x)$.

Case 2: $X = x+1$. Here, the median of the distribution of Y is $\exp(\beta_0)\times\exp\{\beta_1(x+1)\} = \exp(\beta_0)\times\exp(\beta_1 x)\times\exp(\beta_1)$.

Thus, the median of the distribution of Y is *higher* (lower when $\beta_1 < 0$) by the multiplicative factor $\exp(\beta_1)$ when comparing the distributions of Y when $X = x+1$, versus when $X = x$.

You can also interpret the parameters (β_0, β_1) more simply in terms of the usual intercept/slope effects of X on ln(Y), rather than on Y, if you are comfortable with interpreting the transformed variable ln(Y).

5.2.1 Log Transforming Income

The "Charity" data set consists of tax return information and contains variables related to charitable contributions. One variable is adjusted gross income (`Income.AGI`) and another is number of dependents (`DEPS`). A question of interest is, what is the nature of the relationship between the income distributions and number of dependents? Do the distributions shift toward higher incomes, or toward lower incomes, as the number of dependents increases?

Fitting the model using logarithmically transformed adjusted gross income is performed using the following `R` code.

```
charity = read.csv("https://raw.githubusercontent.com/andrea2719/
URA-DataSets/master/charitytax.csv")
attach(charity)
ln.AGI = log(Income.AGI)
fit = lm(ln.AGI ~ DEPS)
summary(fit)
```

Results are as follows:

```
Coefficients:
             Estimate   Std. Error   t value    Pr(>|t|)
(Intercept)  10.51200      0.04226   248.746      <2e-16   ***
DEPS          0.01671      0.01483     1.127        0.26
---
Signif. codes:  0 '***' 0.001 '**' 0.01 '*' 0.05 '.' 0.1 ' ' 1

Residual standard error: 0.4951 on 468 degrees of freedom
Multiple R-squared:  0.002707,  Adjusted R-squared:  0.000576
F-statistic:  1.27 on 1 and 468 DF,  p-value: 0.2603
```

From the output, you can construct two estimated models, one in terms of the original variables, and the other in terms of the transformed Y variable. In terms of the transformed Y variable, $W = \ln(\text{Adjusted Gross Income})$, the estimated conditional mean function is

$$\hat{W} = 10.51200 + 0.01671 \times \text{DEPS}$$

In terms of the untransformed Y variable, the model provides an estimate of the *median* of the distribution of the adjusted gross income (Y) variable:

$$\hat{Y} = \exp(10.51200) \times \exp(0.01671 \times \text{DEPS})$$

With this estimated model, you estimate that the *median* income, Y, of people with, say 4 dependents, is $\exp(0.01671) = 1.01685$ *times higher* than the *median income*, Y, of people with 3 dependents. Thus the model estimates that, for each additional dependent, median adjusted gross income is $100 \times (1.01685 - 1.0)\% = 1.685\%$ higher.

The exact 95% confidence interval for β_1 is obtained as before using `confint(fit)`, which gives

```
                   2.5 %      97.5 %
(Intercept)  10.42896153 10.59504687
DEPS         -0.01242624  0.04585371
```

Thus $-0.01242624 < \beta_1 < 0.04585371$, implying that $\exp(-0.01242624) < \exp(\beta_1)$, $< \exp(0.04585371)$ or $0.9876506 < \exp(\beta_1) < 1.046921$, all with exact 95% confidence (assuming all model assumptions are true for the transformed data). Converting the endpoints to percent changes, you can say that the percent increase in median adjusted gross income associated with one additional dependent is between −1.23494% and +4.6921%. In particular, you cannot tell whether there is an increase or a decrease in the median of the adjusted gross income distribution when you compare people with fewer dependents versus more dependents.

As with transformations in the X variable, when you transform the Y variable to $W = f(Y)$, you can perform all the usual checks of the model assumptions using the (X, W) data. The assumptions should be more reasonable when using the transformed data; if not, do not use that transform! Figure 5.6 shows the (X, Y) and (X, W) (where $W = \ln(Y)$) scatterplots with LOESS smooths for the (`DEPS, AGI`) analysis. Code is as follows:

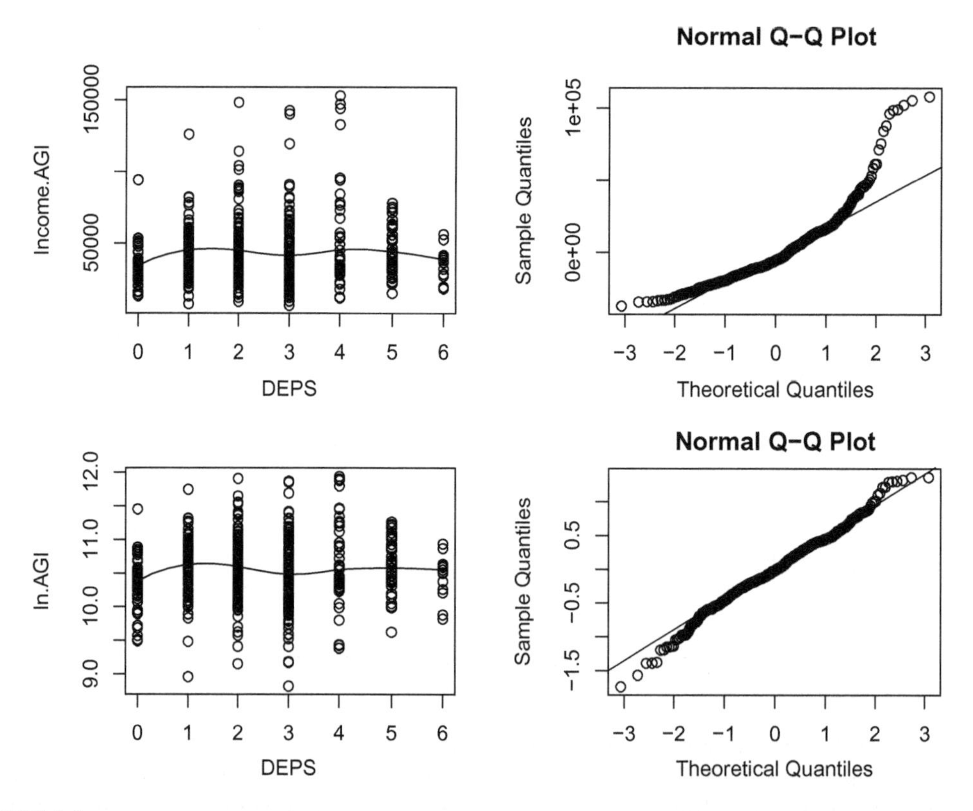

FIGURE 5.6
Scatterplots (left panels) and residual *q-q* plots (right panels) for original Income data versus number of dependents (top panels) and for log-transformed Income data versus number of dependents (bottom panels). The log-transform reduced the deviation from normality in Income distributions.

```
par(mfrow=c(2,2))
plot(DEPS, Income.AGI)
add.loess(DEPS, Income.AGI)
qqnorm(lsfit(DEPS, Income.AGI)$residuals)
qqline(lsfit(DEPS, Income.AGI)$residuals)

ln.AGI = log(Income.AGI)
plot(DEPS, ln.AGI)
add.loess(DEPS, ln.AGI)
qqnorm(lsfit(DEPS, ln.AGI)$residuals)
qqline(lsfit(DEPS, ln.AGI)$residuals)
```

5.2.2 What Should I Use with My Data, ln(*Y*) or No *Y* Transform at All?

Again, the maximized log likelihood is a great statistic for comparing models—higher log likelihood implies a better fit of the model to the data. However, you have to be careful when evaluating the transformations of the Y variable using likelihoods.

Recall that, under the untransformed (classical) model, the maximized log likelihood is

$$LL = -\frac{n}{2}\ln(2\pi) - \frac{n}{2}\ln(\hat{\sigma}^2) - \frac{n}{2}$$

where $\hat{\sigma}^2 = \sum_{i=1}^{n}(y_i - \hat{y}_i)^2/n$, the maximum likelihood estimate of σ^2. Now, suppose you transform Y to $W = aY$, where a is a constant. For example, if you change the units of measurement from dollars to thousands of dollars, then $W = aY = (1/1{,}000)Y$. In this case, the estimated variance of the transformed data W is $a^2 = (1/1{,}000)^2$ times the estimated variance of the untransformed data Y, meaning that the log likelihood of the transformed model is different by $(n/2)\ln(1/1000^2)$ from the log likelihood of the untransformed model.

The following R code illustrates this concept.

```
n = 100; set.seed(12345)
X = runif(n); Y = 20000 + 2000*X + rnorm(n, 0, 500)
par(mfrow=c(1,2)); plot(X,Y)

## W = Y in thousands
W = Y/1000; plot(X,W)

fit.orig  = lm(Y~X); fit.trans = lm(W~X)

LL.orig  = logLik(fit.orig); LL.trans = logLik(fit.trans)

LL.orig; LL.trans
LL.trans - LL.orig
-(n/2)*log(1/1000^2)
```

Figure 5.7 shows the scatterplots of the original and transformed data. It should be clear that the regression model for each case is identical; there really should be no difference between the two.

The output of the R code shows the following:

```
> LL.orig; LL.trans
'log Lik.' -775.5301 (df=3)
'log Lik.' -84.75458 (df=3)
```

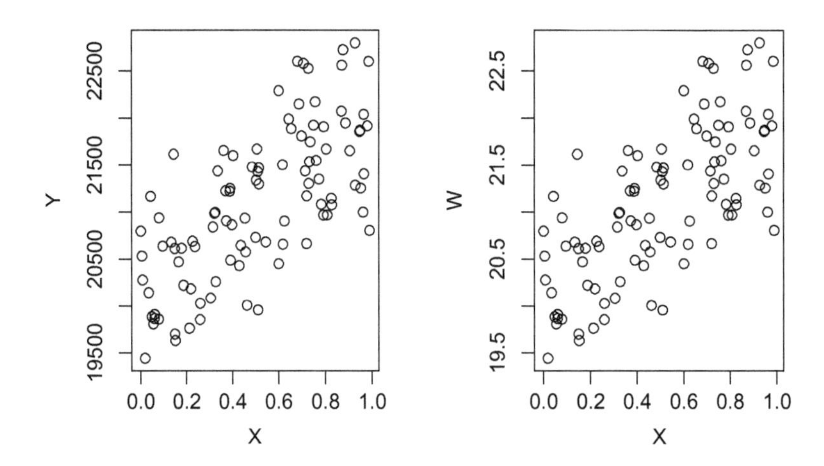

FIGURE 5.7
Scatterplots of original (left panel) and linearly transformed ($W = Y/1000$) dependent variable (right panel). Both meet the regression assumptions equally well.

```
> LL.trans - LL.orig
'log Lik.' 690.7755 (df=3)
> -(n/2)*log(1/1000^2)
[1] 690.7755
```

This output suggests that the transformed model is much better since its log likelihood is 690.7755 higher than the log likelihood for the untransformed data. But there is no essential difference between the two models, as seen in Figure 5.7, so there is a clear problem here with using the likelihood statistics in this way. The resolution to the problem is shown in the box.

When comparing two regression models using likelihood-based statistics, you always must have the same *Y* variable in both. You cannot compare the likelihood of a model for *Y* with the likelihood of a model for $W = f(Y)$.

When you log-transform *Y*, you assume a completely different distribution for the untransformed *Y* variable. Specifically, if you assume that

$$\ln(Y) \mid X = x \sim \mathrm{N}(\mu(x), \sigma^2),$$

then you also assume

$$Y \mid X = x \sim \exp\{\mathrm{N}(\mu(x), \sigma^2)\}$$

The distribution $\exp\{\mathrm{N}(\mu(x), \sigma^2)\}$ is the famous *lognormal* distribution, which is the distribution of a random variable whose logarithm is normally distributed. So when you log-transform *Y*, you assume that $p(y \mid x)$ is a lognormal distribution, rather than a normal distribution, as in the classical model.

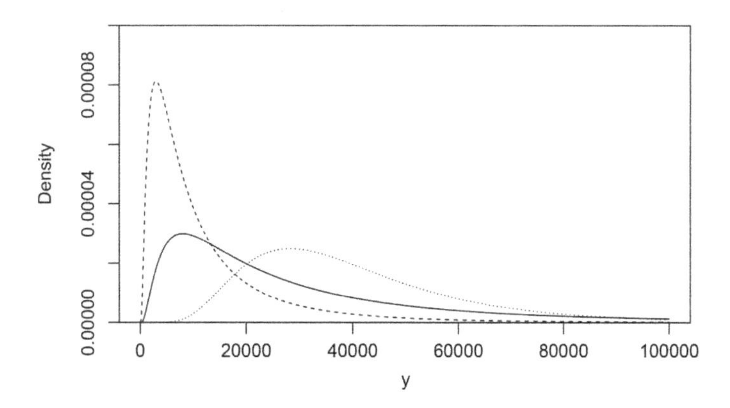

FIGURE 5.8
Examples of lognormal distributions. Solid line: $(\mu,\sigma)=(10,1)$. Dashed line: $(\mu,\sigma)=(9,1)$. Dotted line: $(\mu,\sigma)=(10.5,0.5)$.

The lognormal distribution has parameters μ and σ^2, just like the normal distribution, but these parameters refer to the distribution of ln(Y), not Y. Also, from Jensen's inequality, if μ is the mean of ln(Y), then the mean of $Y=\exp\{\ln(Y)\}$ must be *greater than* $\exp(\mu)$, because exp(·) is a convex function. Specifically, $E(Y)=\exp(\mu)\times\exp(\sigma^2/2)$, which is greater than $\exp(\mu)$ since $\exp(\sigma^2/2)>1$. On the other hand, $\exp(\mu)$ is the *median* of the distribution of Y. Figure 5.8 shows various lognormal distributions for different settings of μ and σ^2.

Probability distributions that are graphed in Figure 5.8 all have the following function form:

$$p(y)=\frac{1}{y\sqrt{2\pi}\sigma}\exp\left[-\frac{\{\ln(y)-\mu\}^2}{2\sigma^2}\right], \text{ for } y>0;\ p(y)=0 \text{ otherwise.}$$

The function form of the lognormal distribution is similar to that of the normal distribution, except there is $\ln(y)$ instead of y in the exponent, and there is an extra y in the denominator. Also, the lognormal distribution only produces positive Y values, whereas the normal distribution produces negative and positive Y values.

Thus, if you assume the classical model $\ln(y)=\beta_0+\beta_1X+\varepsilon$, then you also assume that the distribution of the untransformed $Y \mid X=x$ has the lognormal distribution

$$p(y\mid x,\theta)=\frac{1}{y\sqrt{2\pi}\sigma}\exp\left[-\frac{\{\ln(y)-(\beta_0+\beta_1x)\}^2}{2\sigma^2}\right], \text{ for } y>0;\, p(y\mid x,\theta)=0 \text{ otherwise,}$$

where, as before, the parameter vector is $\theta=(\beta_0,\beta_1,\sigma)$.

The maximum likelihood estimates (MLEs) of the parameters are the same as the MLEs for the classical regression model where you replace Y with $W=\ln(Y)$. But the maximized log likelihood in terms of the untransformed Y is slightly different, because of the extra

"y" in the denominator of the lognormal pdf. The maximized lognormal likelihood, in terms of the raw, untransformed Y variable, is

$$LL_{\text{lognormal}} = -\frac{n}{2}\ln(2\pi) - \frac{n}{2}\ln(\hat{\sigma}_*^2) - \frac{n}{2} - \sum_{i=1}^{n} \ln(y_i)$$

where $\hat{\sigma}_*^2$ is the MLE of σ^2 in the transformed model $\ln(Y) = \beta_0 + \beta_1 X + \varepsilon$. You can compare this log likelihood to the ordinary log likelihood for the classical model $Y = \theta_0 + \theta_1 X + \varepsilon$,

$$LL_{\text{classic}} = -\frac{n}{2}\ln(2\pi) - \frac{n}{2}\ln(\hat{\sigma}^2) - \frac{n}{2}$$

to see which model fits the data better. Of course, higher LL indicates a better fit.

5.2.3 Comparing Log Likelihoods with the Charity Data Set

Here is R code to compare likelihoods for the classical model, which assumes normally distributed Y, with the log-transformed model, which assumes lognormally distributed Y:

```
fit.orig  = lm(Income.AGI ~ DEPS); fit.trans = lm(ln.AGI ~ DEPS)

LL.orig = logLik(fit.orig)
LL.trans = logLik(fit.trans) - sum(log(Income.AGI))
LL.orig; LL.trans; LL.trans - LL.orig
```

The results show that the log likelihood of the original, untransformed model is $LL_{\text{classic}} = -5381.359$ while that of the log-normal model is $LL_{\text{lognormal}} = -5294.951$. Since the log likelihood of the original, untransformed Y data using a lognormal model is 86.4 points higher than the log likelihood of the original untransformed Y data using the classical, normally distributed model, the lognormal model fits the data much better, lending an objective comparison to complement the more informal graphic comparison that favored the lognormal model.

In Section 5.5 below, you will see how to perform similar calculations more easily by using the "boxcox" function in R. You will also see in Chapter 15 that there is R software to perform lognormal regression, as well as regression using a variety of other distributions, and to obtain and compare their log likelihoods automatically.

5.3 The ln(Y) Transformation and Its Use for Heteroscedastic Processes

If the regression model $\ln(Y) = \beta_0 + \beta_1 X + \varepsilon$ is true, then (as discussed above) the conditional mean and variance of the untransformed Y variable are given as follows:

$$\mathrm{E}(Y \mid X = x) = \exp(\beta_0 + \beta_1 x) \times \exp(\sigma^2/2),$$

and

$$\mathrm{Var}(Y \mid X = x) = \{\exp(\sigma^2) - 1\} \times \exp(\sigma^2) \times \exp(2\beta_0 + 2\beta_1 x)$$

Notice that when $\sigma^2 = 0$, then $\{\exp(\sigma^2) - 1\} = 0$, so that $\text{Var}(Y \mid X = x) = 0$, as expected. Notice also that if you assume the lognormal model, you also assume that the variance of $Y \mid X = x$ is *nonconstant*; i.e., heteroscedastic: The formula for $\text{Var}(Y \mid X = x)$ in the lognormal case shows that, if $\beta_1 > 0$, then the variance increases for larger $X = x$.

Figures 5.9 and 5.10 demonstrate the heteroscedasticity of the lognormal regression model, as well as the homoscedasticity of the model for the log-transformed data.

R code for Figures 5.9 and 5.10

```
n = 100; beta0 = -2.00; beta1 = 0.05; sigma = 0.30
set.seed(12345); X = rnorm(n, 70, 10)
lnY = beta0 + beta1*X + rnorm(n, 0, sigma); Y= exp(lnY)
par(mfrow=c(1,2)); plot(X,Y);    abline(v=c(60,80), col="gray")
plot(X, lnY); abline(v=c(60,80), col="gray")
y.seq = seq(0.01,30,.01)
dy1 = dlnorm(y.seq, beta0+beta1*60, sigma)
dy2 = dlnorm(y.seq, beta0+beta1*80, sigma)
par(mfrow=c(1,2))
plot(y.seq, dy1, type="l", xlim = c(0,20), yaxs="i", ylim = c(0, .6),
    ylab="lognormal density", xlab = "Untransformed y")
points(y.seq, dy2, type="l", lty =2)
legend("topright", c("X=60", "X=80"), lty=c(1,2), cex=0.8)
ly.seq = log(y.seq)
dly1 = dnorm(ly.seq, beta0+beta1*60, sigma)
dly2 = dnorm(ly.seq, beta0+beta1*80, sigma)
plot(ly.seq, dly1, type="l", xlim = c(-.2,4), yaxs="i", ylim = c(0,1.6),
    ylab="normal density", xlab = "Log Transformed y")
points(ly.seq, dly2, type="l", lty=2)
legend("topright", c("X=60", "X=80"), lty=c(1,2), cex=0.8)
```

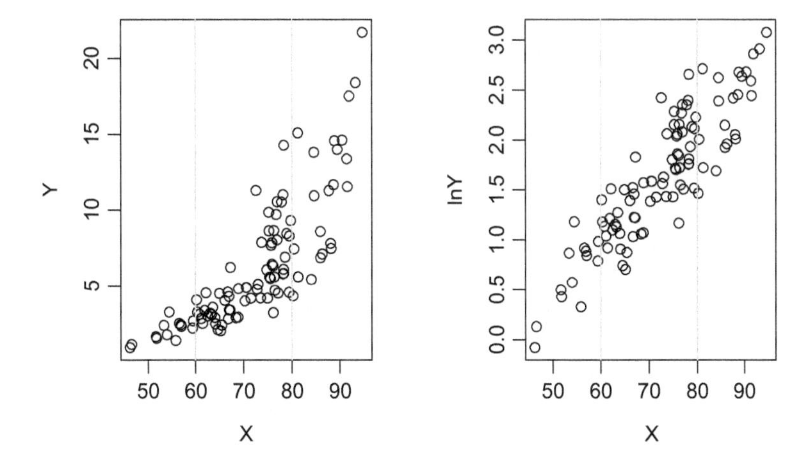

FIGURE 5.9
Data come from the lognormal regression model. Scatterplots are shown for the untransformed (X,Y) data (left panel), and the transformed $(X, \ln(Y))$ data (right panel). Vertical gray lines are given at $X = 60$ and $X = 80$. The variability in Y is greater when $X = 80$ than when $X = 60$, but the variability in $\ln(Y)$ is the same when $X = 80$ as when $X = 60$. The actual distributions in these two cases are shown in Figure 5.10.

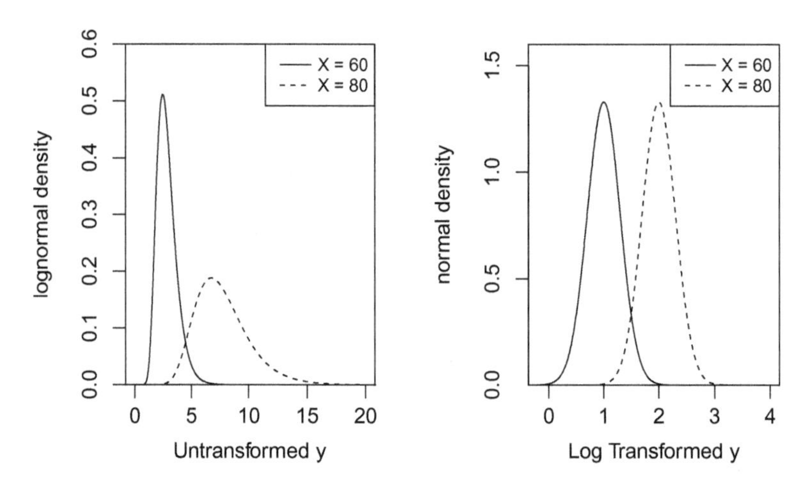

FIGURE 5.10
See Figure 5.9. Left panel shows distributions of Y when $X = 60$ and when $X = 80$; notice that the distribution of $Y \mid X = 80$ has more variance. (The height of the density of $Y \mid X = 80$ is lower for this reason only; recall that the area under a density must be 1.0). On the other hand, the distributions of $\ln(Y)|X = x$ have identical variances.

5.4 An Example Where the Inverse Transformation $1/Y$ Is Needed

Professor Smith collected data on the time it took various computers to perform the same task. He needed to run a massive simulation in a short period of time to meet a deadline for revising a manuscript, so he asked n = 18 graduate students to run some code overnight and send him the results when it was done. Since this was a Monte Carlo simulation, all 18 results were slightly different due to randomness. He then collated all 18 results to get a much larger simulation size and hence more accurate estimates. This allowed him to perform a simulation overnight that otherwise would have taken days to complete.

He was curious as to what factors affected the time it takes for a computer to complete the simulation, so he also had the students record their computer's RAM (in gigabytes) and processor speed (in Gigahertz, or GHz).

One model he used was $Y = \beta_0 + \beta_1 X + \varepsilon$, where Y = time to complete job, and X = Gigabytes RAM (or GB in the code below). However, the results were unsatisfactory: Linearity, constant variance, and normality were clearly violated. He tried using the log-transform on Y, but the results were still not ideal. He then realized that the variable "time to finish the job" could be more directly related to computer performance in its *inverse* transform. After all, time, measured in hours, can be understood as *hours per job*: If a computer took 2 hours to complete the task, then it took 2 hours per 1 job. But the inverse of Y in this example is more directly related to performance: $1/Y = 1/2 = 0.50$ *jobs per hour*. Another computer that took 20 minutes (1/3 hour) to complete the one job would be able to complete $1/(1/3) = 3.0$ *jobs per hour*. Higher jobs per hour clearly indicates a better computer.

With ratio data, the units of measurement are (*a* per *b*), and the inverse transformation often makes sense simply because the measurements become (*b* per *a*), which is just as easy to interpret. For example, a car that gets 30 miles per gallon of gasoline equivalently can be stated to take (1/30) gallons per mile. You could use either measure in a statistical analysis, without question from any critical reviewer—miles per gallon and gallons per mile convey the same information. Which form to use? Simply choose the form that *least violates* the model assumptions.

The following code replicates the analyses shown in Figure 5.6, for these data, but using the $W = 1/Y$ transformation, which he called "speed", because higher values indicate a speedier computer.

R code for Figure 5.11

```
comp = read.table("https://raw.githubusercontent.com/andrea2719/
URA-DataSets/master/compspeed.txt ")
attach(comp)
reg.orig = lm(time ~ GB); summary(reg.orig)

par(mfrow=c(2,2)); plot(GB, time); add.loess(GB, time)
qqnorm(reg.orig$residuals); qqline(reg.orig$residuals)

speed = 1/time; reg.trans = lm(speed ~ GB)
summary(reg.trans)
plot(GB, speed); add.loess(GB, speed)
qqnorm(reg.trans$residuals); qqline(reg.trans$residuals)
```

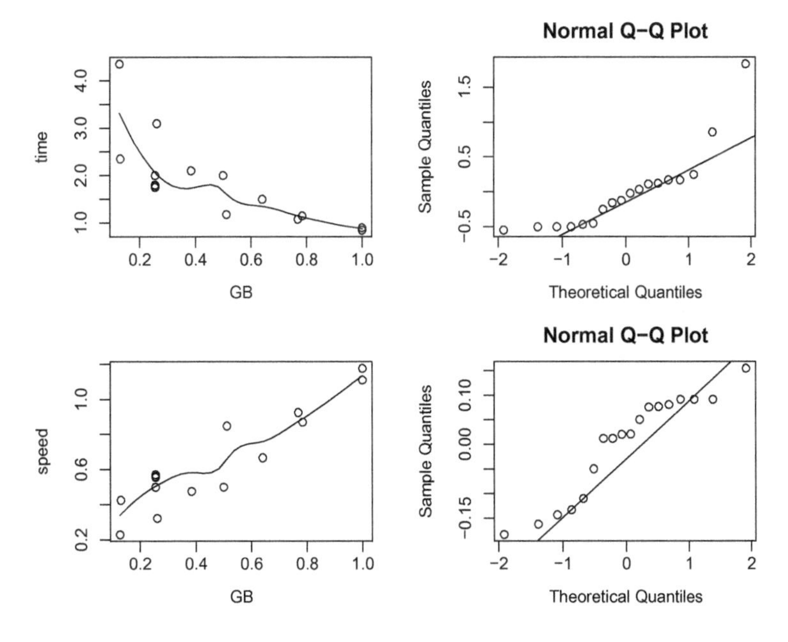

FIGURE 5.11
Scatterplots (left panels) and residual *q*-*q* plots (right panels) for original computer time data versus GB RAM (top panels), and for inverse-transformed time data (speed) versus GB (bottom panels).

While we plan to discuss the R^2 statistic in much more detail in Chapter 8, it is worth noting that the summaries of the two model fits show that the model with transformed Y has a much higher $R^2 = 0.8595$, compared to $R^2 = 0.5366$ for the untransformed Y model. But more importantly, the assumptions are less violated in the transformed model, as shown in Figure 5.11.

Notice in Figure 5.11, top left panel, that there is clear evidence of heteroscedasticity, with higher variance for computers with smaller GB RAM. There also appears to be some curvature in the sense that the curve flattens for large GB. This makes sense because the time cannot go below zero; hence there must be a flattening of the mean function. The normal *q*-*q* plot shows outliers, perhaps more an indication of heteroscedasticity than of non-normality. On the other hand, the heteroscedasticity is clearly lessened when using the $1/Y$ transform, as seen in the lower-left plot. And the normal *q*-*q* plot shows a light-tailed (non-outlier prone) appearance, which is the most benign of non-normal forms. Thus, the inverse Y transformation was a good choice in this example, from multiple standpoints: (i) The assumptions were less violated after inverse transform, (ii) the inverse transform has a simple, practical interpretation (as "speed" here), and (iii) the R^2 statistic is much higher for the transformed model. In the next section we note that, in addition to points (i), (ii) and (iii), the log likelihood is also higher for inverse-transformed Y data than it is for other transformations.

5.5 The Box-Cox Transformation

Recall that an objective way to compare different models is to compare their log likelihoods. Recall also that the appropriate likelihood is always for the *untransformed* Y, which means you need to figure out how the classical model for the transformed data $\left(f(Y) = \beta_0 + \beta_1 X + \varepsilon\right)$ translates into a model $Y \mid X = x \sim p(y \mid x, \theta)$ for the original data. The likelihood you need is then the product of the $p(y_i \mid x_i, \theta)$ terms (n terms).

The Box-Cox method considers the family of transformed models

$$Y^{\lambda} = \beta_0 + \beta_1 X + \varepsilon, \text{ for } \lambda \neq 0,$$

with $\lambda = 0$ given by the logarithmic model

$$\ln(Y) = \beta_0 + \beta_1 X + \varepsilon,$$

Special cases of interest are $\lambda = 1$ (implying no transformation), $\lambda = 0$ (log transform), $\lambda = -1$ (inverse transform), and $\lambda = 0.5$ (square root transform).

Each λ induces a model for the original data of the form $Y \mid X = x \sim p_\lambda(y \mid x,\theta)$, and all of these $p_\lambda(y \mid x,\theta)$ distributions are non-normal except in the untransformed case where $\lambda = 1$. For example, as shown above, the case $\lambda = 0$ implies $p_0(y \mid x,\theta)$ is a lognormal distribution. The maximized log likelihoods for the original Y data, which involve these (non-normal) distributions $p_\lambda(y \mid x,\theta)$, one for each possible λ, can then be compared to select "good" values of λ. You can use the "boxcox" function on a fitted `lm` object in `R` to estimate the "best" λ, but you first need to install the `MASS` library.

Box-Cox analysis for various data sets

```
library(MASS)
par(mfrow=c(2,2))

## Production Cost Data
ProdC = read.table("https://raw.githubusercontent.com/andrea2719/
URA-DataSets/master/ProdC.txt")
fit = lm(Cost ~ Widgets, data=ProdC); boxcox(fit)

## Charity Data: Income versus dependents
charity = read.csv("https://raw.githubusercontent.com/andrea2719/
URA-DataSets/master/charitytax.csv")
fit = lm(Income.AGI ~ DEPS, data=charity); boxcox(fit)

## Computer time versus RAM
comp = read.table("https://raw.githubusercontent.com/andrea2719/
URA-DataSets/master/compspeed.txt")
fit = lm(time ~ GB, data=comp); boxcox(fit)

## Simulated: lambda = 0 (log transform) is known to be correct.
n = 100; beta0 = -2.00; beta1 =  0.05; sigma =  0.30
set.seed(12345)
X = rnorm(n, 70, 10)
lnY = beta0 + beta1*X + rnorm(n, 0, sigma); Y = exp(lnY)
fit = lm(Y ~ X); boxcox(fit)
```

The log likelihoods on the vertical axis of the upper left panel of Figure 5.12 do not correspond directly with the log likelihoods calculated "by hand" above because the boxcox function ignores irrelevant constants (involving terms like n and π) in its calculations of the likelihoods. But differences between log likelihoods are identical when using either way of calculation because the irrelevant constants cancel in the difference calculation.

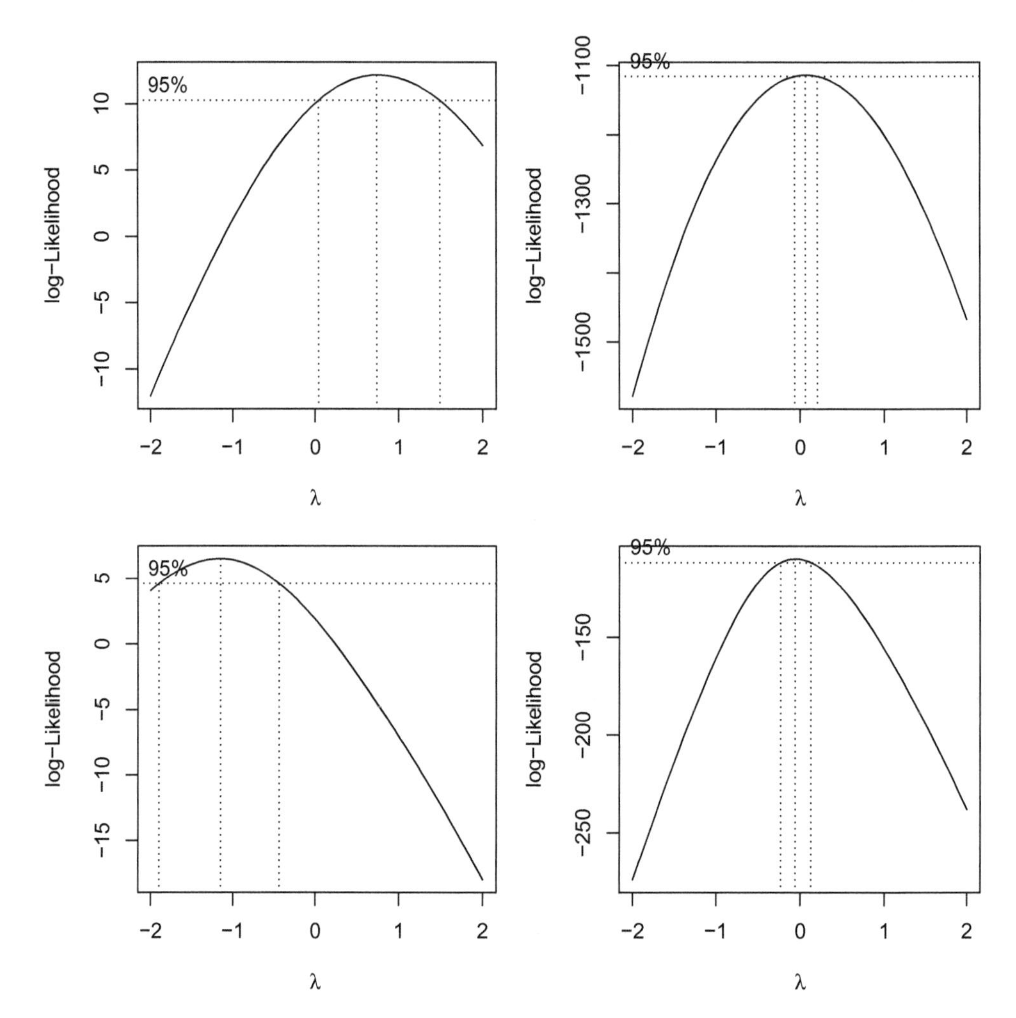

FIGURE 5.12
Box-Cox plots. Upper left, the Production Cost data: The parameter $\lambda = 1$ is inside the 95% confidence interval for possible λ, indicating that no transformation is a reasonable model. Upper right, the Income vs. Dependents analysis: The parameter $\lambda = 0$ is inside the 95% confidence interval for possible λ, indicating that the logarithmic transformation is a reasonable model. Lower left, the computer time versus RAM model: The parameter $\lambda = -1$ is inside the 95% confidence interval for possible λ, indicating that the inverse transformation is a reasonable model; however, $\lambda = 0$ is outside of the confidence interval, meaning that the log-transform is not supportable. Lower right, simulated lognormal data: The parameter $\lambda = 0$ is inside the 95% confidence interval for possible λ indicating that the logarithmic transformation is a reasonable model. In the lower right panel, the logarithmic transform is correct, because the $p(y \mid x, \theta)$ distributions giving the $Y \mid X = x$ data are in fact lognormal distributions, as given by the simulation code.

5.6 Transforming Both Y and X

It should be clear from the examples above that you can transform either Y alone or X alone. But you might have had an idea in your head that if you transform Y, then you should also transform X. In some cases, this is reasonable. After all, if Y and X have a relationship that is close to linear to begin with, then transforming one without the other will automatically introduce severe nonlinear curvature, badly violating the linearity assumption.

For example, consider total crimes (Y) in a U.S. state as it relates to the population covered (X), as reported by the FBI. The following code shows four scatterplots: (X, Y), (X, ln(Y)), (ln(X), Y), and (ln(X), ln(Y)). It is clear, *in this example,* that if you transform one variable, then you must also transform the other.

```
Cr=read.table("https://raw.githubusercontent.com/andrea2719/
URA-DataSets/master/crimes_2012.txt")
attach(Cr); par(mfrow=c(2,2))
plot(Tot_Crimes ~ Pop_Cov)
lcrimes = log(Tot_Crimes); lpop = log(Pop_Cov)
plot(Tot_Crimes~lpop); plot(lcrimes~Pop_Cov); plot(lcrimes~lpop)
```

But it is not always needed to transform X when you transform Y. In the computer example there was no need to transform the X variable, RAM, even though we transformed Y, time, to $1/Y$.

Among all of the possible combinations of X- and Y-transforms, you can compare likelihood-based statistics to see, objectively, which model is best supported by the data. You can also evaluate the models according to how badly violated are assumptions, as discussed in Chapter 4, because higher log likelihood does not necessarily validate the assumptions of the model. Rather, the highest log likelihood among a collection of models that you are considering simply identifies the model that fits the data best among those you have considered. There may be other, better models that you have not considered. Further, even if the correct model is in your set of models, it will not necessarily have the best log likelihood because of randomness: Every random sample gives you a different log likelihood.

If you are considering transformations of the form $(f(X),g(Y))$, simply create the two new variables $U = f(X)$ and $W = g(Y)$, and evaluate the assumptions in terms of the transformed (U,W) data as indicated in Chapter 4. If all looks reasonable, then the classical regression model

$$W = \beta_0 + \beta_1 U + \varepsilon$$

is a reasonable model.

5.7 Elasticity

One important special case of transforming both Y and X is the $\ln(X)$, $\ln(Y)$ transformation as indicated in the lower right corner of Figure 5.13. For this model you specify

$$\ln(Y) = \beta_0 + \beta_1 \ln(X) + \varepsilon$$

and, back-transforming, you get

$$\begin{aligned} Y &= \exp\{\beta_0 + \beta_1 \ln(X) + \varepsilon\} \\ &= \exp(\beta_0) \times \exp\{\beta_1 \ln(X)\} \times \exp(\varepsilon) \\ &= \exp(\beta_0) \times \left[\exp\{\ln(X)\}\right]^{\beta_1} \times \exp(\varepsilon) \\ &= \exp(\beta_0) \times X^{\beta_1} \times \exp(\varepsilon) \end{aligned}$$

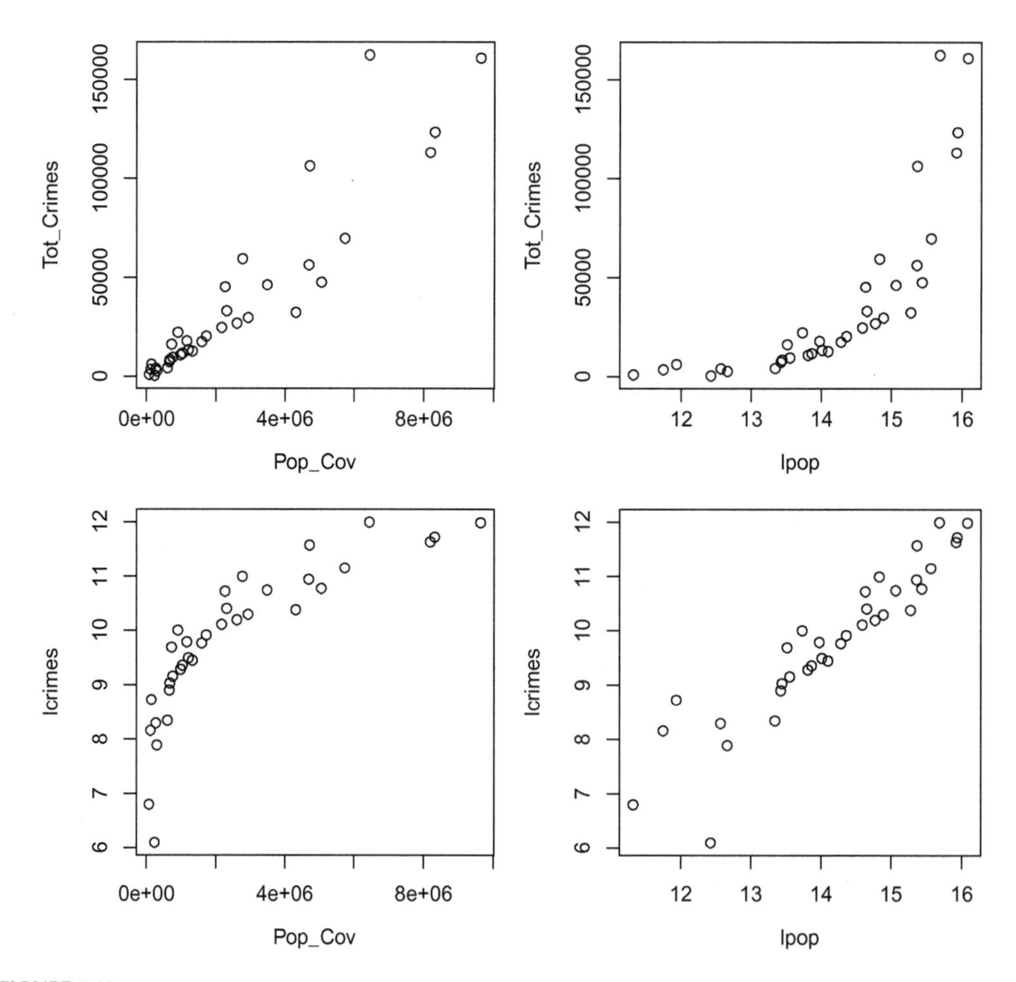

FIGURE 5.13
Upper left, untransformed data. Upper right, ($\ln(X)$, Y) data. Lower left, (X, $\ln(Y)$) data. Lower right, ($\ln(X)$, $\ln(Y)$) data.

Thus, as discussed earlier, the *median* of the distribution of your untransformed $Y \mid X = x$ is equal to $\exp(\beta_0) \times x^{\beta_1}$ when you use the log transformation on both X and Y (assuming that $p(\varepsilon)$ is symmetric).

The coefficient β_1 has a special interpretation in this model. Called *elasticity*, this parameter measures the *percent* increase in the median of the distribution of the untransformed Y variable corresponding to a small *percent* increase in the untransformed X variable. Usually, this is explained as follows "There is a β_1 % increase in the median of Y associated with a 1% increase in X," which is usually close enough, but not precisely correct as we will show below.

First, though, consider the least-squares fit to the $(\ln(X), \ln(Y))$ data in the lower right corner of Figure 5.13: You get the following output from `R` using an `lm` fit:

```
Call:
lm(formula = lcrimes ~ lpop)

Residuals:
     Min        1Q    Median      3Q       Max
-1.91083  -0.16451  -0.06574  0.18266   1.22366

Coefficients:
            Estimate  Std. Error    t value      Pr(>|t|)
(Intercept) -4.85116     1.03321     -4.695      4.52e-05 ***
lpop          1.03531     0.07263     14.256      1.16e-15 ***
---
Signif. codes:  0 '***' 0.001 '**' 0.01 '*' 0.05 '.' 0.1 ' ' 1

Residual standard error: 0.5262 on 33 degrees of freedom
Multiple R-squared:  0.8603,    Adjusted R-squared:  0.8561
F-statistic: 203.2 on 1 and 33 DF,  p-value: 1.164e-15
```

Thus, in terms of the transformed variables V = ln(Population) and W = ln(Total Crimes), the estimated model is

$$\hat{W} = -4.85116 + 1.03531V$$

which gives an estimate of the *mean* of the distribution of ln(Crimes) as a function of ln(Population).

Back-transforming, you get an estimate of the *median* of the distribution of Crimes (Y) as a function of population (X):

$$\hat{Y} = \exp(-4.85116)X^{1.03531}$$

To understand this function clearly, note that it refers to the untransformed variables. Thus, it should provide a good representation of the untransformed data in the upper left panel of Figure 5.13; the following code draws this graph over the scatterplot to verify this assertion.

R code for Figure 5.14

```
par(mfrow=c(1,1))
plot(Tot_Crimes ~ Pop_Cov)
pop.seq = 0:10000000
median.estimate = exp(-4.85116) * pop.seq^1.03531
points(pop.seq, median.estimate, type="l")
```

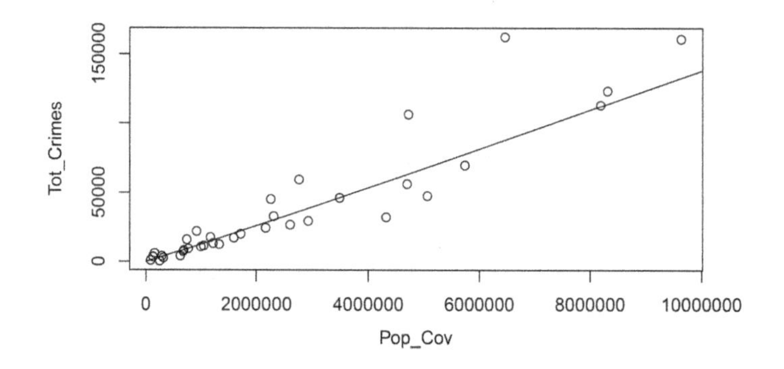

FIGURE 5.14
Graph of untransformed (Population, Crimes) data, with back-transformed fit $\hat{Y} = \exp(-4.85116)X^{1.03531}$ (from the regression of ln(Y) on ln(X)) superimposed. The back-transformed fit is an estimate of the median of the distribution of crimes for a given population size.

In linear models, the coefficient β_1 is interpreted in terms of a one unit increase in X. In the back-transformed (ln(X), ln(Y)) model, for which the median of Y is $\exp(\beta_0) \times x^{\beta_1}$, the coefficient is interpreted in terms of a one *percent* increase in X. Roughly, a one *percent* increase in X corresponds to a β_1 *percent* increase in the median of Y. The statement can be made precise through calculus, where you consider the effect of a δ % increase in X on Y, and let δ tend to zero, but let's just use the approximation.

For example, consider a 1% increase in X = population if the population size is 1,000. Then the new population size is $1000 + (0.01)1000 = 1010$. According to the mathematical theory of elasticity, this should correspond to a 1.03531% increase in $\hat{Y} = \exp(-4.85116)X^{1.03531}$. Let's check:

$$X = 1000 \quad \Rightarrow \quad \hat{Y} = \exp(-4.85116)1000^{1.03531} = 9.979252$$

Now, increase X by 1%:

$$X = 1010 \quad \Rightarrow \quad \hat{Y} = \exp(-4.85116)1010^{1.03531} = 10.08259.$$

The % increase in $\hat{Y}$ is $100(10.08259 - 9.979252)/9.979252\% = 1.035529\%$.

So you can see that the estimated β_1 coefficient, 1.03531, is approximately the percent increase in the median of Y, given a 1% increase in X. Again, the approximation can be made more precise using calculus by considering smaller percent increases in X, but the given interpretation is good enough.

Notice also that, as in the interpretation of β_1 in the ordinary linear model, it does not matter where you start the X value. For example, suppose you start at X = 10,00,000 instead of X = 1,000. Repeating the arithmetic above, you arrive at the same % increase result.

In this example, the fact that β_1 is estimated to be greater than 1.0 means that there is upward (convex) curvature in the function. When $0 < \beta_1 < 1$, the function has a downward (concave) bend, and when $\beta_1 = 1.0$, the function is linear. Thus, it seems that with larger states, there is an "acceleration" of total crimes, since β_1 is estimated to be greater than one. However, the 95% confidence interval for β_1 (obtained using the classical linear model on the (ln(X), ln(Y)) transformed data) is $0.8875545 < \beta_1 < 1.183068$. The interval is not conclusive as to whether $\beta_1 > 1$ or $\beta_1 < 1$, and it also admits the possibility that the linear function (where $\beta_1 = 1$) is a reasonable model.

Exercises

For exercises 1 to 14, use the following data set:

```
Firms = read.csv("https://raw.githubusercontent.com/andrea2719/
URA-DataSets/master/Firms.csv")
```

The data set contains measurements of Income and Assets of various firms. Consider the prediction of Assets (Y) using Income (X).

1. Plot the (X, Y) data. Does it look like typical regression data we have seen so far? How is it different?
2. Find the minimum of the Y values using an `R` function.
3. Ten percent of the firms have Income that is less than or equal to 32. Show this using the quantile function of `R`.
4. Overlay the least-squares line to the scatterplot in 1.
5. Use the least-squares line to predict Y when X = 32.
6. Comment on your answer to 5, in light of your answer to 2.
7. Plot the (ln(X),ln(Y)) data. Does it look more like typical regression data we have seen than the untransformed plot of 1? Why?
8. Overlay the least-squares line that represents the (ln(X), ln(Y)) data upon the scatterplot of the (ln(X), ln(Y)) data.
9. Use the line in 8 to predict ln(Y) when ln(X) = ln(32).
10. Use the prediction in 9 to predict Y when X = 32. Show your math logic.
11. Compare 10 to 5. Which is better, in light of your answer to 2?

12. Express the function which predicts ln(Y) as a linear function of ln(X) in terms of the original (X, Y) variables. See the book for how to back-transform in this way.
13. Overlay the function you found in 12 over the original (X, Y) scatterplot. Interpret the meaning of this function in terms of medians of conditional distributions, as discussed in the book.
14. Show the (X, Y) scatterplot where X is restricted to be less than 100 and where Y is restricted to be less than 500.

6

The Multiple Regression Model

So far, we have considered the simple regression model, $Y \mid X = x \sim p(y \mid x)$, where there is just one X variable. The multiple regression model uses more than one X, labeled as $X_1, X_2, \ldots, X_k$.

First, please understand the following:

There is no "correct" collection of X variables to use in your model.

To elaborate, suppose you want to predict the outcome of giving someone a loan, $Y = 1$ for successful repayment, and $Y = 0$ for default. You may consider different models involving different X variables, such as X_1 = age, X_2 = income, X_3 = assets. In this setting *all* of the following regression models are *correct models*:

- $p(y \mid x_1)$, the model for the distribution of repayment (Y), given only that the customer's age is x_1.
- $p(y \mid x_2)$, the model for the distribution of repayment (Y), given only that the customer's income is x_2.
- $p(y \mid x_3)$, the model for the distribution of repayment (Y), given only that the customer's assets are x_3.
- $p(y \mid x_1, x_2)$, the model for the distribution of repayment (Y), given only that the customer's age is x_1 and his/her income is x_2.
- $p(y \mid x_1, x_3)$, the model for the distribution of repayment (Y), given only that the customer's age is x_1 and his/her assets are x_3.
- $p(y \mid x_2, x_3)$, the model for the distribution of repayment (Y), given only that the customer's income is x_2 and his/her assets are x_3.
- $p(y \mid x_1, x_2, x_3)$, the model for the distribution of repayment (Y), given only that the customer's age is x_1, and his/her income is x_2, and his/her assets are x_3.

  Every single one of these models is correct! Which one you *use* depends on what you might know. If you only know $X_2 = x_2$, then you should use $p(y \mid x_2)$. And if you have no information on any X variable, then you should use the purely probabilistic model for Y given by:

- $p(y)$ itself, the model for the distribution of repayment (Y), given no concomitant X information.

The model $p(y)$ is also correct, and it is equally as important as any of the other models. As you will see in this chapter, the model $p(y)$ is essential for the definition of the famous "R^2" statistic.

There are two main, but somewhat distinct, reasons for including many X variables in multiple regression. One reason concerns *prediction*, the other concerns *causation*. The choice of X variables you might *choose* to use (not that there is any "right" collection) depends partly on whether your goal is to predict or to assess causality.

6.1 Prediction

A major topic within the discipline that has come to be known as "Data Science" is called "Predictive Modeling," which concerns the use of non-linear (and therefore realistic) regression models. Such models include neural networks, random forests, decision trees, and LOESS. In predictive modeling, your aim is to predict a Y, for example, the "Yes/No" outcome of "will this customer repay the loan," as a function of various X variables such as X_1 = age, X_2 = income, X_3 = assets, X_4 = debt, X_5 = credit history, ..., X_k = # of children. The goal is to come up with a function $f(x_1,\ldots,x_k)$ that is as close as possible to $Y \mid X_1 = x_1,\ldots,X_k = x_k$.

In theory, the correct solution is to choose $f(x_1,\ldots,x_k)$ to be the expected value of the conditional distribution of $Y \mid X_1 = x_1,\ldots,X_k = x_k$. It is a mathematical theorem that, *if*

$$f(x_1,\ldots,x_k) = \mathrm{E}(Y \mid X_1 = x_1,\ldots,X_k = x_k),$$

then $\mathrm{E}\{Y - f(X_1,\ldots,X_k)\}^2$ is a minimum. Thus the best predictor of Y, in the sense of minimizing the average squared deviation from the predictor, is the conditional mean function $f(x_1,\ldots,x_k)$, which is almost always a non-linear function of $x_1,\ldots,x_k$.

The proof of this theorem is given below the following example.

6.1.1 Predicting Loan Repayment

Suppose you are a loan officer. A person comes to your office presenting X_1 = 31 years old, X_2 = \$60 K income, X_3 = \$0 assets, X_4 = \$100 K debt, X_5 = 500 credit score, ..., and X_k = 2 children. This person wants a loan. You would like to know whether she or he will be able to repay the loan (Y = 1) or not (Y = 0). Then your best prediction of Y is $\mathrm{E}(Y \mid X_1 = 31,\ldots,X_k = 2)$, which is the average of all potentially observable Y's for people *with exactly the same characteristics* as this person.

If there were many people who might apply for the loan who shared these exact characteristics, and whose loan-repaying outcomes (0's of 1's) were observed, then you could appeal to the Law of Large numbers, getting a good estimate of $\mathrm{E}(Y \mid X_1 = 31,\ldots,X_k = 2)$ by *averaging* the Y values (in this case the Y values are 0's and 1's) for this group of people.

The problem with this solution is that there might not be anyone else in the whole world with exactly this same profile (X_1 = 31 years old, X_2 = \$60 K income, X_3 = \$0 assets, X_4 = \$100 K debt, X_5 = 500 credit score, ..., X_k = 2 children). So, while the solution is simple, *in theory*, you have to use some special methods to estimate $f(x_1,\ldots,x_k)$. Logistic regression, discussed in Chapter 13, is one such method. Neural networks and regression trees are other (more flexible) methods; these are discussed in Chapters 17 and 18.

Putting aside the issue of how to estimate $f(x_1,\ldots,x_k) = \mathrm{E}(Y \mid X_1 = x_1,\ldots,X_k = x_k)$, how do you know that $f(x_1,\ldots,x_k)$ actually *is* the best predictor of Y? The answer lies in the famous "Law of Total Expectation," which is actually a mathematical theorem, not just a mild suggestion or ugly rule of thumb.

Law of Total Expectation

Suppose (V, W) are random variables. Let $\mathrm{E}(W \mid V = v) = f(v)$. Then

$$\mathrm{E}(W) = \mathrm{E}_V\{f(V)\}$$

In simple, informal words, the Law of Total Expectation states that "the (weighted) average of the within-group averages is equal to the global average."

It bears repeating that the best way to understand abstract concepts is by using simulation because you can attach actual numbers to everything.

6.1.2 Simulation Demonstrating the Law of Total Expectation

Suppose $Y \mid X = x$ is produced by the (normal) N($0.3x$, $(0.04x)^2$) distribution, and X itself is produced by the (uniform) U(40, 120) distribution. Figure 6.1 is a graph of n = 300 (x, y) data points produced by this model.

In this model, the conditional mean of Y is $f(x) = 0.3x$, and the conditional variance of Y is $\mathrm{Var}(Y \mid X = x) = (0.04x)^2$, both of which are visible in Figure 6.1. The Law of Total Expectation states that the unconditional (or *marginal*) expected value of Y (that is, the expected value of Y without regard for X) is given by

$$\mathrm{E}(Y) = \mathrm{E}_X(0.3X),$$

where X itself is distributed as U(40,120).

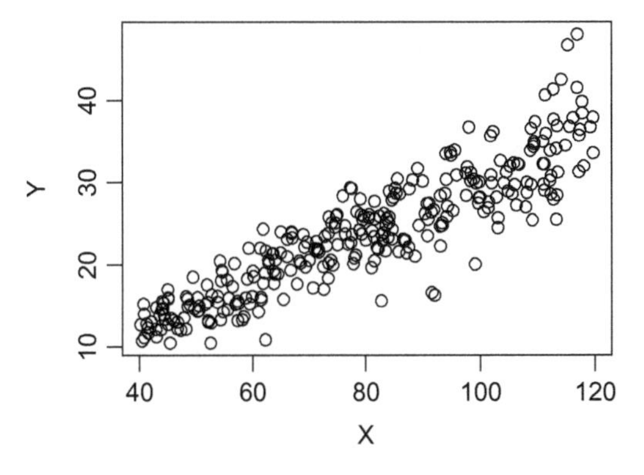

FIGURE 6.1
Graph of n = 300 data points produced by the model $X \sim$ U(40,120), $Y \mid X = x \sim$ N($0.3x$, $(0.04x)^2$).

Here is R code to illustrate the Law of Total Expectation in this context.

```
nsim = 10000000
X = runif(nsim, 40,120)
Y = 0.3*X + rnorm(nsim, 0, 0.04*X)
Cond.Mean = 0.3*X
Cond.Var = (0.04*X)^2

## Estimated unconditional expected value of Y
mean(Y)
## Estimated expected value of the conditional mean
mean(Cond.Mean)
```

Results are as follows:

```
> ## Estimated unconditional expected value of Y
> mean(Y)
[1] 24.00024
> ## Estimated expected value of the conditional mean
> mean(Cond.Mean)
[1] 24.00005
```

The difference between 24.00024 and 24.00005 is due to simulation error. The corresponding true quantities are equal, by the Law of Total Expectation.

(The result of the simulation also can be shown mathematically using the model equation $Y = 0.3X + \varepsilon$, which implies $\mathrm{E}(Y) = \mathrm{E}(0.3X) + \mathrm{E}(\varepsilon) = \mathrm{E}(0.3X)$. Since X is uniformly distributed between 40 and 120, its mean is 80, so $\mathrm{E}(0.3X) = 0.3(80) = 24.0$. But the simulation makes the Law of Total Expectation easier to understand.)

There are several important applications of the Law of Total Expectation. The first is:

Application 1 of the Law of Total Expectation

The best predictor of $Y \mid X = x$ is the conditional mean function $f(x)$.

We will now prove this result. The idea is to consider any other function $h(X)$ that you can use to predict Y, then find the mean squared prediction error using $h(X)$, then show that the mean squared prediction error using $f(X)$ is smaller. Note that

$$\mathrm{E}\{Y - h(X)\}^2 = \mathrm{E}\{(Y - f(X)) + (f(X) - h(X))\}^2,$$

then, by the algebra $(a + b)^2 = a^2 + b^2 + 2ab$; and by the linearity property of $\mathrm{E}(\cdot)$, we have that

$$\mathrm{E}\{Y - h(X)\}^2 = \mathrm{E}\{(Y - f(X))^2\} + \mathrm{E}\{(f(X) - h(X))^2\} + 2\mathrm{E}\{(Y - f(X))(f(X) - h(X))\}$$

Now, consider the last term in the above sum and let $g(x) = \mathrm{E}\{(Y - f(X))(f(X) - h(X)) \mid X = x\}$.

Then, applying the fact that x is constant, and the linearity property of expectation

$$g(x) = (f(x) - h(x))\mathrm{E}\{(Y - f(X)) \mid X = x\}$$
$$= (f(x) - h(x))\{\mathrm{E}(Y \mid X = x) - f(x)\}$$

Then, by definition of $f(x)$

$$g(x) = (f(x) - h(x))\{f(x) - f(x)\}$$
$$= 0$$

From the Law of Total Expectation, we know that $\mathrm{E}\{(Y - f(X))(f(X) - h(X))\} = \mathrm{E}\{g(X)\}$; but from the demonstration above, for every $X = x$, we know that $g(x) = 0$; hence $g(X) = 0$ and $\mathrm{E}\{g(X)\} = 0$. Thus

$$\mathrm{E}\{Y - h(X)\}^2 = \mathrm{E}\{(Y - f(X))^2\} + \mathrm{E}\{(f(X) - h(X))^2\} + 2\mathrm{E}\{(Y - f(X))(f(X) - h(X))\}$$
$$= \mathrm{E}\{(Y - f(X))^2\} + \mathrm{E}\{(f(X) - h(X))^2\} + 0$$
$$\geq \mathrm{E}\{(Y - f(X))^2\}$$

Hence, no matter what function $h(X)$ you use to predict Y, the mean squared prediction error will always be greater than or equal to the mean squared prediction error of the conditional mean function $f(X)$. This proves that the conditional mean function $f(X)$ minimizes mean squared prediction error.

A second application of the Law of Total Expectation is that additional variables always improve the best predictor, or at least, never make it worse.

Application 2 of the Law of Total Expectation

Additional predictor (X) variables either improve or do not change the best predictor of Y.

To elaborate, let $\boldsymbol{X}$ be a set of k explanatory variables, and let U be an additional variable. Let $f(\boldsymbol{x})$ be defined, as above, as $f(\boldsymbol{x}) = \mathrm{E}(Y \mid \boldsymbol{X} = \boldsymbol{x})$, and let $g(\boldsymbol{x}, u) = \mathrm{E}(Y \mid \boldsymbol{X} = \boldsymbol{x}, U = u)$. Then

$$\mathrm{E}\{Y - g(\boldsymbol{X}, U)\}^2 \leq \mathrm{E}\{Y - f(\boldsymbol{X})\}^2$$

The proof of this result is essentially the same as what was just proven: Let $X = (\boldsymbol{X}, U)$. The conditional mean function $g(\boldsymbol{X}, U)$ minimizes the mean squared prediction error among all possible functions $h(\boldsymbol{X}, U)$. Since $f(\boldsymbol{X})$ is an example of such an $h(\cdot)$ function, it follows that $h(\boldsymbol{X}, U)$ has mean squared prediction error that is less than or equal to that of $f(\boldsymbol{X})$.

Again, to understand technical concepts such as this in a simple way, you can use simulation. Consider modifying the simulation above, shown after Figure 6.1, by including a "number of children" variable, U, independent of income. Suppose also that the true model for house expense is $Y = 0.3X + 2U + \varepsilon$, where $\varepsilon \mid X = x, U = u \sim \text{N}(0, (0.04x)^2)$, suggesting that people spend more on housing when they have more children. Also, suppose the Poisson(1.2) model for U = Number of children, implying an average of 1.2 children. Here,

$$g(x,u) = \text{E}(Y \mid X = x, U = u) = 0.3x + 2u$$

and, using independence,

$$\begin{aligned} f(x) = \text{E}(Y \mid X = x) &= \text{E}(0.3X + 2U + \varepsilon \mid X = x) \\ &= 0.3x + \text{E}(2U \mid X = x) + \text{E}(\varepsilon \mid X = x) \\ &= 0.3x + 2(1.2) + 0 \\ &= 2.4 + 0.3x \end{aligned}$$

Application 2 simply states in this case that the prediction $0.3X + 2U$ has smaller average squared difference from Y than does the prediction $2.4 + 0.3X$. This result should make intuitive sense because the number of children, U, conveys information about housing expense, Y. Here is the simulation code to verify the intuition:

```
nsim = 10000000
X = runif(nsim, 40,120); U = rpois(nsim, 1.2)
Y = 0.3*X + 2*U + rnorm(nsim, 0, 0.04*X)

## Predictors of Y
Cond.Mean.X = 0.3*X + 2.4; Cond.Mean.X.U = 0.3*X + 2*U

## Mean squared difference from predictor to Y
mean( (Y - Cond.Mean.X)^2  ); mean( (Y - Cond.Mean.X.U)^2)
```

The results show, as expected, that using both variables results in smaller mean squared prediction error.

```
> mean( (Y - Cond.Mean.X)^2  )
[1] 15.89988
> mean( (Y - Cond.Mean.X.U)^2)
[1] 11.09123
```

Another application of the Law of Total Expectation is the famous "Law of Total Variance," which expresses the marginal variance of Y, $\text{Var}(Y)$, as the sum of the variance of the conditional mean function, plus the expected value of the conditional variance function.

Application 3 of the Law of Total Expectation:

The Law of Total Variance

Let $\mathrm{E}(Y \mid X = x) = f(x)$, and let $\mathrm{Var}(Y \mid X = x) = v(x)$. Then

$$\mathrm{Var}(Y) = \mathrm{Var}\{f(X)\} + \mathrm{E}\{v(X)\}$$

The Law of Total Variance is proven using similar mathematical tricks as shown above, which you can easily find online.

Self-study question: Find a mathematical proof of the Law of Total Variance online. How is the Law of Total Expectation used in the proof?

To understand the Law of Total Variance, see Figure 1.7 again (again!) Think of the possible values X (= hours studying) in that graph as random, produced by a distribution $p(x)$. As shown in the graph, for each such $X = x$, there is a distribution $p(y \mid x)$; four of these are shown in that graph. Each of these distributions has a mean, $f(x)$, and a variance, $v(x)$.

Considering the totality of all possible $X = x$ values, the mean function $f(X)$ is a random variable, and the variance function $v(X)$ is also a random variable. The law of total variance states that the marginal variance of Y, which you can visualize as the variation in Figure 1.7 along the "Exam Score" axis, without regard for the explanatory variable, is equal to the variance of the conditional means, $\mathrm{Var}\{f(X)\}$, plus the expected value of the conditional variances, $\mathrm{E}\{v(X)\}$.

And again, the concept is perhaps easiest to understand via simulation.

6.1.3 Simulation Demonstrating the Law of Total Variance

Consider again the example where Y = housing expense is related to X = income via the conditional model $Y \mid X = x \sim \mathrm{N}(0.3x, (0.04x)^2)$. Further, as before, $X \sim \mathrm{U}(40,120)$. In this model, the conditional mean of Y is $f(x) = 0.3x$, and the conditional variance of Y is $(0.04x)^2$, both of which are visible in Figure 6.1. The Law of Total Variance states that the marginal variance of Y (that is, the variance of Y without regard for X) is given by

$$\mathrm{Var}(Y) = \mathrm{Var}(0.3X) + \mathrm{E}\left\{(0.04X)^2\right\}.$$

There is code shown above to illustrate the Law of Total Expectation via simulation for this model; the following code modifies that code to illustrate the Law of Total Variance.

```
nsim = 10000000
X = runif(nsim, 40,120); Y = 0.3*X + rnorm(nsim, 0, 0.04*X)
Cond.Mean = 0.3*X; Cond.Var = (0.04*X)^2

## Estimated unconditional variance of Y
var(Y)
```

```
## Estimated variance of the conditional mean
var(Cond.Mean)
## Estimated expected value of the conditional variance
mean(Cond.Var)
## Verifying the Law of Total Variance
var(Cond.Mean) + mean(Cond.Var)
```

Results are as follows:

```
> ## Estimated unconditional variance of Y
> var(Y)
[1] 59.08472
>
> ## Estimated variance of the conditional mean
> var(Cond.Mean)
[1] 47.99515
>
> ## Estimated expected value of the conditional variance
> mean(Cond.Var)
[1] 11.094
>
> ## Verifying the Law of Total Variance
> var(Cond.Mean) + mean(Cond.Var)
[1] 59.08915
```

The difference between 59.08472 and 59.08915 is due to simulation error. The corresponding true quantities are mathematically equal, by the Law of Total Variance.

Self-study question: Using the mathematics of expectation, what are the true values that are estimated in the simulation code above?

One more application of the Law of Total Expectation is the definition of the true R^2 statistic. In cases where the classical model is true, then the statistic takes a special form, which you have already seen in the R output from the "lm" function, labeled "Multiple R-squared." The R^2 statistic refers to the proportion of the marginal variance in Y that is explained by the conditional mean function, $f(X)$.

The definition of the true R^2 statistic comes from the Law of Total Variance, which in turn comes from the Law of Total Expectation. Recall that the Law of Total Variance states that $\mathrm{Var}(Y) = \mathrm{Var}\{f(X)\} + \mathrm{E}\{v(X)\}$. The true R^2 is the proportion of the variance in Y that is explained by X.

Application 4 of the Law of Total Expectation:
The true R^2 statistic

$$\Omega^2 = \mathrm{Var}\{f(X)\} / \mathrm{Var}(Y)$$

In the simulation above involving only X = income, the estimate of $\mathrm{Var}\{f(X)\}$ was given as 47.99515 and the estimate of $\mathrm{Var}(Y)$ was given as 59.08472. Thus, $47.99515 / 59.08472 = 81.2\%$

of the marginal variance in Y is explained by the true regression function, i.e., by the conditional mean function that relates housing expense to income.

Since $\text{Var}(Y) = \text{Var}\{f(X)\} + \text{E}\{v(X)\}$, we have $\text{Var}\{f(X)\} = \text{Var}(Y) - \text{E}\{v(X)\}$, giving (by simple substitution into the equation above) another way to write the true R^2 statistic:

$$\Omega^2 = 1 - \text{E}\{v(X)\} / \text{Var}(Y)$$

Using common words, this statistic is the average (proportional) reduction of the marginal variance in Y that you achieve when you predict Y using X. You also get 81.2% in the example above using this form of the true R^2, and the interpretation is that the marginal variance of Y is reduced by an estimated 81.2% (on average) when you use X to predict Y.

To summarize this section so far:

- The best function of the $\mathbf{X}$ data that you can use to predict Y is $f(\mathbf{X})$, the conditional mean function.
- Adding additional predictors U cannot make the performance of the predictor $g(\mathbf{X}, U)$ worse; i.e., the mean squared prediction error cannot increase.
- The Law of Total Variance states that the variance of Y is equal to the variance of the conditional mean function $f(\mathbf{X})$ plus the expected value of the conditional variance $v(\mathbf{X})$. Hence the proportion of (unconditional variance) in Y that is explained by the predictor $f(\mathbf{X})$ is given by $\Omega^2 = \text{Var}\{f(\mathbf{X})\} / \text{Var}(Y)$; this is the true R^2 statistic.

In practice, the conditional mean function $f(\mathbf{X})$ is unknown and must be estimated. Once you estimate this function as $\hat{f}(\mathbf{X})$, the randomness of the data plays a role in the estimate's accuracy because each estimated parameter is also random. In such cases, it does not always help to include additional variables in the model (like the variable called "U" above), because the randomness of the estimates corresponding to the additional predictor variables adds variability over and above the theoretical reduction in variance as illustrated in the simulation study above. This additional variability incurred by the random parameter estimates can make the predictions worse with additional predictor variables, even when those variables are truly related to the response variable Y.

For example, the simulation study above showed that Y = housing expense is better predicted by using X = income and U = Number of dependents, than by using X = income alone. However, the predictions using X = income *alone* might be better *when you have to estimate the true regression function*. Intuitively, this can easily happen if your estimates of the true parameters are poor, such as when you have a small sample size. We will return to this issue later when we discuss variable selection in Chapter 11, where we also discuss the related concept called "overfitting."

But rather than simply obtaining *predictions* of Y, scientists and researchers are often more interested in understanding what *causes* Y. Even an excellent predictive model is not necessarily good for inferring causality, as the following section shows.

6.2 Why Prediction Is Different from Causation?

You cannot prove causation rigorously by using observational data. Causation is difficult to establish, and usually requires context-based arguments in addition to statistical analysis. While you would probably like to conclude that your favorite X variable "causes" your Y, you will ordinarily be safer to conclude that your favorite X variable "is associated with" Y. The following example highlights the difference.

6.2.1 Does Eating Ice Cream Cause You to Drown?

Consider the (standard) example where the response variable is Y = Drowning Deaths in a week in the U.S.A., and X = Ice Cream Sales that same week. There is an association between Y and X. A scatterplot of such data might look as shown in Figure 6.2.

From the data shown in Figure 6.2, there is clearly a *predictive* relationship: When Ice Cream Sales are higher, the distribution of Drownings is shifted toward higher numbers. Thus, you can use X = Ice Cream Sales to obtain improved predictions of Y = Drownings.

Putting a little more structure into the model, suppose the linearity assumption is met so that the means of the distributions $p(y \mid x)$ fall on a straight line, and write the following predictive model:

$$Y = \beta_0 + \beta_1 X + \varepsilon,$$

where Y = Drownings, and X = Ice Cream Sales. Using this model, you can logically conclude the following:

Predictive association interpretation

Consider two large sets of potentially observable weeks, one where Ice Cream Sales is equal to 250 and the other where Ice Cream Sales is equal to 350. Then the mean of the Drownings variable is $100\beta_1$ more in the second scenario than in the first.

The predictive association interpretation differs from the causal interpretation. The following interpretation uses a correct definition of causality, but causality *does not logically follow* from the predictive model.

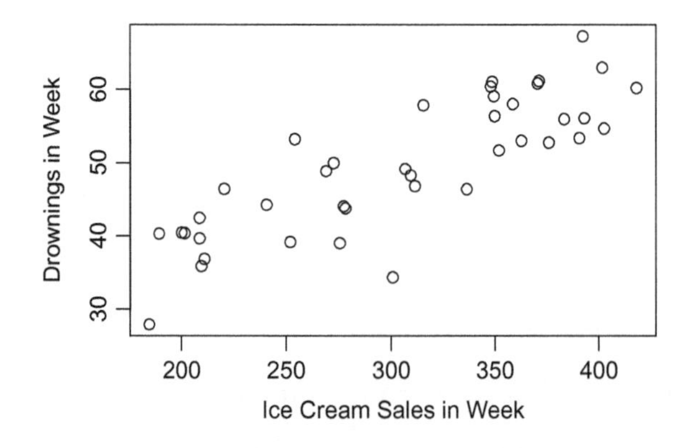

FIGURE 6.2
Plot of hypothetical data showing number of drownings in a week as related to the specific amount of ice cream sales in that week.

Causal interpretation (does *not* follow logically from the model)

Consider a large set of potentially observable weeks where Ice Cream Sales is equal to 250. Now, imagine *intervening* in Nature to change Ice Cream Sales to 350 *for that same collection of weeks, while holding everything else (except drownings) fixed.* Then the mean of the Drownings variable is $100\beta_1$ more in the second scenario than in the first.

The causal interpretation is not allowed by the predictive model, because in the data that you can actually observe, the two groups of weeks where Sales = 350 and where Sales = 250 do not have all other variables fixed. In particular, it is likely that in weeks where the Sales variable is higher, temperature will also be higher. Thus, in Figure 6.2, we are likely seeing the effect of Temperature (with higher temperatures, there will be more people enjoying water recreation, which in turn implies more drownings), rather than the effect of Ice Cream Sales.

Besides prediction accuracy, another reason you might wish to use multiple X variables is to make the interpretation of the parameter β_1 closer to the causal interpretation. Such additional X variables are called *confounding variables,* and when included in the model, they are called *control variables.*

Definition of a confounding variable

A confounding variable is a variable, say Z, external to the model involving Y and X, that is related to both Y and X and thus makes causal assessment of the effect of X on Y dubious.

With multiple regression, you can get closer to the causal interpretation by controlling potential *confounding* variables. If you include Z = (average weekly temperature) in the model, then you have the following multiple regression model:

$$Y = \gamma_0 + \gamma_1 X + \gamma_2 Z + \delta.$$

This multiple regression model also allows predictive interpretations, but the predictive interpretation is now slightly closer to the correct causal interpretation.

Predictive association interpretation using the "held fixed" concept

Consider two large sets of potentially observable weeks, one where Ice Cream Sales is equal to 250 and Temperature is equal to 25, and the other where Ice Cream Sales is equal to 350 and Temperature is equal to 25. Then the mean of the Drownings variable is $100\beta_1$ more in the second scenario than in the first.

Note that the Temperature variable is *held fixed* in this interpretation. It does not matter at what value it is held fixed, whether 25 (degrees Celsius) or any other number. Assuming $E(Y \mid X = x, Z = z) = \gamma_0 + \gamma_1 x + \gamma_2 z$, the same interpretation of γ_1 applies for all possible $Z = z$.

Note also that this interpretation is still associative rather than causal. However, it is closer to causal because at least we are holding one relevant variable fixed. We would expect the value of γ_1 to be nearly 0.0 in this analysis because we really believe that there is no causal link between Drownings and Ice Cream Sales. Rather, we believe there is a predictive, but non-causal, association between Ice Cream Sales and Drownings because both variables are positively related to Temperature. If the Ice Cream Sales variable still had a coefficient that was substantially different from 0.0 in the model that included both Temperature and Sales, we would suspect some variable other than Temperature is also confounding the results. But we might be wrong. Maybe all that ice cream pulls bodies to the bottom of the ocean!

Finally, note that we used β_1 in the single-variable model and γ_1 in the two-variable model because the coefficient of Ice Cream Sales has very different interpretations in those two models. It is not always necessary to flip from one Greek letter to another, as long as you understand that the meaning of the parameter changes from model to model. But it is safer to do so, just so that you are not confused. A fact about regression in general, which applies to multiple regression as well as regression with transformed variables as you saw in Chapter 5, is given in the following set-off box:

The interpretation of a parameter is entirely dependent upon the particular model in which the parameter appears.

The interpretation of a parameter changes whenever you bring an additional X variable into the regression model. With more X variables, more is "held fixed," hence the resulting parameter has an interpretation which becomes different, and closer to causal.

As an alternative to finding additional X variables and using them in the model, you can use an *instrumental variable* to estimate causal effects. This method is controversial. The appendix of this chapter describes the instrumental variable method and its drawbacks.

6.3 The Classical Multiple Regression Model and Interpretation of Its Parameters

Similar to the simple regression model, the classical multiple regression model specifies the conditional distributions that produce all observations Y_i, $i = 1,2,\ldots,n$, in the data set, as well as infinitely many potentially observable observations *not* in the data set. But with multiple regression, these conditional distributions refer to *combinations* of the multiple X variables.

The classical multiple regression model is given as follows:

The classical multiple regression model

$$Y_i \mid X_{i1} = x_{i1}, X_{i2} = x_{i2}, \ldots, X_{ik} = x_{ik} \sim_{\text{ind}} \text{N}\left(\beta_0 + \beta_1 x_{i1} + \beta_2 x_{i2} + \ldots + \beta_k x_{ik}, \sigma^2\right)$$

The case where $k = 2$ is instructive. Whereas in simple linear regression (where $k = 1$) the conditional mean function is a *line* that can be shown in a 2-dimensional graph, the multiple regression case where $k = 2$ gives a conditional mean function that is a *plane* that can be shown in a 3-dimensional graph. In Figure 6.3, each point of the *plane* corresponds to the mean of the distribution of Y for a given *combination* $X_1 = x_1$, $X_2 = x_2$. The following `R` code and resulting 3-D graph show this function.

`R` code for Figure 6.3

```
x = seq(1,2,.05)
x1 = rep(x, each = 21)
x2 = rep(x,21)
EY  = -1 + 6*x1 + 10*x2

library(lattice)
newcols <- colorRampPalette(c("grey90", "grey10"))
wireframe(EY ~ x1*x2,
  xlab = "X1", ylab = "X2",
  main = "Multiple Regression Function", xlim = c(1,2), ylim = c(1,2),
  drape = TRUE,
  colorkey = FALSE,
  col.regions=newcols(100),
  scales = list(arrows=FALSE,cex=.8,tick.number = 5),
  screen = list(z = -60, x = -60))
```

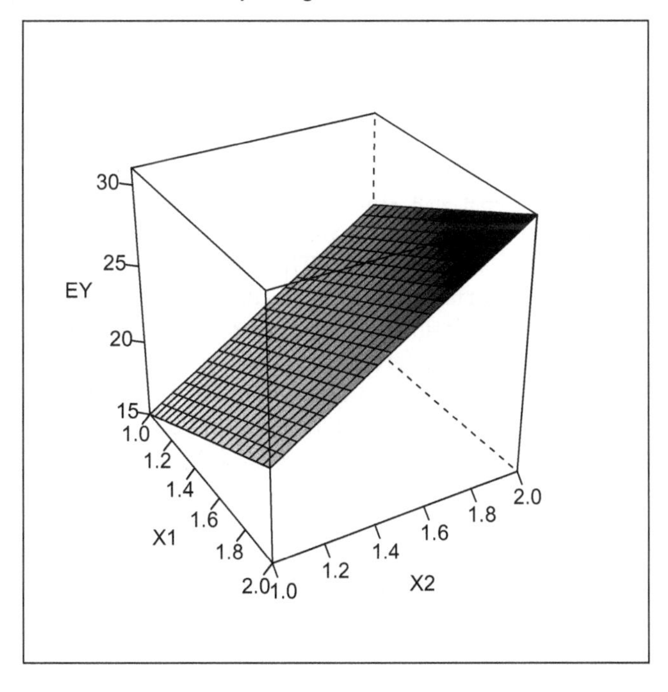

FIGURE 6.3
The conditional mean function $E(Y \mid X_1 = x_1, X_2 = x_2) = -1 + 6x_1 + 10x_2$.

The classical multiple regression model states that for every combination of values $X_1 = x_1$, $X_2 = x_2$, ..., $X_k = x_k$, the data Y is produced from a normal distribution that is specific to that combination of X values. Further, the model supposes that these distributions all have the same variance, and that their means all fall precisely on a function $f(x_1, x_2, \ldots, x_k) = \beta_0 + \beta_1 x_1 + \beta_2 x_2 + \ldots + \beta_k x_k$, for some values of the β parameters. Finally, the model assumes that the observations Y_i, $i = 1, 2, \ldots, n$, are produced (conditional on the X's) independently of one another from normal distributions.

The parameters of the classical multiple regression model are β_0, β_1, β_2, ..., β_k, and σ, and are interpreted as follows:

Interpretation of the parameters in the classical multiple regression model, in case where there are no functional relationships among the x's

1. β_0 is the mean of the potentially observable Y when $X_1 = 0$, $X_2 = 0$, ..., $X_k = 0$.
2. β_1 is interpreted as follows: Consider two different cases of potentially observable data: One where $X_1 = x_1 + 1$, $X_2 = x_2$, ..., $X_k = x_k$, and the other where $X_1 = x_1$, $X_2 = x_2$, ..., $X_k = x_k$. Then β_1 is the difference between the means of the potentially observable Y data for these two cases. In other words, β_1 is the difference in the means of the distributions of Y associated with a one unit difference in X_1, with all other X variables held fixed.
3. β_2 is interpreted similarly as β_1, interchanging X_1 with X_2 in the description above, and similarly for all the rest of the β's.
4. σ^2 is the variance of the potentially observable Y values where $X_1 = x_1$, $X_2 = x_2$, ..., $X_k = x_k$, and for any other setting of the X's, e.g., $X_1 = z_1$, $X_2 = z_2$, ..., $X_k = z_k$, because the distributions are all assumed to have the same variance. Thus, σ is the common standard deviation of each of these (infinitely many) conditional distributions.

Notice the phrase "in the case where there are no functional relationships among the X's." Examples where there are functional relationships among the X's include the quadratic model $E(Y \mid X = x) = \beta_0 + \beta_1 x + \beta_2 x^2$, and the interaction model $E(Y \mid X_1 = x_1, X_2 = x_2) = \beta_0 + \beta_1 x_1 + \beta_2 x_2 + \beta_3 x_1 x_2$. The reason for the "no functional relationship" stipulation among the X variables is that you cannot increase one X and hold the rest fixed when there is a functional relationship. You can still interpret the parameters in such models, but the interpretations are different than those given above.

The following `R` code reads the "Grade Point Average" data (introduced in Chapter 1), and performs multiple regression to predict Y = Grade Point Average (GPA) for a graduate student, as a function of X_1 = GMAT score, and X_2 = Degree Program (1 = PhD, 0 = Masters).

```
grades = read.table("https://raw.githubusercontent.com/andrea2719/
URA-DataSets/master/gpa_gmat.txt")
Y = grades$gpa
X1 = grades$gmat
# A 1/0 variable; 1 = PhD, 0 = Masters
X2 = ifelse(grades$degree =="P",1,0)
head(data.frame(X1, X2, Y))
```

TABLE 6.1

The Classical Multiple Regression Model as it Applies to Modeling Y = GPA as a Function of X_1 = GMAT and X_2 = Degree Program (Master's or PhD). Each Value in the "Y" Column is Assumed to be Produced by the $N(\beta_0 + \beta_1 x_1 + \beta_2 x_2, \sigma^2)$ Distribution when you use the Classical Regression Model

	X_1	X_2	Y	
1	480	1	3.842	$\leftarrow N(\beta_0 + \beta_1(480) + \beta_2(1), \sigma^2)$
2	510	0	3.454	$\leftarrow N(\beta_0 + \beta_1(510) + \beta_2(0), \sigma^2)$
3	500	0	3.750	$\leftarrow N(\beta_0 + \beta_1(500) + \beta_2(0), \sigma^2)$
4	460	0	3.000	$\leftarrow N(\beta_0 + \beta_1(460) + \beta_2(0), \sigma^2)$
5	630	1	3.430	$\leftarrow N(\beta_0 + \beta_1(630) + \beta_2(1), \sigma^2)$
6	680	1	3.550	$\leftarrow N(\beta_0 + \beta_1(680) + \beta_2(1), \sigma^2)$
...	...	...	...	...
494	610	0	3.866	$\leftarrow N(\beta_0 + \beta_1(610) + \beta_2(0), \sigma^2)$

The assumptions of the multiple regression model are essential to interpret the parameters; these are explained as shown in Table 6.1.

As stated in "Interpretation of Parameters ..." above, β_0 is the mean of the distribution that produces GPA when both GMAT and Degree Program are 0. Since GMAT cannot be zero, this interpretation is an extrapolation to what would happen is the GMAT scale were extended to include the score 0. In such a case, β_0 is the mean of the distribution of GPA for hypothetical Master's students who have GMAT = 0 (i.e., both X_1 *and* X_2 are 0). However, when the range of the X data cannot include zero, we generally avoid interpreting the intercept this way. Instead, you can think of the intercept simply as a "fitting constant," a term that is needed in the model to adjust the overall function up or down as needed to fit the potentially observable data.

To interpret β_1, consider two different types of potentially observable students: (i) PhD students with GMAT = 601 and (ii) PhD students with GMAT = 600. According to the model, the means of the GPA distributions in these two cases are, respectively, $\beta_0 + \beta_1(601) + \beta_2(1)$, and $\beta_0 + \beta_1(600) + \beta_2(1)$. The mean of the GPA distribution in case (i) is β_1 higher than the mean of the GPA distribution in case (ii). The same conclusion applies if we fix attention to Master's students, where $X_2 = 0$, and the same conclusion applies if we compare students whose GMATs are, say 501 and 500. So β_1 is the difference in the means of the GPA distributions associated with a one unit difference in GMAT scores, holding Degree Program fixed.

To interpret β_2, consider two different types of potentially observable students: (i) PhD students with GMAT = 600 and (ii) Master's students with GMAT = 600. The means of the GPA distributions in these two cases are, respectively, $\beta_0 + \beta_1(600) + \beta_2(1)$, and $\beta_0 + \beta_1(600) + \beta_2(0)$. The mean of the GPA distribution in case (i) is β_2 higher than the mean of the GPA distribution in case (ii). The same conclusion applies if we compare Master's and Ph. D. students whose GMATs are, say 500. So β_2 is the difference in the means of the GPA distributions associated with a one-unit difference in Degree Program, holding GMAT fixed.

In the case of β_2, you should interpret a "one-unit difference in Degree Program" more directly because Degree Program can only be 0 (Master's) or 1 (PhD): A one-unit difference in degree program simply refers to the difference between Master's and PhD students. So in this example, β_2 is the difference in the means of the GPA distributions, comparing Ph. D. students with Master's students, holding GMAT fixed.

To interpret σ, consider any group of potentially observable students having fixed X_1 and X_2, such as Ph. D. students with GMAT = 600. The parameter σ^2 is the variance of the distribution of potentially observable GPAs in this group. Since this parameter is assumed constant for all such groups, σ^2 is also the variance of the distribution of potentially observable GPAs among Ph. D. students with GMAT = 700, and σ^2 is also the variance of the distribution of potentially observable GPAs among Master's students with GMAT = 500.

You can estimate these parameters using `R` as follows, obtaining maximum likelihood (OLS) estimates of the β's and the "unbiased version" of the estimate of σ. Note that you supply the multiple X variables to the `lm` function separated by the "+" sign.

```
> summary(lm(Y~X1+X2))

Call:
lm(formula = Y ~ X1 + X2)

Residuals:
     Min       1Q   Median       3Q      Max
-1.13932 -0.20578  0.01747  0.22294  0.62498

Coefficients:
             Estimate Std. Error t value Pr(>|t|)
(Intercept) 2.7506999  0.1191639  23.083  < 2e-16 ***
X1          0.0013572  0.0002156   6.296  6.8e-10 ***
X2          0.1793805  0.0503072   3.566 0.000398 ***
---
Signif. codes:  0 '***' 0.001 '**' 0.01 '*' 0.05 '.' 0.1 ' ' 1

Residual standard error: 0.2852 on 491 degrees of freedom
Multiple R-squared:  0.1142,    Adjusted R-squared:  0.1106
F-statistic: 31.64 on 2 and 491 DF,  p-value: 1.187e-13
```

So $\hat{\beta}_1 = 0.00136$ is the estimated mean GPA difference associated with one unit higher GMAT, for students in the same degree program (i.e., holding Degree Program fixed), and $\hat{\beta}_2 = 0.179$ is the estimated mean GPA difference between Ph. D. and Master's students who have the same GMAT score (i.e., holding GMAT score fixed). Also, $\hat{\sigma} = 0.2852$ is the estimated standard deviation of GPAs for students who are in the same degree program and have the same GMAT score. For example, the estimated standard deviation of GPA in the potentially observable cohort of PhD students who have GMAT = 700 is 0.2852, and the estimated standard deviation of GPA in the potentially observable cohort of Master's students who have GMAT = 710 is also 0.2852.

Finally, $\hat{\beta}_0 = 2.75$ is the estimate of the mean GPA for Master's students having GMAT = 0; this is a correct interpretation, but not very useful because GMAT cannot be 0.

Appendix A: Use of Instrumental Variables to Estimate Causal Effect

As indicated in this chapter, use of additional potential confounding variables in the regression model makes the interpretation of the β for an X variable of interest closer to causal, because more is "held fixed." An alternative approach, one that does not require you to identify potential confounders at all, is to use an *instrumental variable*. There are many controversial aspects of this methodology; therefore, we will write the results using theorems, so that it is clear precisely what is being assumed and what logically, factually, follows from those assumptions.

A.1 Foundations

Estimands

$$\mu_x = \mathrm{E}(X);\; \mu_y = \mathrm{E}(Y);\;\; \sigma_x^2 = \mathrm{Var}(X) = \mathrm{E}(X-\mu_x)^2;\;\; \sigma_y^2 = \mathrm{Var}(Y) = \mathrm{E}(Y-\mu_y)^2$$

$$\sigma_{xy} = \mathrm{Cov}(X,Y) = \mathrm{E}\{(X-\mu_x)(Y-\mu_y)\};\;\; \rho_{xy} = \mathrm{Corr}(X,Y) = \sigma_{xy}/(\sigma_x\sigma_y)$$

Estimators

$$\hat{\mu}_x = (1/n)\sum_{i=1}^{n} X_i;\;\; \hat{\sigma}_x^2 = \{1/(n-1)\}\sum_{i=1}^{n}(X_i - \hat{\mu}_x)^2$$

$$\hat{\sigma}_{xy} = \{1/(n-1)\}\sum_{i=1}^{n}(X_i-\hat{\mu}_x)(Y_i-\hat{\mu}_y);\; \hat{\rho}_{xy} = \hat{\sigma}_{xy}/(\hat{\sigma}_x\hat{\sigma}_y)$$

The focus of this appendix will be on the consistent estimation of causal effects. Unbiasedness is not nearly as important as consistency, as most estimators we use are biased, such as the ordinary sample standard deviation, and most maximum likelihood estimators. The more important issue is whether an estimate becomes closer to its estimand with a larger sample size n. If not, it is an inconsistent estimator.

Definition of Consistency: An estimator $\hat{\theta}$ of an estimand θ is **consistent** if $\mathrm{plim}_{n\to\infty}\,\hat{\theta} = \theta$.

By the Law of Large Numbers, averages are consistent estimators of expected values. Thus we must find expected values in this document. The Law of Total Expectation, defined earlier in this chapter, gives us the first result.

Lemma 1: The law of total expectation

Let (X, Y) be jointly distributed. Let $f(X,Y)$ be a "standard" type of function involving ratios, exponents, polynomials, etc. Let $\mathrm{E}\{f(X,Y) \mid X = x\} = g(x)$. Then $\mathrm{E}\{f(X,Y)\} = \mathrm{E}\{g(X)\}$.

(More briefly, $\mathrm{E}\{f(X,Y)\} = \mathrm{E}_X[\mathrm{E}_Y\{f(X,Y) \mid X\}]$.)

Proof of Lemma 1

It's not proven here. But it appears in many probability texts; Wikipedia reports Theorem 34.4 of Billingsley (1995).

Another needed result:

Lemma 2: The additivity property of covariance

Let Y, $X_1, \ldots, X_k$ be random variables with finite variance. Then $\text{Cov}(Y, \Sigma_i X_i) = \Sigma_i \text{Cov}(Y, X_i)$.

Proof of Lemma 2

Denote $E(Y)$ as μ_y and denote $E(X_i)$ as μ_i. Note that, by definition, linearity property of expectation, and algebra

$$\begin{aligned}\text{Cov}(Y, \Sigma_i X_i) &= \text{E}\{(Y - \mu_y)\Sigma_i(X_i - \mu_i)\} \\ &= \Sigma_i \text{E}\{(Y - \mu_y)(X_i - \mu_i)\} \\ &= \Sigma_i \text{Cov}(Y, X_i) \qquad \blacksquare\end{aligned}$$

The first theorem assumes the conditional mean form of the regression model, which states that the means of the conditional distributions lie precisely on a line with slope β_1. It then establishes the connection between the covariance σ_{xy} and β_1. Non-normal distributions are allowed.

Theorem 6.1: Relationship Between the Covariance and the Regression Coefficient

Assume $(X, Y) \sim p(x, y)$, with $\text{Var}(X) < \infty$, $\text{Var}(Y) < \infty$, and with $\text{E}(Y \mid X = x) = \beta_0 + \beta_1 x$. Then $\sigma_{xy} = \beta_1 \sigma_x^2$.

Proof of Theorem 6.1

By definition, $\sigma_{xy} = \text{E}\{(X - \mu_x)(Y - \mu_y)\}$. By Lemma 1,

$$\begin{aligned}\sigma_{xy} &= \text{E}_X \text{E}_Y \{(X - \mu_x)(Y - \mu_y) | X\} \\ &= \text{E}_X \{(X - \mu_x)(\beta_0 + \beta_1 X - \beta_0 - \beta_1 \mu_x)\} \\ &= \text{E}_X \{\beta_1 (X - \mu_x)^2\} \\ &= \beta_1 \text{E}_X \{(X - \mu_x)^2\} \\ &= \beta_1 \sigma_x^2 \qquad \blacksquare\end{aligned}$$

Theorem 6.2: Consistency of Variance and Covariance Estimates

Assume the pairs (U_i, V_i) are independent and identically distributed (iid) from $p(u,v)$, where $\text{Var}(U_i) < \infty$ and $\text{Var}(V_i) < \infty$. Then $\hat{\sigma}_u^2, \hat{\sigma}_v^2$, and $\hat{\sigma}_{uv}$ are consistent estimates of σ_u^2, σ_v^2, and σ_{uv}, respectively.

Sketch of proof of Theorem 6.2 (can be made more rigorous)

Follows from the Law of Large Numbers as applied to the quantities $\{(U_i - \mu_u)^2\}, \{(V_i - \mu_v)^2\}$, $\{U_i\}$, and $\{V_i\}$; also from the fact that $\{(n-1)/n\} \rightarrow 1$ ■

Theorem 6.3: Consistent estimation of the regression coefficient in terms of variance and covariance

Assume the pairs (X_i, Y_i) are iid from $p(x,y)$, where $p(x,y)$ satisfies the linear conditional mean assumption and the finite variances assumption of Theorem 6.1. Then $\text{plim}_{n\to\infty} \hat{\sigma}_{xy}/\hat{\sigma}_x^2 = \beta$.

Proof of Theorem 6.3

Applying Theorem 6.2 and also the fact that smooth functions of convergent variables also converge, $\text{plim}_{n\to\infty} \hat{\sigma}_{xy}/\hat{\sigma}_x^2 = \sigma_{xy}/\sigma_x^2$. Applying Theorem 6.1, $\sigma_{xy}/\sigma_x^2 = \beta_1$. Hence, $\text{plim}_{n\to\infty} \hat{\sigma}_{xy}/\hat{\sigma}_x^2 = \beta_1$ ■

As seen earlier in this book, the estimator $\hat{\sigma}_{xy}/\hat{\sigma}_x^2$ is also the usual OLS estimator. So, in addition to the result shown above that the estimator $\hat{\sigma}_{xy}/\hat{\sigma}_x^2$ is consistent, it is also unbiased, and BLUE, and it is also the maximum likelihood estimator under the classical model. It's a pretty nice estimator!

But the problem, as noted earlier in this chapter, is that the β_1 in the conditional mean expression $\text{E}(Y \mid X_1 = x_1) = \beta_0 + \beta_1 x_1$ is not the causal effect of X_1. It can easily happen that X_1 is correlated with some confounding variable X_2, and that the real effect on Y is from X_2, not X_1.

For an example, consider the Y = Computer Processing Speed, and X_1 = Computer RAM example introduced in Chapter 5. In an observational study using many different student's computers, the coefficient β_1 in the model $\text{E}(\text{Speed} \mid \text{RAM} = x) = \beta_0 + \beta_1 x$ is not the causal effect of increasing the computer's RAM. The reason is that RAM is correlated with excluded variables such as X_2 = "GHz": In the given observational study, students whose machines have higher RAM also tend to have higher GHz because better computers tend to have higher values of both variables. The coefficient β_1 is correctly interpreted as the mean difference between Speed reported by students whose computer has RAM 1.5 (say) and Speed reported by students whose computer has RAM 0.5. But this value β_1 can be positive even if RAM had no effect whatsoever because in the group of potentially observable students with RAM = 1.5 there is generally higher GHz than in the group of potentially observable students with RAM = 0.5.

So, while the regression coefficient β_1 is correctly interpreted as a mean difference between the means of Y for two groups defined by X values that differ by 1.0, it cannot be interpreted as a causal effect that would result from actually manipulating the X value. In the computer case, it is easy to understand how to perform such manipulation in order to assess the causal effect: Simply open up the computer and insert an additional memory chip.

The assumption $\mathrm{Cov}(X,\varepsilon)=0$

In many sources, you will see that $\mathrm{Cov}(X,\varepsilon)=0$ is stated as one of the assumptions of the regression model. But as long as the regression coefficients β_j are interpreted in terms of means of conditional distributions, then this assumption is automatically satisfied, and therefore need not be stated as an assumption. The following theorem shows why.

Theorem 6.4: X and ε are Uncorrelated in the Conditional Mean Model

Assume the conditional mean model as in Theorem 6.1: $(X,Y)\sim p(x,y)$, with $\mathrm{Var}(X)<\infty$, $\mathrm{Var}(Y)<\infty$, and with $\mathrm{E}(Y\mid X=x)=\beta_0+\beta_1 x$. Define $\varepsilon=Y-(\beta_0+\beta_1 X)$. Then $\mathrm{Cov}(X,\varepsilon)=0$.

Proof of Theorem 6.4

Constants do not affect variance and covariance. Hence we can eliminate β_0 and write $\mathrm{Cov}(X,\varepsilon)=\mathrm{Cov}(X,Y-\beta_1 X)=\mathrm{Cov}(X,Y)-\beta_1\mathrm{Cov}(X,X)$, by Lemma 2 and the linearity property of expectation. But $\mathrm{Cov}(X,X)=\mathrm{Var}(X)=\sigma_x^2$; the result $\mathrm{Cov}(X,\varepsilon)=0$ then follows directly from Theorem 6.1. ■

So, the assumption $\mathrm{Cov}(X,\varepsilon)=0$ does not need to be stated, because it is automatically true in the conditional mean model. But just for the sake of argument, what if you had a model $Y=\gamma_0+\gamma_1 X+\delta$, where $\mathrm{Cov}(X,\delta)\neq 0$? Such a model could not possibly be the conditional mean model, because as shown in Theorem 6.4, the conditional mean model forces $\mathrm{Cov}(X,\delta)=0$. So the parameters γ_0 and γ_1 cannot be the ordinary regression parameters in such a model, which is why they are given symbols other than β_0 and β_1. But suppose you want to estimate this model in the ordinary way as $\hat{\sigma}_{xy}/\hat{\sigma}_x^2$. What do you think might happen? It should be no surprise that the result is inconsistent for γ_1. The reason is clear: The parameter γ_1 is different from β_1, and the OLS estimator $\hat{\beta}_1=\hat{\sigma}_{xy}/\hat{\sigma}_x^2$ is consistent for β_1. Thus, the OLS estimator $\hat{\beta}_1=\hat{\sigma}_{xy}/\hat{\sigma}_x^2$ is necessarily inconsistent for γ_1. The specific degree of inconsistency is quantified in terms of $\sigma_{x\delta}\equiv\mathrm{Cov}(X,\delta)$, as shown in the following Theorem.

Theorem 6.5: Quantifying the degree of inconsistency of the OLS estimator when X is correlated with ε

Assume (X_i,Y_i) are iid from $p(x,y)$, with finite variances. Let γ_0, γ_1 be any constants such that $\mathrm{Cov}\{X,Y-(\gamma_0+\gamma_1 X)\}\neq 0$. Then $\mathrm{plim}_{n\to\infty}\hat{\sigma}_{xy}/\hat{\sigma}_x^2=\gamma_1+\sigma_{x\delta}/\sigma_x^2$.

Proof of Theorem 6.5

As a consequence of Theorem 6.2, $\mathrm{plim}_{n\to\infty}\hat{\sigma}_{xy}/\hat{\sigma}_x^2=\sigma_{xy}/\sigma_x^2$. But $\sigma_{xy}=\mathrm{Cov}(X,Y)=\mathrm{Cov}(X,\gamma_0+\gamma_1 X+\delta)=\gamma_1\sigma_x^2+\sigma_{x\delta}$, implying the result. ■

A.2 The Causal Model

If you want the coefficient of X to be a causal effect, you need a different model than the conditional mean model, one that includes many more X's. Here is one:

$$Y = \gamma_0 + \gamma_1 X_1 + \gamma_2 X_2 + \ldots + \gamma_k X_k + \varepsilon'$$

In this model:

- X_1 is your main X variable of interest, the one whose causal effect on Y you wish to measure.
- X_2 through X_k are all other (usually unmeasured) variables that also causally affect Y. In this model, changes (manipulations) in X_1 cause changes in the distribution of Y when all other possible causal variables $X_2 - X_k$ are held fixed.
- ε' is a random error term. This term might be identically zero, in which case the causal model is a deterministic model, and this does not change any of the arguments below. Otherwise, with enough X's, it is reasonable to assume that this term is uncorrelated with everything; e.g., ε' might be subatomic quantum noise.

You can re-arrange the causal model as follows:

$$Y = \gamma_0 + \gamma_1 X + \delta$$

where $X = X_1$ and

$$\delta = \gamma_2 X_2 + \ldots + \gamma_k X_k + \varepsilon'$$

In this model, $\mathrm{Cov}(X,\delta) \neq 0$. Instead, $\mathrm{Cov}(X,\delta) = \sum_{j=2}^{k} \gamma_j \mathrm{Cov}(X, X_j)$. Thus, applying Theorem 6.5, the OLS estimate $\hat{\sigma}_{xy} / \hat{\sigma}_x^2$ is inconsistent for γ_1, with probability limit $\gamma_1 + \sum_{j=2}^{k} \gamma_j \mathrm{Cov}(X, X_j) / \sigma_x^2$.

A.3 The Instrumental Variable Method

The goal is to come up with an estimator whose probability limit is γ_1. If you could measure all the relevant unobserved confounders $X_2,\ldots,X_k$, then the simple OLS multiple regression estimate of γ_1 in model (1) would do the trick. But you usually cannot. And even if you could, there might be hundreds of such variables, and you would not want to run OLS with so many predictors. What to do? Try to find an **instrumental variable**.

Consider the model $Y = \gamma_0 + \gamma_1 X + \delta$, where γ_1 is the causal effect of X. An **instrumental variable** is a variable Z such that:

1. Z is correlated with X (preferably reasonably strongly correlated), and
2. Z is uncorrelated with δ

The instrumental variable (IV) estimator of γ_1

The instrumental variable estimator of γ_1 is given by

$$\hat{\gamma}_1 = \frac{\hat{\sigma}_{zy}}{\hat{\sigma}_{zx}}$$

Theorem 6.6: Consistency of the IV Estimator

Assume the data pairs (X_i, Y_i, Z_i) are sampled iid from $p(x, y, z)$, with all variances finite. Assume in addition that Z is an instrumental variable to the causal model $Y = \gamma_0 + \gamma_1 X + \delta$. Then $\hat{\gamma}_1 = \hat{\sigma}_{zy} / \hat{\sigma}_{zx}$ is a consistent estimator of γ_1.

Proof of Theorem 6.6

From Theorem 6.2, $\hat{\sigma}_{zy}$ is a consistent estimator of $\mathrm{Cov}(Z, Y)$. But $\mathrm{Cov}(Z, Y) = \mathrm{Cov}(Z, \gamma_0 + \gamma_1 X + \delta) = \gamma_1 \mathrm{Cov}(Z, X) + \mathrm{Cov}(Z, \delta) = \gamma_1 \mathrm{Cov}(Z, X)$.

Also from Theorem 6.2, $\hat{\sigma}_{zx}$ is a consistent estimator of $\mathrm{Cov}(Z, Y)$. The conclusion of Theorem 6.6 follows immediately since the ratio of consistent estimators is itself a consistent estimator. ■

Discussion of the IV estimation result:

To perform this analysis, the crucial assumption is that you can find a Z that is uncorrelated with δ in the model $Y = \gamma_0 + \gamma_1 X + \delta$. This is challenging! Since $\mathrm{Cov}(X, \delta) = \sum_{j=2}^{k} \gamma_j \mathrm{Cov}(X, X_j)$, either you (a) find a Z so that all the correlations $\mathrm{Corr}(Z, X_j)$ are 0, for all other causal variables X_j $(j > 1)$, or (b) hope to find a Z so that the terms $\gamma_j \mathrm{Cov}(Z, X_j)$, $2, \ldots, k$, somehow all add to 0. Condition (b) seems impossible to justify *a priori*, so you are left with (a). Now, how do you do that? One way that works is that if Z is randomly assigned by the experimenter. In that case, mathematically, all terms $\mathrm{Corr}(Z, X_j)$ are in fact zero for all X's that precede the random assignment. The sample correlations will, of course, be different from zero, but that's what the standard errors account for. The main concern is whether the true correlations are zero, and if they are then the IV estimator is consistent.

Other than randomization, you need to find a Z that is truly uncorrelated with *all potential confounders*. Because this requirement cannot be tested using data, it will ordinarily be a subjective assumption. Thus, instrumental variable methods are questionable.

Reference

Billingsley, P. (1995). *Probability and measure*. New York, NY: John Wiley & Sons.

Exercises

1. Hans goes to college! He was just accepted to Calisota Tech University (CTU) University as an entering Freshman, with X_1 = High school GPA = 3.4, X_2 = SAT Verbal = 600, X_3 = SAT Math = 480. He is also X_4 = Male. In words (not numbers of symbols), what is the best predictor of Y = Hans' GPA at graduation, given the above X information? (Incorporate CTU University in your answer, as well as the given X data.)
2. A true regression model is $Y \mid X = x \sim \text{N}(10 + 3x, x^2)$. Assume that $X \sim \text{N}(2, 0.5^2)$.
   a. Apply the Law of Total Expectation to find E(Y). As part of your answer, (i) identify $f(X)$, and (ii) apply the linearity property of expectation.
   b. Apply the law of Total Variance to find Var(Y).
   c. Find the true R^2 value using your answer to 2.b.
   d. Report a scatterplot of data produced by the model above, and relate the scatterplot to the true R^2 value in 2.c.
3. Consider the computer speed example

   ```
   comp = read.table("https://raw.githubusercontent.com/andrea2719/
   URA-DataSets/master/compspeed.txt")
   ```

   with Y = speed (= 1/time to complete) and X = RAM. The model is $Y \mid X = x \sim \text{N}(\beta_0 + \beta_1 x, \sigma^2)$.
   a. Give the predictive interpretation of β_1. (Not $\hat{\beta}_1$).
   b. Give the causal interpretation of β_1, and explain why the causal interpretation is not allowed in this example.
   c. Unlike the case of the ice cream sales example, the required "intervention" of the causal interpretation is physically possible in this example, because you can add RAM to a computer. Explain how you could perform another experiment, with intervention, to find the causal effect of RAM on Speed.
4. Use the computer speed example again, but use the model where Y = speed (= 1/time to complete), X_1 = RAM, and X_2 = GHz. The model is $Y \mid X_1 = x_1, X_2 = x_2 \sim \text{N}(\beta_0 + \beta_1 x_1 + \beta_2 x_2, \sigma^2)$.
   a. Give the predictive interpretation of β_1. (Not $\hat{\beta}_1$). Make it clear that this answer is different from your answer to 3.a.
   b. Why is the interpretation you gave in 4.a. closer to the causal interpretation?
5. Consider the charitable contributions data set

   ```
   charity = read.csv("https://raw.githubusercontent.com/andrea2719/
   URA-DataSets/master/charitytax.csv")
   ```

   Let Y = ln(Charitable.Cont) (logarithm of charitable contributions) X_1 = DEPS (dependents), X_2 = ln(Income.AGI) (logarithm of adjusted gross income).
   a. Fit the regression model and report the values of the four estimated parameters.
   b. Interpret the true parameters that correspond to the estimates of 5.a. (Do not back-transform, just interpret in terms of ln(Charitable.Cont), DEPS, and ln(Income.AGI).)

c. Find the 95% confidence interval for the true parameter β_1 using "confint" on your `lm` object.
d. From the confidence interval in 5.c., β_1 is more likely to be positive rather than negative. Supposing that β_1 is truly greater than 0, explain why β_1 does not have a causal interpretation. Use the "intervention" concept in your answer, while referring specifically to Charitable contributions, number of dependents, and Income.

7

Multiple Regression from the Matrix Point of View

In the case of simple regression, you saw that the OLS estimate of slope has a simple form: It is the estimated covariance of the (X, Y) distribution, divided by the estimated variance of the X distribution, or $\hat{\beta}_1 = \hat{\sigma}_{xy} / \hat{\sigma}_x^2$. There is no such simple formula in multiple regression. Instead, you must use matrix algebra, involving matrix multiplication and matrix inverses. If you are unfamiliar with basic matrix algebra, including multiplication, addition, subtraction, transpose, identity matrix, and matrix inverse, you should take some time now to get acquainted with those particular concepts before reading on. (Perhaps you can locate a "matrix algebra for beginners" type of web page.)

Done? Ok, read on.

Our first use of matrix algebra in regression is to give a concise representation of the regression model. Multiple regression models refer to n observations and k variables, both of which can be in the thousands or even millions. The following matrix form of the model provides a very convenient shorthand to represent all this information.

$$\boldsymbol{Y} = \mathbf{X}\boldsymbol{\beta} + \boldsymbol{\varepsilon}$$

This concise form covers all the n observations and all the X variables (k of them) in one simple equation. Note that there are boldface non-italic terms and boldface italic terms in the expression. To make the material easier to read, we use the convention that boldface means a matrix, while boldface italic refers to a *vector*, which is a matrix with a single column. Thus $\boldsymbol{Y}$, $\boldsymbol{\beta}$, and $\boldsymbol{\varepsilon}$, are vectors (single-column matrices), while $\mathbf{X}$ is a matrix having multiple columns.

To understand this model, consider your data set, which has the structure shown in Table 7.1.

You can relate the matrices in the model $\boldsymbol{Y} = \mathbf{X}\boldsymbol{\beta} + \boldsymbol{\varepsilon}$ easily to the data set shown in Table 7.1 as follows:

The $\boldsymbol{Y}$ vector is the list of all the Y_i values:

$$\boldsymbol{Y} = \begin{bmatrix} Y_1 \\ Y_2 \\ \vdots \\ Y_n \end{bmatrix}$$

The $\mathbf{X}$ matrix is the array of all the X_{ij} values, with an additional column of 1's to account for the intercept β_0:

$$\mathbf{X} = \begin{bmatrix} 1 & X_{11} & X_{12} & \ldots & X_{1k} \\ 1 & X_{21} & X_{22} & \ldots & X_{2k} \\ \vdots & \vdots & \vdots & \ddots & \vdots \\ 1 & X_{n1} & X_{n2} & \ldots & X_{nk} \end{bmatrix}$$

TABLE 7.1

Structure of the Multiple Regression Data Set

Obs	X_1	X_2	...	X_k	Y
1	X_{11}	X_{12}	...	X_{1k}	Y_1
2	X_{21}	X_{22}	...	X_{2k}	Y_2
...	...	...	...	...	...
n	X_{n1}	X_{n2}	...	X_{nk}	Y_n

The β vector contains all the unknown β parameters, including the intercept:

$$\beta = \begin{bmatrix} \beta_0 \\ \beta_1 \\ \vdots \\ \beta_k \end{bmatrix}$$

The matrix product $\mathbf{X}\beta$ gives the multiple regression function:

$$\mathbf{X}\beta = \begin{bmatrix} 1 & X_{11} & X_{12} & \dots & X_{1k} \\ 1 & X_{21} & X_{22} & \dots & X_{2k} \\ \vdots & \vdots & \vdots & \vdots & \vdots \\ 1 & X_{n1} & X_{n2} & \dots & X_{nk} \end{bmatrix} \begin{bmatrix} \beta_0 \\ \beta_1 \\ \vdots \\ \beta_k \end{bmatrix} = \begin{bmatrix} 1 \times \beta_0 + X_{11} \times \beta_1 + \dots + X_{1k} \times \beta_k \\ 1 \times \beta_0 + X_{21} \times \beta_1 + \dots + X_{2k} \times \beta_k \\ \vdots \\ 1 \times \beta_0 + X_{n1} \times \beta_1 + \dots + X_{nk} \times \beta_k \end{bmatrix}$$

Finally, the ε vector contains the differences from $\boldsymbol{Y}$ to the regression function $\mathbf{X}\beta$:

$$\varepsilon = \boldsymbol{Y} - \mathbf{X}\beta$$

7.1 The Least Squares Estimates in Matrix Form

One use of matrix algebra is to display the model for all n observations and all X variables succinctly as shown above. Another use is to identify the OLS estimates of the β's. There is simply no way to display the OLS estimates other than by using matrix algebra, as follows:

$$\hat{\beta} = (\mathbf{X}^{\mathrm{T}}\mathbf{X})^{-1}\mathbf{X}^{\mathrm{T}}Y.$$

(The "T" symbol denotes transpose of the matrix.) To see why the OLS estimates have this matrix representation, recall that in the simple, classical regression model, the maximum likelihood (ML) estimates must minimize the sum of squared "errors" called SSE. The same is true in multiple regression: The ML estimates must minimize the function

$$\mathrm{SSE}(\beta_0, \beta_1, \dots, \beta_k) = \sum_{i-1}^{n} \{y_i - (\beta_0 + \beta_1 x_{i1} + \dots + \beta_k x_{ik})\}^2$$

In the case of two X variables ($k = 2$), you are to choose $\hat{\beta}_0$, $\hat{\beta}_1$, and $\hat{\beta}_2$ that define the *plane*, $f(x_1, x_2) = \hat{\beta}_0 + \hat{\beta}_1 x_1 + \hat{\beta}_2 x_2$, such as the one shown in Figure 6.3, that minimizes the sum of squared vertical deviations from the 3-dimensional *point cloud* $(x_{i1}, x_{i2}, y_i), i = 1, 2, \ldots, n$. Figure 7.1 illustrates the concept.

R code for Figure 7.1

```
X1 = runif(20,1,2); X2 = runif(20,1,2)
Y = -1 + 2*X1 + 4*X2 + rnorm(20,0,2)

fit <- lm(Y ~ X1 + X2)
pch = ifelse(fit$residuals < 0, "-", "+")

library(scatterplot3d)
s3d <-scatterplot3d(X1,X2,Y, pch=pch, cex.symbols=1.6, highlight.3d= FALSE,
   type="h", main="3-D Scatter and Fitted Regression Plane")
s3d$plane3d(fit, draw_polygon = TRUE)
```

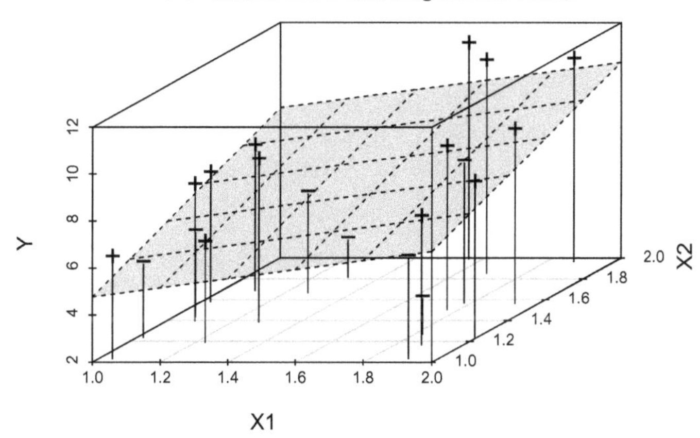

FIGURE 7.1
Three-dimensional scatterplot of $n = 20$ data values (x_{i1}, x_{i2}, y_i), along with the least-squares plane that minimizes the sum of squared vertical deviations from y_i values to the plane. Positive deviations (or residuals) are shown as "+", and negative deviations (or residuals) are shown as "–."

Thus, in the case $k = 2$, the least-squares plane is the one plane (out of infinitely many possible planes) whose coefficients minimize $\text{SSE}(\beta_0, \beta_1, \beta_2)$. For $k > 2$, you can visualize the plane essentially the same way as shown in Figure 7.1, just realize that the plane exists in higher-dimensional space, which is not as easy to visualize, and it is called a *hyperplane*.

To find the best-fitting (hyper)plane, you need to (i) take derivatives of the $\text{SSE}(\beta_0, \beta_1, \ldots, \beta_k)$ function, successively with respect to each $\beta_0, \beta_1, \ldots, \beta_k$, (ii) set these derivatives to zero, and (iii) solve for the values $\hat{\beta}_0, \hat{\beta}_1, \ldots, \hat{\beta}_k$ that determine the minimizing hyperplane. Performing steps (i) and (ii) and algebraically simplifying yields the following system of $(k + 1)$ equations:

Equation 0: $\partial \text{SSE}/\partial\beta_0 = 0 \;\Rightarrow\; n\beta_0 + \Sigma x_{i1}\beta_1 + \;\ldots\; + \Sigma x_{ik}\beta_k = \Sigma y_i$

Equation 1: $\partial \text{SSE}/\partial\beta_1 = 0 \;\Rightarrow\; \Sigma x_{i1}\beta_0 + \Sigma x_{i1}^2\beta_1 + \ldots + \Sigma x_{i1}x_{ik}\beta_k = \Sigma x_{i1}y_i$

… … … … … … … … …

Equation k: $\partial \text{SSE}/\partial\beta_k = 0 \;\Rightarrow\; \Sigma x_{ik}\beta_0 + \Sigma x_{ik}x_{i1}\beta_1 + \;\ldots\; + \Sigma x_{ik}^2\beta_k = \Sigma x_{ik}y_i$

The values of $\beta_0, \beta_1, \ldots, \beta_k$ that solve these (k+1) equations are the OLS estimates, $\hat{\beta}_0, \hat{\beta}_1, \ldots, \hat{\beta}_k$, which comprise the vector $\hat{\beta}$. You can write these (k+1) equations in matrix form as

$$(\mathbf{X}^{\mathrm{T}}\mathbf{X})\hat{\beta} = \mathbf{X}^{\mathrm{T}}\mathbf{Y}$$

Self-study question: How do you get the matrix equation $(\mathbf{X}^{\mathrm{T}}\mathbf{X})\hat{\beta} = \mathbf{X}^{\mathrm{T}}\mathbf{Y}$ from Equation 0, Equation 1, …, Equation k?

If $(\mathbf{X}^{\mathrm{T}}\mathbf{X})$ has an inverse, then the linear equations have the solution $\hat{\beta} = (\mathbf{X}^{\mathrm{T}}\mathbf{X})^{-1}\mathbf{X}^{\mathrm{T}}\mathbf{Y}$, since

$$\begin{aligned}
& (\mathbf{X}^{\mathrm{T}}\mathbf{X})\hat{\beta} = \mathbf{X}^{\mathrm{T}}\mathbf{Y} \\
\Rightarrow\quad & (\mathbf{X}^{\mathrm{T}}\mathbf{X})^{-1}(\mathbf{X}^{\mathrm{T}}\mathbf{X})\hat{\beta} = (\mathbf{X}^{\mathrm{T}}\mathbf{X})^{-1}\mathbf{X}^{\mathrm{T}}\mathbf{Y} \\
\Rightarrow\quad & \mathbf{I}\,\hat{\beta} = (\mathbf{X}^{\mathrm{T}}\mathbf{X})^{-1}\mathbf{X}^{\mathrm{T}}\mathbf{Y} \\
\Rightarrow\quad & \hat{\beta} = (\mathbf{X}^{\mathrm{T}}\mathbf{X})^{-1}\mathbf{X}^{\mathrm{T}}\mathbf{Y}
\end{aligned}$$

In the above algebra, there are references to matrix inverses (having the "−1" superscript) and the identity matrix ($\mathbf{I}$) that we assume you have already learned about before reading this far in the chapter.

7.2 The Regression Model in Matrix Form

The model representation $\mathbf{Y} = \mathbf{X}\beta + \varepsilon$ is not complete because it states nothing about the assumptions. The following expression is a complete representation of the classical model; notice how simple the model looks when expressed in matrix form.

The classical model in matrix form

$$\mathbf{Y} \mid \mathbf{X} = \mathbf{x} \sim \mathrm{N}_n(\mathbf{x}\beta, \sigma^2\mathbf{I})$$

Here, the $\mathbf{X} = \mathbf{x}$ condition refers to a specific realized matrix $\mathbf{x}$ of the *random matrix* $\mathbf{X}$ and is a simple generalization of the $X = x$ condition we have used repeatedly to its matrix form. The matrix $\mathbf{X}$ contains potentially observable (random) X values, as well as fixed values for any non-random X data. The first column of $\mathbf{X}$ is ordinarily the column of 1's needed to capture the intercept term β_0, and this column is not random.

In Appendix A of Chapter 1, we introduced the *bivariate normal distribution*, which is a distribution of two variables. Here, the symbol "$N_n(\mathbf{x}\boldsymbol{\beta}, \sigma^2\mathbf{I})$" refers to a *multivariate normal distribution*. The "*n*" subscript identifies that it is a distribution of the *n* variables $Y_1, Y_2, \ldots, Y_n$. The $\mathbf{x}\boldsymbol{\beta}$ term refers to the mean vector of the distribution, and the term $\sigma^2\mathbf{I}$ refers to its covariance matrix (explained in detail below).

All assumptions in the classical regression model are embodied in the concise matrix form of the model: The correct functional specification assumption is embodied in the mean vector ($\mathbf{x}\boldsymbol{\beta}$) specification, the constant variance and independence assumptions are implied by specification of $\sigma^2\mathbf{I}$ as covariance matrix, as will be described below, and the normality assumption is embodied in the multivariate normal specification.

A covariance matrix is a matrix that contains all the variances and covariances among a set of random variables. For example, if (W_1, W_2, W_3) are jointly distributed random variables, then the covariance matrix of $\mathbf{W} = (W_1, W_2, W_3)$ is given by

$$\mathrm{Cov}(W) = \begin{bmatrix} \mathrm{Var}(W_1) & \mathrm{Cov}(W_1, W_2) & \mathrm{Cov}(W_1, W_3) \\ \mathrm{Cov}(W_2, W_1) & \mathrm{Var}(W_2) & \mathrm{Cov}(W_2, W_3) \\ \mathrm{Cov}(W_3, W_1) & \mathrm{Cov}(W_3, W_2) & \mathrm{Var}(W_3) \end{bmatrix}$$

Notice that the row/column combination tells you which pair of variables are involved, or which variable is involved in the case of the diagonal elements. Note also that the covariance of a variable with itself is just the variance of that variable, which explains why the variances are on the diagonal of the covariance matrix.

The assumption that $\mathrm{Cov}(\mathbf{Y} \mid \mathbf{X} = \mathbf{x}) = \sigma^2\mathbf{I}$ unpacks as follows:

$$\mathrm{Cov}(\mathbf{Y} \mid \mathbf{X} = \mathbf{x}) = \sigma^2\mathbf{I} = \sigma^2 \begin{bmatrix} 1 & 0 & \cdots & 0 \\ 0 & 1 & \cdots & 0 \\ \vdots & \vdots & \ddots & \vdots \\ 0 & 0 & \cdots & 1 \end{bmatrix} = \begin{bmatrix} \sigma^2 & 0 & \cdots & 0 \\ 0 & \sigma^2 & \cdots & 0 \\ \vdots & \vdots & \ddots & \vdots \\ 0 & 0 & \cdots & \sigma^2 \end{bmatrix}$$

In particular, the assumption states that the potentially observable *Y* data are conditionally uncorrelated (since the covariances are all zeros) and that the potentially observable *Y* data all have the same conditional variance (since the diagonal elements are all the same number, σ^2).

The normality assumption is needed for exact inferences as described in Chapter 3. But as described in Chapter 2, normality is not necessary for some of the good results that you get in regression. In particular, it was stated that, under the conditions of the Gauss-Markov model, which drops the normality assumption, the OLS estimates are unbiased. The standard errors likewise do not depend on the normality assumption. Here is a relaxed form of the model, where normality is dropped, but the same conditional mean and covariance forms remain.

The Gauss-Markov model in matrix form

$$\mathbf{Y} \mid \mathbf{X} = \mathbf{x} \sim (\mathbf{x}\boldsymbol{\beta}, \sigma^2\mathbf{I})$$

7.3 Unbiasedness of the OLS Estimator $\hat{\beta}$ Under the Gauss-Markov Model

Strangely, it is easier to prove the unbiasedness of the estimators $\hat{\beta}$ using their matrix form. We will assume the Gauss-Markov model, thus the mathematical theorem is stated as follows: *If* the data are produced by the Gauss-Markov model, *then* the OLS $\hat{\beta}$ is an unbiased estimator of $\boldsymbol{\beta}$. Since the Gauss-Markov model includes the (normality-assuming) classical model as a special case, the proof of unbiasedness of the estimators $\hat{\beta}$ in the Gauss-Markov model implies, *a fortiori*, unbiasedness of the estimators $\hat{\boldsymbol{\beta}}$ in the classical model as well.

There are two parts to the unbiasedness argument. The first is that the estimates are unbiased, *conditional* on the observed values of the random X variables. The second is to note that, by the law of total expectation, the estimates are also unbiased when considered over all possible samples of random X data.

7.3.1 Unbiasedness of the OLS Estimates $\hat{\beta}$ Conditional on the X Data

The **X** matrix has random and fixed elements; the first column of the ones, for example, contains fixed (non-random) elements. Also, if there are X variables that are fixed in advance of observing the Y data, as occurs for example in designed experiments and in some kinds of stratified sampling, then these X variables are also fixed, not random.

Let $\mathbf{X} = \mathbf{x}$ denote a particular realization of the **X** matrix. This is the **X** data you have in your current study, for example. We are going to sneak in a condition here: We will consider only $\mathbf{X} = \mathbf{x}$ such that $(\mathbf{x}^T\mathbf{x})$ is invertible. This is not a major assumption, because if this matrix was not invertible, there would be no unique OLS estimates to speak of, so you would not be worried about unbiasedness. In the proofs below, the condition "$\mathbf{X} = \mathbf{x}$" will always be interpreted as "$\mathbf{X} = \mathbf{x}$ where $(\mathbf{x}^T\mathbf{x})$ is invertible," but it will not be explicitly displayed that way in order to save space. Here is the proof of conditional unbiasedness:

$$
\begin{aligned}
&\mathrm{E}(\hat{\beta} \mid \mathbf{X} = \mathbf{x}) \\
&= \mathrm{E}\{(\mathbf{X}^T\mathbf{X})^{-1}\mathbf{X}^T Y \mid \mathbf{X} = \mathbf{x}\} && \text{(substitution)} \\
&= \mathrm{E}\{(\mathbf{X}^T\mathbf{X})^{-1}\mathbf{X}^T(\mathbf{X}\beta + \varepsilon) \mid \mathbf{X} = \mathbf{x}\} && \text{(substitution)} \\
&= \mathrm{E}\{(\mathbf{X}^T\mathbf{X})^{-1}(\mathbf{X}^T\mathbf{X}\beta + \mathbf{X}^T\varepsilon) \mid \mathbf{X} = \mathbf{x}\} && (\mathbf{A}(\mathbf{B}+\mathbf{C}) = \mathbf{AB} + \mathbf{AC}) \\
&= \mathrm{E}\{(\mathbf{X}^T\mathbf{X})^{-1}(\mathbf{X}^T\mathbf{X})\beta + (\mathbf{X}^T\mathbf{X})^{-1}\mathbf{X}^T\varepsilon) \mid \mathbf{X} = \mathbf{x}\} && (\mathbf{A}(\mathbf{B}+\mathbf{C}) = \mathbf{AB} + \mathbf{AC}) \\
&= \mathrm{E}\{\mathbf{I}\beta + (\mathbf{X}^T\mathbf{X})^{-1}\mathbf{X}^T\varepsilon) \mid \mathbf{X} = \mathbf{x}\} && (\mathbf{A}^{-1}\mathbf{A} = \mathbf{I}) \\
&= \mathrm{E}\{\beta + (\mathbf{X}^T\mathbf{X})^{-1}\mathbf{X}^T\varepsilon) \mid \mathbf{X} = \mathbf{x}\} && (\mathbf{IA} = \mathbf{A}) \\
&= \beta + \mathrm{E}\{(\mathbf{X}^T\mathbf{X})^{-1}\mathbf{X}^T\varepsilon) \mid \mathbf{X} = \mathbf{x}\} && \text{(linearity property of expectation)} \\
&= \beta + (\mathbf{x}^T\mathbf{x})^{-1}\mathbf{x}^T\,\mathrm{E}(\varepsilon \mid \mathbf{X} = \mathbf{x}) && \text{(applying the condition } \mathbf{X} = \mathbf{x}\text{, and} \\
& && \text{the linearity property of expectation)} \\
&= \beta + (\mathbf{x}^T\mathbf{x})^{-1}\mathbf{x}^T\,\mathrm{E}(Y - \mathbf{X}\beta \mid \mathbf{X} = \mathbf{x}) && \text{(substitution)} \\
&= \beta + (\mathbf{x}^T\mathbf{x})^{-1}\mathbf{x}^T\{\mathrm{E}(Y \mid \mathbf{X} = \mathbf{x}) - \mathbf{x}\beta\} && \text{(applying the condition } \mathbf{X} = \mathbf{x}\text{, and} \\
& && \text{the linearity property of expectation)} \\
&= \beta + (\mathbf{x}^T\mathbf{x})^{-1}\mathbf{x}^T(\mathbf{x}\beta - \mathbf{x}\beta) && \text{(by the G-M model assumption)} \\
&= \beta + 0 = \beta && \text{(since } \mathbf{A0} = \mathbf{0}\text{)}
\end{aligned}
$$

As always, unbiasedness does *not* say that $\hat{\beta}$ is equal to β. Rather, the *random* $\hat{\beta}$ differs from the *fixed* β by a *random deviation vector*, seen in the equation $\hat{\beta} = \beta + (\mathbf{X}^{\mathrm{T}}\mathbf{X})^{-1}\mathbf{X}^{\mathrm{T}}\varepsilon$ in the seventh line of the demonstration above. As proven, the random deviation vector, $(\mathbf{X}^{\mathrm{T}}\mathbf{X})^{-1}\mathbf{X}^{\mathrm{T}}\varepsilon$, has elements that are neither systematically above nor below zero, on average.

The simulation that gave Figure 3.1 shows the fixed-*X* unbiasedness of the parameters $\hat{\beta}$: In that simulation, the "Widgets" data were fixed for every sample, at the specific values:

```
Widgets = c(1500,800,1500,1400,900,800,1400,1400,1300,1400,700,1000,1200,
1200,900,1200,1700,1600,1200,1400,1400,1000,1200,800,1000,1400,1400,1500,
1500,1600,1700,900,800,1300,1000,1600,900,1300,1600,1000)
```

These values are the same as in the original data set. When the *Y* values are simulated at random, using these fixed *X* values, from the model $Y = 55 + 1.5X + \varepsilon$, where $\varepsilon \sim \mathrm{N}(0,\ 250^2)$, it was seen that the random estimates $\hat{\beta}$ were not systematically different from the target $\beta = (55,\ 1.5)^{\mathrm{T}}$.

However, the conditional-*X* presentation seems to lack applicability. Certainly, in different data sets, you would expect that the *Y* data (Cost), will differ. But you should also expect that the *X* data (Widgets), would differ in different samples. So it makes more sense to investigate unbiasedness from a *random-X* point of view, where every possible sample has a different collection of *X* data, just as what you expect in real life. The random-*X* viewpoint is more relevant and natural in the Production Cost example, and in most examples involving observational data.

7.3.2 Unbiasedness of the OLS Estimates $\hat{\beta}$, *not* Conditional on the Values of the X Data

Unconditional unbiasedness is actually very easy to prove. Recall the law of total expectation: $\mathrm{E}_V\{\mathrm{E}(U \mid V)\} = \mathrm{E}(U)$. Applying this to the OLS estimator $\hat{\beta}$, let $\mathrm{E}(\hat{\beta} \mid \mathbf{X} = \mathbf{x}) = f(\mathbf{x})$. We learned in the previous section that $f(\mathbf{x}) = \beta$; i.e., it does not matter what is the specific value of the matrix $\mathbf{X}$, the estimator is still conditionally unbiased, $f(\mathbf{X}) = \beta$ for every possible matrix $\mathbf{X}$. Hence, by the law of total expectation,

$$\mathrm{E}(\hat{\beta}) = \mathrm{E}_{\mathbf{X}}\{f(\mathbf{X})\} = \mathrm{E}_{\mathbf{X}}(\beta) = \beta,$$

since β is a constant. In other words, for any specific realization $\mathbf{X} = \mathbf{x}$ of the *X* data, the expected value of $\hat{\beta}$ is β. Thus it makes sense that when you average β (a constant) over all possible values of $\mathbf{X}$, you still get β, because the average of a constant, is just that constant.

Self-study question: What do you get when you take the average of 12, 12, 12, 12, and 12? How does that relate to the unbiasedness of the OLS estimates, *not* conditional on the **X** data?

To demonstrate unbiasedness via simulation in the random-*X* case, you need to specify a distribution to produce the *X* data, then use the two-step process shown in Table 1.5 of Chapter 1. First, you generate an *X* value, $X = x$, from a distribution $p(x)$; then you generate a value of *Y* from the distribution $p(y \mid x)$. Every simulation will, therefore, have different

X data, unlike the simulation shown in Chapter 3, leading to Figure 3.1, where the $\mathbf{X}$ data were fixed for all simulated data sets. In the following code, the model $p(x)$ used to produce the X = Widgets data is the N(1200, 300^2) model, with X values rounded to the nearest 100 to mimic the type of data for X = Widgets that was actually observed in the real study. The model parameters $(\beta_0, \beta_1, \sigma)$ are as specified in the original simulation, with $(\beta_0, \beta_1, \sigma) = (55, 1.5, 250)$.

R code for Figure 7.2

```
Nsim = 100000; b1.ols = numeric(Nsim)
for (i in 1:Nsim) {
   Widgets.sim = round(rnorm(40, 1200, 300), -2)
   Cost.sim = 55 + 1.5*Widgets.sim + rnorm(40,0,250)
   b1.ols[i] = lm(Cost.sim ~ Widgets.sim)$coefficients[2]   }
hist(b1.ols, freq=F, breaks=100, main="",xlab = expression("OLS
estimate of" ~beta[1]))
abline(v=1.5, lwd=2.5)
```

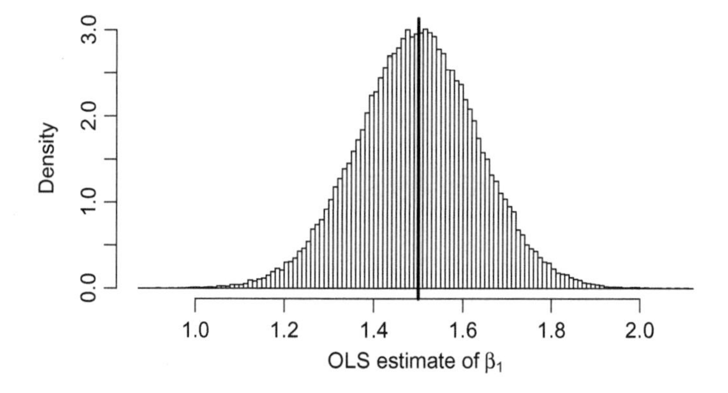

FIGURE 7.2
Histogram of 10,000 estimates $\hat{\beta}_1$, each calculated from a sample of 40 observations from a random-X model: The Y are simulated from the N($55+1.5X$, 250^2) model, where the X data themselves are simulated from a rounded-off normal distribution. The vertical line is the target, $\beta_1 = 1.5$. The estimates are neither systematically larger nor systematically smaller than the target, as expected when the estimator is unbiased.

7.4 Measurement Error

Very often, the variables X and Y that you might use are imperfectly measured. For example, you might wish to assess the effect of a company's accounting mistakes on their performance, but people do not like to admit mistakes, and also people might not even know that they have made mistakes. Thus, you cannot measure "Accounting Mistakes" directly. Instead, you decide to use a proxy variable, such as X = CEO's Guess of Accounting Mistakes. Clearly, the CEO's guess is not the same as the actual number of mistakes, but they should be correlated.

Letting X_A = Actual Mistakes and X_G = CEO's Guess, it should be clear that the conditional distribution $p(y \mid X_G = x)$ is different from the conditional distribution $p(y \mid X_A = x)$, simply because X_G and X_A are different variables. The results of the previous section show that the OLS estimates based on the data (X_G, Y) give unbiased estimates for the model $\boldsymbol{Y} = \mathbf{X}_G\boldsymbol{\beta}_G + \boldsymbol{\varepsilon}_G$ (under the Gauss-Markov assumptions), and that the OLS estimates based on the data (X_A, Y) give unbiased estimates for the model $\boldsymbol{Y} = \mathbf{X}_A\boldsymbol{\beta}_A + \boldsymbol{\varepsilon}_A$ (again, under the Gauss-Markov assumptions). However, it should be no surprise that the data (X_G, Y) give *biased* estimates for the model $\boldsymbol{Y} = \mathbf{X}_A\boldsymbol{\beta}_A + \boldsymbol{\varepsilon}_A$, simply because $\boldsymbol{\beta}_G$ and $\boldsymbol{\beta}_A$ are different. Usually, the effect of this bias is toward zero; that is, the estimated effect (slope) based on (X_G, Y) will be closer to zero, the R^2 statistic will be smaller, and the p-value larger than when you use (X_A, Y).

The following simulation illustrates this bias in the case where the X_A variable is "Widgets" and the X_G variable is a guess at the number of Widgets, e.g., by a plant supervisor. The simulation assumes that X_G differs from X_A by random error, with $X_G = X_A + \delta$. The term δ is called the "measurement error," and is assumed to be independent of X_A and ε_A. The code is similar to the code that produced Figure 7.2 above, but with the addition of normally distributed measurement error δ.

R code for Figure 7.3

```
Nsim = 100000; b1.ols.XG = numeric(Nsim); b1.ols.XA = numeric(Nsim)
for (i in 1:Nsim) {
  Widgets.A = round(rnorm(40, 1200, 300), -2) #Actual Widgets
  delta = rnorm(40, 0 , 100) # Measurement error
  Widgets.G = Widgets.A + delta # Guess of Widgets
  Cost = 55 + 1.5*Widgets.A + rnorm(40,0,250)
  b1.ols.XA[i] = lm(Cost ~ Widgets.A)$coefficients[2]
  b1.ols.XG[i] = lm(Cost ~ Widgets.G)$coefficients[2] }

par(mfrow=c(2,1))
hist(b1.ols.XA, freq=F, breaks=100, xlim=c(1,2),
  xlab="OLS estimate with true X", main="")
abline(v=1.5, lwd=2, lty=2)

hist(b1.ols.XG, freq=F, breaks=100, xlim = c(1,2),
  xlab="OLS estimate with measurement error in X", main="")
abline(v=1.5, lwd=2, lty=2)
```

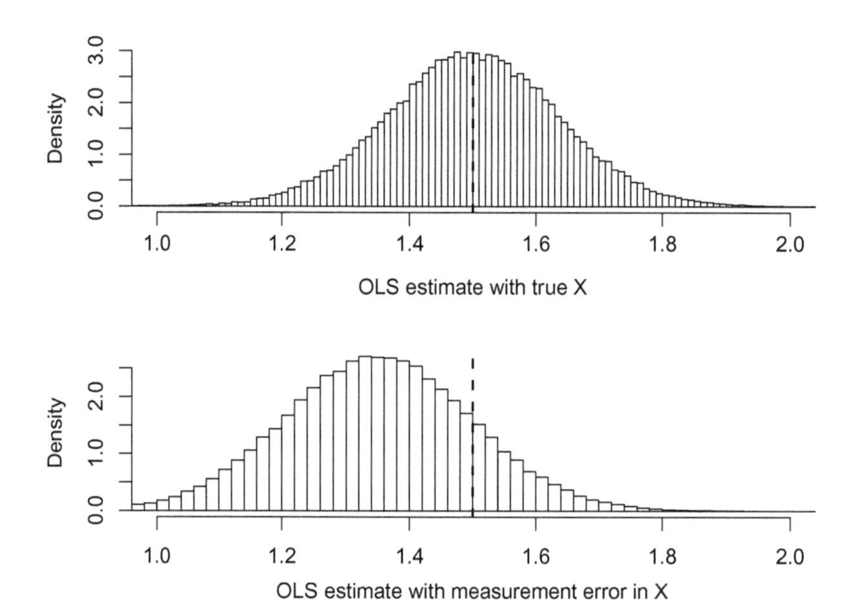

FIGURE 7.3
Histograms of estimated slope estimates under the cases where the X variable is measured without error (top panel), and where the X variable is measured with error (bottom panel). Dashed lines are the true slope (1.5) of the model using the true X.

Figure 7.3 shows that, if you use the Widget Guess in your regression model, then your estimated slope tends to be smaller than the true slope of the (Widgets, Cost) relationship. On the other hand, if you use the Widget Guess in your regression model, then your estimated slope is an unbiased estimate of the true slope of the (Widget Guess, Cost) relationship.

Some sources erroneously state that an assumption of regression models is that the X variables are measured without error. But this assumption is not needed. When you have measurement error, the estimates are unbiased for the parameters of the relationship between the measured variable and Y. There is no reason to expect that the estimates should be unbiased for a regression relationship where a different variable is used. In other words, the estimated slope of a (X, Y) data set is a biased estimate of the slope of a (Z, Y) relationship.

7.5 Standard Errors of OLS Estimates

Recall from Section 3.5 that the conditional (on $\mathbf{X} = \mathbf{x}$) variance of $\hat{\beta}_1$ is, in the case of simple regression where there is only one X variable, given by

$$\mathrm{Var}(\hat{\beta}_1 \mid \mathbf{X} = \mathbf{x}) = \frac{\sigma^2}{(n-1)s_x^2}$$

Taking the square root gives you the conditional standard deviation, and replacing σ with $\hat{\sigma}$ gives you the standard error:

$$s.e.(\hat{\beta}_1 \mid \mathbf{X} = \mathbf{x}) = \frac{\hat{\sigma}}{s_x\sqrt{n-1}}$$

Usually, the $\mathbf{X} = \mathbf{x}$ condition is not displayed explicitly, but clearly this standard error formula is conditional on the observed values of $\mathbf{X}$ because there is a sample estimate, s_x, in the formula.

In multiple regression where there is more than one X variable, the standard error formula cannot be displayed in such a simple form. Matrix methods are needed instead. To get these standard errors, we first need the variance of an estimated β. Strangely, it turns out that it is easier to compute the *entire covariance matrix* of the estimated vector $\hat{\boldsymbol{\beta}}$, then pick the variances out from the diagonal elements of the covariance matrix. To do this, we need to invoke a famous result, a generalization of the "linearity property of variance" to covariance matrices. The result is as follows:

Linearity property of covariance matrices

Let $\boldsymbol{W}$ be a random $(p \times 1)$ vector and let $\mathbf{A}$ be a fixed (not random) $(q \times p)$ matrix. Then

$$\mathrm{Cov}(\mathbf{A}\boldsymbol{W}) = \mathbf{A}\mathrm{Cov}(\boldsymbol{W})\mathbf{A}^{\mathrm{T}}$$

To apply this result to regression, note that $\hat{\boldsymbol{\beta}} = (\mathbf{X}^{\mathrm{T}}\mathbf{X})^{-1}\mathbf{X}^{\mathrm{T}}\boldsymbol{Y}$. So let $\mathbf{A} = (\mathbf{X}^{\mathrm{T}}\mathbf{X})^{-1}\mathbf{X}^{\mathrm{T}}$ and let $\boldsymbol{W} = \boldsymbol{Y}$. Conditional on $\mathbf{X} = \mathbf{x}$, $\mathbf{A}$ is constant and $\boldsymbol{Y}$ has covariance matrix $\sigma^2\mathbf{I}$ under the Gauss-Markov model (normality is not needed here). Then

$$\mathrm{Cov}(\hat{\boldsymbol{\beta}} \mid \mathbf{X} = \mathbf{x}) = \mathrm{Cov}((\mathbf{X}^{\mathrm{T}}\mathbf{X})^{-1}\mathbf{X}^{\mathrm{T}}\boldsymbol{Y} \mid \mathbf{X} = \mathbf{x}) \quad \text{(by substitution)}$$

$$= (\mathbf{x}^{\mathrm{T}}\mathbf{x})^{-1}\mathbf{x}^{\mathrm{T}}(\sigma^2\mathbf{I})((\mathbf{x}^{\mathrm{T}}\mathbf{x})^{-1}\mathbf{x}^{\mathrm{T}})^{\mathrm{T}} \quad \text{(by the linearity property of covariance)}$$

$$= \sigma^2(\mathbf{x}^{\mathrm{T}}\mathbf{x})^{-1}\mathbf{x}^{\mathrm{T}}((\mathbf{x}^{\mathrm{T}}\mathbf{x})^{-1}\mathbf{x}^{\mathrm{T}})^{\mathrm{T}}$$

(since constants can be multiplied in any order, and since multiplying matrices by the identity matrix leaves them, unchanged)

$$= \sigma^2(\mathbf{x}^{\mathrm{T}}\mathbf{x})^{-1}\mathbf{x}^{\mathrm{T}}\mathbf{x}\{(\mathbf{x}^{\mathrm{T}}\mathbf{x})^{-1}\}^{\mathrm{T}}$$

(since $(\mathrm{AB})^{\mathrm{T}} = \mathrm{B}^{\mathrm{T}}\mathrm{A}^{\mathrm{T}}$ and since the transpose of a transpose is the original matrix)

$$= \sigma^2(\mathbf{x}^{\mathrm{T}}\mathbf{x})^{-1}\mathbf{x}^{\mathrm{T}}\mathbf{x}(\mathbf{x}^{\mathrm{T}}\mathbf{x})^{-1} \quad \text{(since the } (\mathbf{x}^{\mathrm{T}}\mathbf{x})^{-1} \text{ matrix is symmetric)}$$

$$= \sigma^2(\mathbf{x}^{\mathrm{T}}\mathbf{x})^{-1}\mathbf{I} \quad \text{(by definition of matrix inverse)}$$

$$= \sigma^2(\mathbf{x}^{\mathrm{T}}\mathbf{x})^{-1}$$

(because multiplying by $\mathbf{I}$ leaves the matrix unchanged)

Thus the conditional (on $\mathbf{X} = \mathbf{x}$) covariance matrix of the random estimates $\hat{\boldsymbol{\beta}}$ is as follows.

Conditional covariance matrix of the OLS estimates

$$\mathrm{Cov}(\hat{\boldsymbol{\beta}} \mid \mathbf{X} = \mathbf{x}) = \sigma^2(\mathbf{x}^{\mathrm{T}}\mathbf{x})^{-1}$$

As the derivation followed from the assumption of the Gauss-Markov model, thus it is important to re-state that the above formula requires the Gauss-Markov assumptions, homoscedasticity, and independence in particular. As we will see in later chapters, other formulas are needed when these assumptions are violated.

The standard errors that are reported by almost all regression software use the conditional-x formulation. As discussed previously, this usually does not limit the generalizability of the results, because the conditional-x results generalize to the unconditional-X case through the Law of Total Expectation.

Letting $\mathbf{C} = (\mathbf{x}^{\mathrm{T}}\mathbf{x})^{-1}$, the first diagonal element of $\mathbf{C}$ corresponds to the intercept, β_0, the second diagonal element corresponds to the first slope, β_1, and the last diagonal (the $(k+1)^{\mathrm{st}}$ diagonal element) corresponds to the slope β_k. So let the diagonal elements of $\mathbf{C} = (\mathbf{x}^{\mathrm{T}}\mathbf{x})^{-1}$ be denoted as $c_{00}, c_{11}, \ldots, c_{kk}$ to underscore the correspondence with $\beta_0, \beta_1, \ldots, \beta_k$. Then the variances of the OLS estimates are given by:

$$\mathrm{Var}(\hat{\beta}_j \mid \mathbf{X} = \mathbf{x}) = \sigma^2 c_{jj},\ j = 0,1,\ldots,k.$$

Taking the square root gives you the conditional standard deviation, and replacing σ with its estimate gives the standard errors:

Standard errors of the OLS estimates

$s.e.(\hat{\beta}_j) = \hat{\sigma}\sqrt{c_{jj}},\ j = 0,1,\ldots,k$, where c_{jj} is the $(j+1)^{\mathrm{st}}$ diagonal element of $(\mathbf{x}^{\mathrm{T}}\mathbf{x})^{-1}$.

Strangely, the standard error formula looks simpler in the case of multiple regression than with simple regression! That's one (among many) of the beauties of using matrix algebra in regression analysis.

To demonstrate the matrix form of the standard error formula using `R`, consider the "Computer Speed" data introduced in Section 5.4 of Chapter 5. There are three variables in the data set, "time," "GHz," and "GB." As discussed in Section 5.4, the transformation $Y = 1/\mathrm{time}$ is reasonable. The other variables will be used in raw form, $X_1 = \mathrm{GHz}$ and $X_2 = \mathrm{GB}$. We will now show how to obtain the $\hat{\beta}_j$ and $s.e.(\hat{\beta}_j)$ in matrix form, and note that those are identical to what is given in the standard `lm` output.

Please run the following code line by line to see how the matrix calculations correspond precisely to the `lm` output. Note that `R` uses `%*%` for matrix multiplication, `t()` for transpose, and `solve()` for inverse.

`R` code to show matrix calculations used in regression

```
compS = read.table("https://raw.githubusercontent.com/andrea2719/
URA-DataSets/master/compspeed.txt")
attach(compS); Y = 1/time; X1 = GHz; X2 = GB
fit = lm(Y ~ X1 + X2); summary(fit)
```

(Continued)

```
n = length(Y); Intcpt = rep(1, n)
X = as.matrix(cbind(Intcpt, X1, X2))
XTX = t(X) %*% X; XTX
XTX.inv = solve(XTX); XTX.inv
## This XTX inverse matrix is called "cov.unscaled"
## in the fitted lm object:
summary(fit)$cov.unscaled

XTY = t(X) %*% as.matrix(Y)
beta.hat = XTX.inv %*% XTY; beta.hat
## Compare with the lm results:
fit$coefficients

## Sum of squared residuals
Resid = Y - X %*% beta.hat
SS.Resid = t(Resid) %*% Resid
sigma.hat = sqrt(SS.Resid/(n-3))
sigma.hat
## Compare with the lm result:
summary(fit)$sigma

## Standard Errors
Cov.beta.hat = as.numeric(sigma.hat^2) * XTX.inv
Cov.beta.hat
standard.errors = sqrt(diag(Cov.beta.hat))
standard.errors
## Compare with the lm result:
summary(fit)$coefficients[,2]
```

The code demonstrates that the output of `lm` corresponds exactly with the matrix formulations.

Self-study question: Where does the vector $\hat{\beta}$ appear in the output from the code?
Self-study question: How do you interpret the parameter β_1 in the model that is estimated using the code?

7.6 Application of the Theory: The Graduate Student GPA Data Analysis, Revisited

This chapter has a number of theoretical results. How do these results apply to you, the student or researcher who wants to interpret the results of your data analysis? Glad you asked! Let's apply the results to the analysis of the data introduced in Chapter 6, which were used to predict GPA as a function of GMAT and degree program. To make the results easier to read, we will use the variable names "GMAT" and "PhD" (1 = PhD, 0 = Masters) rather than "X1" and "X2" as in Chapter 6.

Here is how the concepts presented in this chapter apply to this concrete situation.

- All estimates and standard errors are matrix functions of the observed data set, as described and calculated above.
- The fitted function is (Predicted GPA) = 2.7506999 + 0.0013572 × GMAT + 0.1793805 × PhD. This function defines the plane that minimizes the sum of squared vertical deviations from individual GPA values to the plane. Figure 7.4 is obtained using the same code that produced Figure 7.1.
- The processes at work that gave these data on n = 494 students' GPAs *could have* given rise to a completely different set of n = 494 GPAs, even with exact same PhD and GMAT data values as in the current data set. These other data are potentially observable only and do not refer to specific, existing other students. These other possible data simply reflect other possibilities that might have occurred at that particular point in time and place. Unbiasedness of the parameter estimates means that, while the estimates will be different for every other data set, on average they will be neither systematically above nor below the targets β_0, β_1 and β_2 that govern the production of the GPA data. In other words, unbiasedness implies that your estimates, 2.7506999, 0.0013572, and 0.1793805, are randomly sampled values from distributions whose means are precisely β_0, β_1 and β_2, respectively.
- The same conclusion regarding unbiasedness holds when you imagine the other data sets all having different PhD and GMAT data (the random-X viewpoint). While this way of looking at the other data sets makes it easier to view them as simply belonging to a different set of n = 494 students, it is still best not to think about it that way, because there never existed another set of 494 students coming from the same processes that produced these students. Rather, again, you should view these other possible data sets as *potentially observable*, just as in the fixed-X viewpoint.
- Again assuming the data-generating processes just described are well modelled via the classical model, then the standard deviations of all the other parameter

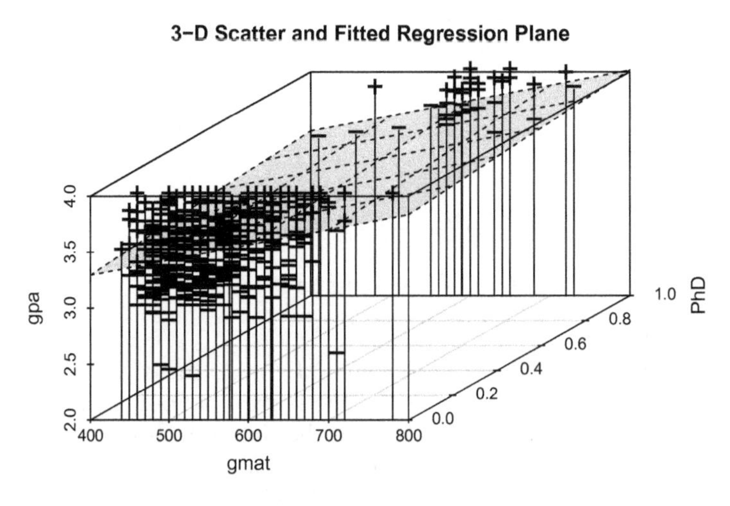

FIGURE 7.4
Best-fitting plane through the (PhD, GMAT, GPA) point cloud. GPAs below the plane (negative residuals) are shown as "–", GPAs above the plane (positive residuals) are shown as "+."

estimates you would get from all these other data sets, assuming the same PhD and GMAT data for all data sets (the conditional-x framework), would be approximately 0.1191639363, 0.0002155794, and 0.0503072073. Thus, since data values from a distribution are typically within ± two standard deviations of the mean, and because the means of the distributions of the estimated β's are in fact the true β's (by unbiasedness), you can expect, for example, that the true β_2 (measuring the true mean difference between GPA's of PhD and Masters student who share a common GMAT) will be within the range $0.1793805 \pm 2(0.0503072073)$. In other words, you can claim confidently that $0.07876609 < \beta_2 < 0.2799949$ (grade points). Under the assumptions of the classical model, an exact 95% confidence version of this interval uses the T distribution to get the multiplier rather than using 2.0; the more precise interval is $0.0805365726 < \beta_2 < 0.27822450$.

- 95% of data sets in the conditional-x samples will have the true β_2 inside of similarly constructed intervals; the same conclusion holds in the unconditional case because of the Law of Total Expectation: Over all possible random-X samples, the average coverage level is the average of the conditional coverage levels 95%, 95%, …, etc. Because the average of a constant is just that constant, the interpretation of "95%" holds in both the fixed-X and random-X frameworks.

Exercises

1. Consider the following data,

```
sales = read.csv("https://raw.githubusercontent.com/andrea2719/
URA-DataSets/master/sales.csv")
```

but model SALES as a function of the *inverse* of RATE and PPG (price per gallon of gas). Using that model (and only that model), graph the estimated regression function as shown in Figure 7.1, in the transformed (inverse of RATE, non-transformed PPG) units, with data points overlaid, using "+" for points above the function and "–" for points below the function.

2. See the "`R` code to show matrix calculations used in regression" in Chapter 7. Perform a similar analysis using the data and model of problem 1.
3. What is the effect of having more measurement error? To study the effect of having more measurement error, change the measurement error code for Figure 7.3 from `delta = rnorm(40, 0, 100)` to `delta = rnorm(40, 0, 200)`. Then (i) explain why that change means there is more measurement error, (ii) repeat the analysis of the text, and (iii) compare the results of your analysis to what is in the book to answer the original question.
4. What is the effect of having less measurement error? To study the effect of having less measurement error, change the measurement error in code for Figure 7.3 from `delta = rnorm(40, 0, 100)` to `delta = rnorm(40, 0, 50)`. Then (i) explain why that change means there is less measurement error, (ii) repeat the analysis of the text, and (iii) compare the results of your analysis to what is in the book to answer the original question.

8

R-Squared, Adjusted R-Squared, the F Test, and Multicollinearity

This chapter discusses the additional output in the regression analysis, from the context of multiple regression in the classical model. It also discusses multicollinearity, its effects, and remedies.

8.1 The *R*-Squared Statistic

Recall that the true R^2 statistic was introduced in Chapter 6 as

$$\Omega^2 = 1 - \mathrm{E}\{v(X)\} / \mathrm{Var}(Y),$$

where $v(x)$ is the conditional variance of Y given $X = x$, written as $v(x) = \mathrm{Var}(Y \mid X = x)$.

The number Ω^2 is a measure of how well the "X" variable(s) predict(s) your "Y" variable. You can understand this concept in terms of *separation* of the distributions $p(y \mid X = x)$, for the two cases (i) X = a "low" value, and (ii) X = a "high" value. When these distributions are well-separated, then X is a good predictor of Y.

For example, suppose the true model is

$$Y = 6 + 0.2X + \varepsilon,$$

where $X \sim \mathrm{N}(20, 5^2)$ and $\mathrm{Var}(\varepsilon) = \sigma^2$. Then $\mathrm{Var}(Y) = 0.2^2 \times 5^2 + \sigma^2 = 1 + \sigma^2$, and $v(x) = \mathrm{Var}(Y \mid X = x) = \sigma^2$, implying that $\Omega^2 = 1 - \sigma^2 / (1 + \sigma^2) = 1 / (1 + \sigma^2)$ is the true R^2.

Three cases to consider are (i) $\sigma^2 = 9.0$, implying a low $\Omega^2 = 0.1$, (ii) $\sigma^2 = 1.0$, implying a medium value $\Omega^2 = 0.5$, and (iii) $\sigma^2 = 1/9$, implying a high $\Omega^2 = 0.9$. In all cases, let's say a "low" value of X is 15.0, one standard deviation below the mean, and a "high" value of X is 25.0, one standard deviation above the mean.

Now, when $X = 15$, the distribution $p(y \mid X = 15)$ is the $\mathrm{N}(9.0, \sigma^2)$ distribution; and when $X = 25$, the distribution $p(y \mid X = 25)$ is the $\mathrm{N}(11.0, \sigma^2)$ distribution. Figure 8.1 displays these distributions for the three cases above, where the true R^2 is either 0.1, 0.5, or 0.9 (which happens in this study when σ^2 is either 9.0, 1.0, or 1/9). Notice that there is greater *separation* of the distributions $p(y \mid x)$ when the true R^2 is higher.

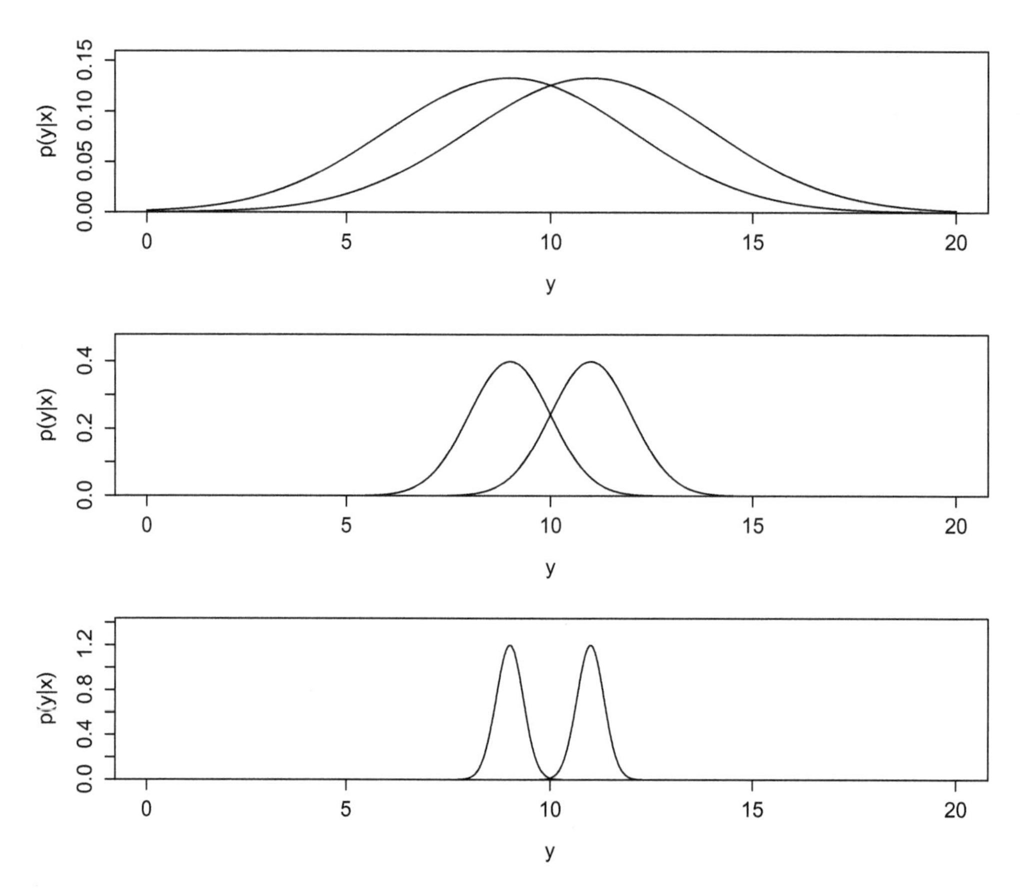

FIGURE 8.1
Separation of distributions $p(y \mid X = \text{low})$ (left distributions) and $p(y \mid X = \text{high})$ (right distributions) in cases where the true R^2 is 0.1 (top panel), 0.5 (medium panel) and 0.9 (bottom panel). In all cases "X = low" and "X = high" refer to an X that is either one standard deviation below or above the mean.

In the case of the classical regression model as shown in Figure 8.1, the conditional variance $\text{Var}(Y \mid X = x) = v(x)$ is a constant σ^2 and does not depend on $X = x$. The unconditional variance is $\text{Var}(Y) = \sigma_Y^2$, so the true R^2 statistic, *in the classical regression model*, is

$$\Omega^2 = 1 - \sigma^2 / \sigma_Y^2$$

Also in the classical regression model, the maximum likelihood estimate of σ^2 is $\hat{\sigma}^2 = \text{SSE} / n$, where $\text{SSE} = \sum_{i=1}^{n} (y_i - \hat{y}_i)^2$, is the sum of squared vertical deviations from y_i values to the fitted OLS function. Further, the maximum likelihood estimate of σ_Y^2 is $\hat{\sigma}_Y^2 = \text{SST} / n$, where $\text{SST} = \sum_{i=1}^{n} (y_i - \bar{y})^2$, is the "total" sum of squared vertical deviations from y_i values to the flat line where $y = \bar{y}$. Using the maximum likelihood estimates of conditional and unconditional variance, you get the maximum likelihood estimate of the true *R*-squared statistic.

Maximum likelihood estimate of Ω^2

$$R^2 = 1 - (\text{SSE} / n) / (\text{SST} / n) = 1 - \text{SSE} / \text{SST}$$

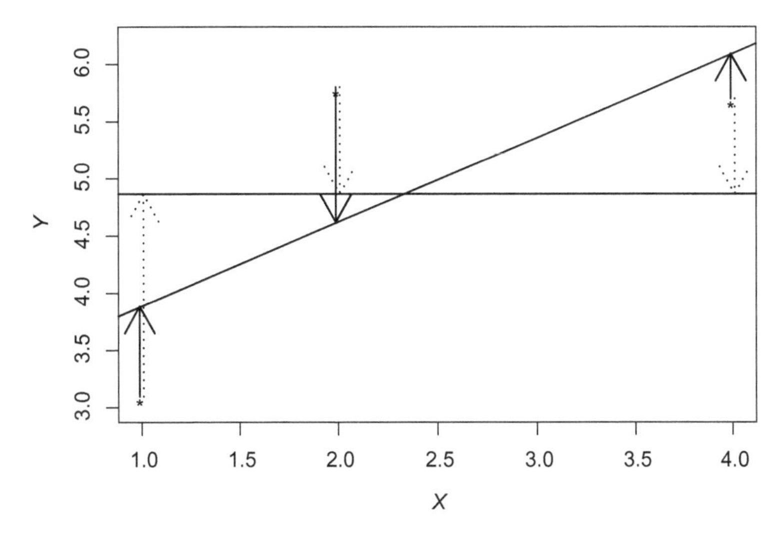

FIGURE 8.2
Scatterplot of $n = 3$ (x, y) data points (1, 3.1), (2, 5.8), and (4, 5.7), indicated by *'s. The horizontal line is the $y = \bar{y}$ line, and the diagonal line is the least-squares line. Vertical deviations from the $y = \bar{y}$ line are shown as dotted arrows; SST is the sum of these squared deviations. Vertical deviations from the least-squares line are shown as solid arrows; SSE is the sum of these squared deviations. The R^2 statistic equals $1 - \text{SSE} / \text{SST} = 1 - 0.461 = 0.539$.

See Figure 8.2 to understand the R^2 statistic.

In Chapter 5, you saw models with different transformations in the X variable. The model with the highest maximized log likelihood was the one with the smallest estimated conditional variance SSE/n, hence it was also the model with the smallest SSE since n is always the same when considering different models for the same data set. Also, SST is always the same when considering different models for the same data set, because SST does not involve the predicted values from the model. Thus, among the different models having different transformations of the X variable (but no Y transformations), the model with the highest log likelihood corresponds precisely to the model with the highest R^2 statistic.

While it is mathematically factual that $0 \le R^2 \le 1.0$, there is no "Ugly Rule of Thumb" for how large an R^2 statistic should be to be considered "good." Rather, it depends on norms for the given subject area: In finance, any non-zero R^2 for predicting stock returns is interesting, because the efficient markets hypothesis states that the true R^2 is zero in this case. In chemical reaction modeling, the outputs are essentially deterministic functions of the inputs, so an R^2 statistic that is less than 1.0, e.g., 0.99, may not be good enough because it indicates faulty experimental procedures. With human subjects and models to predict their behavior, the R^2 statistics are typically less than 0.50 because people are, well, people. We have our own minds and are not robots that can be pigeon-holed by some regression model.

Our advice is to rely less on R^2, and more on the separation of distributions as seen in Figure 8.1. When we get to more complex models, the usual R^2 statistic becomes less interpretable, and in some cases it is non-existent. But you always will have conditional distributions $p(y \mid x)$, so you should always graph those distributions as shown in Figure 8.1 to see how well your X predicts your Y.

8.2 The Adjusted *R*-Squared Statistic

Recall that, in the classical model, $\Omega^2 = 1 - \sigma^2/\sigma_Y^2$, and that the standard R^2 statistic replaces the two variances with their maximum likelihood estimates. Recall also that maximum likelihood estimates of variance are biased. With a larger number of predictor variables (i.e., larger k), the estimate $\hat{\sigma}^2 = \text{SSE}/n$ becomes increasingly biased downward, implying in turn that the ordinary R^2 becomes increasingly biased upward.

Replacing the two variances with their unbiased estimates gives the adjusted R^2 statistic:

$$R_a^2 = 1 - \{\text{SSE}/(n-k-1)\}/\{\text{SST}/(n-1)\}$$

The adjusted R^2 statistic is still biased as an estimator of $\Omega^2 = 1 - \sigma^2/\sigma_Y^2$ because of Jensen's inequality, but it is less biased than the ordinary R^2 statistic. You can interpret the adjusted R^2 statistic in the same way as the ordinary one.

Which estimate is best, adjusted R^2 or ordinary R^2? You guessed it: Use simulation to find out. Despite its reduced bias, the adjusted R^2 is not necessarily closer to the true Ω^2, as simulations will show. In addition, the adjusted R^2 statistic can be less than 0.0, which is clearly undesirable. The ordinary R^2, like the estimand Ω^2, always lies between 0 and 1 (inclusive).

The following R code locates these R^2 statistics in the `lm` output, and computes them "by hand" as well, using the model where Car Sales is predicted using a quadratic function of Interest Rate.

```
CarS = read.table("https://raw.githubusercontent.com/andrea2719/
URA-DataSets/master/CarS.txt")
attach(CarS); Y = NSOLD; X1 = INTRATE; X2 = X1^2
fit = lm(Y ~ X1 + X2); summary(fit)

## By hand calculations of R-squared statistics
SST = sum( (Y-mean(Y))^2 )
SSE = sum(fit$residuals^2)
R.squared = 1 - SSE/SST
n = length(NSOLD)
R.squared.adj = 1 - (SSE/(n-3))/(SST/(n-1))
R.squared; R.squared.adj
```

The summary of the fit shows the following R^2 and adjusted R^2 statistic:

```
Multiple R-squared:  0.5154,    Adjusted R-squared:  0.5071
F-statistic: 62.22 on 2 and 117 DF,  p-value: < 2.2e-16
```

The "by hand" calculations agree:

```
> R.squared; R.squared.adj
[1] 0.5154053
[1] 0.5071217
```

8.3 The *F* Test

See the `R` output a few lines above: Underneath the R^2 statistic is the *F*-statistic. This statistic is related to the R^2 statistic in that it is also a function of SST and SSE (review Figure 8.2). It is given by

$$F = \{(\text{SST} - \text{SSE})/k\} / \{\text{SSE}/(n-k-1)\}$$

If you add the line `((SST-SSE)/2)/(SSE/(n-3))` to the `R` code above, you will get the reported *F*-statistic, although with more decimals: `62.21945`.

With a little algebra, you can relate the *F*-statistic directly to the R^2 statistic, showing that for fixed k and n, larger R^2 corresponds to larger *F*:

$$F = \{(n-k-1)/k\} \times R^2 / (1-R^2)$$

Self-study question: Why is the equation relating *F* to R^2 true?

The *F*-statistic is used to test the *global null hypothesis* $H_0 : \beta_1 = \beta_2 = \ldots = \beta_k = 0$, which states that *none of* the regression variables $X_1, X_2, \ldots,$ or X_k is related to *Y*. Under the classical model where $H_0 : \beta_1 = \beta_2 = \ldots = \beta_k = 0$ is true, the *F*-statistic has a precise and well-known distribution.

Distribution of the *F*-statistic under the classical model where $\beta_1 = \beta_2 = \ldots = \beta_k = 0$

$$F \sim F_{k,n-k-1},$$

where $F_{k,n-k-1}$ is the *F* distribution with k numerator degrees of freedom and $n-k-1$ denominator degrees of freedom.

Recall also that the degrees of freedom for "error," *dfe*, was given by $dfe = n - k - 1$; in the Car Sales data this value is $120 - 2 - 1 = 117$, as shown in the `lm` output above. The numerator degrees of freedom, k (= 2 in the example from above), is sometimes called the "model" degrees of freedom, hence symbolized as *dfm*, because it represents the flexibility (freedom) of the model. For example, the class of quadratic regression models, which have $dfm = 2$, is *more flexible* than the class of linear models, which have $dfm = 1$. We can say that the quadratic class is "more flexible" because quadratic models allow curvature, and because linear models lie within the class of quadratic models (where $\beta_2 = 0$).

When $H_0 : \beta_1 = \beta_2 = \ldots = \beta_k = 0$ is true, the theoretical R^2 statistic is exactly $\Omega^2 = 0$. And when H_0 is false, some of the *X* variables are related to *Y*, giving larger values of R^2, hence larger *F*-statistics. Unlike the *T*-test for testing individual regression coefficients discussed in Section 3.9, the *p*-value for testing $H_0 : \beta_1 = \beta_2 = \ldots = \beta_k = 0$ via the *F* test considers the extreme values of *F* to be *only the large values*, not both the large and the small ones. Smaller *F* values are expected under H_0; hence small values of *F* are not "extreme." The *p*-value is, therefore, the probability in the right tail only of the $F_{k,n-k-1}$ distribution.

To understand the F-statistic and the "chance only" model where $\beta_1 = \beta_2 = \ldots = \beta_k = 0$, you should use simulation. (As always!)

8.3.1 Simulation Study to Understand the *F* Statistic

```
CarS = read.table("https://raw.githubusercontent.com/andrea2719/
URA-DataSets/master/CarS.txt")
attach(CarS); X1 = INTRATE; X2 = X1^2
n = length(X1)
Y.sim =15 +0*X1 +0*X2 +rnorm(n,0,2) ## Notice the 0's: Null model is true
fit = lm(Y.sim ~ X1 + X2); summary(fit)
```

The code above generates data Y that is unrelated to either X_1 or X_2; in other words, the null hypothesis $H_0 : \beta_1 = \beta_2 = 0$ is in fact true. From the code above, we got $F = 1.415$ (yours will vary by randomness). To understand what is the range of possible F values that are explained by chance alone, you need to repeat this simulation many (ideally, infinitely many) times. We will now simulate many data sets under both the null and non-null (or *alternative*) models, save the F values, draw the null histogram and overlay the theoretically correct $F_{dfm,dfe}$ density, and then indicate how the non-null F values appear.

R code for Figure 8.3

```
Nsim = 10000
Fsim.null = numeric(Nsim)
Fsim.alt  = numeric(Nsim)

for (i in 1:Nsim) {
  Y0.sim =15 +0*X1 +0*X2 +rnorm(n,0,2) ## Null model is true
  Y1.sim  = 15 - 0.15*X1 + .003*X2 + rnorm(n,0,2) ##Non-null model
  fit.null = lm(Y0.sim ~ X1 + X2)
  fit.alt  = lm(Y1.sim  ~ X1 + X2)
  Fsim.null[i] = summary(fit.null)$fstatistic[1]
  Fsim.alt[i]  = summary(fit.alt)$fstatistic[1]
  }

par(mfrow=c(3,1))
hist(Fsim.null, breaks=100, freq=F, main="", xlab="F value")

hist(Fsim.null, breaks=100, freq=F, main="", xlab="F value")
flist = seq(0,15,.01)
fdist = df(flist,2,117)
crit = qf(.95,2,117)
points(flist, fdist, type="l")
abline(v=crit, col="gray")

hist(Fsim.null, breaks=100, freq=F, main="", xlab="F value")
points(flist, fdist, type="l")
abline(v=crit, col="gray")
hist(Fsim.alt,breaks=100, freq=F, add=T, lty=2, border="gray")
```

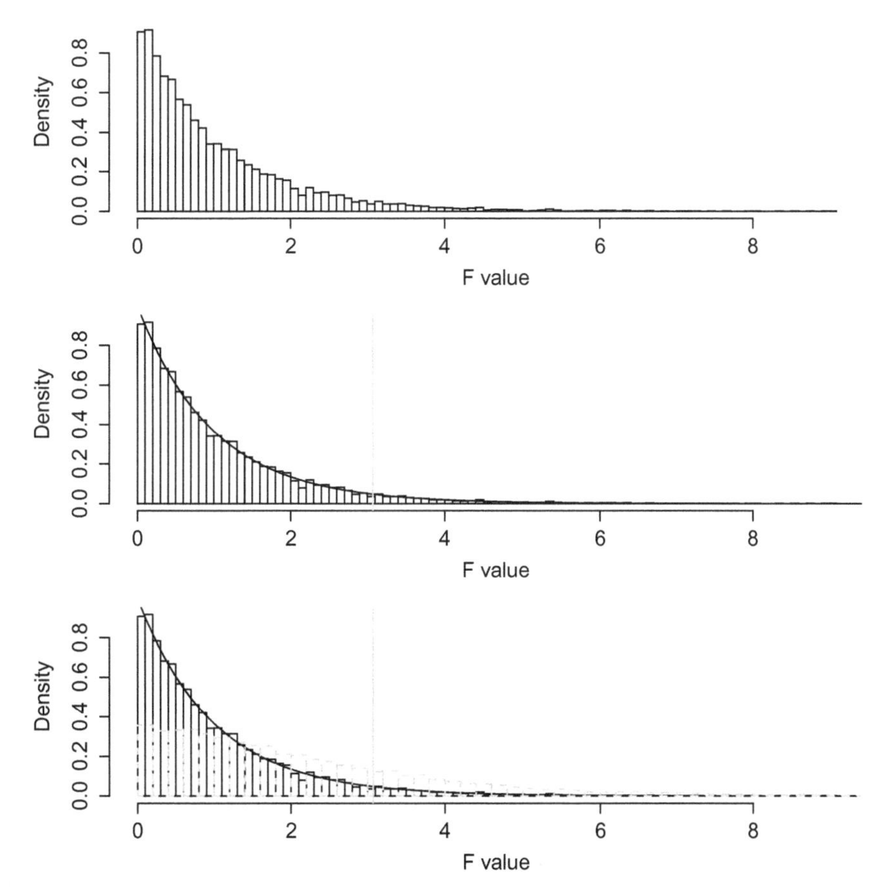

FIGURE 8.3
Top panel: Histogram of 10,000 simulated *F*-statistics under the null model. Middle panel: Same as top panel but with the theoretically correct $F_{2,117}$ distribution overlaid (solid curve), as well as its 0.95 quantile 3.074 (vertical line). Bottom panel: Same as middle panel, but with histogram of 10,000 simulated *F*-statistics under an alternative model superimposed (histogram in dotted lines).

The observed *F*-statistic from the original data was 62.22. As seen in Figure 8.3, this value is off the chart, and hence is nearly impossible to observe by chance alone when $\beta_1 = \beta_2 = 0$. The *p*-value is calculated from the solid $F_{2,117}$ curve shown in the middle and bottom panels of Figure 8.3; it is the area under that curve beyond 66.22, and is calculated in `R` as `1-pf(66.48768,2, 117)`, giving the value 0. But the `lm` output shows "`p-value: < 2.2e-16`," which is actually more correct than "*p*-value = 0." A true *p*-value can never be mathematically zero, but if the probability is infinitesimally small, then the computer might report it as 0 due to computer limitations. For example, the output "`p-value: < 2.2e-16`" indicates that the computer cannot accurately calculate and/or display probabilities that are smaller than 2.2×10^{-16}.

Because an *F* value such as 62.22 is nearly impossible under the model where $\beta_1 = \beta_2 = 0$, it is logical to conclude that ($\beta_1 = \beta_2 = 0$) is not true. In other words, it is logical to conclude that either $\beta_1 \neq 0$, or $\beta_2 \neq 0$, or that both β_1 and β_2 differ from 0. Be careful, though: The *F* test is not specific. A large *F* value does *not* tell you that *both* parameters differ from zero, nor can it identify *which* parameter differs from zero. It can only tell you that *at least one* parameter (either β_1 or β_2) differs from 0.

8.4 Multicollinearity

Multicollinearity (MC) refers to the X variables being "collinear" to varying degrees. In the case of two X variables, X_1 and X_2, collinearity means that the two variables are close to linearly related. A "perfect" multicollinearity means that they are perfectly linearly related. See Figure 8.4.

Often, "multicollinearity" with just two X variables is called simply "collinearity." Figure 8.4, right panel, illustrates the meaning of the term "collinear."

With more X variables, it is not so easy to visualize multicollinearity. But if one of the X variables, say X_j, is closely related to all the other X variables via

$$X_j \cong a_0X_0 + a_1X_1 + \ldots + a_{j-1}X_{j-1} + a_{j+1}X_{j+1} + \ldots + a_kX_k$$

then there is multicollinearity. And if the "$\cong$" is, in fact, an "=" in the equation above, then there is a *perfect* multicollinearity. (Note that the variable X_0 is the intercept column having all 1's).

A perfect multicollinearity causes the $\mathbf{X}^{\mathrm{T}}\mathbf{X}$ matrix to be non-invertible, implying that there are no unique least-squares estimates. Equations 0 through k shown in Section 7.1 can still be solved for estimates of the β's, but some equation or equations will be redundant with others, implying that there are infinitely many solutions for $\hat{\beta}_0, \hat{\beta}_1, \ldots,$ and $\hat{\beta}_k$. Thus the effects of the individual X_j variables on Y are not identifiable when there is a perfect multicollinearity.

To understand the notion that there can be an infinity of solutions for the estimated β's, consider the case where there is only one X variable. A perfect multicollinearity, in this case, means that $X_1 = a_0X_0$, so that the X_1 column is all the same number, a_0. Figure 8.5 shows how data might look in this case, where $x_i = 10$ for every $i = 1, \ldots, n$, and also shows several possible least-squares fits, all of which have the same minimum sum of squared errors.

A similar phenomenon happens with the case of two X variables as shown in the right panel of Figure 8.4: There are an infinity of three-dimensional planar functions (review

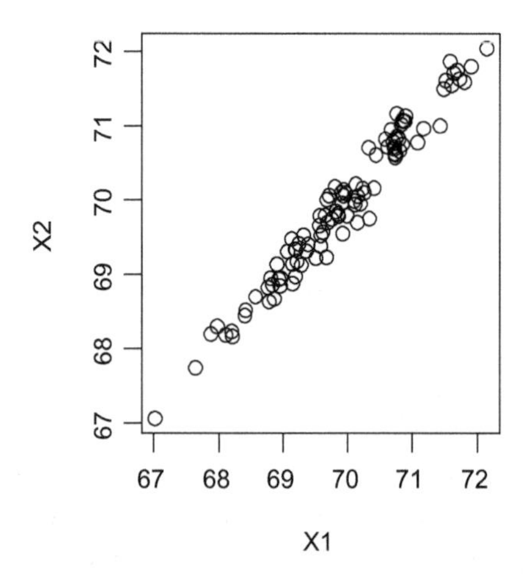

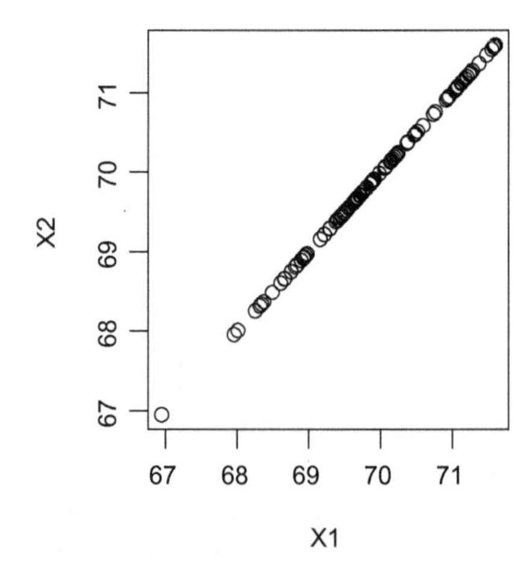

FIGURE 8.4
Left panel: Collinear X variables having correlation 0.976. Right panel: Perfectly collinear X variables having correlation 1.0.

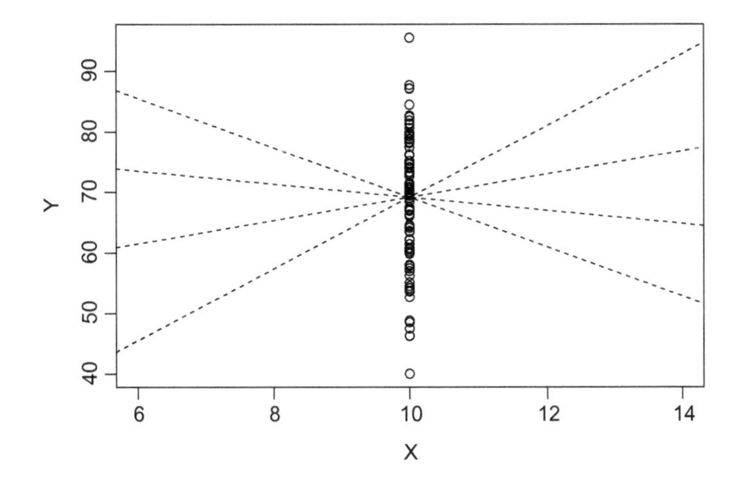

FIGURE 8.5
Scatterplot of data where $x_i = 10$ for all $i = 1,\ldots,n$. There are infinitely many least-squares lines, some of which are shown as dashed lines, all of which provide the minimum SSE. Here, the X column of data is $X_1 = (10,10,\ldots,10)$, which is perfectly collinear with the intercept column $X_0 = (1,1,\ldots,1)$.

Figure 7.1 in Chapter 7) of X_1 and X_2 that minimize the SSE, all of which planes contain the two-dimensional line shown in the right panel of Figure 8.4.

Using R, you will get one of these infinitely many estimated planar functions. For example, the R code below generates perfectly collinear (X_1, X_2) data, then generates Y data from these X data that satisfy all regression assumptions.

```
set.seed(12345)
X1 = rnorm(100)
X2 = 2*X1 -1       # Perfect collinearity
Y = 1 + 2*X1 + 3*X2 + rnorm(100,0,1)
summary(lm(Y~X1+X2))
```

This code produces the following output:

```
Call:
lm(formula = Y ~ X1 + X2)

Residuals:
     Min       1Q   Median       3Q      Max
-2.20347 -0.60278 -0.01114  0.61898  2.60970

Coefficients: (1 not defined because of singularities)
            Estimate Std. Error t value Pr(>|t|)
(Intercept) -1.97795    0.10353  -19.11   <2e-16 ***
X1           8.09454    0.09114   88.82   <2e-16 ***
X2                NA         NA      NA       NA
---
Signif. codes:  0 '***' 0.001 '**' 0.01 '*' 0.05 '.' 0.1 ' ' 1

Residual standard error: 1.011 on 98 degrees of freedom
Multiple R-squared:  0.9877,    Adjusted R-squared:  0.9876
 F-statistic:  7888 on 1 and 98 DF,  p-value: < 2.2e-16
```

Notice the "NA" for the coefficient of `X2`. Recognizing that $\mathbf{X}^T\mathbf{X}$ is not invertible, and hence that there are infinitely many solutions for the estimated β's, the `lm` function in `R` simply assigned $\hat{\beta}_2 = 0$ and estimated β_1. Note also the comment "`Coefficients: (1 not defined because of singularities)`." In matrix algebra, "singular" means "not invertible"; the comment lets you know that `R` recognizes that $\mathbf{X}^T\mathbf{X}$ is not invertible.

To visualize the infinity of solutions for the regression plane in the example above, have a look at the 3-D representation of the data just simulated in Figure 8.6 below. In that graph, there are infinitely many planes that will separate the positive and negative residuals as shown—some are steeper on one side of the vertical sheet where the data lie, some are steeper on the other side of the sheet.

For yet another way to understand that you cannot estimate the coefficients uniquely when there is perfect multicollinearity, recall that β_2 is the difference between the means of the distributions of potentially observable Y values in two groups:

Group 1: $X_1 = x_1, \;\; X_2 = x_2$

Group 2: $X_1 = x_1, \;\; X_2 = x_2 + 1$

However, if X_2 is perfectly related to X_1, it is impossible to increase X_2 while leaving X_1 fixed. This fact provides an intuitive explanation for why it is impossible to estimate the effect of larger X_2 when X_1 is held fixed.

The intuitive logic that you cannot estimate the effect of increasing X_2 while X_1 is held constant for the case of perfectly collinear X variables, also explains the problem with near-perfect collinearity, as shown in the left panel of Figure 8.4. Since the data are so closely related, there is very little variation in X_2 when you fix X_1, say, by drawing a vertical line over any particular value of X_1. Recall also that, to estimate the effect of an X variable on Y, you need variation in that X variable. The relevant variation in the case of multiple

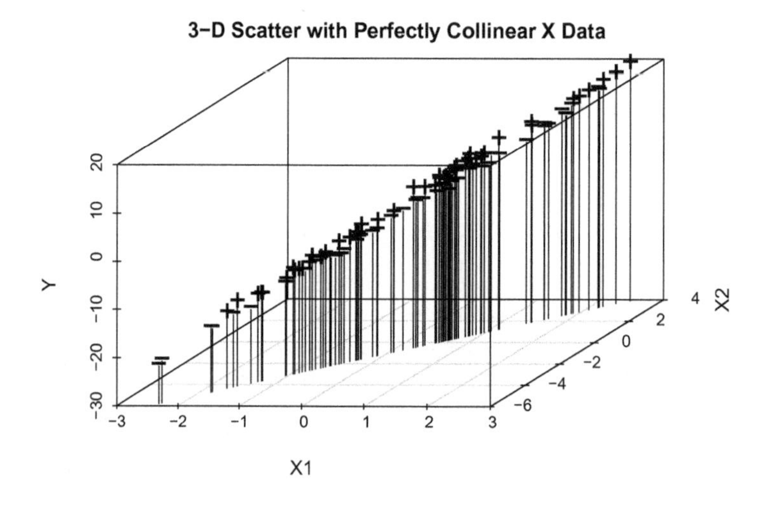

FIGURE 8.6
Three-D scatterplot of data where the X variables are perfectly collinear. The data lie in a vertical sheet above the line of collinearity on the (X1, X2) plane. There are infinitely many planes, all having the same minimum SSE, for which the points indicated "+" points are above, and the points indicated by "−" are below.

regression, where you are estimating the effect of an X variable holding the other variables fixed, is exactly the variation in that X variable where the other variables are fixed. If there is little such variation, as shown in the left panel of Figure 8.4, you will get unique estimates of the β's, but they will be relatively imprecise estimates because, again, there is so little relevant variation in the X data.

Therefore, the *main problem* with multicollinearity is that the estimates of the β's are relatively (relative to the case where the X variables are unrelated) imprecisely estimated. This imprecision manifests itself in higher standard errors of the estimated β's.

There is a simple formula to explain how multicollinearity affects the standard errors of the estimated β's: Recall from Chapter 7, Section 3, that

$$\text{s.e.}(\hat{\beta}_j) = \hat{\sigma}\sqrt{c_{jj}},\ j = 0,1,\ldots,k$$

In simple regression, where there is just one X variable, this expression reduces to the form you saw in Chapter 3,

$$\text{s.e.}(\hat{\beta}_1) = \frac{\hat{\sigma}}{s_x\sqrt{n-1}}$$

Some fairly complicated matrix algebra gives the following representation of the standard errors for the multiple regression case:

$$\text{s.e.}(\hat{\beta}_j) = \hat{\sigma}\sqrt{c_{jj}} = \frac{\hat{\sigma}}{s_{x_j}\sqrt{n-1}} \times \left(\frac{1}{1-R_j^2}\right)^{1/2}$$

Here, R_j^2 is the R-squared statistic that you get by regressing X_j on *all other* X variables. Higher R_j^2 is an indication of more extreme multicollinearity, which implies less precision in the estimate $\hat{\beta}_j$.

Two important special cases are (1) $R_j^2 = 0$, in which case the standard error formula for $\hat{\beta}_j$ is exactly as given in the simple regression where there is only one X variable, and (2) $R_j^2 \to 1$, in which case the standard error tends to infinity. Such behavior is expected because when X_j is increasingly related to the other X variables, there is less and less relevant variation in X_j when all other X variables are held fixed.

The term $1/(1-R_j^2)$ is called the *variance inflation factor* (VIF) because it measures how much larger is the variance of $\hat{\beta}_j$ due to multicollinearity. By the same token, $\{1/(1-R_j^2)\}^{1/2}$ can be called a *standard error inflation factor.*

8.4.1 The Effects of Multicollinearity on the *T* Statistics

This simulated data set shows what happens with highly MC X data. Note that the model has an F-statistic that is far in the right tail of the distribution (p-value = 2.306×10^{-8}), but neither X variable has a large T statistic: The MC between X_1 and X_2 makes it difficult to assess the effect of X_1 on Y when X_2 is held fixed, and vice versa.

```
set.seed(12345)
x1 = rep(1:10, each=10)
x2 = x1 + rnorm(100, 0, .05)   # X2 differs only slightly from X1
```

```
## You can see the collinearity in the graph:
plot(x1,x2)

## The true model has beta0 = 7, beta1=1, and beta2 = 1,
## with all assumptions satisfied.
y = 7 + x1 + x2 + rnorm(100,0,10)
dat.high.mc = data.frame(y,x1,x2)

high.mc = lm( y ~ x1 + x2, data = dat.high.mc)
summary(high.mc)
```

The output is as follows:

```
Call:
lm(formula = y ~ x1 + x2, data = dat.high.mc)

Residuals:
     Min       1Q   Median       3Q      Max
-21.4159  -5.9119   0.0834   5.5034  26.4582

Coefficients:
            Estimate Std. Error t value Pr(>|t|)
(Intercept)    5.909      2.191   2.697  0.00825 **
x1           -16.712     18.362  -0.910  0.36500
x2            18.953     18.332   1.034  0.30376
---
Signif. codes:  0 '***' 0.001 '**' 0.01 '*' 0.05 '.' 0.1 ' ' 1

Residual standard error: 10.14 on 97 degrees of freedom
Multiple R-squared:  0.3041,    Adjusted R-squared:  0.2898
F-statistic:  21.2 on 2 and 97 DF,  p-value: 2.306e-08
```

In the example above, the R_j^2 statistics, obtained via `summary(lm(X1~X2))` and `summary (lm(X2~X1))`, are 0.9996306 and 0.9996306, respectively, implying standard error inflation factors $1/(1-0.9996306)^{1/2} = 52.02973$. Thus, the standard errors, 18.362 and 18.332, are 52.02973 times larger than they would have been had the variables been uncorrelated. A slight modification of the simulation model to keep all the same (same n, same σ, nearly the same variances of X_1 and X_2) except with uncorrelated X variables verifies that result.

```
set.seed(12345)
x1 = rep(1:10, each=10)
x2 = rep(1:10, 10)

## You can see the lack of collinearity in the graph:
plot(x1,x2)

## The true model has beta0 = 7, beta1=1, and beta2 = 1,
## with all assumptions satisfied.
y = 7 + x1 + x2 + rnorm(100,0,10)
dat.no.mc = data.frame(y,x1,x2)
```

```
no.mc = lm( y ~ x1 + x2, data = dat.no.mc)
summary(no.mc)
```

The output is as follows:

```
Call:
lm(formula = y ~ x1 + x2, data = dat.no.mc)

Residuals:
    Min      1Q  Median      3Q     Max
-27.296  -8.823   2.274   6.425  20.775

Coefficients:
            Estimate Std. Error t value Pr(>|t|)
(Intercept)   9.6931     3.2288   3.002  0.00341 **
x1            1.2958     0.3894   3.328  0.00124 **
x2            0.6604     0.3894   1.696  0.09311 .
---
Signif. codes:  0 '***' 0.001 '**' 0.01 '*' 0.05 '.' 0.1 ' ' 1

Residual standard error: 11.18 on 97 degrees of freedom
Multiple R-squared:  0.1257,    Adjusted R-squared:  0.1077
F-statistic: 6.974 on 2 and 97 DF,  p-value: 0.001479
```

The R^2 statistics relating X_1 to X_2 and X_2 to X_1 are both 0.0 in this second example. The standard error multiplier in the first example with the high multicollinearity was 52.02973; checking, we see that $52.02973 \times 0.3894 = 20.26$, reasonably close to the standard errors 18.36 and 18.33 in the original analysis with multicollinear variables. Differences are explained by randomness in the simulations.

Regression models that are estimated using MC data can still be useful. There is no absolute requirement that MC be below a certain level. In fact, in some cases it is strongly recommended that highly correlated variables be retained in the model. For example, in many cases you should include the linear term in a quadratic model, even though the linear and quadratic terms are highly correlated. This is called the "Variable Inclusion Principle"; more on this in the next chapter. It is most important that you simply recognize the effects of multicollinearity, which are (i) high variances of parameter estimates, (ii) tenuous parameter interpretations, and (iii) in the extreme case of perfect multicollinearity, non-existence of unique least-squares estimates.

It makes no sense to "test" for MC in the usual hypothesis testing "null vs. alternative hypothesis" sense. But when correlations between the X variables are extremely high (e.g., many greater than 0.9) or variance inflation factors are very high (e.g., greater than 9.0; implying a standard error inflation factor greater than 3.0), you might consider MC to be problematic.

When you have extreme multicollinearity, here is a list of possible actions you might take. But first, please realize the following statement:

You do not necessarily have to do anything at all about multicollinearity.

But in all cases, you should diagnose the degree of multicollinearity among the X variables. You should also recognize the effects of MC, which are (i) large standard errors, (ii) tenuous parameter interpretations, and (iii) non-existence of unique OLS estimates in the case of perfect MC.

8.4.2 Possible Actions to Take with Multicollinear *X* Variables

1. In some cases, you might avoid using MC variables using the following strategies:
   a. Drop less important and/or redundant X variables.
   b. Combine X variables into an index. For example, if X_1, X_2 and X_3 are all measuring the same thing, then you might use their sum or average in the model in place of the original three X variables.
   c. Use principal components to reduce the dimensionality of the X variables (this is discussed in Multivariate Analysis courses).
   d. Use "common factors" (or latent variables), to represent the correlated X variables, and fit a structural equations model relating the response Y to these common factors (also discussed in Multivariate Analysis courses).
   e. Use ratios in "size-related" cases. For example, if you have the two firm-level variables X_1 = Total Sales and X_2 = Total Assets in your model, they are bound to be highly correlated. So you might use the two variables X_1 = (Total Assets)/(Total Sales) and X_2 = (Total Sales) (perhaps in log form) in your model instead of the two variables (Total Sales) and (Total Assets).
2. In some cases, you *must leave* multicollinear variables in the model. These cases include
   a. Predictive Multicollinearity: Two variables can be highly correlated, but both are essential for predicting Y. When you leave one or the other out of the model, you get a much poorer model.
   b. Variable Inclusion Principles: Whenever you include higher-order terms in a model, you should also include the implied lower order terms. For example, if you include X^2 in the model, then you should also include X. But X and X^2 are highly correlated. Nevertheless, both X and X^2 should be used in model, despite the fact that they are highly correlated, for reasons we will give in the next chapter.
   c. Research Hypotheses: Your main research hypothesis is to assess the effect of X_1, but you recognize that the effect of X_1 on Y might be confounded by X_2. If this is the case, you are simply stuck with including both X_1 and X_2 in your model.
3. Either redesign your study or collect more data to reduce MC.
   a. If you have the opportunity to select the (X_1, X_2) values, then you should attempt to do so in a way that makes those variables as uncorrelated as possible. For example, (X_1, X_2) might refer to two process inputs, each either "Low" or "High," and you should select them in the arrangement (L,L), (L,H), (H,L), (H,H), with equal numbers of runs at each combination, to ensure that X_1 and X_2 are uncorrelated.
   b. The main problem resulting from MC is that the standard errors are large. You can make standard errors smaller by collecting a larger sample size n.

Exercises

1. See the "by hand" calculation in Chapter 8 right before Section 8.3. Perform a similar analysis using the "Production Cost" data set to verify that the "by hand" calculations of R^2 and adjusted R^2 agree precisely with those in the lm output. Here is the data set:

```
ProdC = read.table("https://raw.githubusercontent.com/andrea2719/
URA-DataSets/master/ProdC.txt")
```

2. What happens to the distribution of the *F*-statistic when the β's become closer to zero? The (intuitive) answer: The distribution gets closer to the distribution of the *F*-statistic when the β's are equal to zero. Change the "R Code for Figure 8.3" to demonstrate this concept. Use graphs based on the bottom panel only. Be sure to explain how your graphs show the concept.
3.
   a. Perform a simulation study with one *Y* and one *X* where the *p*-value for testing $H_0 : \beta_1 = 0$ is greater than 0.05. In your code, pick a β_1 that is not equal to 0. Use "`set.seed`" so that your results are replicable.
   b. For the simulation data of 3.a, show (to as many decimals as you can get from the computer; do not copy-and-paste) that the *p*-values for the *F* test and for the *T* test are identical.
   c. Modify the β_1 of your code of 3.a, using the same `set.seed` and all else, to the point where the *p*-value is between 0.001 and 0.05, and repeat 3.b.
4. Often, multicollinearity can be reduced through appropriate transformations. Consider the following analysis.

```
census = read.csv("https://raw.githubusercontent.com/andrea2719/
URA-DataSets/master/census.csv")
attach(census)
crime.rate = nCrimes/pop

# Original analysis
fit1= lm(crime.rate ~ pop + nDoctors + nBeds + civil + income + area)
library(car)
vif(fit1)
```

   a. Calculate the VIF's "by hand" in terms of R^2 statistics and verify that the results are as given by the code.
   b. Using those R^2 statistics, explain what each of the VIFs tells you about the relationships between the variables.
   c. Repeat the analysis of 4.b after making the transformations indicated in the following code.

```
# Revised analysis
pop.log = log(pop)
docs.per = nDoctors/pop
beds.per = nBeds/pop
civ.per = civil/pop
inc.per = income/pop
area.log = log(area)

fit2= lm(crime.rate ~ pop.log + docs.per + beds.per + civ.per + inc.per
+ area.log)
vif(fit2)
```

9

Polynomial Models and Interaction (Moderator) Analysis

In Chapter 4, we introduced the quadratic model as a device to test for curvature in the conditional mean function. You could also use this model as a *bona fide* model to predict Y, although we warned you that it can have poor properties at the extremes of the X data. On the other hand, a benefit of using the quadratic model is that it provides a simple estimate of the x that gives the highest or lowest value of $E(Y \mid X = x)$.

When extended to multiple regression, quadratic models give you a conditional mean function that is a *response surface*, rather than a rigid plane as shown in Figure 6.3. Such models are useful because they are more flexible than rigid linear and planar functions, and are therefore more realistic. Nature does not particularly care about rigid linear and planar functions; Nature will put the conditional means wherever Nature wants to. Thus, models with greater flexibility, such as quadratic and other polynomial models, will better approximate Nature's whims.

The general polynomial function of one X variable is

$$f(x) = \beta_0 + \beta_1 x + \beta_2 x^2 + \beta_3 x^3 + \ldots + \beta_d x^d$$

The value d is called the *degree* of the polynomial. In most cases, you should choose $d = 2$ (no higher degree than quadratic), although you occasionally will need $d = 3$ (cubic) and higher-order polynomial models because the quadratic model is not flexible enough. For example, in Moreau et al. 2005, the authors developed and validated a *cubic* function of a person's age to predict their risk of death after suffering severe burn injury. They showed that the quadratic function simply was not flexible enough to accurately model the effect of age on mortality risk, especially when considering the wide age spectrum from infants to elderly patients, while a cubic model gave sufficient additional flexibility.

The general polynomial model of a single X variable extends to multiple X variables as well, but the additional wrinkle is that there are *interaction* terms in the model. The most general second-degree polynomial function in two variables, $X_1 = x_1$ and $X_2 = x_2$, is given by

$$f(x_1, x_2) = \beta_0 + \beta_1 x_1 + \beta_2 x_1^2 + \beta_3 x_2 + \beta_4 x_2^2 + \beta_5 x_1 x_2.$$

In this expression, the β terms are constants to be estimated using the data. The term "$x_1 x_2$" is called the "interaction term." The *degree* of a polynomial function in multiple X variables is either the *maximum exponent* or the *maximum sum of exponents in an interaction term,* whichever is larger. Thus, there is no term like $x_1^2 x_2$ in a second-degree polynomial function: The degree of the polynomial term $x_1^2 x_2 = x_1^2 x_2^1$ is $2 + 1 = 3$, so if that term were included in $f(\cdot)$, then $f(\cdot)$ would be a *third-degree* polynomial function.

The classical interaction model, used for "moderator analysis," where the effect of a variable X_1 on Y depends on the value of (or is moderated by) another variable X_2, is just the special case of the second-degree polynomial function that excludes the pure quadratic terms:

$$f(x_1, x_2) = \beta_0 + \beta_1 x_1 + \beta_2 x_2 + \beta_3 x_1 x_2.$$

Polynomial functions (including the interaction model, or moderation model) are discussed and applied in this chapter.

9.1 The Quadratic Model in One *X* Variable

The simplest of polynomial models is the simple quadratic model,

$$f(x) = \beta_0 + \beta_1 x + \beta_2 x^2$$

These models are quite flexible, see Figure 9.1 for various examples.

R code for Figure 9.1

```
x = seq(0.2, 10, .1)
EY1 = -3 + 0.3*x + 0.1*x^2; EY2 =  2 - 0.9*x + 0.3*x^2
EY3 = -1 + 3.0*x - 0.4*x^2; EY4 =  1 + 1.2*x - 0.1*x^2
plot(x, EY1, type="l", lty=1, ylab = "E(Y | X=x)")
points(x, EY2, type="l", lty=2); points(x, EY3, type="l", lty=3)
points(x, EY4, type="l", lty=4)
legend(0,10, c("b0     b1     b2 ", "-3.0    0.3    0.1","
  2.0    -.9    0.3", "-1.0   3.0    -.4", " 1.0    1.2    -.1"), lty =
  c(0,1,2, 3,4))
```

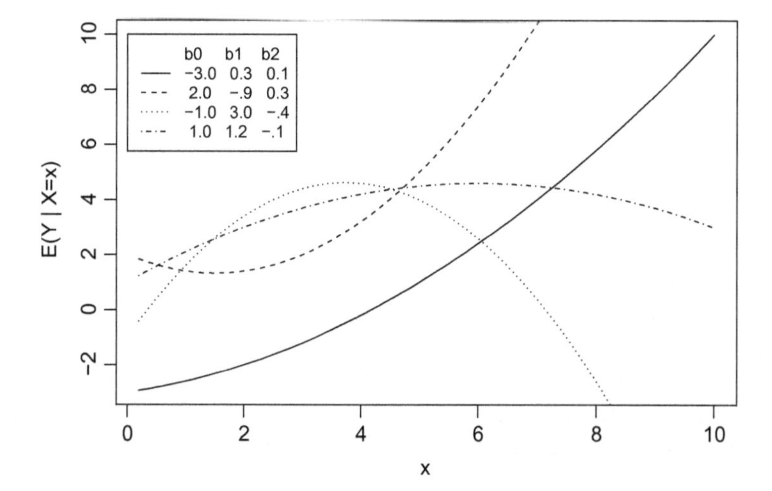

FIGURE 9.1
Quadratic functions $\beta_0 + \beta_1 x + \beta_2 x^2$, using four different settings of $(\beta_0, \beta_1, \beta_2)$ as shown in the legend.

As it is the case for all models in this chapter, the "β" coefficients cannot be interpreted in the way discussed in Chapter 8, where you increase the value of one X variable while keeping the others fixed, because there are functional relationships among the various terms in the model. Specifically, in the example of a quadratic polynomial function, you cannot increase x^2, while keeping x fixed. But you can still interpret the parameters by understanding the graphs in Figure 9.1. In particular, β_2 measures curvature: When $\beta_2 < 0$, there is concave curvature, when $\beta_2 > 0$ there is convex curvature, and when $\beta_2 = 0$, there is no curvature. Further, the larger the $|\beta_2|$, the more extreme is the curvature.

The intercept term β_0 has the same meaning as before: It is the value of $f(x)$ when $x = 0$. This interpretation is correct but, as always, it is a not useful interpretation when the range of the x data does not cover 0. Still, the coefficient is needed in the model as a "fitting constant," which adjusts the function up or down as needed to match the observable data.

To interpret β_1, note that it *is* possible to increase x by 1 when x^2 is fixed, but the only way that can happen is when you move from $x = -0.5$ to $x = +0.5$. Consider the solid graph shown in Figure 9.1: Here, $f(x) = -3 + 0.3x + 0.1x^2$, so that $f(-0.5) = -3 + 0.3(-0.5) + 0.1(-0.5)^2 = -3.125$, and $f(+0.5) = -2.825$; these values differ by exactly 0.3, the coefficient β_1 that multiplies x. While this math gives a correct way to interpret β_1 in the quadratic model, it is not useful if the range of the X data does not cover zero.

Incorrect interpretations of the linear term β_1 include "β_1 is the effect of x by itself" and "β_1 is the degree of linearity." The former is wrong because you cannot isolate the effect of x from x^2 in the quadratic model; again, because they are functionally related. The latter interpretation is wrong because the degree of linearity (or lack of curvature) is measured by β_2, not β_1.

Like the intercept term β_0, there is usually no point in testing the hypothesis that the coefficient β_1 of the linear term in a quadratic function is equal to zero because the parameter by itself has little meaning. Nevertheless, it is necessary to keep it in the model, regardless of its "significance," or lack thereof (as determined by its p-value). See the "Variable Inclusion Principle" discussed in Section 9.4 of this chapter.

One important application of β_1 in the quadratic model is its use in identifying the x value that gives the maximum or minimum of $f(x)$. Using calculus, $f'(x) = \beta_1 + 2\beta_2 x = 0$ implies that $x = -\beta_1/(2\beta_2)$ is the x value that maximizes or minimizes the function. For the quadratic functions shown in Figure 9.1, the x values maximizing (or minimizing) each of the four curves are, respectively, –1.5 (the minimum occurs outside the range of the graph for the solid curve), 1.5, 3.75, and 6.0. Thus, it is best to think of β_1 as simply a parameter that, along with β_2, governs the shape of the quadratic function, and that can be used to identify the optimal x location.

For example, have a look at Figure 1.16 again. The LOESS fit shows curvature that might be modeled well as a quadratic function. Such a quadratic function will give you a simple estimate of the X value (product complexity) that maximizes average consumer preference (Y); namely, $x_{\max} = -\hat{\beta}_1/(2\hat{\beta}_2)$. This value is identified and graphed using the following code.

R code for Figure 9.2

```
cp = read.table("https://raw.githubusercontent.com/andrea2719/
URA-DataSets/master/complex.txt")
attach(cp)
plot(Complex, jitter(Pref), ylab="Consumer Preference",
   xlab="Product Complexity")

Comp2 = Complex^2
fit = lm(Pref ~ Complex + Comp2)
b0 = fit$coefficients[1]; b1 = fit$coefficients[2];
  b2 = fit$coefficients[3]

comp.seq = seq(0,40,.1)
yhat = b0 + b1*comp.seq + b2*comp.seq^2
points(comp.seq, yhat, type="l", col= "gray", lwd=2)
Comp.opt = -b1/(2*b2); Comp.opt
abline(v = Comp.opt, col="gray", lwd=2)
```

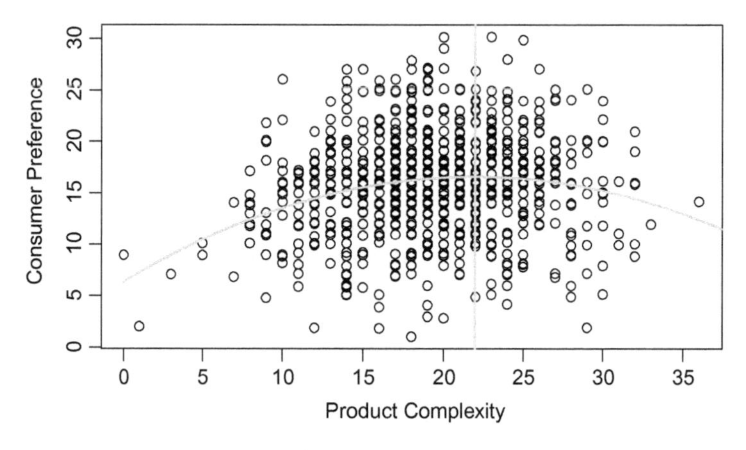

FIGURE 9.2
Product complexity data, with fitted quadratic function (curve) and location of optimal value of Product Complexity (vertical line at Complexity = 21.96809).

The fitted quadratic function is $\hat{y} = 6.341194 + 0.931838x - 0.021209x^2$, where Y = Consumer Preference and X = Product Complexity. The fact that the coefficient of the quadratic term is negative ($\hat{\beta}_2 = -0.021209$) implies that the fitted function is concave; this is seen clearly in Figure 9.2. The coefficient of the linear term, $\hat{\beta}_1 = 0.931838$, can be interpreted as an estimate of the mean preference for cases where Complexity = 0.5, minus the mean preference for cases where Complexity = −0.5, but, since the Complexity range does not cover 0, this interpretation is not useful. A more useful interpretation of $\hat{\beta}_1$ is that, along with $\hat{\beta}_2$, it gives you an estimate of the Product Complexity that yields maximum average Preference: $\hat{x}_{opt} = -\hat{\beta}_1/(2\hat{\beta}_2) = 21.96809$, also shown in Figure 9.2. Of course, 21.96809 is just an estimate of the true value $x_{opt} = -\beta_1/(2\beta_2)$; Bayesian methods based on drawing simulations from the posterior distribution of $(\beta_0, \beta_1, \beta_2, \sigma)$ provide a very simple way to construct an interval range (called a *credible interval* in the Bayesian parlance) of plausible values for x_{opt}.

9.2 The Quadratic Model in Two or More *X* Variables

Just as it is the case that quadratic functions in a single *X* variable are more realistic than rigid linear functions, quadratic functions of several *X* variables are more realistic than rigid planar functions. Like linear functions are special cases of quadratic functions in a single variable, planar functions of multiple variables are special cases of quadratic functions of several variables. Thus, as is the case with simple regression, you can test the hypothesis that the response function is planar by fitting the general quadratic function and testing that the coefficients of non-linear terms are all simultaneously zero. You can do this by using an *F* test; more details on this particular *F* test are given in Chapter 10.

The general quadratic response surface as given in the introduction to this chapter is $f(x_1, x_2) = \beta_0 + \beta_1 x_1 + \beta_2 x_1^2 + \beta_3 x_2 + \beta_4 x_2^2 + \beta_5 x_1 x_2$. An example for particular choices of the coefficients β is shown in Figure 9.3. Notice that the response function is curved, not planar, and is, therefore, more realistic. The term "response surface" is sometimes used instead of "response function" in the case of two or more *X* variables; the "surface" term is explained by the appearance of the graph of the function in Figure 9.3.

In addition to modeling and testing for curvature in higher dimensional space, quadratic models are also useful for identifying an optimal *combination* of *X* values that maximizes or minimizes the response function; see the "`rsm`" package of `R` for more information. While quadratic models are more flexible (and therefore more realistic) than planar models, they can have poor extrapolation properties and are often less realistic than the similarly flexible, curved class of response surfaces known as *neural network regression models*. In Chapter 17, we compare polynomial regression models with neural network regression models.

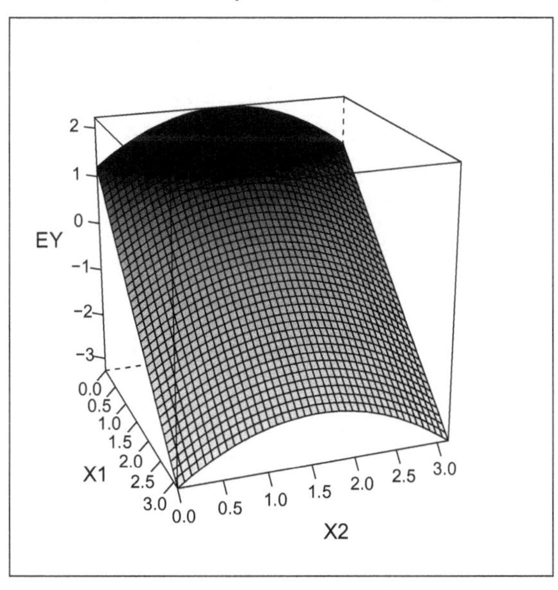

FIGURE 9.3
A quadratic response surface of the form $f(x_1, x_2) = \beta_0 + \beta_1 x_1 + \beta_2 x_1^2 + \beta_3 x_2 + \beta_4 x_2^2 + \beta_5 x_1 x_2$.

9.3 Interaction (or Moderator) Analysis

The commonly-used interaction model is a special case of the general quadratic model, involving the interaction term but no quadratic terms. When performing interaction analysis, you typically will assume the following conditional mean function:

$$E(Y \mid X_1 = x_1, X_2 = x_2) = \beta_0 + \beta_1 x_1 + \beta_2 x_2 + \beta_3 x_1 x_2$$

A slight modification of the Product Complexity example provides a case study in which interaction is needed. Suppose you measure Y = Intent to Purchase a Luxury Product, say expensive jewelry, using a survey of consumers. You also measure the attractiveness (X_1) of a web design used to display and promote the product, say measured in a scale from 1 to 10, with 10 = most attractive design, and the person's income (X_2) in a scale from 1 to 5, with 5 = most wealthy.

Figure 9.4 shows an example of how this conditional mean function might look. Like the quadratic response surface, it is a curved function in space, not a plane. But note in particular that the effect of X_1, Attractiveness of Web Design, on Y = Intent to Purchase, depends on the value of X_2, Income: For consumers with the lowest income, $X_2 = 1$, the slice of the surface corresponding to $X_2 = 1$ is nearly flat as a function of X_1 = Attractiveness of Web Design. That is to say, for people with the lowest income, Attractiveness of Web Design

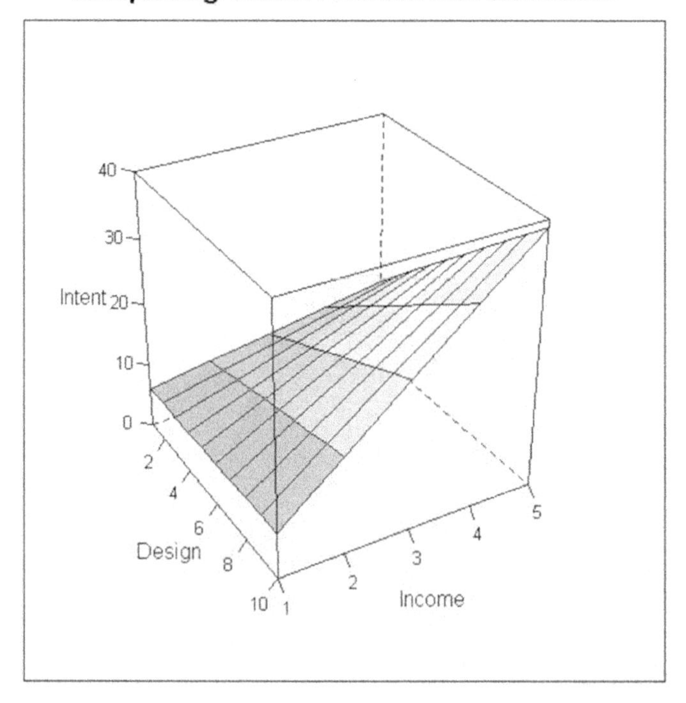

FIGURE 9.4
Conditional mean function showing interaction. The function is not a plane; rather, it is a twisted plane. Because of this twisting nature, the effect of Design on Intent is flat when Income = 1, but there is a large effect of Design on Intent when Income = 5.

has little effect on Intent to Purchase this luxury product. No surprise! They do not have enough money to purchase luxury items, so the web design is mostly irrelevant to them. On the other hand, for people with the highest income (X_2 = 5), the slice of the surface corresponding to $X_2 = 5$ increases substantially as a function of X_1 = Attractiveness of Web Design. Thus, this single model states both (i) that Attractiveness of Web Design (X_1) has little effect on Intention to Purchase a Luxury Product for people with little money, and (ii) that Attractiveness of Web Design (X_1) has a substantial effect on Intention to Purchase a Luxury Product for people with lots of money.

The interaction model is called a "moderation" model because it states that the effect of X_1 on Y is "moderated" by the value of X_2. If X_2 is low, then X_1 has one effect on Y. If X_2 is high, then X_1 has a different effect on Y.

The function shown in Figure 9.4 is $\mathrm{E}(Y \mid X_1 = x_1, X_2 = x_2) = 5 - 0.6x_1 + x_2 + 0.7x_1x_2$. The "slice" of this graph corresponding to low income is

$$\mathrm{E}(Y \mid X_1 = x_1, X_2 = 1) = 5 - 0.6x_1 + 1 + 0.7x_1(1) = 6 + 0.1x_1.$$

Thus the increase in E(Y) in this slice as x_1 ranges from 1 to 10 is 0.9, barely visible in the graph.

On the other hand, for consumers with high income, the "slice" of this graph is

$$\mathrm{E}(Y \mid X_1 = x_1, X_2 = 5) = 5 - 0.6x_1 + 5 + 0.7x_1(5) = 10 + 2.9x_1.$$

Thus, the increase in E(Y) in this slice as x_1 ranges from 1 to 10 is equal to 26.1, which is clearly visible in the graph.

9.3.1 Path Diagrams

Often, regression models are displayed as path diagrams. For example, the classical simple regression model where the mean of Y is a function of a single variable X_1 is often displayed in a path diagram as follows:

$$X_1 \longrightarrow Y$$

The classical multiple regression model where the mean of Y is a function of two variables, X_1 and X_2, with no interaction, is often displayed as follows:

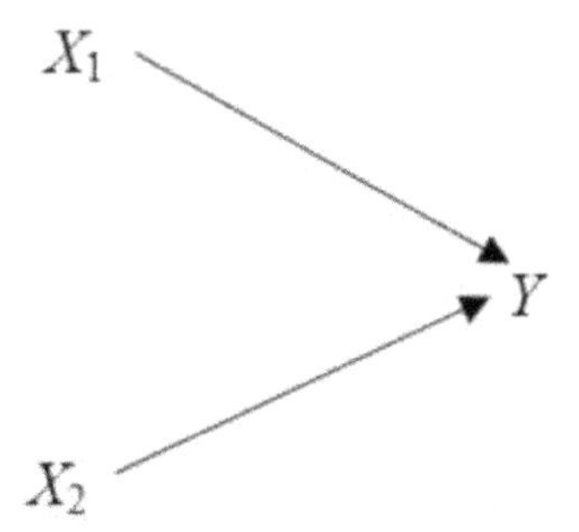

The arrows in the path diagram refer to the effects of the variable in question on Y, as measured by its coefficient. In the interaction model, the coefficient of the variable X_1 changes, depending on the value of the moderator variable X_2. In the example above, the coefficient of X_1 was 0.1 when $X_2 = 1$, and was 2.9 when $X_2 = 5$. Thus, moderator variables are often displayed in a path diagram as "having an effect on the effect" of the primary variable, like this:

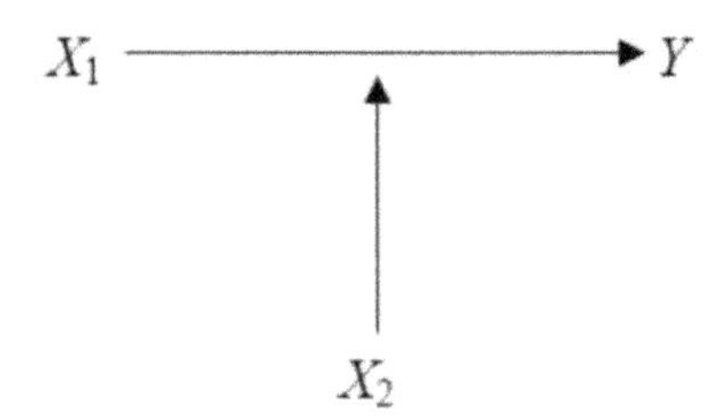

9.3.2 Parameter Interpretation in Interaction Models

Parameter interpretation in interaction models is complicated by the fact that the X variables in the model are functionally related to each other. Thus, the standard "increase X_1 by +1 while holding all other terms fixed" interpretation cannot be used, because you cannot increase X_1 by +1 while simultaneously keeping both X_2 and X_1X_2 fixed.

However, you can increase X_1 while holding X_2 fixed. Consider two groups of potentially observable data, one where $X_1 = x_1$ and $X_2 = x_2$, and the other where $X_1 = x_1 + 1$ and $X_2 = x_2$. The interpretation of the parameters then follows logically from the model, as in all other cases you have seen (and will see): *If* the no interaction model is truly the data-generating process, *then* the following conclusions are mathematically true.

Group 1: $X_1 = x_1$, $X_2 = x_2$: Here, $\mathrm{E}(Y \mid X_1 = x_1, X_2 = x_2) = \beta_0 + \beta_1 x_1 + \beta_2 x_2 + \beta_3 x_1 x_2$
Group 2: $X_1 = x_1 + 1$, $X_2 = x_2$: Here,

$$\begin{aligned}\mathrm{E}(Y \mid X_1 = x_1, X_2 = x_2) &= \beta_0 + \beta_1(x_1 + 1) + \beta_2 x_2 + \beta_3(x_1 + 1)x_2 \\ &= \beta_0 + \beta_1 x_1 + \beta_2 x_2 + \beta_3 x_1 x_2 + (\beta_1 + \beta_3 x_2) \\ &= (\text{Group 1 mean}) + (\beta_1 + \beta_3 x_2)\end{aligned}$$

Thus, in the interaction model, the coefficient β_1 does *not* measure the effect of increasing X_1 when X_2 is held constant because there is a different effect for every possible $X_2 = x_2$ (namely, $\beta_1 + \beta_3 x_2$). You can interpret β_1 as the increase in E(Y) per unit increase in X_1 when $X_2 = 0$, but now we are back to the interpretation that is practically useless when the X_2 data do not cover 0.0. Still, just like the intercept term β_0 is necessary in regression, regardless of its interpretability or "significance," the β_1 term is necessary in the interaction model, regardless of its interpretability or "significance." See Section 9.4 of this chapters for elaboration on this concept.

A clearly incorrect interpretation of β_1 that is sometimes given is "the effect of X_1 by itself," which is clearly wrong because the X_1 and X_2 must be considered simultaneously in this model. As just shown, β_1 is only the effect of X_1 when $X_2 = 0$ in the interaction model.

Like the intercept β_0, the parameters β_1 and β_2 in the interaction model are simply "necessary terms," and need not be interpreted further. Instead, you can interpret the model by first identifying two cohorts, one with a "low value" and the other with a "high value" of the moderator variable. (As in the interpretation of the true R^2 statistic given by Figure 8.1, "low" and "high" can refer to values that are a standard deviation from the mean.) Then you can calculate and graph the resulting estimated mean functions. Finally, you can interpret the model using those graphs and equations in terms of the subject matter, just as we did above for the made-up example showing how Income moderates the effect of Attractiveness of Web Page on Intent to Purchase a Luxury Item.

Here is a real, published example that illustrates the preferred method of analyzing interaction (moderating) effects.

9.3.3 Effect of Misanthropy on Support for Animal Rights: The Moderating Effect of Idealism

The paper "Misanthropy, idealism, and attitudes toward animals," (Wuensch et al. 2002) develops theory and analyzes data showing that animal rights support is related to misanthropy for non-idealistic students, but that animal rights support is not strongly related to misanthropy for idealist students. They developed and validated a survey instrument to measure three variables, Y = Support for Animal Rights, X_1 = Misanthropy, and X_2 = Idealism, as averages of responses (typically in the 1,2,3,4,5 scale) to three sets of questions on the survey. As averages, these data are more continuous than the 1,2,3,4,5 scale itself, and hence better modeled using normal distributions than are the actual 1,2,3,4,5 data.

The model that allows the effect of X_1 on Y to depend on the value of X_2 (i.e., the moderation model) is the interaction model

$$Y_i \mid X_{i1} = x_{i1}, X_{i2} = x_{i2} \sim_{\text{ind}} \text{N}(\beta_0 + \beta_1 x_{i1} + \beta_2 x_{i2} + \beta_3 x_{i1} x_{i2}, \sigma^2)$$

We estimate this model using `lm`, and also include graphical assessments of the linearity, constant variance, and normality assumptions as shown in Chapter 4. Notice that we use the predicted values $\hat{y}_i$ in these diagnostic analyses, rather than the x_i variables, as is appropriate for multiple regression.

R code for assessing assumptions of the interaction model in animal rights study

```
AnR = read.csv("https://raw.githubusercontent.com/andrea2719/
URA-DataSets/master/AnimalRights.csv")
attach(AnR)
Y = AnRights; X1 = Misanthropy; X2 = Idealism
X1X2 = X1*X2   # The interaction term

fit = lm(Y ~ X1 + X2 + X1X2)
summary(fit)

yhat = fit$fitted.values; resid = fit$residuals

par(mfrow=c(2,2)); par(mar=c(4, 4, 1, 1))
plot(yhat,Y)   # Graph 1: Correct functional spec
abline(lsfit(yhat,Y))
add.loess(yhat,Y, col="gray")
plot(yhat, resid)  # Graph 2: Correct functional spec
abline(h=0)
add.loess(yhat, resid, col="gray")
plot(yhat, abs(resid))  # Graph 3: Constant variance
add.loess(yhat, abs(resid), col="gray")
qqnorm(resid, main="")  # Graph 4: Normality
qqline(resid, col="gray")
```

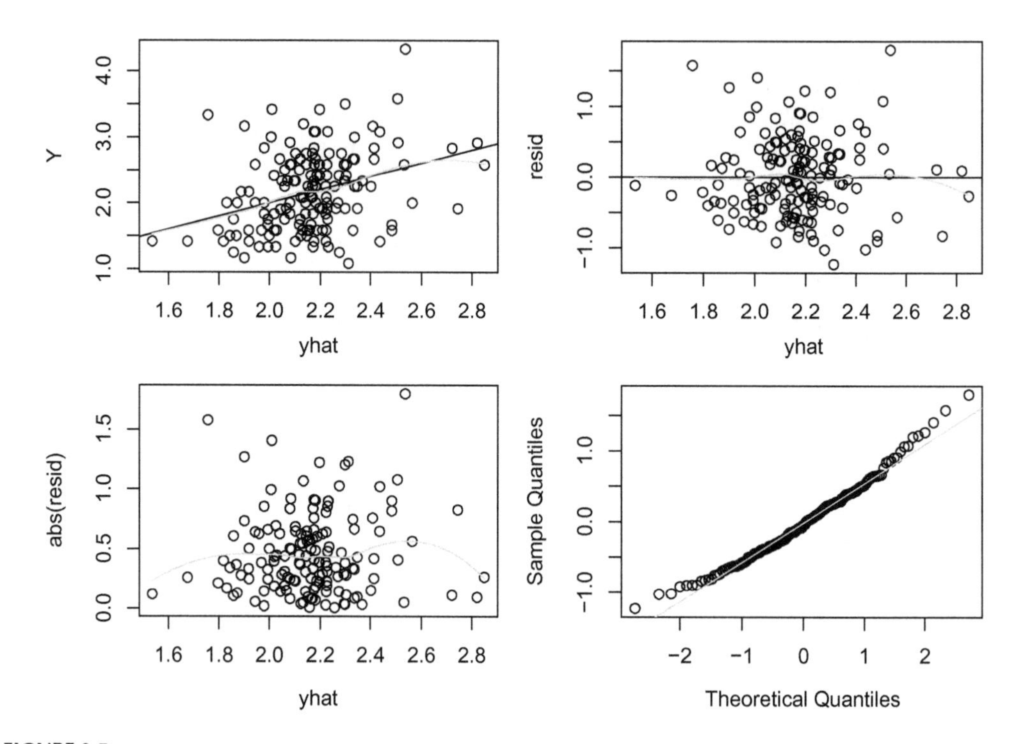

FIGURE 9.5
Graphical assessments of model assumptions in the interaction model. Top two panels assess the correct functional specification assumption: Slight curvature is noted but does not appear to be extreme. Bottom left panel assesses homoscedasticity; no pronounced trend in variability is seen. Bottom right panel assesses normality; the residuals show slight positive skew, but it also does not appear extreme.

According to Figure 9.5, the classical regression model with interaction appears reasonable. The results of the analysis as given by `summary(fit)` are as follows:

```
Call:
lm(formula = Y ~ X1 + X2 + X1X2)

Residuals:
     Min       1Q   Median       3Q      Max
-1.22820 -0.39732 -0.04499  0.35286  1.79772

Coefficients:
            Estimate Std. Error t value Pr(>|t|)
(Intercept)  -0.8182     1.1054  -0.740   0.4603
X1            1.1600     0.4542   2.554   0.0117 *
X2            0.6415     0.2954   2.171   0.0315 *
X1X2         -0.2438     0.1228  -1.986   0.0488 *
---
Signif. codes:  0 '***' 0.001 '**' 0.01 '*' 0.05 '.' 0.1 ' ' 1

Residual standard error: 0.5713 on 150 degrees of freedom
Multiple R-squared:  0.113,    Adjusted R-squared:  0.0953
F-statistic: 6.372 on 3 and 150 DF,  p-value: 0.0004291
```

Conclusions are as follows:

- An estimated 11.3% of the variance in Y is explained by the conditional mean function $E(Y \mid X_1 = x_1, X_2 = x_2) = \beta_0 + \beta_1 x_1 + \beta_2 x_2 + \beta_3 x_1 x_2$ ($R^2 = 0.113$).
- At least one of β_1, β_2 or β_3 is not zero ($F_{3,150} = 6.372$, $p = 0.0004291$).
- The interaction term is barely "significant," using the historical criteria ($T_{150} = -1.986$, $p = 0.0488$); which means that the moderating effect of X_2 on the effect of X_1 that is observed in the data is not easily explained by chance alone, where "chance alone" refers to a data-generating process that truly has no moderating effect.
- The other β coefficients are "significant" but not worth mentioning because their interpretations, by themselves, are not interesting. It does not matter whether they are "significant" or not. It only matters that the size of the β_3 coefficient is meaningfully large because the point of the analysis is to estimate the moderating effect of X_2 on the effect of X_1.

To understand the moderating effect of X_2 on the effect of X_1, the best approach to use (i) algebra with the model equation to show the different effects of X_1 on Y in cases of low and high X_2, and then (ii) display the results of the algebra graphically.

First, the algebra. The fitted interaction model is

$$\hat{Y} = -0.8181990 + 1.1600136X_1 + 0.6415269X_2 - 0.2438042X_1X_2.$$

The whole idea behind moderator analysis is that the effect of a main variable, X_1, on the response Y, depends on the value of the moderator variable, X_2. To display these different effects, pick two values of the moderator, one "low" and one "high." You might think the min and max of the moderator would suffice for this purpose, but they do not. The reason is that the min and max are not "typical" low and high values, they are the extremes and thus are usually *atypical* values. So instead, you can pick values one standard deviation from the mean, as in Figure 8.1. You can also pick percentiles—popular choices are the 10th and 90th percentiles and the 25th and 75th percentiles. We will pick the 10th and 90th percentiles of the Idealism measure (X_2) to represent "typical" low and high values of Idealism.

In `R`, you can use `quantile(X2, c(.10,.90))` to get these values; they are 3.0 and 4.4 respectively. (Recall that Idealism is the average of questionnaire items all measured on the 1,2,3,4,5 scale, so the averages are similarly between 1 and 5.) Now, for "Low" Idealism, the effect of Misanthropy (X_1) on Support for Animal Rights (Y) is estimated as

$$\begin{aligned}
\hat{Y} &= -0.8181990 + 1.1600136X_1 + 0.6415269(3.0) - 0.2438042X_1(3.0) \\
&= \{-0.8181990 + 0.6415269(3.0)\} + \{1.1600136 - 0.2438042(3.0)\}X_1 \\
&= 1.106382 + 0.428601X_1
\end{aligned}$$

The effect of Misanthropy (X_1) on support for Animal Rights (Y) is estimated to be 0.428601 when Idealism is low (specifically, when Idealism is at the 10th percentile of the Idealism distribution).

Now, for "High" Idealism, the effect of Misanthropy (X_1) on Support for Animal Rights (Y) is estimated as

$$\hat{Y} = -0.8181990 + 1.1600136X_1 + 0.6415269(4.4) - 0.2438042X_1(4.4)$$

$$= \{-0.8181990 + 0.6415269(4.4)\} + \{1.1600136 - 0.2438042(4.4)\}X_1$$

$$= 2.004519 + 0.08727512X_1$$

The effect of Misanthropy (X_1) on support for Animal Rights (Y) is estimated to be 0.08727512 when Idealism is high (specifically, when Idealism is at the 90th percentile of the Idealism distribution).

Comparing these two models confirms the researchers' hypothesis that Support for Animal Rights has little relation to Misanthropy among idealistic students, but is more strongly related to Misanthropy among non-idealistic students.

Graphs are nearly always better than equations. Simply display these two equations on the same set of axes; the conclusions described above from the algebra are clearly visible. First, note that the X_1 variable ranges from 1 to 4, so the graphs will use that range. Here is `R` code to do it.

`R` code for Figure 9.6

```
X1.seq = seq(1,4,1)
X2.low = quantile(X2,.10)
X2.high = quantile(X2,.90)
b = fit$coefficients; b0 = b[1]; b1 = b[2]; b2 = b[3]; b3 = b[4]
Y.low.X2 = b0 + b1*X1.seq + b2*X2.low + b3*X1.seq*X2.low
Y.high.X2 = b0 + b1*X1.seq + b2*X2.high + b3*X1.seq*X2.high

plot(X1.seq, Y.low.X2, type="l", xlab = "Misanthropy",
  ylab = "Predicted Animal Rights Support")
points(X1.seq, Y.high.X2, type="l", lty=2)
legend(2.7, 2.0, c("Low Idealism", "High Idealism"), lty=c(1,2))
```

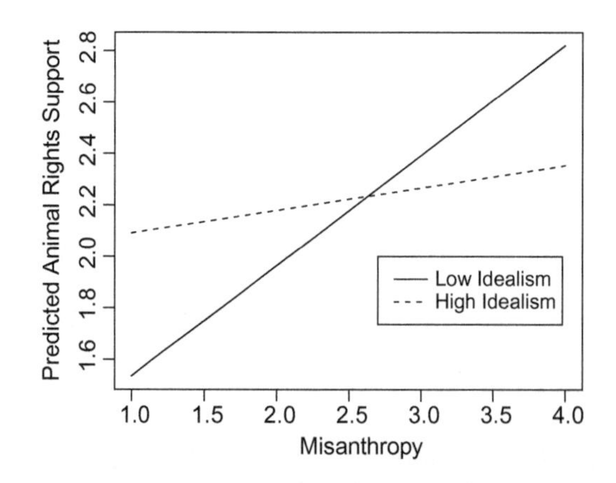

FIGURE 9.6
Graph showing estimated effects of Misanthropy on Support for Animal Rights, among people who are at the 10th percentile of Idealism (solid line), and among people who are at the 90th percentile of Idealism (dashed line). Higher slope indicates a greater effect.

Compare Figure 9.6. with Figure 9.4: All we are doing in Figure 9.6 is graphing "slices" of the 3-D response surface corresponding to low and high values of X_2, and then graphing these slices as 2-D functions, which are easier to interpret than 3-D functions. The same "slice" trick can be used in cases where you have more than two *X* variables as well: Just hold all the other *X* variables constant (say, at their mean values), calculate the two functions algebraically as shown above, draw a graph like Figure 9.6, and interpret the results in terms of the subject matter.

Before leaving interaction models, we need to set off in a box what is perhaps the most common misinterpretation of interaction:

Interaction between X_1 and X_2 has *nothing whatsoever to do with* correlation between X_1 and X_2

Researchers often state "X_1 and X_2 interact" when they mean "X_1 and X_2 are correlated," and vice versa. But *correlation* and *interaction* are completely unrelated concepts. The easiest way to understand the distinction is to recognize that *interaction* between X_1 and X_2 actually involves *three* variables: X_1, X_2 and *Y*. Specifically, X_1 and X_2 interact if the *effect* of X_1 on *Y depends on* the value of X_2. On the other hand, *correlation* between X_1 and X_2 involves no third variable.

9.4 The Variable Inclusion Principle

Earlier in this chapter and in prior chapters, we alluded to this concept, which is stated specifically as follows:

The variable inclusion principle

Include all lower-order terms related to higher-order terms in a polynomial regression model (regardless of "significance" or lack thereof.)

Recall that the "order" of a polynomial term is either the exponent or sum of the exponents. Lower order terms are those terms in the same variable or variables, with smaller order.

Here are some applications of the variable inclusion principle:

- **Quadratic term**, x^2. Lower order terms are $x^1 = x$ and $x^0 = 1$. The variable inclusion principle states that if you have x^2 in the model, then you should also include x (the linear term) and 1 (the intercept).
- **Cubic term**, x^3. Lower order terms are x^2, $x^1 = x$, and $x^0 = 1$. The variable inclusion principle states that if you have x^3 in the model, then you should also include x^2 (the quadratic term), x (the linear term), and 1 (the intercept).
- **Interaction term**, x_1x_2. Write this term as $x_1^1x_2^1$. Lower order terms are $x_1^1x_2^0 = x_1$, $x_1^0x_2^1 = x_2$, and $x_1^0x_2^0 = 1$. The variable inclusion principle states that if you have x_1x_2 in the model, then you should also include x_1, x_2 and 1 (the intercept).
- **Linear term**, x. The only lower-order term is $x^0 = 1$ (the intercept). The variable inclusion principle states that you should always put an intercept in the model.

You will see examples in the literature where researchers do not adhere to this principle, so it is important to understand the logic behind the principle as clearly as possible. The logic is this:

If you violate the variable inclusion principle, then the coefficient of the higher-order term does not measure what you want it to measure.

What do you want these coefficients of higher-order terms to measure?

- In the case of the quadratic model $f(x) = \beta_0 + \beta_1 x + \beta_2 x^2$, you want the coefficient β_2 of x^2 to measure curvature. After all, the only reason you would use a quadratic term is to model curvature of the conditional mean function.
- In the case of the cubic model $f(x) = \beta_0 + \beta_1 x + \beta_2 x^2 + \beta_3 x^3$, you want the coefficient β_3 of x^3 to measure curvature that cannot be modeled with a simple quadratic model.
- In the case of the interaction model $f(x_1, x_2) = \beta_0 + \beta_1 x_1 + \beta_2 x_2 + \beta_3 x_1 x_2$, you want the coefficient β_3 of $x_1 x_2$ to measure interaction. Specifically, you want β_3 to be zero when there is no interaction.
- In the case of the simple linear model $f(x) = \beta_0 + \beta_1 x$, you want the coefficient β_1 of x to measure the effect of X on Y.

Let's consider the last bullet first, as it is the easiest.

9.4.1 Why You Should Always Include the Intercept Term

Consider the following data:

```
x = c(72, 72, 70, 69, 72, 64, 72, 69, 69, 67, 70, 76, 71, 72)   # and
y = c(6, 11,  7, 10, 12, 17, 10, 16,  9,  8,  5, 10,  7, 10)
```

You can estimate the standard regression model as follows:

```
fit1 = lm(y~x); summary(fit1)
```

This gives the following output:

```
Call:
lm(formula = y ~ x)

Residuals:
    Min      1Q  Median      3Q     Max
-5.0138 -2.9042  0.2055  2.4299  5.5474

Coefficients:
            Estimate Std. Error t value Pr(>|t|)
(Intercept)  40.7266    23.1458   1.760    0.104
x            -0.4388     0.3287  -1.335    0.207

Residual standard error: 3.34 on 12 degrees of freedom
Multiple R-squared:  0.1293,    Adjusted R-squared:  0.0567
F-statistic: 1.781 on 1 and 12 DF,  p-value: 0.2067
```

Notice that the intercept term is "insignificant" with $p = 0.104$. Some regression sources erroneously state that "insignificant" terms should be dropped from the model. Further, some researchers will do anything to "prove" their theories, including playing around with models until they get the result they want. Suppose you are one of these unscrupulous researchers. In an attempt to make your X variable "significant," you try the no-intercept model. After all, you have seen in one of these dubious regression sources that "insignificant" variables should be removed. So you decide to let your dubious sources provide your justification for removing the intercept term.

Here is what happens when you do that. In the `lm` function of `R`, just put a "`-1`" in the model term to remove the intercept. (The logic for the "`-1`" is that the minus sign "–" suggests "remove," and the "1" refers to the intercept X variable, $x^0 = 1$). Here is the code:

```
fit2 = lm(y ~ x-1)
summary(fit2)
```

The output is as follows:

```
Call:
lm(formula = y ~ x - 1)

Residuals:
    Min      1Q  Median      3Q     Max 
-4.7468 -2.3924 -0.3038  0.8291  8.0886 

Coefficients:
  Estimate Std. Error t value Pr(>|t|)    
x  0.13924    0.01366   10.19 1.44e-07 ***
---
Signif. codes:  0 '***' 0.001 '**' 0.01 '*' 0.05 '.' 0.1 ' ' 1

Residual standard error: 3.599 on 13 degrees of freedom
Multiple R-squared:  0.8888,	Adjusted R-squared:  0.8802 
F-statistic: 103.9 on 1 and 13 DF,  p-value: 1.441e-07
```

Suddenly, the results look great! You have "proven" your theory: X is "significantly" ($p = 1.4 \times 10^{-7}$) related to Y with a positive coefficient. The R^2 also statistic looks great (but actually is meaningless here). The main problem is shown in Figure 9.7: Notice that the no-intercept model (which actually assumes that the intercept is equal to 0) shows a positive effect of X on Y while the model with intercept shows a negative effect. Which is right? Here's a little secret: We simulated the data in such a way that there really is a negative effect of X on Y. So the no-intercept model "decisively proved" $\beta_1 > 0$ (according to the p-value 1.4×10^{-7}) when, in reality, $\beta_1 < 0$.

This analysis provides evidence supporting the point that we gave above: If you violate the variable inclusion principle, then the coefficient of the higher-order term does not measure what you want it to measure. Here, leaving out the intercept term (thus violating the inclusion principle) means that the slope does not measure the true effect of X on Y.

Let's consider the first bullet now.

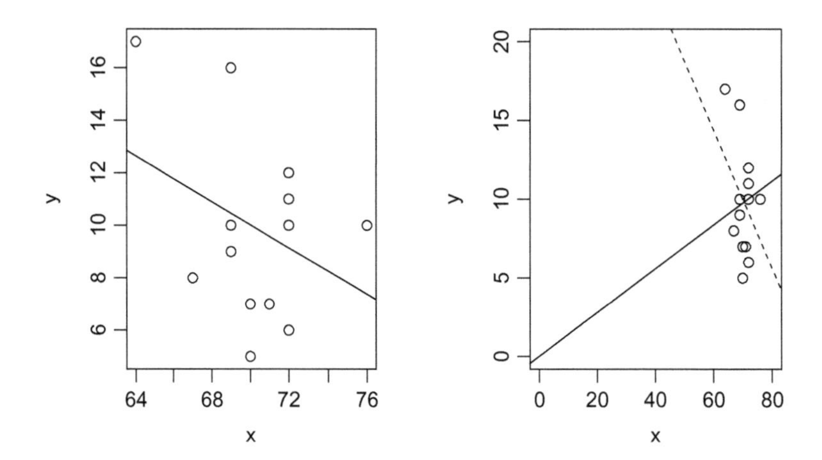

FIGURE 9.7
Graphs of the intercept and no-intercept models. Left panel: Fitted model with an intercept term. Right panel: Same data, different graph scale. Fitted model with intercept (same as left panel, decreasing line, now dotted line) as well as fitted model without intercept term (increasing line, assumes intercept is 0).

9.4.2 Why You Should Include the Linear Term in a Quadratic Model

Please run the following `R` code.

```
## Code to illustrate the danger of violating the inclusion principle
## Note that Y does not depend on X squared.

set.seed(12345)
x = rnorm(100, 10, 1)
y = 5 + 2*x + rnorm(100,0,2)
x.sq = x**2

# In the model where the inclusion principle is not followed, the
# quadratic term is "significant," incorrectly indicating curvature.

inclusion.model      =  lm(y ~ x + x.sq)
summary(inclusion.model)

no.inclusion.model  =  lm(y ~ x.sq)
summary(no.inclusion.model)
```

The model is truly linear, and not curved, in the simulation above. In the model that obeys the inclusion criterion, the quadratic term is correctly deemed "insignificant" ($p = 0.525$). The linear term is also "insignificant" ($p = 0.161$), so one might be tempted to remove it from the model, thus violating the inclusion principle. In the resulting model that does not obey the inclusion criterion, where the linear term is excluded, the quadratic term is highly "significant" ($p < 2\times10^{-16}$).

Again, if you violate the variable inclusion principle, then the coefficient of the higher-order term does not measure what you want it to measure. Here, leaving out the linear term (thus violating the inclusion principle) means that the coefficient of the quadratic term does not measure curvature. Rather, since the linear term was excluded, the quadratic term simply acts as a surrogate for the linear term. Instead of measuring only curvature, the coefficient of the quadratic term also measures the linear effect, if you violate the variable inclusion principle.

9.4.3 Why You Should Include the Linear Terms in an Interaction Model

Again, please run the following R code.

```
## Note that there is no interaction.
set.seed(12345)
x1 = rnorm(100, 10, 1)
x2 = 15 + 2*rnorm(100)
y = 5 + 2*x1 + x2 + rnorm(100)
x1.x2 = x1*x2

##  In the model where the inclusion principle is not followed,
## the interaction term is "significant," incorrectly indicating
## that the effect of X1 on Y depends on X2.
inclusion.model     = lm(y ~ x1 + x2 + x1.x2)
summary(inclusion.model)

no.inclusion.model = lm(y ~ x1 +      x1.x2)
summary(no.inclusion.model)
```

The model is truly one with no interaction in the simulation above. In the model that obeys the inclusion criterion, the interaction term is correctly deemed "insignificant" ($p = 0.3908$). The x2 term is also "insignificant" ($p = 0.3139$), so one might be tempted to remove it from the model, thus violating the inclusion principle. In the resulting model that does not obey the inclusion criterion, where the linear term is excluded, the interaction term is highly "significant" ($p < 2 \times 10^{-16}$).

If you violate the variable inclusion principle, then the coefficient of the higher-order term does not measure what you want it to measure. Here, leaving out the x2 term (thus violating the inclusion principle) means that the coefficient of the interaction term x1.x2 does not measure interaction. Rather, since the x2 term was excluded, the x1.x2 interaction term simply acts as a surrogate for the x2 linear term. Instead of measuring only interaction, the coefficient of the interaction term also measures the linear effect of x2, if you violate the variable inclusion principle.

References

Moreau, A. R., Westfall, P. H., Cancio, L. C., & Mason, A. D. (2005). Development and validation of an age-risk score for mortality prediction following thermal injury. *Journal of Trauma, Injury, Infection & Critical Care, 58*(5), 967–972.

Wuensch, K. L., Jenkins, K. W., & Poteat, G. M. (2002). Misanthropy, idealism and attitudes towards animals. *Anthrozoös, 15*(2), 139–149.

Exercises

1. Consider the following data with SALES as a function of RATE with no transformations.

```
sales = read.csv("https://raw.githubusercontent.com/andrea2719/
URA-DataSets/master/sales.csv")
```

Graph the scatterplot, and overlay the (i) linear, (ii) quadratic, (iii) cubic, (iv) quartic, and (v) quintic fitted functions on the same plot using different line types. See here https://stackoverflow.com/a/26959959/6945791 for an excellent post showing how to make this easy. Report the SSE and R^2 for each of the 5 models and comment.

2. Consider the "charity" data set graphed in Figure 5.6, read as follows:

```
charity = read.csv("https://raw.githubusercontent.com/andrea2719/
URA-DataSets/master/charitytax.csv")
```

In that data set are variables CHARITY, INCOME and DEPS, among others. CHARITY is the natural logarithm of the person's charitable contributions, INCOME is the natural logarithm of their adjusted gross income, and DEPS is the number of dependents. LEAVE the log-transformed variables "as is" throughout your analyses; do not back-transform.

For this problem, you will consider the additive and interaction models, which are, respectively, (i.) $\text{CHARITY} = \beta_0 + \beta_1\text{INCOME} + \beta_2\text{DEPS} + \varepsilon$ and (ii.) $\text{CHARITY} = \beta_0 + \beta_1\text{INCOME} + \beta_2\text{DEPS} + \beta_3\text{INCOME} \times \text{DEPS} + \varepsilon$

a. Find the 10th and 90th percentiles of INCOME. Call these numbers INC.L and INC.H.

b. Using the estimated model (i.), find the equation of the estimated mean value of CHARITY as a function of DEPS when INCOME = INC.L.

c. Using the estimated model (i.), find the equation of the estimated mean value of CHARITY as a function of DEPS when INCOME = INC.H.

d. Graph the two functions in 2.b and 2.c on the same axes, with different line types. Include a legend.

e. Using the estimated model (ii.), find the equation of the estimated mean value of CHARITY as a function of DEPS when INCOME = INC.L.

f. Using the estimated model (ii.), find the equation of the estimated mean value of CHARITY as a function of DEPS when INCOME = INC.H.

g. Graph the two functions in 2.e and 2.f on the same axes, with different line types. Include a legend.

h. Is the interaction effect shown in 2.g explainable by chance alone? (You may use the p-value to answer).

10

ANOVA, ANCOVA, and Other Applications of Indicator Variables

An *indicator variable* is an X variable that takes only two values, either 0 or 1. You have already seen an indicator variable in the prediction of GPA as a function of GMAT and degree program (Master's or PhD) in Chapter 6. The "`degree`" variable is actually character in the data set, with values "`M`" and "`P`." Obviously, you cannot put "`M`" and "`P`" into a regression model since they are not numbers. One use of indicator variables is to re-code such nominal data numerically as 0 or 1, so that you can use them in a regression model.

Indicator variables are very useful, with a wide variety of applications, as we show in this chapter. We will start as simply as possible, and progress to more complex applications.

10.1 Using a Single Indicator Variable to Represent a Single Nominal Variable Having Two Levels (Two-Sample Comparison)

First, notice that we snuck in a term, "level," into the title of this section. The term "level" means "distinct value." So, the "`degree`" variable in the GPA data set discussed in Chapter 6 has two levels, "`M`" and "`P`." Another example is gender: If a survey question asks the respondent to identify their gender, the resulting Gender variable has two levels, "Female" and "Male." In data where each observation is a firm, or company (called firm-level data analysis), you might have publicly held and privately held firms in your data set. This information might be recorded in a variable called PUBLIC, with values "Yes" and "No." In this case, the PUBLIC variable has two levels, "Yes" and "No." If ethnicity is recorded in a survey, there will usually be more than two possibilities. Suppose there are 7 distinct ethnicities a person may identify with, including an "Other" category. Then the ETHNICITY variable will have 7 levels.

Consider the GPA prediction example, using only the "`degree`" variable to predict it. For now, let's code "PhD" as 1 and "Master's" as 0. (You can code it the other way, too, as (0,1). We will do that right after we do it this way so we can compare results.) Call the resulting (1,0) indicator variable "PHD."

The classical regression model states as follows:

$$\text{GPA} \mid \text{PHD} = x \sim \text{N}(\beta_0 + \beta_1 x, \sigma^2)$$

Equivalently,

$$\text{GPA} = \beta_0 + \beta_1 \text{PHD} + \varepsilon, \quad \text{where } \varepsilon \sim \text{N}(0, \sigma^2)$$

Drawing a scatterplot is always the best first step in the analysis of regression data; in this case, it appears as shown in Figure 10.1, with the least-squares line superimposed.

R code for Figure 10.1

```
ba = read.table("https://raw.githubusercontent.com/andrea2719/
URA-DataSets/master/gpa_gmat.txt")
attach(ba)
PHD = ifelse(degree=="P",1,0)
plot(PHD, gpa); abline(lsfit(PHD, gpa))
```

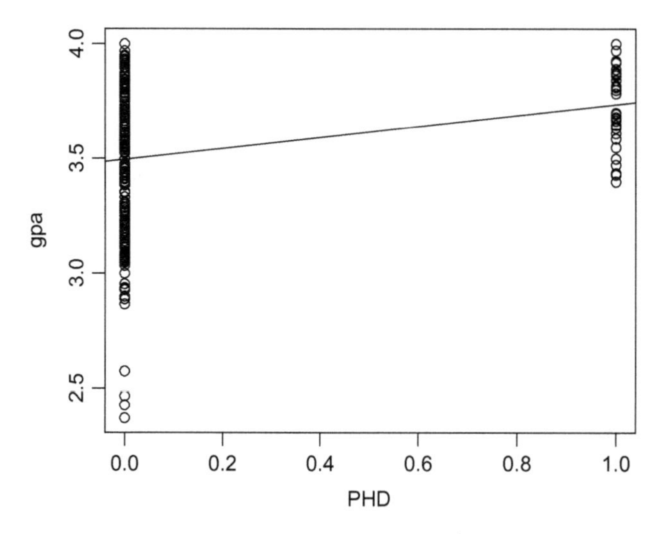

FIGURE 10.1
Scatterplot of the (PHD, gpa) data, with least-squares line superimposed. The indicator variable "PHD" is 1.0 for PhD students, and 0.0 for Master's students. Usually, the line is not displayed in plot with indicator variable, since there is no value between 0 and 1. The line is included here to establish the connection with typical regression models.

As always, the regression model is a model that specifies $p(y \mid x)$, the distribution of the *potentially observable* values of Y when $X = x$. In this case, there are only two possible X values, and unlike previous examples where the X data were more continuous, here you can visualize the Y data from these distributions very clearly in Figure 10.1. There is one distribution $p(y \mid x)$ of *potentially observable* GPAs for Master's students (where PHD = 0), and there is another distribution $p(y \mid x)$ of *potentially observable* GPAs for PhD students (where PHD = 1). Figure 10.1 shows reasonably large samples of *actually observed* GPA data that come from these two distributions of *potentially observable* GPAs.

In this chapter, you will see repeatedly a special technique for interpreting parameters in models where there is an indicator variable.

Technique for interpreting parameters in models with indicator variables

Always write down the separate models for the groups represented by the indicator variables.

To illustrate the technique for interpreting parameters of the GPA model, you should write down the model separately for the two groups represented by the PHD variable.

Group 1: PHD = 0 (Master's students)

$$\begin{aligned} \text{GPA} &= \beta_0 + \beta_1(0) + \varepsilon \\ &= \beta_0 + \qquad \varepsilon \end{aligned}$$

Group 2: PHD = 1 (PhD students)

$$\begin{aligned} \text{GPA} &= \beta_0 + \beta_1(1) + \varepsilon \\ &= \beta_0 + \beta_1 + \quad \varepsilon \end{aligned}$$

Since the ε terms have mean zero, from these models you can conclude:

- β_0 is the mean of the distribution of the *potentially observable* GPAs for Master's students.
- $\beta_0 + \beta_1$ is the mean of the distribution of the *potentially observable* GPAs for PhD students.
- Hence, β_1 is the *difference* between the mean of the *potentially observable* GPAs for PhD students and the mean of the *potentially observable* GPAs for Master's students.

When there is a single indicator variable in the regression model, the least-squares estimates of the means are identical to the sample averages of the data. Specifically, the command `mean(gpa[PHD==1])` gives the ordinary average of the GPAs for PhD students, 3.736306. The corresponding average of the Master's students' GPAs is obtained as `mean(gpa[PHD==0])`, giving 3.49621. Using the regression analysis, the commands `fit = lm(gpa ~ PHD); summary(fit)` gives the results:

```
Coefficients:
            Estimate Std. Error t value Pr(>|t|)
(Intercept)  3.49621    0.01384 252.577  < 2e-16 ***
PHD          0.24010    0.05128   4.682 3.67e-06 ***
```

From the output, the estimated mean GPA for PhD students is $3.49621 + 0.24010 = 3.73631$, and the estimated mean GPA for Master's students is 3.49621, as shown by the ordinary averages. The value 0.24010 is likewise the difference between the ordinary averages from the PhD and Master's students. The predictions from the indicator variable regression model correspond exactly to the averages of the actual data.

Self-study question: How does the correspondence between the predicted values and the simple averages relate to the definition of the regression function as *means* of *conditional distributions*?

Because β_1 is the *difference* between the mean of the *potentially observable* GPAs for PhD students and the mean of the *potentially observable* GPAs for Master's students, the null model where $\beta_1 = 0$ states that these means are identical. As shown in the `lm` output above, there is a 0.2401 difference between the observed GPA averages, and this difference is 4.682 standard errors from 0. Such a difference is not easily explained by chance alone $(p = 3.67 \times 10^{-6})$ when the null model is true, hence the observed difference is not just a chance (idiosyncratic) difference. Instead, it reflects a systematic tendency for PhD students to have higher grade point averages than Master's students.

You get a better quantification of this systematic difference by using the confidence interval. With `confint(fit)`, you get $0.1393487 < \beta_1 < 0.3408432$, implying that the mean of *potentially observable* GPA values for PhD students is somewhere between 0.1393487 and 0.3408432 *higher than* the mean of *potentially observable* GPA values for Master's students.

10.1.1 Does It Matter Whether the Indicator Variable Is Coded as 1,0 vs. 0,1?

It does not matter in the sense that you can code it either way; neither is preferred. It does matter in the sense that the parameter interpretations will all be reversed. If you code Master's as 1 and PhD as 0, then β_0 becomes the mean for PhD students, and β_1 is the difference between Master's mean and PhD mean. Here is the output:

```
Coefficients:
             Estimate Std. Error t value Pr(>|t|)
(Intercept)  3.73631    0.04937  75.676  < 2e-16 ***
Masters     -0.24010    0.05128  -4.682 3.67e-06 ***
```

Now, the intercept 3.73631 is the mean of the observed GPA for PhD students, and the difference between Master's and PhD means is –0.24010. It's exactly the same information as before, just labeled differently. Further, using the `confint` function, you get $-0.3408432 < \beta_1 < -0.1393487$, meaning that the mean of *potentially observable* GPAs for Master's students is somewhere between 0.1393487 and 0.3408432 *lower than* the mean of *potentially observable* GPAs for PhD students. Again, this information is exactly as before, just from a different point of view.

The analysis above that compares Master's and PhD students is identical to the *two-sample t test*, which we hope you have seen in a prior course. In `R`, you can perform the two-sample t test as follows:

```
t.test(gpa ~ degree, var.equal=T)
```

The output provides identical $|T|$ statistic and confidence interval as the regression analysis:

```
        Two Sample t-test

data:  gpa by degree
t = -4.6824, df = 492, p-value = 3.672e-06
alternative hypothesis: true difference in means is not equal to 0
95 percent confidence interval:
 -0.3408432 -0.1393487
sample estimates:
mean in group M mean in group P
       3.496210        3.736306
```

A happy conclusion is that you really do not need to bother with the two-sample t test at all—you can do it all via regression with an indicator variable. Still, you should know what the "two-sample t test" refers to: It refers to the comparison of averages from two different (independent) samples. Assumptions for the two-sample t test are identical to the assumptions for the regression model with an indicator variable: The assumptions are independence, common variance, and normality. Linearity need not be mentioned because the linearity assumption is actually true: Since there are only two possible values of X (0 and 1), the means of the distributions $p(y \mid X = 0)$ and $p(y \mid X = 1)$ *must* fall perfectly on the line. So, there is no need to mention the linearity assumption when performing regression analysis with a single indicator variable.

10.2 Using Multiple Indicator Variables to Represent a Single Nominal Variable Having Three or More Levels (ANOVA)

"Location, location, location!" is what the real estate folks tell you. The price of a home is strongly affected by where the house is located.

Consider the following data set on house prices.

```
house = read.csv("https://raw.githubusercontent.com/andrea2719/
URA-DataSets/master/house.csv", header=T); attach(house)
```

Two variables in the data set are "`sell`" and "`location`"; respectively the selling price (in $1,000s) of a home, and its location. The "Location" variable has five levels, either location A, B, C, D, or E.

Whenever you have a nominal variable, you should always check the sample sizes within all of the levels, because small sample sizes will give imprecise estimates. By entering the command `table(location)`, you get the following result:

```
location
 A  B  C  D  E
26 15  4  9 10
```

Notice that there are only 4 homes in location "C." Thus, the average of potentially observable house price in location C is estimated less accurately than it is in the other locations.

The goal is to model $Y \mid X = x$, selling price of a home, as a function of X, location. As always in regression, you will specify a model $p(y \mid X = x)$ for every $X = x$; which, in this case, means you need to specify the models $p(y \mid X = A)$, $p(y \mid X = B)$, $p(y \mid X = C)$, $p(y \mid X = D)$, and $p(y \mid X = E)$.

A problem is that location is character data (A, B, C, D, and E). You cannot use the classical regression model to specify these distributions, because that model states that the means of the distributions are a linear function of $X = x$, and here, the specific values of X are not numbers, they are letters. You might consider recoding the locations as a new variable U, where $U = 1$ for location A, $U = 2$, for location B, ..., $U = 5$ for location E, and then perform the regression of Y on U, but that would presume a perfect linear function of mean values for the different locations. This really makes no sense at all—the locations are not even ordinal, in the sense that A is somehow "less than" B, which is in turn "less

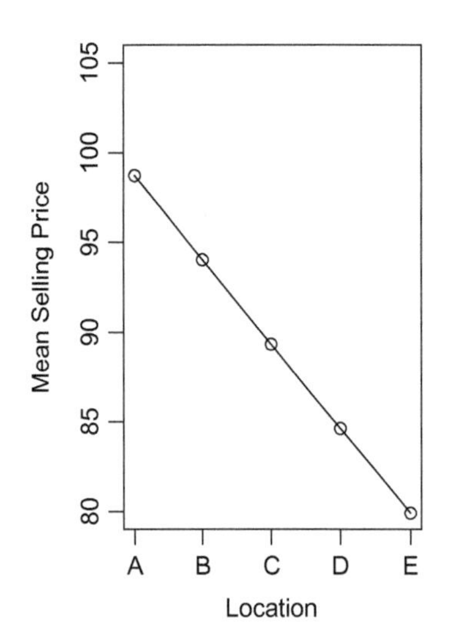

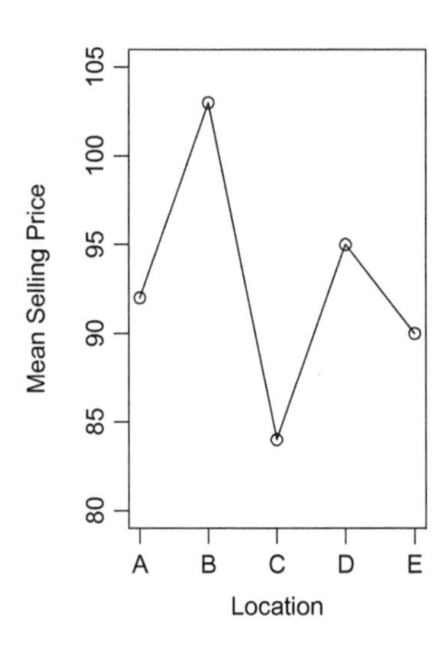

FIGURE 10.2
Mean price of a house ($1,000s) as a function of location. Left panel: A linear model assuming that the Location variable is coded as 1,2,3,4,5; obviously a bad model. Right panel: A more realistic model in which the means for each location can be any numbers. (In both graphs, there is no need to connect the dots; it just makes the graphs easier to read.)

than" C, and so forth. Rather, they are just five locations, and the mean selling prices could be anything for each location, with no connection (linear or otherwise) between the means for different locations. Figure 10.2 shows graphs of the mean function assuming such a *U* variable, and the more realistic case where the mean selling prices can be any numbers.

So, how do you model "Location" so that the mean selling price can be anything at all in each location, as shown in the right panel of Figure 10.2? Simple. Define *four* indicator variables for the locations, leaving one out. Specifically, let:

XA = 1 if location A, 0 otherwise
XB = 1 if location B, 0 otherwise
XC = 1 if location C, 0 otherwise
XD = 1 if location D, 0 otherwise

What happened to location E? As in the case of PhD/Master's example above, where the (0,1) and (1,0) codings were shown to be equivalent, the model here is also equivalent no matter which category you "leave out." Only the parameter interpretations change.

Here is the `R` code to create these indicator variables.

```
XA = ifelse(location=="A",1,0)
XB = ifelse(location=="B",1,0)
XC = ifelse(location=="C",1,0)
XD = ifelse(location=="D",1,0)
```

These variables are now four new columns in your data set, named `XA`, `XB`, `XC`, and `XD`. Each of these columns has values 0 or 1. In `R`, these variables are *all* included in the `lm` call:

```
Fit.ind = lm(sell ~ XA + XB + XC + XD)
```

The model fit by R corresponds to the mathematical model:

$$\text{Price} = \beta_0 + \beta_1\text{XA} + \beta_2\text{XB} + \beta_3\text{XC} + \beta_4\text{XD} + \varepsilon$$

You should interpret the β parameters by separating the potentially observable data into the five groups indicated by the "location" variable.

Group 1: Location = A, Price = $\beta_0 + \beta_1(1) + \beta_2(0) + \beta_3(0) + \beta_4(0) + \varepsilon = (\beta_0 + \beta_1) + \varepsilon$
Group 2: Location = B, Price = $\beta_0 + \beta_1(0) + \beta_2(1) + \beta_3(0) + \beta_4(0) + \varepsilon = (\beta_0 + \beta_2) + \varepsilon$
Group 3: Location = C, Price = $\beta_0 + \beta_1(0) + \beta_2(0) + \beta_3(1) + \beta_4(0) + \varepsilon = (\beta_0 + \beta_3) + \varepsilon$
Group 4: Location = D, Price = $\beta_0 + \beta_1(0) + \beta_2(0) + \beta_3(0) + \beta_4(1) + \varepsilon = (\beta_0 + \beta_4) + \varepsilon$
Group 5: Location = E, Price = $\beta_0 + \beta_1(0) + \beta_2(0) + \beta_3(0) + \beta_4(0) + \varepsilon = \beta_0 + \varepsilon$

Since the ε terms have mean zero, from these models you can conclude:

- β_0 is the mean of the distribution of the *potentially observable* Prices in Location E.
- $\beta_0 + \beta_1$ is the mean of the distribution of the *potentially observable* Prices in Location A.
- $\beta_0 + \beta_2$ is the mean of the distribution of the *potentially observable* Prices in Location B.
- $\beta_0 + \beta_3$ is the mean of the distribution of the *potentially observable* Prices in Location C.
- $\beta_0 + \beta_4$ is the mean of the distribution of the *potentially observable* Prices in Location D.

Notice that these five mean values can be any numbers whatsoever. Unlike the linear model shown in the left panel of Figure 10.2, the indicator variable model places no constraints whatsoever on the five mean values.

The following four bullet points give the interpretations of the individual β values.

- β_1 is equal to the mean of the *potentially observable* Prices in location A, *minus* the mean of the *potentially observable* Prices in location E.
- β_2 is equal to the mean of the *potentially observable* Prices in location B, *minus* the mean of the *potentially observable* Prices in location E.
- β_3 is equal to the mean of the *potentially observable* Prices in location C, *minus* the mean of the *potentially observable* Prices in location E.
- β_4 is equal to the mean of the *potentially observable* Prices in location D, *minus* the mean of the *potentially observable* Prices in location E.

This is another case where the least-squares estimates of the means are identical to the ordinary data averages, as shown by the following R code:

```
# Ordinary averages of the observed data:
mean(house$sell[location=="A"])
mean(house$sell[location=="B"])
mean(house$sell[location=="C"])
mean(house$sell[location=="D"])
mean(house$sell[location=="E"])
```

This code gives you the averages 104.1115, 82.38667, 87.75, 82.3, and 86.17. The estimated means using the indicator variable regression model are identical to the ordinary averages of the data, which you can see using the following code:

```
Fit.ind = lm(sell ~ XA + XB + XC + XD, data=house)
b = Fit.ind$coefficients
b0 = b[1]; b1 = b[2]; b2 = b[3]; b3 = b[4]; b4 = b[5]
b0+b1; b0+b2; b0+b3; b0+b4; b0
```

You can view the regression results using `summary(Fit.ind)`, with output as follows:

```
Coefficients:
            Estimate Std. Error t value Pr(>|t|)
(Intercept)   86.170      4.841  17.800  < 2e-16 ***
XA            17.942      5.696   3.150  0.00257 **
XB            -3.783      6.250  -0.605  0.54725
XC             1.580      9.057   0.174  0.86210
XD            -3.870      7.034  -0.550  0.58425
---
Signif. codes:  0 '***' 0.001 '**' 0.01 '*' 0.05 '.' 0.1 ' ' 1

Residual standard error: 15.31 on 59 degrees of freedom
Multiple R-squared:  0.3184,    Adjusted R-squared:  0.2722
F-statistic:  6.89 on 4 and 59 DF,  p-value: 0.0001277
```

From this output, you can conclude:

- The estimated mean for location E is 86.170.
- The mean for location A is estimated to be 17.942 higher than the mean for location E.
- The mean for location B is estimated to be 3.783 lower than the mean for location E.
- The mean for location C is estimated to be 1.580 higher than the mean for location E.
- The mean for location D is estimated to be 3.870 lower than the mean for location E.
- The standard deviation of potentially observable prices in each location is estimated to be 15.31.
- The R^2 statistic is 0.3184, implying the 31.84% of the variation in house prices is explained by locational differences in price.

The F test has a special interpretation here. The null model states that

$$\text{Price} = \beta_0 + (0)\text{XA} + (0)\text{XB} + (0)\text{XC} + (0)\text{XD} + \varepsilon = \beta_0 + \varepsilon$$

In other words, the null model states that all five locations have the same mean house price, β_0. If the null model is true, then the differences between the observed average house prices for the five different locations are *explained by chance alone.*

The result of the F test is $F_{4,59} = 6.89$ (p = 0.0001277), indicating that the differences between the observed average prices are not easily explained by chance alone, under a model where the true mean prices are identical. Hence, you can conclude that the true mean prices are not all identical.

While the F test rules out the hypothesis that the location means are identical, it does not tell you which means differ from which other means. You can have a large F when four of the locations have identical means, and the fifth location's mean differs. To identify which locations differ from which other locations, you can examine the individual T tests, as follows:

- The mean for location A differs from the mean for location E by an amount that is not easily explained by chance alone ($p = 0.00257$).
- The difference between location B and location E means is explainable by chance ($p = 0.54725$).
- The difference between location C and location E means is explainable by chance ($p = 0.86210$).
- The difference between location D and location E means is explainable by chance ($p = 0.58425$).

There are several other comparisons to be made here, such as A vs. B, B vs. C, A vs. C, etc. (Actually, there are 10 total such comparisons.) The appropriate method for making all such comparisons simultaneously is called "Tukey's Method"; see Bretz et al. 2010.

Finishing up, here is the code to draw the graph of mean house prices for this example.

R code for Figure 10.3

```
sell = c(b0+b1, b0+b2, b0+b3, b0+b4, b0)
X = 1:5
plot(X, sell, xaxt="n", xlab = "Location", ylim = c(80,105),
ylab="Mean Selling Price")
axis(1, at = 1:5, lab=c("A", "B", "C", "D", "E"))
points(X, sell, type="l")
```

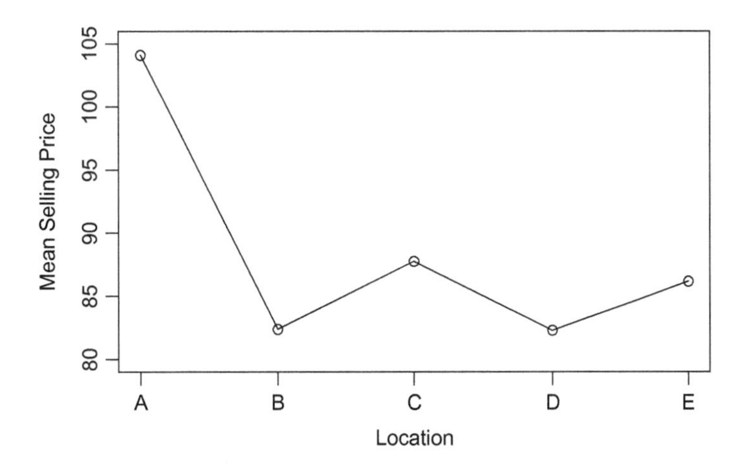

FIGURE 10.3
Estimated mean house prices as a function of location. (Note: Connecting the dots is not needed, it just makes the graph easier to read.)

As before, it does not matter which location you leave out, the model predictions are the same; only the parameter estimates will differ because the intercept will refer to a different category than location E. The R software is smart enough to do this analysis without requiring you to create the indicator variables manually. That's a good thing because if you had, say, 100 different locations rather than 5, the coding would be extremely tedious! Without creating indicator variables at all, you can make R perform the same analysis simply by specifying

```
Fit.ind1 = lm(sell ~ location, data=house)
summary(Fit.ind1)
```

The resulting output is:

```
Coefficients:
             Estimate Std. Error t value Pr(>|t|)
(Intercept)  104.112      3.002  34.678  < 2e-16 ***
locationB    -21.725      4.963  -4.377 4.99e-05 ***
locationC    -16.362      8.222  -1.990  0.05123 .
locationD    -21.812      5.920  -3.684  0.00050 ***
locationE    -17.942      5.696  -3.150  0.00257 **
---
Signif. codes:  0 '***' 0.001 '**' 0.01 '*' 0.05 '.' 0.1 ' ' 1

Residual standard error: 15.31 on 59 degrees of freedom
Multiple R-squared:  0.3184,    Adjusted R-squared:  0.2722
F-statistic:  6.89 on 4 and 59 DF,  p-value: 0.0001277
```

Notice that R created four indicator variables for you, but left-out location "A" rather than location "E." While the two estimated models are the same in terms of R^2 and F statistics, the parameter estimates differ, because now the intercept is the mean house price for location "A," and all other parameters are differences from location "A." As you can see in Figure 10.3, all locations have smaller mean prices than location "A," explaining why all the estimated coefficients are negative.

Why do you need to leave one category out at all? The answer is given in Chapter 8. If you have perfectly collinear variables, then the $\mathbf{X}^T\mathbf{X}$ matrix is non-invertible, and there are infinitely many solutions to the equations $(\mathbf{X}^T\mathbf{X})\boldsymbol{b} = \mathbf{X}^T\mathbf{Y}$. If you created indicator variables XA, XB, XC, XD, and XE in the house example, and then included them all in the model, you would have that $XA + XB + XC + XD + XE = 1$, implying that every indicator variable is a perfect linear function of the others.

Self-study question: Why is it true that $XA + XB + XC + XD + XE = 1$?

To summarize, when you create indicator variables from a nominal variable, you must

- Leave out one category to ensure unique least-squares estimates.
- Interpret the intercept as the mean of Y for the left-out category.
- Interpret all indicator variable coefficients in terms of differences from the left-out category.
- Understand that the predicted values, the R^2 statistic, and the F statistic will be the same, no matter which category you leave out.

As a final comment, the analysis of the housing price data is an example of a "one-way ANOVA," or "one-way analysis of variance," which refers to a comparison of two or more independent groups. (The two-sample t test is a special case of one-way ANOVA.) The F statistic from the one-way ANOVA, which you can obtain from `R` using `summary(aov(sell ~ location, data=house))`, is exactly the same F statistic that you get using the regression model with indicator variables $\left(F_{4,59} = 6.89\right)$.

The term "analysis of variance," for which ANOVA is the acronym, is somewhat of a misnomer. The main purpose of ANOVA is not to analyze variances, it is to analyze mean values. The reason "variance" is mentioned is that larger F and R^2 values occur with ANOVA data when there is high *variance between* the groups; i.e., when the groups differ greatly.

A happy consequence of the correspondence between ANOVA and regression is that you really do not need to bother with ANOVA at all—you can do it all with regression using indicator variables, e.g., using the `lm` software of `R`. On the other hand, you should know what one-way ANOVA refers to: It refers to the comparison of means of (independent) data in different groups. Because the models are identical, the assumptions of the one-way ANOVA model are also identical to those of the regression model: They are independence, constant variance, and normality. The linearity assumption (or more precisely, the *correct functional specification assumption*) of the regression model is automatically true when a single nominal predictor variable is represented by indicators, since the model places no constraints on the mean values (see the right panel of Figure 10.2).

Self-study question: What is the meaning of the word "constraint" in the previous sentence?

10.3 Using Indicator Variables and "Ordinary" X Variables in the Same Model (ANCOVA)

You have already seen a model with both an indicator variable and an "ordinary" X variable. It is what we used to predict GPA as a function of GMAT and degree program (Master's or PhD) in Chapter 6. The "`degree`" variable was modeled as an indicator (1/0) variable, and the "GMAT" variable is what we mean by an "ordinary" X variable. The GPA variable was modeled as follows:

$$\text{GPA} = \beta_0 + \beta_1\text{PHD} + \beta_2\text{GMAT} + \varepsilon$$

The conditional mean function implied by this model is the plane $\beta_0 + \beta_1\text{PHD} + \beta_2\text{GMAT}$; the least-squares estimate of this plane, along with the actual data values, are shown in Figure 7.4. As shown above in this chapter, the best way to understand models with indicator variables is to separate them by the possible levels of the qualitative variable that is modeled via the indicator variable. This method also helps you to understand the meaning of the plane shown in Figure 7.4.

Group 1: PHD = 0 (Master's students)

$$\begin{aligned}\text{GPA} &= \beta_0 + \beta_1\left(0\right) + \beta_2\text{GMAT} + \varepsilon \\ &= \beta_0 + \beta_2\text{GMAT} + \varepsilon\end{aligned}$$

Group 2: PHD = 1 (PhD students)

$$\begin{aligned} \text{GPA} &= \beta_0 + \beta_1(1) + \beta_2\text{GMAT} + \varepsilon \\ &= (\beta_0 + \beta_1) + \beta_2\text{GMAT} + \varepsilon \end{aligned}$$

Since the ε terms have mean zero, you can conclude from these models that:

- β_0 is the mean of the distribution of the *potentially observable* GPAs for Master's students having GMAT = 0 (a correct but essentially useless interpretation since GMAT cannot equal 0).
- $\beta_0 + \beta_1$ is the mean of the distribution of the *potentially observable* GPAs for PhD students having GMAT = 0 (again, correct but essentially useless).
- β_2 is the difference between the means of the *potentially observable* GPAs for Master's students in two groups, one having GMAT = $x + 1$, the other having GMAT = x. Since the slope β_2 is the same in both models, β_2 is also the difference between the means of the *potentially observable* GPAs for PhD students in two groups, one having GMAT = $x + 1$, the other having GMAT = x.
- β_1 is the difference between the mean of the *potentially observable* GPAs for PhD students and the mean of *potentially observable* GPAs for Master's students who all share the same GMAT score. Unlike the two intercepts, this parameter has a useful interpretation because the GMAT score can be set at any relevant value like 500, or 600. The interpretation is not restricted to the irrelevant case where GMAT = 0.

Three-dimensional (3-D) plots are more difficult to understand than two-dimensional (2-D) plots. Rather than use the 3-D graph shown in Figure 7.4, it is better to estimate the model and display the two linear functions shown above in Group 1 and Group 2 on the same 2-D plot; see Figure 10.4.

R code for Figure 10.4

```
grades = read.table("https://raw.githubusercontent.com/andrea2719/
URA-DataSets/master/gpa_gmat.txt")
attach(grades)
fit = lm(gpa ~ degree + gmat)
b = fit$coefficients; b0 = b[1];b1 = b[2]; b2 = b[3]
pch = ifelse(degree == "P", 17, 1)
plot(gpa ~ gmat, data=grades, pch = pch, xlim = c(400,800),
     xlab = "GMAT Score", ylab = "GPA", cex=.8)
gmat.list = seq(400,800,10)
PhD.line = (b0 + b1) + b2*gmat.list
Masters.line = b0 + b2*gmat.list
points(gmat.list, PhD.line, type="l", lwd=2)
points(gmat.list, Masters.line, type="l", lty=2, lwd=2)
legend(692, 3.1, c("PhD", "Masters"), lty = c(1,2), lwd=c(2,2))
```

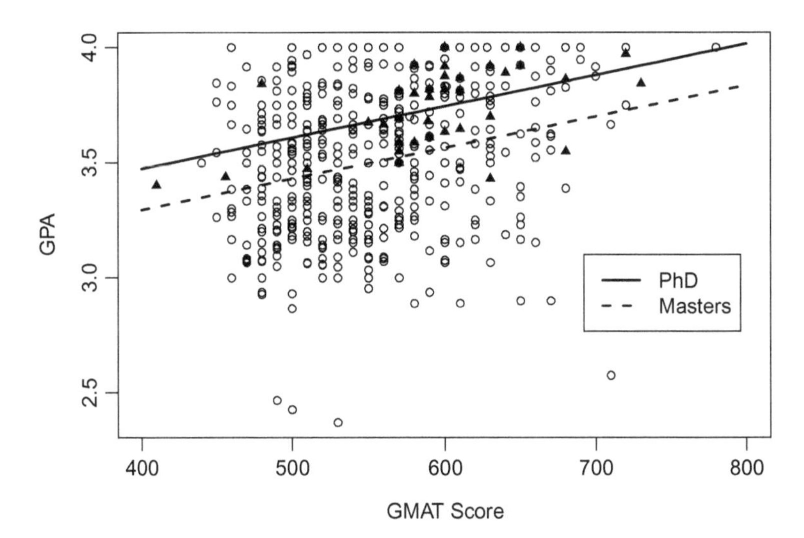

FIGURE 10.4
Scatterplot of (GMAT, GPA) data, showing PhD students as solid triangles and Master's students as hollow circles. Fitted functions are from the indicator variable model, without interaction.

Figure 10.4 is a two-dimensional representation of Figure 7.4. In particular, the lines shown in Figure 10.4 are the "slices" of the plane shown in Figure 7.4 corresponding to PhD students (where PHD = 1 in Figure 7.4) and Master's students (where PHD = 0 in Figure 7.4).

The summary of the fitted model graphed in Figure 10.4 is given as follows:

```
Call:
lm(formula = gpa ~ degree + gmat)

Residuals:
     Min       1Q   Median       3Q      Max
-1.13932 -0.20578  0.01747  0.22294  0.62498

Coefficients:
              Estimate Std. Error t value Pr(>|t|)
(Intercept) 2.7506999  0.1191639  23.083  < 2e-16 ***
degreeP     0.1793805  0.0503072   3.566 0.000398 ***
gmat        0.0013572  0.0002156   6.296  6.8e-10 ***
---
Signif. codes:  0 '***' 0.001 '**' 0.01 '*' 0.05 '.' 0.1 ' ' 1

Residual standard error: 0.2852 on 491 degrees of freedom
Multiple R-squared:  0.1142,    Adjusted R-squared:  0.1106
F-statistic: 31.64 on 2 and 491 DF,  p-value: 1.187e-13
```

Figure 10.4 greatly facilitates interpretation of the estimated coefficients shown in this output:

- The dashed line extended to GMAT = 0 intercepts the GPA axis at $\hat{\beta}_0 = 2.7506999$ (grade points).
- The slopes of both lines are identical: $\hat{\beta}_2 = 0.0013572$ (grade points).
- The vertical distance between the two lines, for any fixed value of GMAT Score, is $\hat{\beta}_1 = 0.1793805$ (grade points).

In addition, you may conclude as follows:

- The R^2 statistic is 0.1142, implying the 11.42% of the variation in grade point averages is explained by the combined effects of GMAT score and degree program (Master's or PhD).
- The F statistic is $F_{2,491} = 31.64$, too large to be explained by a data-generating process where degree program and GMAT are both unrelated to GPA. Under such a data-generating process, there is a 1.2×10^{-13} probability of observing and F statistic as large as 31.64. Thus, you can rule out chance, and conclude that either degree program, or GMAT, or both, is (are) related to GPA.
- Neither of the estimated coefficients for GMAT and degree program can be explained easily by chance alone (p-values 6.8×10^{-10} and 0.000398, respectively). Thus, you may logically conclude that, for fixed GMAT, degree program is related to GPA; and for fixed degree program, GMAT is related to GPA.

We showed previously in this chapter that the predicted values of Y corresponded to simple data averages when the model contained indicators for a single nominal X variable. That was unusual! It never happened in any of the other regression examples. And you can see that it does not happen here, either: Have a look at Figure 10.4 again. The student with the largest GMAT (780) is clearly visible in the upper right portion of the graph. Since there is only one such student, the average GPA for students in the data set having GMAT = 780 is simply that student's GPA, which is 4.0. But the regression model predicts GPA for such students much lower than 4.0, as can be seen by the dashed line in Figure 10.4.

Further, the two lines in Figure 10.4 are not even the least-squares lines that you would get if you performed the regressions separately for the Master's and PhD students. Such separate analyses would not constraint the slopes to be equal, and therefore the estimated slopes of the two functions will be different. In such a case, you would say that the effect of GMAT on GPA depends on degree program, which can also be stated as "the effect of GMAT on GPA is moderated by degree program." As we described in Chapter 9, you can easily estimate a moderation effect by including an interaction term in your model.

10.4 Interaction Between Indicator Variables and "Ordinary" X Variables (ANCOVA with Interaction)

Figure 10.5 shows the estimated least-squares lines for the GPA prediction case when the data are separated by students.

R code for Figure 10.5

```
PHD = grades[degree=="P",]; MAS = grades[degree=="M",]
abline(lsfit(PHD$gmat, PHD$gpa), lty=2, col="gray")
abline(lsfit(MAS$gmat, MAS$gpa), lty=2, col="gray")
```

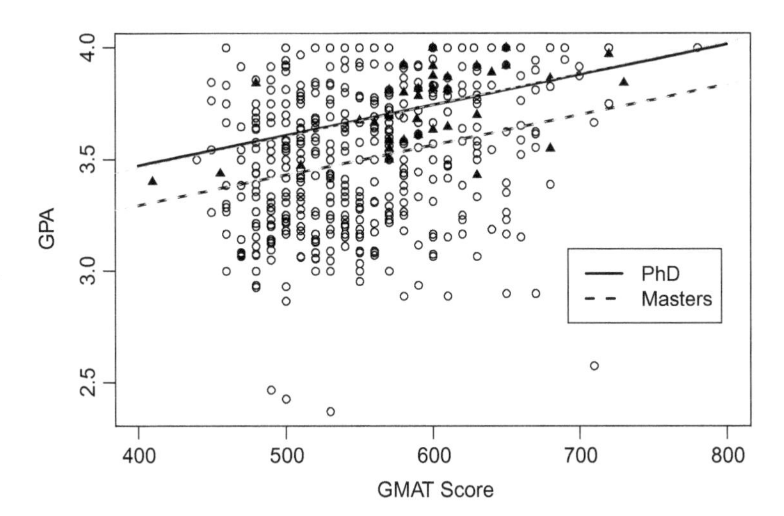

FIGURE 10.5
Identical to Figure 10.4, but with least-squares lines for separate data sets superimposed as dashed lines.

From Figure 10.5, it appears that the original model, which forced equal slopes for the two groups, is perfectly adequate because when we separate the students into the two groups and estimate the lines separately, they have nearly equal slopes. It does not have to be this way, though. As an extreme example, if the effect of GMAT was estimated to be positive for Master's students but negative for PhD student, then we certainly would not want to assume a common slope model.

As we saw in the first two indicator variable applications (two-sample t test and ANOVA), the estimates of the model parameters correspond to simple averages of subsets of the data. Here, we will see something similar when we estimate the interaction model: The estimated regression lines from the single interaction model estimated using the combined data set correspond precisely to the least-squares estimates shown in Figure 10.5 on the two subsets (one for PhD students, one for Master's students). The interaction model is estimated using a product term, as in Chapter 9, as follows:

```
fit.int = lm(gpa ~ degree + gmat + degree*gmat)
summary(fit.int)
```

The output is as follows:

```
Call:
lm(formula = gpa ~ degree + gmat + degree * gmat)

Residuals:
     Min       1Q   Median       3Q      Max
-1.13852 -0.20587  0.01738  0.22278  0.62453

Coefficients:
              Estimate Std. Error t value Pr(>|t|)
(Intercept)  2.753e+00  1.244e-01  22.138  < 2e-16 ***
degreeP      1.430e-01  4.694e-01   0.305    0.761
gmat         1.352e-03  2.251e-04   6.007 3.69e-09 ***
degreeP:gmat 6.154e-05  7.905e-04   0.078    0.938
---
Signif. codes:  0 '***' 0.001 '**' 0.01 '*' 0.05 '.' 0.1 ' ' 1
```

```
Residual standard error: 0.2855 on 490 degrees of freedom
Multiple R-squared:  0.1142,    Adjusted R-squared:  0.1088
F-statistic: 21.05 on 3 and 490 DF,  p-value: 7.634e-13
```

To interpret the results, note that the theoretical model being estimated here is:

$$\text{GPA} = \beta_0 + \beta_1\text{PHD} + \beta_2\text{GMAT} + \beta_3\text{PHD} \times \text{GMAT} + \varepsilon$$

As always, the best way to understand parameters in an indicator variable model is to separate the model by levels of the nominal variable that is represented by the indicator variable. Here the nominal variable is "degree," and its levels are PhD and Master's.

Group 1: PHD = 0 (Master's students)

$$\begin{aligned}\text{GPA} &= \beta_0 + \beta_1(0) + \beta_2\text{GMAT} + \beta_3(0)\text{GMAT} + \varepsilon \\ &= \beta_0 + \beta_2\text{GMAT} + \varepsilon\end{aligned}$$

Group 2: PHD = 1 (PhD students)

$$\begin{aligned}\text{GPA} &= \beta_0 + \beta_1(1) + \beta_2\text{GMAT} + \beta_3(1)\text{GMAT} + \varepsilon \\ &= (\beta_0 + \beta_1) + (\beta_2 + \beta_3)\text{GMAT} + \varepsilon\end{aligned}$$

Notice that these two models can have any intercepts and any slopes whatsoever, unlike the no-interaction model that constrained the slopes to be equal.

As it turns out, the estimated models obtained using the entire data set are mathematically identical to the least-squares estimates obtained by separating the data into the two groups, as you will see when you run the following code. Please run the following code now.

```
## Estimated regression lines using data subsets.

# Master's students
lsfit(gmat[degree=="M"], gpa[degree=="M"])$coefficients

#PhD students
lsfit(gmat[degree=="P"], gpa[degree=="P"])$coefficients

## Estimated regression lines using the combined
## data and interaction model

PHD = ifelse(degree=="P",1,0)
fit.int = lm(gpa ~ PHD + gmat + gmat*PHD)
b = fit.int$coefficients
b0 = b[1]
b1 = b[2]
b2 = b[3]
b3 = b[4]
```

```
# Master's students
c(b0,b2)

# PhD students
c(b0+b1, b2+b3)
```

You can interpret the parameters of the interaction model as follows. Notice that the first two interpretations are identical to the no-interaction model for the same data as discussed above.

- β_0 is the mean of the distribution of the *potentially observable* GPAs for Master's students having GMAT = 0.
- $\beta_0 + \beta_1$ is the mean of the distribution of the *potentially observable* GPAs for PhD students having GMAT = 0.
- β_2 is the difference between the means of the *potentially observable* GPAs for Master's students in two groups, one having GMAT $= x + 1$, the other having GMAT $= x$. Unlike the no-interaction model, in which slopes for the two models were constrained to be equal, this interpretation only applies to Master's students in this model because of the interaction term.
- $\beta_2+\beta_3$ is the difference between the means of the *potentially observable* GPAs for PhD students in two groups, one having GMAT $= x + 1$, the other having GMAT $= x$.

As a consequence, you may conclude the β_1 is the difference between intercepts of the two models. Unlike the case with the no-interaction model, the parameter β_1 is no longer interesting because its interpretation requires GMAT = 0. Also, β_3 is the difference between the slopes of the two models. This parameter, on the other hand, *is* interesting, because if it is large enough, you should use the separate slope model. If it is small, then the common slope model may be adequate.

Repeating the output from `summary(fit.int)` above:

```
Coefficients:
              Estimate Std. Error t value Pr(>|t|)
(Intercept)  2.753e+00  1.244e-01  22.138  < 2e-16 ***
degreeP      1.430e-01  4.694e-01   0.305    0.761
gmat         1.352e-03  2.251e-04   6.007 3.69e-09 ***
degreeP:gmat 6.154e-05  7.905e-04   0.078    0.938
```

Consider the last line, which concerns the interactions term, β_3. The PhD slope is estimated to be 6.154×10^{-5} higher than the slope for Master's students; this is a small difference but is visible in the dashed lines of Figure 10.5. However, this difference is only 0.078 standard errors from zero, and is thus, explainable by chance alone. Of course, that is not to say that the slopes are the same.

Arguably, the PhD slope should actually be *less than* the Master's slope, because the PhD data are closer to the upper boundary of 4.0. Since the data are close to the boundary, there logically must be a flattening of the function relative to that of the Master's students, whose data are generally farther from the boundary. So, while the difference between estimated slopes is explainable by chance, the true slopes really are different, and arguably in the reverse direction as suggested by the data.

So, should you use the separate slope model (the interaction model) or the common slope model (the no-interaction model)? In reality, there is always interaction, just like there is always curvature. The reason that there is always interaction is the same reason that there is always heteroscedasticity, nonlinearity, and non-normality. But, despite the fact that there are always violations of model assumptions, you often can use the model anyway; review Figure 1.13 to understand this concept anew. Thus, despite existence of interaction between variables, you can often use a no-interaction model anyway. Please look at Figure 10.5 again: In it, you see that the simpler no-interaction model, while wrong, is quite adequate because it differs so little from the interaction model. Plus, the no-interaction model has the advantage that it is simpler, requiring fewer parameters. Further, the parameters of the no-interaction model are easier to interpret than those of the interaction model. Overall, the no-interaction model is easily preferred here, despite being wrong.

The best answer to the question "which model should I use?" involves comparing graphs such as shown in Figure 10.5, using subject matter considerations as discussed in the preceding two paragraphs, and using simulation to resolve any remaining issues. You should not rely on *p*-values, because *p*-values do not address the question, "Which model should I use?" They only address the question, "Are the deviations we see in our data explainable by chance alone?"

10.4.1 Does Location Affect House Price, Controlling for House Size?

Even though the realtors say "location, location, location!", the observed effects of location on house price might simply be due to the fact that bigger homes tend to be in some locations. After all, square footage is a strong determinant of house price. To compare prices in different locations for homes of the same size, simply add "sqfeet" to the model like this:

```
house = read.csv("https://raw.githubusercontent.com/andrea2719/
URA-DataSets/master/house.csv", header=T)
attach(house)
fit.main = lm(sell ~ location + sqfeet, data=house)
summary(fit.main)
```

The results are as follows:

```
Coefficients:
             Estimate Std. Error t value Pr(>|t|)
(Intercept)  25.898669   5.060777   5.118 3.67e-06 ***
locationB   -21.106407   2.152655  -9.805 6.41e-14 ***
locationC   -21.431288   3.579304  -5.988 1.43e-07 ***
locationD   -24.846429   2.574269  -9.652 1.13e-13 ***
locationE   -27.304759   2.538505 -10.756 1.94e-15 ***
sqfeet        0.041224   0.002578  15.993  < 2e-16 ***
---
Signif. codes:  0 '***' 0.001 '**' 0.01 '*' 0.05 '.' 0.1 ' ' 1

Residual standard error: 6.638 on 58 degrees of freedom
Multiple R-squared:  0.874,     Adjusted R-squared:  0.8631
F-statistic: 80.47 on 5 and 58 DF,  p-value: < 2.2e-16
```

In this analysis, 87.4% of the variation in price is explained by the combined effects of location and square footage, much more than the 31.8% that was explained by location alone. This result is expected because, again, square footage is a strong determinant of house price. The model estimates that, for a fixed location, an additional square foot of house costs an additional 0.041224 (thousands of dollars) on average. And the negative coefficients of the "location" indicator variables suggest that, for any fixed square footage, homes in location "A" cost more on average than homes in any other location. Thus, the location differences discovered earlier cannot be explained simply by differences in home size for the different locations.

Whenever possible, you should use graphs to clarify such complex information.

R code for Figure 10.6

```
b = fit.main$coefficients
b0 = b[1];b1 = b[2]; b2 = b[3]; b3 = b[4]; b4 = b[5]; b5 = b[6]
pch = as.numeric(location)
plot(sell ~ sqfeet, data=house, pch=pch)
sqf = seq(1000,3000,10)
A.line = b0 + b5*sqf
B.line = b0 + b1 +  b5*sqf
C.line = b0 + b2 +  b5*sqf
D.line = b0 + b3 +  b5*sqf
E.line = b0 + b4 +  b5*sqf
points(sqf, A.line, type="l", lty=1)
points(sqf, B.line, type="l", lty=2)
points(sqf, C.line, type="l", lty=3)
points(sqf, D.line, type="l", lty=4)
points(sqf, E.line, type="l", lty=5)
legend("bottomright",c("Region A","Region B","Region C","Region D",
"Region E"), lty = 1:5, pch=1:5)
```

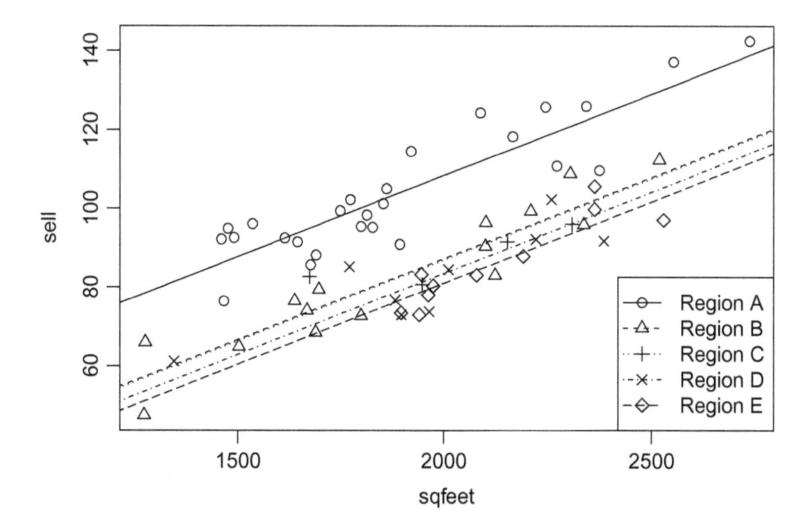

FIGURE 10.6
Housing price as a function of square footage for different locations, with fitted values from the no-interaction (parallel slope) model.

Notice in Figure 10.6 that location "A" has higher estimated mean sales than for any other location, at fixed value of square footage. For example, when sqfeet = 2,000, the estimated mean sales for location A is 108.3465, which is 21.1064 higher than the estimated mean sales for location B when sqfeet = 2,000, corresponding to $\hat{\beta}_1 = -21.106407$ in the fitted model summary above.

The graphs in Figure 10.6 show parallel slopes, because the effect of home size on price is assumed to be the same for all locations. In reality, it is impossible that all these slopes are exactly the same. As shown above, the way to assess whether an assumption (constraint) is reasonable is simply to fit the model that allows deviations from the assumption, and then to compare the results.

To allow separate effects of sqfeet on house price for the different locations, simply include interaction terms. Here, since there are four indicator variables, you must interact all four of them with sqfeet. Fortunately, `R` does this for you automatically—you can do it all inside the `lm` function, as follows:

```
fit.int = lm(sell ~ location + sqfeet + location*sqfeet , data=house)
summary(fit.int)
```

The resulting fit shows an additional four parameters, one for each indicator variable:

```
Coefficients:
                    Estimate Std. Error t value Pr(>|t|)
(Intercept)        24.212444   7.352927   3.293  0.00175 **
locationB         -23.599025  11.496279  -2.053  0.04496 *
locationC          17.405228  29.655472   0.587  0.55971
locationD          -9.371093  16.956577  -0.553  0.58278
locationE         -35.232771  22.573505  -1.561  0.12441
sqfeet              0.042113   0.003813  11.044 1.81e-15 ***
locationB:sqfeet    0.001331   0.005978   0.223  0.82461
locationC:sqfeet   -0.019278   0.014630  -1.318  0.19316
locationD:sqfeet   -0.007885   0.008565  -0.921  0.36135
locationE:sqfeet    0.003637   0.010699   0.340  0.73524
---
Signif. codes:  0 '***' 0.001 '**' 0.01 '*' 0.05 '.' 0.1 ' ' 1

Residual standard error: 6.695 on 54 degrees of freedom
Multiple R-squared:  0.8807,     Adjusted R-squared:  0.8608
F-statistic: 44.28 on 9 and 54 DF,  p-value: < 2.2e-16
```

Not much has been gained here: The R^2 statistic shows 88.1% of the variance in price is explained by this model, compared to 87.4% without the interaction terms. Further, the adjusted R^2, which is less biased, is actually *smaller* for this model with interaction terms (86.1% versus 86.3% for the no-interaction model), but the adjusted R^2 statistic is not the best one to use for comparing models. Better statistics for model comparison are given in the next chapter.

The additional parameters in the interaction model are differences of slopes for the given location compared to the "left-out" location, location A in this case. For example, location B is estimated to have a 0.001331 higher slope than location A. But, as always, it is best to draw graphs to facilitate understanding.

R code for Figure 10.7

```
b = fit.int$coefficients
b0 = b[1];b1 = b[2]; b2 = b[3]; b3 = b[4]; b4 = b[5]; b5 = b[6]
b6 = b[7]; b7 = b[8]; b8 = b[9]; b9 = b[10]
pch = as.numeric(location)
plot(sell ~ sqfeet, pch=pch, data=house)
sqf = seq(1000,3000,10)
A.line = b0 + b5*sqf
B.line = b0 + b1 +  (b5+b6)*sqf
C.line = b0 + b2 +  (b5+b7)*sqf
D.line = b0 + b3 +  (b5+b8)*sqf
E.line = b0 + b4 +  (b5+b9)*sqf
points(sqf, A.line, type="l", lty=1)
points(sqf, B.line, type="l", lty=2)
points(sqf, C.line, type="l", lty=3)
points(sqf, D.line, type="l", lty=4)
points(sqf, E.line, type="l", lty=5)
legend("bottomright", c("Location A","Location B","Location C",
"Location D","Location E"), lty = 1:5, pch=1:5)
```

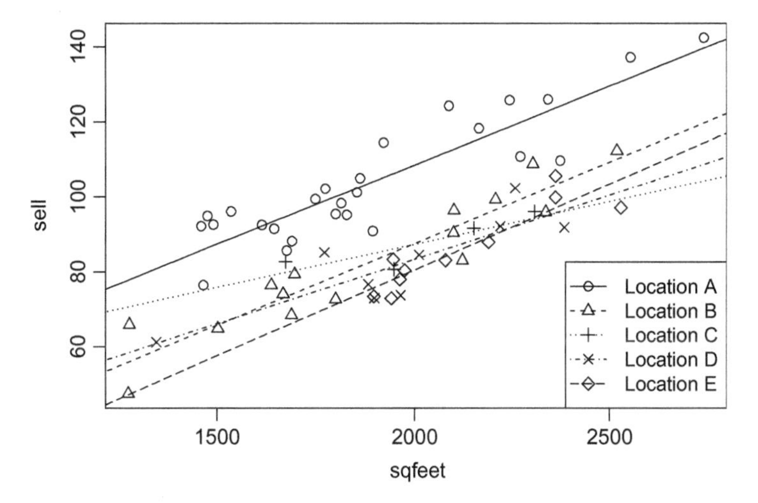

FIGURE 10.7
Housing price as a function of square footage for different locations, with fitted values from the interaction (separate slopes) model.

Notice that the lines in Figure 10.7 all have different slopes. As was the case with the grade point average example, these lines are identical to what you would get if you created five subsets of data, one for each location, and then performed ordinary linear regressions to predict price in each subset as a function of sqfeet.

Self-study question: How do you use R to check that the two methods described in the previous sentence will give you the same five regression lines?

Notice in particular that the location C subset seems to have a slope that is dramatically different than the others. But this is no surprise because, as we saw above, there are only four homes in location C (indicated by "+" symbols in Figures 10.6 and 10.7), so the estimate of the least-squares line is not as accurate in that location as it is in other locations.

Finally, the analyses performed in this section and in the previous section are called "ANCOVA" or "Analysis of Covariance." A happy conclusion is that you do not need to use any special methods to perform ANCOVA, you can just use regression with indicator variables and "ordinary" X variables. But you should at least know that ANCOVA refers to comparison of data in different groups (like ANOVA), but where an "ordinary" X variable (called a "covariate" in this context) is used in the model as well. But other than terminology, the ANCOVA model, assumptions, test statistics, and results are identical to what you get using regression with indicator variables.

Like ANOVA is a misnomer in that the methodology does not refer to analyzing variances, ANCOVA is also a misnomer in that the methodology does not refer to analyzing covariances. In some cases, you really do want to analyze variances and covariances by using your regression data; we give such methods in Chapter 12.

10.5 Full Model versus Restricted Model *F* Tests

As we have mentioned repeatedly, tests of hypotheses are not the best way to evaluate models and assumptions. However, the F test that was introduced in Chapter 8 is so common in the history of ANOVA, ANCOVA, and regression that we would be remiss not to mention it.

Models such as those shown in Figures 10.7 and 10.6 are often compared by using the F test, which is a test to compare "full" versus "restricted" classical regression models. (For models other than the classical regression model, full/restricted model comparison is more commonly done using the likelihood ratio test, which is used starting in Chapter 12 of this book.)

In the usual regression analysis, a full model typically has the form:

$$Y = \beta_0 + \beta_1 X_1 + \beta_2 X_2 + \ldots + \beta_k X_k + \varepsilon$$

Here, the parameters β_0, β_1, $\beta_2, \ldots$, and β_k are *unconstrained*; that is, each parameter can possibly take any value whatsoever between $-\infty$ and ∞, and the value that one β parameter takes is not dependent on (or constrained by) the value that any other β parameter takes.

A restricted model is the same model, but with *constraints* on the parameters. The most common restrictions are constraints such as $\beta_1 = \beta_2 = 0$, although other constraints such as $\beta_2 = 1$, or $\beta_1 - \beta_2 = 0$, or $\beta_0 + 15\,\beta_2 = 100$ are also possible.

The separate slope model graphed in Figure 10.7 is a full model relative to the restricted model that constrains all the interaction β's to be zero, shown in Figure 10.6. The F test can be used to compare these models. To construct the F test, let SSE_F denote the error sum of squares in the full model, and let SSE_R denote the error sum of squares in the restricted model. It is a mathematical fact that

$$SSE_F \le SSE_R$$

This is an important point, so it bears more explanation. Recall $\text{SSE}(b_0, b_1, \ldots, b_k) = \sum_{i=1}^{n}\{y_i - (b_0 + b_1 x_{i1} + \cdots + b_k x_{ik})\}^2$. Now, the *least-squares* algorithm tells you that SSE_F is the *minimum* value of $\text{SSE}(b_0, b_1, \ldots, b_k)$ for *all possible combinations* $(b_0, b_1, \ldots, b_k)$.

In the restricted model, some of the b values are constrained, e.g., to 0. Therefore, the *set of possible combinations* $\{b_0, b_1, \ldots, b_k\}$ in the restricted model is a *subset* of the set of possible combinations $\{b_0, b_1, \ldots, b_k\}$ in the unrestricted model. Thus, SSE_R is the minimum of $\text{SSE}(b_0, b_1, \ldots, b_k)$ over a set of values $\{b_0, b_1, \ldots, b_k\}$ that is a *subset* of the unrestricted set. The minimum of a set of values has to be as small, or smaller than the minimum of any subset, right? That fact proves $\text{SSE}_\text{F} \le \text{SSE}_\text{R}$.

For example, the minimum of the set of numbers {7, 4, 3, 5, 0, 4, 4, 3, 5, 4} is zero. Now, pick a subset of that set, like the first four: {7, 4, 3, 5}. The minimum of the subset (the restricted set) is three, which is larger than zero.

Thus, even when the restricted model is the true model, the full model will *appear* to fit better (or at least no worse), because $\text{SSE}_\text{F} \le \text{SSE}_\text{R}$.

Self-study question: How does the fact that $\text{SSE}_\text{F} \le \text{SSE}_\text{R}$ imply mathematically that the R^2 statistic ordinarily will be higher, and never lower, for the full model, even when the restricted model is true?

If the restricted model is true, *then* the improvement of the fit of the full model, as measured by $\text{SSE}_\text{R} - \text{SSE}_\text{F}$, is *explained by chance alone.* The F statistic is a re-packaging of this difference into a form that has a known distribution under the chance-only (restricted) model.

The *F* statistic for comparing a full versus restricted (null) model

$$F = \frac{(\text{SSE}_\text{R} - \text{SSE}_\text{F}) / (dfe_\text{R} - dfe_\text{F})}{\text{SSE}_\text{F} / dfe_\text{F}}$$

The main component of this statistic is the term $\text{SSE}_\text{R} - \text{SSE}_\text{F}$ in the numerator, which tells you that the F statistic will be big if the restricted model fits the data much worse than the unrestricted model.

Again, the chance-only (null) model is the restricted model. It is a mathematical theorem that, *if* the data are in fact produced by the restricted model, with all classical model assumptions true, *then* distribution of the F statistic is given as follows:

Null distribution of the *F* statistic

$$F \sim F_{dfe_\text{R} - dfe_\text{F},\, dfe_\text{F}}$$

If the restricted (null) model is false, then you will see larger values of $\text{SSE}_\text{R} - \text{SSE}_\text{F}$, making the distribution of the F statistic shifted to the right of the $F_{dfe_\text{R} - dfe_\text{F},\, dfe_\text{F}}$ distribution. Thus, the p-value for the test is the area under the $F_{dfe_\text{R} - dfe_\text{F},\, dfe_\text{F}}$ distribution *to the right* of the observed F statistic.

For example, let the full model be the interaction model shown in Figure 10.7, and let the restricted model be the no-interaction model shown in Figure 10.6. The restriction that is placed on the full model is the restriction that all the β's corresponding to interaction terms are zero, i.e., that $\beta_6 = \beta_7 = \beta_8 = \beta_9 = 0$.

As shown in the outputs above, the error degrees of freedom are $dfe_R = 58$ and $dfe_F = 54$; hence the numerator degrees of freedom are $dfe_R - dfe_F = 4$, which is equal to the number of constraints. The error sum of squares are not given in the output, but you can obtain them as sums of squared residuals:

```
SSE.F = sum(fit.int$residuals^2)
SSE.R = sum(fit.main$residuals^2)
```

This gives the values 2,420.42 and 2,555.744, respectively. Thus, the F statistic is

```
F = ((SSE.R - SSE.F)/(58-54))/(SSE.F/54)
```

which gives you $F = 0.7547747$. The degrees of freedom for the F distribution are 4 and 54, so the p-value is `1-pf(F, 4,54)`, which gives you $p = 0.5592687$.

Thus, the differences between the estimated slopes seen in Figure 10.7 are explainable by chance alone: If the data are produced by a model where the slopes were in fact equal, but an interaction model is estimated, then there will be random differences in the estimated slopes that are explained by chance alone. The differences in the estimated slopes as large as seen in Figure 10.7 are in the range of such chance-only differences, meaning that the differences seen in the slopes of Figure 10.7 are explainable by chance alone.

As you might expect, there is a simpler way to get that F statistic using `R`.

10.5.1 Computing the *F* Statistic to Compare Full and Restricted Models

First, fit both models via `lm`, resulting in fitted `lm` objects called (say) `fit.Restricted` and `fit.Full`. Then enter the command `anova(fit.Restricted, fit.Full)`.

The fitted models above were called `fit.main` and `fit.int`; the command `anova(fit.main, fit.int)`results in the following output:

```
Analysis of Variance Table

Model 1: sell ~ location + sqfeet
Model 2: sell ~ location + sqfeet + location * sqfeet

  Res.Df    RSS Df Sum of Sq      F Pr(>F)
1     58 2555.7
2     54 2420.4  4    135.32 0.7548 0.5593
```

Notice that we already got all these numbers in the "by hand" calculations: In the "`Res.Df`" column you see the error degrees of freedom, $dfe_R = 58$ and $dfe_F = 54$. In the "`RSS`" column you see the sum of squared residuals $SSE_R = 2555.744$ and $SSE_F = 2420.42$ calculated "manually" above. In the "`Df`" column you see the numerator degrees of freedom for the F statistic, specifically $dfe_R - dfe_F = 58 - 54 = 4$. In the "`Sum of Sq`" column you see the difference in SSE's, $SSE_R - SSE_F = 2555.744 - 2420.42 = 135.32$. Finally, the "`F`" and "`Pr(>F)`" columns give the F-statistic and p-value calculated manually above.

To understand the interpretation that the differences in slopes are *explainable by chance alone*, you first need to understand what *explained by chance alone* means. As always, simulation helps you enormously to understand this concept.

10.5.2 Simulation to Understand the Null (Chance-Only) Model

First, we will create a simulation model to produce data that looks like the housing data, but with equal slopes. We use the estimated coefficients and estimated conditional standard deviation in the restricted model summary above to suggest true parameter values to use in the simulation, but the actual values are not important. What is important is that the data are simulated according to the classical model with identical slopes.

```
house = read.csv("https://raw.githubusercontent.com/andrea2719/
URA-DataSets/master/house.csv", header=T)
attach(house)
n = nrow(house)
locB = (location=="B"); locC = (location=="C")
locD = (location=="D"); locE = (location=="E")
sell.sim = 26 -21*locB - 21*locC - 25*locD -27*locE +
.04*sqfeet + rnorm(n,0,6.6)
fit.Full = lm(sell.sim ~ location + sqfeet + location*sqfeet)
fit.Rest = lm(sell.sim ~ location + sqfeet)
anova(fit.Rest, fit.Full)
```

In the simulated data, the restricted model really is the true model: Look at the line in the code where "`sell.sim`" is defined, and you will see no interaction terms. Nevertheless, if you fit the full model with interaction terms, the sum of squared errors is necessarily smaller for the full model, as proven above with the "constraint/subset" analysis. Here is the output:

```
Analysis of Variance Table

Model 1: sell.sim ~ location + sqfeet
Model 2: sell.sim ~ location + sqfeet + location * sqfeet
  Res.Df    RSS Df Sum of Sq      F Pr(>F)
1     58 2503.7
2     54 2328.9  4    174.77 1.0131 0.4089
```

The error sum of squares is 174.77 smaller for the full model, but according to the F test, this difference is explainable by chance. However, we know that, in this case, the difference is *explained* by chance, because the data come from a process where there is no interaction (see the "`sell.sim`" line of code again.) Even if the p-value were less than 0.05 (which would, in fact, happen in 1 out of 20 simulations of the type just performed), we would still know that the difference is *explained* by chance alone.

If this simulation were repeated 1,000's of times, and the F statistics were stored and then shown in a histogram, the mathematical theory states that this histogram will approximate the $F_{4,54}$ distribution. The mathematical theory is right (it always is!) The following simulation shows the results.

R code for Figure 10.8

```
NSIM = 10000; F.sim = numeric(NSIM)
for (i in 1:NSIM) {
sell.sim = 26 -21*locB - 21*locC - 25*locD -27*locE +
.04*sqfeet + rnorm(n,0,6.6)
fit.Full = lm(sell.sim ~ location + sqfeet + location*sqfeet)
fit.Rest = lm(sell.sim ~ location + sqfeet)
F.sim[i] = anova(fit.Rest, fit.Full)$F[2]
 }

F.list = seq(0,5,.001); F.dist = df(F.list,4,54)
plot(F.list, F.dist, type="l", xlab = "Null F Value",)
 ylab = "F(4,54) density", ylim = c(0,.9), yaxs="i")
hist(F.sim, breaks=50, add=T, freq=FALSE, lty=2)
abline(v = qf(.95, 4,54), lwd=2)
```

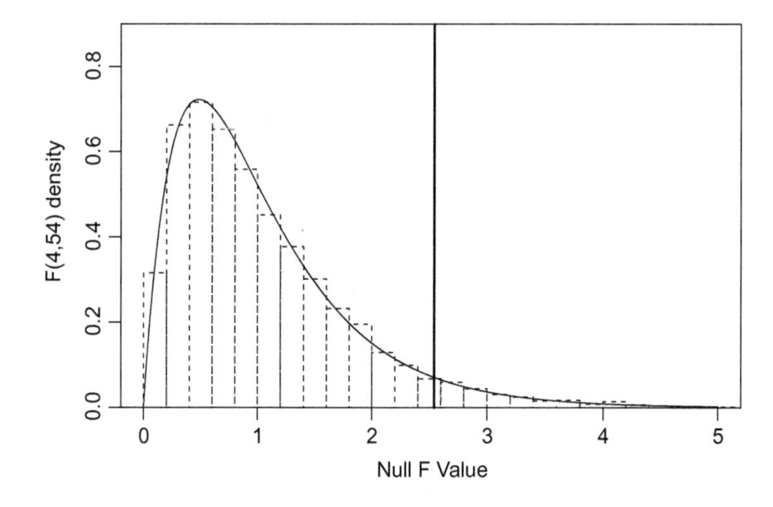

FIGURE 10.8
The $F_{4,54}$ density, (solid line), with 0.95 quantile, 2.543, shown as vertical line. The histogram (dashed lines) is constructed using 10,000 simulated F statistics that come from the null model. If an F value from a real study is less than 2.543 ($p > 0.05$), then it is within the range of F values that are typical under the null model. If an F value from a real study is more than 2.543 ($p < 0.05$), then it is within a range of F values that are not typical under the null model.

You can specify a collection of full/restricted models and test them all simultaneously using the ANOVA function in R. For example, consider the following nested sequence of models, each of which is a restricted model relative to the next model in the list.

Model 1: $\text{Price} = \beta_0 + \beta_1\text{sqfeet} + \varepsilon$

Model 2: $\text{Price} = \beta_0 + \beta_1\text{sqfeet} + \beta_2\text{XB} + \beta_3\text{XC} + \beta_4\text{XD} + \beta_5\text{XE} + \varepsilon$

Model 3: $\text{Price} = \beta_0 + \beta_1\text{sqfeet} + \beta_2\text{XB} + \beta_3\text{XC} + \beta_4\text{XD} + \beta_5\text{XE}$
$\quad + \beta_6\text{XB}\times\text{sqfeet} + \beta_7\text{XC}\times\text{sqfeet} + \beta_8\text{XD}\times\text{sqfeet} + \beta_9\text{XE}\times\text{sqfeet} + \varepsilon$

You can test all these simultaneously as follows:

```
model1 = lm(sell ~ sqfeet, data=house)
model2 = lm(sell ~ sqfeet + location, data=house)
model3 = lm(sell ~ sqfeet + location + sqfeet*location, data=house)
anova(model1, model2, model3)
```

This gives the following output:

```
Analysis of Variance Table

Model 1: sell ~ sqfeet
Model 2: sell ~ sqfeet + location
Model 3: sell ~ sqfeet + location + sqfeet * location
  Res.Df     RSS Df Sum of Sq       F Pr(>F)
1     62 11241.3
2     58  2555.7  4    8685.6 48.4442 <2e-16 ***
3     54  2420.4  4     135.3  0.7548 0.5593
---
Signif. codes:  0 '***' 0.001 '**' 0.01 '*' 0.05 '.' 0.1 ' ' 1
```

The bottom F-test, with $F = 0.7548$, tests the interaction model relative to the restriction that the slopes are truly equal, as already done. The second-to-bottom F-test, with $F = 48.4442$, tests whether there are location differences relative to the restricted model (model 1) that states there are no locational differences.

To interpret the statistic $F = 48.4442$, suppose (hypothetically) that the data are produced by pricing processes where there really are no location effects, only sqfeet effects. Still, when the model having location effects is fit to the data, the results will show non-zero estimates of location differences that are explained by chance alone (i.e., by the idiosyncratic nature of the particular observed data set). However, since the F value, 48.4442, is so large, it is difficult to explain the observed location differences in this way. Thus, you can rule out chance as a possible explanation for the observed location differences, and conclude that at least two of the locations differ in mean selling price.

So, what should you do with the "insignificant" interaction terms? According to the F test, they are collectively explainable by chance alone, so should they be dropped from the model? After all, there is no violation of the inclusion principle if you do drop them, since they are the highest-order terms.

Contrary to what you might have heard or read in other sources, the F test does not give you a recipe for action. "Explainable by chance alone" is quite different from "explained by chance alone." Just because terms are explainable by chance alone does not mean they are truly zero. Despite p-value based "significance," or lack thereof, you know for an absolute fact that slopes are really all different in reality—why would Nature choose to make all those slopes exactly the same? The real question is, how different are they?

Further, the graphs of the interaction model as shown in Figure 10.7 are interesting, and they do tell you what is in the data. There may be interesting information in the estimated model. Of course, if the different slope estimates are explainable by chance alone, then you have to interpret such graphs more cautiously.

As mentioned on Chapter 6, sometimes you might choose to exclude variables that you know are really related to Y (such as interaction effects) because including such variables adds more error to the predictions, in the form of parameter estimates ($\hat{\beta}$'s) that are in error (i.e., different from the true β's). But to decide whether to include such variables, you have to compare the predictive ability of the model with and without them. Such comparisons are best done using penalized likelihoods or cross-validation, to be discussed in Chapter 11. The F test (and any other test for that matter) is simply the wrong tool to decide whether to keep variables in a model.

Do not decide whether to keep variables in a model based solely on the F test, T test, or any other p-value yielding test.

These tests tell you whether results are explainable by chance alone, or are not easily explained by chance alone, and that is all. Tests do not address the question, "Should the variable be kept in the model?"

You might think that these F tests are redundant with the information already contained in the ordinary regression output. Look at the coefficients for the interaction terms in the full model for example: They are all "insignificant," based on their p-values, corroborating the small F-statistic 0.7548. And look at the common slope model: All four of the location difference estimates are "significant," based on their p-values, corroborating the large F-statistic 48.4442 shown just above. Isn't it sufficient to just look at the parameter estimates and their p-values, rather than bother with the full model/restricted model F tests?

The answer is, "No." For one thing, what if some of the tests are "significant" and some not? How many have to be "significant" in order to call the global test "significant"? In particular, p-values are uniformly distributed when the data are produced by the restricted model, so with 20 tests from the restricted model, you *expect* one of the p-values to be less than 0.05, explained purely by chance.

Further, it can easily happen that all of the coefficients of the indicator variables are "insignificant," yet the global test is "highly significant." This can happen, for example, when the reference category has a small sample size. In that case, the comparisons against the reference category might all have large p-values. But there may be big differences among the remaining categories, which will be correctly detected by the full model/restricted model F statistic.

It can also happen that the individual $|T|$ statistics are small, yet the global F statistic is large, when the X variables are multicollinear (this case is called a *predictive* multicollinearity).

Finally, note that the "ordinary" F test printed in bottom of the `lm` output, and that was discussed in Chapter 8, is also an example of the full model/restricted model test. The full model is $Y = \beta_0 + \beta_1 X_1 + \cdots + \beta_k X_k + \varepsilon$; the restricted model is $Y = \beta_0 + \varepsilon$, and the restriction is that $\beta_1 = \beta_2 = \cdots = \beta_k = 0$. The least-squares estimate of β_0 in the restricted model is just the average of the Y data, so the "SST" term used in the F statistic that was discussed in Chapter 8 is just the error sum of squares for this particular restricted model, which has only the intercept term, β_0.

10.6 Two Nominal Variables (Two-Way ANOVA)

Have a look at the grade point average data set again:

```
grades = read.table("https://raw.githubusercontent.com/andrea2719/
URA-DataSets/master/gpa_gmat.txt")
names(grades)
```

TABLE 10.1

College Majors Corresponding to the Numerical Codes of the "Major" Variable in the Grade Point Average Data Set

Major	Degree Program
2	MS Accounting
66	MS Finance
70	General Business
114	MS Management Information Systems
115	MS Production and Operations Management
118	MS Marketing
203	Business Administration

This gives you:

```
[1] "grad"    "gpa"     "gmat"    "major"   "degree" "sex"     "ethnic"
```

Suppose you want to predict Y = GPA as a function of the two nominal variables "Major" and "Sex." Since these are nominal variables, you should first look at the data to see how many observations there are for the different (Major, Sex) combinations, by using the "table" function `table(grades$sex, grades$major)`, which gives you:

```
     2  66  70 114 115 118 203
  F 25   0  87   0   0   0  11
  M 36  10 232   4   2   3  84
```

The categories of the "Major" variable are coded numerically, with values 2, 66, ..., 203, but there are actual degree programs attached to these numerical codes, as shown in Table 10.1.

Notice from use of the "`table`" function above that there are few observations, and no females at all, in the MS programs except for the MS Accounting program. We will drop those data, but this is not an absolute requirement. Mainly, we will drop these data to make the main points more easily. Empty cells such as (Female, MS Finance) are allowed, but add a layer of complication.

So we will first subset the data as follows.

```
grades1 = subset(grades, major %in% c(2,70,203))
attach(grades1)
```

Note that the "Major" variable is numeric, but the numbers do not have numeric meaning; they are just labels. This gives us a great opportunity to demonstrate when and why you need to use the "`as.factor`" function in R. First, fit the model involving major only, using the R commands `Fit.n = lm(gpa ~ major); summary(Fit.n)`. This yields the following output:

```
 Call:
lm(formula = gpa ~ major)

Residuals:
     Min       1Q   Median       3Q      Max
-1.14214 -0.23614  0.02444  0.25444  0.51009
```

```
Coefficients:
             Estimate Std. Error t value Pr(>|t|)
(Intercept) 3.4892303  0.0238190  146.49   <2e-16 ***
major       0.0003415  0.0002218    1.54    0.124
---
Signif. codes:  0 '***' 0.001 '**' 0.01 '*' 0.05 '.' 0.1 ' ' 1

Residual standard error: 0.2985 on 473 degrees of freedom
Multiple R-squared:  0.004988,  Adjusted R-squared:  0.002884
F-statistic: 2.371 on 1 and 473 DF,  p-value: 0.1243
```

This output provides estimates to predict GPA as a function of Major, where "Major" is in its numeric form, with values 2, 70 and 203. The left panel of Figure 10.9 displays this estimated model. This model is clearly nonsense because again, there is no numeric meaning to the "Major" variable.

R code for Figure 10.9

```
par(mfrow=c(1,2))
plot(major, gpa, xlab = "Major", ylab="GPA")
abline(lsfit(major,gpa))
major.123 = ifelse(major == 2,1, ifelse(major==70,2,3))
plot(major.123, gpa, xaxt="n", xlab = "Major", ylab="GPA")
axis(1, at = 1:3, lab=c("002", "070", "203"))
Fit.f = lm(gpa ~ as.factor(major))
b = Fit.f$coefficients
b0 = b[1]; b1 = b[2]; b2 = b[3]
ybars = c(b0, b0+b1, b0+b2)
points(1:3, ybars, type="l")
points(1:3, ybars, pch=4, cex=2)
```

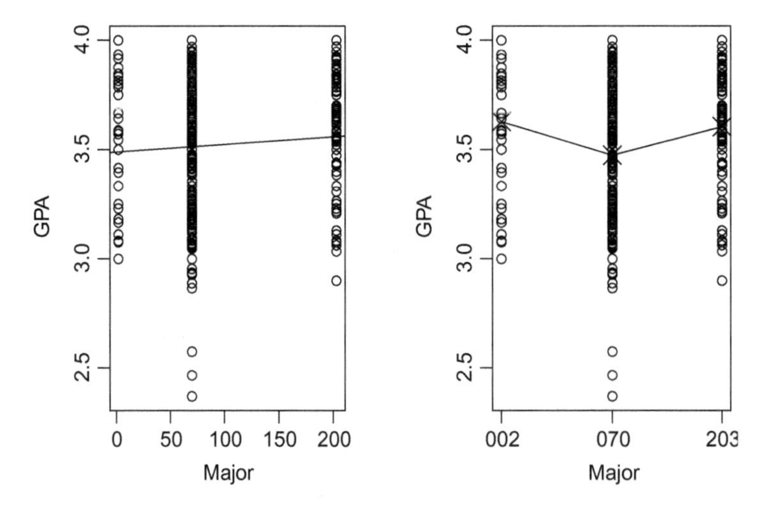

FIGURE 10.9
Left panel: Ordinary linear regression to predict GPA as a function of Major, where Major is considered as ordinary numeric data. Right panel: Indicator variable regression model to predict GPA as a function of Major, where Major is considered as character (nominal) data, a result of using the `as.factor` function in R. Ordinary averages in right panel are indicated by "×."

Rather than assume a linear trend as shown in the left panel of Figure 10.9, you should use indicator variables for two of the majors. As discussed above with the ANOVA models, the indicator variable model will allow each of the three majors to have any mean whatsoever, with no linear constraint. You can estimate this model easily by replacing "`major`" in the `lm` model statement with "`as.factor(major)`" as shown in the `Fit.f` object in the code above. The "`as.factor`" function changes the internal representation of the object in `R` from numeric to character, and then `R` automatically creates the needed indicator variables, as shown below:

```
Fit.f = lm(formula = gpa ~ as.factor(major))
summary(Fit.f)
```

Output:

```
Coefficients:
                     Estimate Std. Error t value Pr(>|t|)
(Intercept)           3.62726    0.03741  96.966  < 2e-16 ***
as.factor(major)70   -0.15382    0.04083  -3.767 0.000186 ***
as.factor(major)203  -0.02361    0.04794  -0.493 0.622576
---
Signif. codes:  0 '***' 0.001 '**' 0.01 '*' 0.05 '.' 0.1 ' ' 1

Residual standard error: 0.2922 on 472 degrees of freedom
Multiple R-squared:  0.04859,   Adjusted R-squared:  0.04456
F-statistic: 12.05 on 2 and 472 DF,  p-value: 7.845e-06
```

Notice that the model F statistic is now "significant," indicating that the estimated differences between means of the GPA variable for the different majors are not easily explained by chance alone; whereas the F statistic was "insignificant" in the (incorrect) model where Major was treated as numeric.

To summarize:

When to use "as.factor"

Use "`as.factor`" to create indicator variables out of a numeric variable.

It won't hurt to use `as.factor` if the variable in question is already character data; in this case, you will get the same results with or without using `as.factor`. But if the numeric data are just labels, then you *must* use `as.factor`. In some cases, you can also use `as.factor` with ordinal numeric data, as described in Section 10.7.

Although very important, this discussion of "`as.factor`" was really just a digression. The main point of this section concerns what you need to do when you have two nominal variables, such as Major and Sex. So, here is the first analysis. To make the output look a little nicer, we will use `as.factor` outside the `lm` statement:

```
major = as.factor(major)
fit1 = lm(gpa ~ sex + major)
summary(fit1)
```

The output is as follows:

```
Coefficients:
            Estimate Std. Error t value Pr(>|t|)
```

```
(Intercept)  3.71877    0.04063   91.533   < 2e-16 ***
sexM        -0.15505    0.03039   -5.102 4.89e-07 ***
major70     -0.13256    0.04000   -3.314 0.000992 ***
major203     0.02198    0.04756    0.462 0.644134
---
Signif. codes:  0 '***' 0.001 '**' 0.01 '*' 0.05 '.' 0.1 ' ' 1

Residual standard error: 0.2847 on 471 degrees of freedom
Multiple R-squared:  0.09841,   Adjusted R-squared:  0.09267
F-statistic: 17.14 on 3 and 471 DF,  p-value: 1.407e-10
```

The theoretical model corresponding to this fit is as follows, an example of a *two-way ANOVA model without interaction:*

$$\text{GPA} = \beta_0 + \beta_1\text{Male} + \beta_2\text{Major.70} + \beta_3\text{Major.203} + \varepsilon$$

Notice that all the X variables in the model are indicator variables.

You learned earlier in this chapter that the best way to understand the β coefficients in a model with indicator variables is to separate the model by the different levels of the nominal variable. The same is true here, except there are now two nominal variables (Major and Sex), so the best way to interpret the coefficients is to separate the model by different *combinations* of values of the two nominal variables, as shown in Table 10.2.

From Table 10.2, you get the following interpretation of the parameters in the model.

- β_0 is the mean GPA for females in major 002.
- β_1 is the difference of mean GPA, male minus female, *within any fixed major.*
- β_2 is the difference of mean GPA, major 070 minus major 002, *within any fixed sex.*
- β_3 is the difference of mean GPA, major 203 minus major 002, *within any fixed sex.*

Self-study question: How does Table 10.2 follow logically from the model? And how do the given interpretations of the parameters shown in the bullet points logically follow from Table 10.2?

In the interpretations given above, the term "mean GPA" refers to the longer but more accurate phrase, "mean of the distribution of potentially observable GPA values" that we have used previously.

This model is an example of a "no-interaction model." As you saw in Chapter 9 with the analysis of moderator variables, no-interaction models predict that the effect of one variable does not depend on the value of the other. In particular, according to the model

TABLE 10.2

True Mean Values, According to the Two-way ANOVA Model Without Interaction

	Major 002	Major 070	Major 203
Male	$\beta_0 + \beta_1$	$\beta_0 + \beta_1 + \beta_2$	$\beta_0 + \beta_1 + \beta_3$
Female	β_0	$\beta_0 + \beta_2$	$\beta_0 + \beta_3$

above, the effect of the Sex variable is always β_1, no matter what is the major. As with all no-interaction models, this one presumes that the response functions are parallel, as shown in Figure 10.10.

R code for Figure 10.10

```
Major = rep(c("2", "70", "203") , each = 2)
Sex = rep(c("M", "F"), 3)
plot.dat = data.frame(Major, Sex)
names(plot.dat) = c("major", "sex")
pred.fit1 = predict(fit1, plot.dat)
interaction.plot(Major, Sex, pred.fit1, type="b",
ylab = "GPA", ylim = c(3.4,3.8))
```

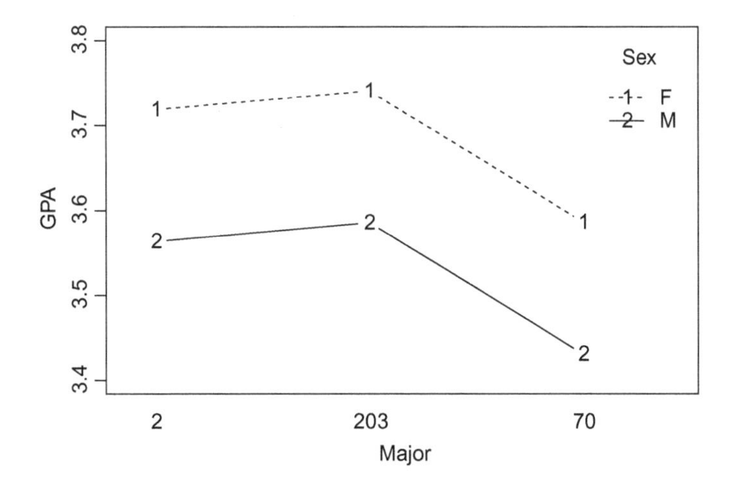

FIGURE 10.10
Estimated mean GPAs for (Major, Sex) combinations, assuming a no-interaction model.

Notice that the estimated means are constrained by the no-interaction model to have a parallel pattern such as what is shown in Figure 10.10. In particular, the model suggests that men are, on average, worse than women by 0.15505 grade points *within every single major.*

Certainly, the averages of the raw data will not obey such a constrained pattern. For instance, as seen in Figure 10.10, the estimated mean GPA for females in Major category "2" is about 3.72. But the raw average is different from 3.72 that is predicted by the no-interaction regression model: The code

```
mean(gpa[major=="2" & sex =="F"])
```

produces 3.62808 as the actual average of the GPAs for the 25 females enrolled in major "2."

When you include interaction terms, the estimated means using the regression model correspond precisely with the actual data averages. Recall from above that you must first re-code the "Major" variable as a factor using `major = as.factor(major)`. The interaction model is then specified using

```
fit2 = lm(gpa ~ sex + major + sex*major)
summary(fit2)
```

The output is as follows:

```
Coefficients:
                Estimate Std. Error t value Pr(>|t|)
(Intercept)     3.628080   0.056738  63.945   <2e-16 ***
sexM           -0.001386   0.073856  -0.019   0.9850
major70        -0.015034   0.064375  -0.234   0.8154
major203        0.106556   0.102642   1.038   0.2997
sexM:major70   -0.190566   0.082016  -2.324   0.0206 *
sexM:major203 -0.146751   0.117171  -1.252   0.2110
---
Signif. codes:  0 '***' 0.001 '**' 0.01 '*' 0.05 '.' 0.1 ' ' 1

Residual standard error: 0.2837 on 469 degrees of freedom
Multiple R-squared:  0.1087,    Adjusted R-squared:  0.09918
F-statistic: 11.44 on 5 and 469 DF,  p-value: 1.968e-10
```

The theoretical model corresponding to this fit is an example of a *two-way ANOVA model with interaction*, given as

$$\begin{aligned} \text{GPA} = &\beta_0 + \beta_1 \text{Male} + \beta_2 \text{Major.70} + \beta_3 \text{Major.203} \\ &+ \beta_4 \text{Male} \times \text{Major.70} + \beta_5 \text{Male} \times \text{Major.203} + \varepsilon \end{aligned}$$

Note that all X terms in the model are indicator variables or products of indicator variables.

Again, to interpret coefficients in model with indicator variables, you must separate the models by levels of the nominal variable(s) that are represented by the indicator variables, as shown in Table 10.1. Table 10.3 shows the case for the model that involves interaction terms.

The model parameters are somewhat more complex to interpret in this case. From Table 10.3, you can deduce the following facts:

- β_0 is the mean GPA for females in Major 002.
- β_1 is the difference of mean GPA, male minus female, *only for Major* 002.
- β_2 is the difference of mean GPA, major 070 minus major 002, *only for Females.*
- β_3 is the difference of mean GPA, major 203 minus major 002, *only for Females.*

TABLE 10.3

True Mean Values, According to the Two-Way ANOVA Model with Interaction

	Major 002	Major 070	Major 203
Male	$\beta_0 + \beta_1$	$\beta_0 + \beta_1 + \beta_2 + \beta_4$	$\beta_0 + \beta_1 + \beta_3 + \beta_5$
Female	β_0	$\beta_0 + \beta_2$	$\beta_0 + \beta_3$

- β_4 is a "difference of differences" that reflects the non-parallel nature of the interaction model: It is the difference between the Male/Female difference in Major 070 ($\beta_1 + \beta_4$) and the Male/Female difference in Major 002 (β_1).
- β_5 is another "difference of differences" parameter: It is the difference between the Male/Female difference in Major 203 ($\beta_1 + \beta_5$) and the Male/Female difference in Major 002 (β_1).

Interaction (moderation) models predict that the effect of one variable *does* depend on the value of the other. In particular, you know from Table 10.3 that the Sex effect depends on Major: The Sex effect is β_1 in Major 002, $\beta_1 + \beta_4$ in Major 070, and $\beta_1 + \beta_5$ in Major 203. Figure 10.11 shows the estimated mean values using this model.

R code for Figure 10.11

```
pred.fit2 = predict(fit2, plot.dat)
interaction.plot(Major, Sex, pred.fit2, type="b", ylab = "GPA",
ylim = c(3.4,3.8))
```

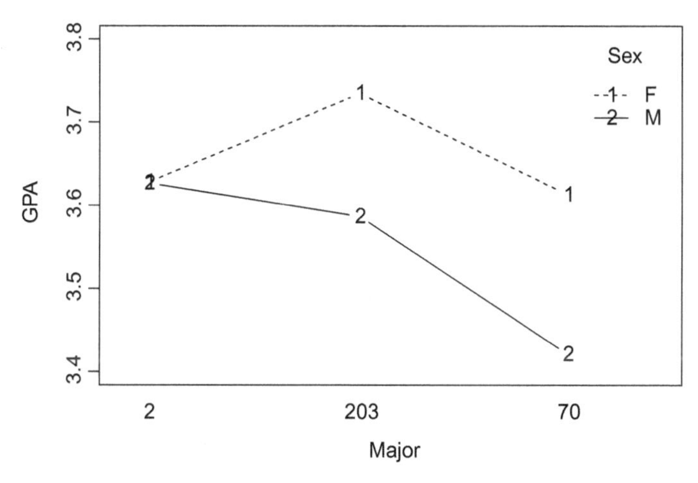

FIGURE 10.11
Estimated mean GPA for (Major, Sex) combinations, assuming the two-way ANOVA model with interaction.

Notice in Figure 10.11 that the estimated means are not constrained in any way, unlike Figure 10.10, in which they were constrained to lie on parallel lines. In particular, the estimated mean values using the regression model with interactions correspond exactly to the sample means in the subsets. As shown in Figure 10.11, the model predicts that the mean for Women in Major 002 is 3.62808, which is exactly the average GPA for the 25 women in that category.

Self-study question: How can you show that all of the six regression predictions in Figure 10.11 correspond precisely to simple averages of the six relevant subsets of the data?

Because the no-interaction model is a restricted model relative to the interaction model, you can test whether the deviations from parallelism shown in Figure 10.11 are explainable by chance alone, where "chance alone" refers to a data-generating process where the slopes are truly parallel. In other words, if, hypothetically, there was no interaction in reality between Sex and Major, is it likely to see non-parallel patterns such as shown in Figure 10.11 by chance alone? You can answer that question by using the *F* test. Simply use the following command:

```
anova(fit1,fit2)
```

The result is:

```
Analysis of Variance Table

Model 1: gpa ~ sex + major
Model 2: gpa ~ sex + major + sex * major
  Res.Df    RSS Df Sum of Sq      F  Pr(>F)
1    471 38.180
2    469 37.745  2   0.43501 2.7026 0.06807 .
---
Signif. codes:  0 '***' 0.001 '**' 0.01 '*' 0.05 '.' 0.1 ' ' 1
```

Thus, you can conclude that the difference between the model fits (as measured by SSE difference) is explainable by chance alone $\left(F_{2,469} = 2.7026, p = 0.06807\right)$.

This result may seem to contradict the summary `lm` output from `fit2` above, where one of the interaction terms was "significant" ($p = 0.0206$):

```
sexM:major70   -0.190566   0.082016   -2.324   0.0206 *
```

Here are some comments about this seeming contradiction:

- It's not really a contradiction. The global *F* test conclusion that the interaction effects are *explainable by chance alone* does not mean that there are no interaction effects. It just means that the test failed to detect evidence of interactions.
- The global *F* test tends to be less powerful in cases where many of the effects tested are small and one is big. In this case, the small effects essentially "dilute" the effects of the large effects, making the *F* statistic smaller.
- Because of the above bullet, you might think it is better not to use the *F* test and just use the individual *T* tests instead. The problem with this is that when you look at multiple *T* tests, the chance of observing a $p < 0.05$ result by chance alone is inflated because *p*-values are uniformly distributed under the null model.

What to do in this case? Use the "insignificant" *F* test or the "significant" *T* test? It really doesn't matter much since (i) "insignificance" of any test for interaction does not imply that there is no interaction, and (ii) tests should not be used for anything other than quantifying hypothetical chance effects. You know, *a priori*, that the interaction effects are real, because there is no reason that Nature would choose to put all those mean values on parallel lines, such as shown in Figure 10.10. Rather than use the *F* or *T* tests to decide on your model, you should draw the graphs shown in Figures 10.10 and 10.11. The purpose of the *F* and *T* tests is to aid your interpretation of the graphs: You should not place too much emphasis on a seeming pattern that can be explained by chance alone.

As discussed, previously multiple full model/restricted model tests can be performed simultaneously. In two-way ANOVA examples such as the one here, there are two distinct sequences that you might consider.

10.6.1 Nested Model Sequence, Version 1

Model 1.0: GPA $= \beta_0 + \varepsilon$

Model 1.1: GPA $= \beta_0 + \beta_1$Male $+ \varepsilon$

Model 1.2: GPA $= \beta_0 + \beta_1$Male $+ \beta_2$Major.070 $+ \beta_3$Major.203 $+ \varepsilon$

Model 1.3: GPA $= \beta_0 + \beta_1$Male $+ \beta_2$Major.070 $+ \beta_3$Major.203 $+ \beta_4$Male $\times$ Major.070 $+ \beta_5$Male $\times$ Major.203 $+ \varepsilon$

To test this sequence, simply fit the various models. Model 0 specifies the model where neither Sex nor Major has any effect of GPA and is fit using `fit1.0 = lm(gpa ~ 1)`. You know how to fit the others. You can test them all simultaneously via

```
anova(fit1.0, fit1.1, fit1.2, fit1.3)
```

which yields

```
Analysis of Variance Table

Model 1: gpa ~ 1
Model 2: gpa ~ sex
Model 3: gpa ~ sex + major
Model 4: gpa ~ sex + major + sex * major
  Res.Df    RSS Df Sum of Sq       F    Pr(>F)
1    474 42.347
2    473 40.421  1   1.92652 23.9382 1.370e-06 ***
3    471 38.180  2   2.24095 13.9226 1.337e-06 ***
4    469 37.745  2   0.43501  2.7026   0.06807 .
---
Signif. codes:  0 '***' 0.001 '**' 0.01 '*' 0.05 '.' 0.1 ' ' 1
```

It's best to interpret these sequences from the bottom up:

- The interaction effects are explainable by chance alone, assuming a model where there are no interaction effects $\left(F_{2,469} = 2.7026, p = 0.06807\right)$.
- The effect of Major is not easily explained by chance alone, assuming a model with only sex effects $\left(F_{2,471} = 13.9226, p = 1.337 \times 10^{-6}\right)$.
- The effect of Sex is not easily explained by chance alone, assuming a model where nothing affects GPA $\left(F_{1,473} = 23.9382, p = 1.370 \times 10^{-6}\right)$.

An alternative sequence is given as follows.

10.6.2 Nested Model Sequence, Version 2

Model 2.0: GPA $= \beta_0 + \varepsilon$

Model 2.1: GPA $= \beta_0 + \beta_2$Major.070 $+ \beta_3$Major.203 $+ \varepsilon$

Model 2.2: $\text{GPA} = \beta_0 + \beta_1\text{Male} + \beta_2\text{Major.070} + \beta_3\text{Major.203} + \varepsilon$

Model 2.3: $\text{GPA} = \beta_0 + \beta_1\text{Male} + \beta_2\text{Major.070} + \beta_3\text{Major.203} + \beta_4\text{Male} \times \text{Major.070} + \beta_5\text{Male} \times \text{Major.203} + \varepsilon$

The only difference between the two nested model sequences is in models 2.1 and 1.1. You might like to test models 1.1 and 2.1 in the same sequence, but you cannot because neither is a full model relative to the other. With sequences of *F* tests, you must ensure that every model in the sequence is nested in the next one.

You can test these all simultaneously via

```
anova(fit2.0, fit2.1, fit2.2, fit2.3)
```

which yields the following output:

```
Analysis of Variance Table

Model 1: gpa ~ 1
Model 2: gpa ~ major
Model 3: gpa ~ sex + major
Model 4: gpa ~ sex + major + sex * major
  Res.Df    RSS Df Sum of Sq       F    Pr(>F)
1    474 42.347
2    472 40.289  2   2.05772 12.7842 3.926e-06 ***
3    471 38.180  1   2.10976 26.2150 4.466e-07 ***
4    469 37.745  2   0.43501  2.7026   0.06807 .
---
Signif. codes:  0 '***' 0.001 '**' 0.01 '*' 0.05 '.' 0.1 ' ' 1
```

Interpretations are as follows, from bottom up:

- The interaction effects are explainable by chance alone, assuming a model where there are no interaction effects $\left(F_{2,469} = 2.7026, p = 0.06807\right)$.
- The effect of Sex is not easily explained by chance alone, assuming a model with only Major effects $\left(F_{1,471} = 26.2150, p = 4.466 \times 10^{-7}\right)$.
- The effect of Major is not easily explained by chance alone, assuming a model where nothing affects GPA $\left(F_{2,472} = 12.7842, p = 3.926 \times 10^{-6}\right)$.

The *F* statistics for Sex and Major differ in the two sequences because there are different null models used as references. But the conclusions are essentially the same: There are indeed Sex and Major effects on GPA.

Finally, note that the methods in this section are called "two-way ANOVA," or "two-way Analysis of Variance." The "two-way" refers to the fact that there are two nominal variables, so that the means can be arranged in two-way tables such as shown above in Tables 10.2 and 10.3. A happy conclusion is that you do not need to bother with specialized software for two-way ANOVA; you can do it all with the `lm` function in R. But you should know what two-way ANOVA refers to: It is a model for the distribution of a *Y* variable as a function of two nominal variables.

You can extend the two-way ANOVA easily to three-way and higher-way ANOVAs: A three-way ANOVA is a model for the distribution of a *Y* variable as a function of three nominal variables; four-way ANOVA refers to four nominal variables, etc. You can also have interaction terms in such models (the number of possible interaction terms increases greatly as the degree of the ANOVA increases), and you can have covariates in all higher-degree ANOVA models, with or without interactions involving the covariates. For example, you could fit a model that is called a "four-way ANOVA with up to three-way interactions, as well as two covariates with all (main effect)*covariate interactions," or you could fit the same model and call it a "regression model using indicator variables." The latter is much simpler, and covers all the various special cases. You can show the specific details of the model by writing down its mathematical form, or by showing its `lm` syntax. All ANOVA and ANCOVA models use the same `lm` that is used for regression; no specialized software or other `R` function is needed.

10.7 Additional Applications of Indicator Variables

Indicator variables are incredibly useful, with many applications other than the various flavors of ANOVAs and ANCOVAs described above. In this final section, we will show you a few of them.

10.7.1 Piecewise Linear Regression; Regime Analysis

Usually, it makes sense to model $E(Y \mid X = x)$ as a continuous function of x, but there are cases where a discontinuity is needed. For a hypothetical example, suppose people with less than \$250,000 income are taxed at 28%, and those with \$250,000 or more are taxed at 34%. Then a regression model to predict Y = Charitable Contributions will likely have a discontinuity at X = 250,000, as shown in Figure 10.12.

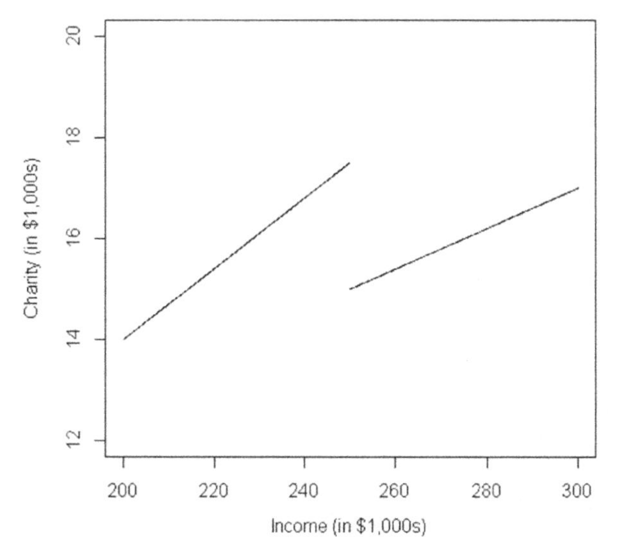

FIGURE 10.12
Hypothetical mean Charitable Contributions as a function of Income, assuming the tax rate jumps at \$250,000.

If you wanted to estimate the model shown in Figure 10.12, you would first create an indicator variable that is 0 for Income < 250, otherwise 1, like this:

```
Ind = ifelse(Income < 250, 0, 1)
```

Then you would include that variable in a regression model, with interactions, like this:

$$\text{Charity} = \beta_0 + \beta_1\text{Income} + \beta_2\text{Ind} + \beta_3\text{Income} \times \text{Ind} + \varepsilon$$

How can you understand this model? Once again, you must separate the model into the various subgroups. Here there are models in this example:

Group 1: Income < 250

$$\begin{aligned}\text{Charity} &= \beta_0 + \beta_1\text{Income} + \beta_2(0) + \beta_3\text{Income} \times (0) + \varepsilon \\ &= \beta_0 + \beta_1\text{Income} + \varepsilon\end{aligned}$$

Group 2: Income ≥ 250

$$\begin{aligned}\text{Charity} &= \beta_0 + \beta_1\text{Income} + \beta_2(1) + \beta_3\text{Income} \times (1) + \varepsilon \\ &= (\beta_0 + \beta_2) + (\beta_1 + \beta_3)\text{Income} + \varepsilon\end{aligned}$$

Thus, β_0 and β_1 are the intercept and slope of the model when Income < 250, while $(\beta_0 + \beta_2)$ and $(\beta_1 + \beta_3)$ are the intercept and slope of the model when Income ≥ 250.

You can test the hypothesis that there is a discontinuity at the break point. Figure 10.12 shows a break, but you can hypothesize that the lines connect, and you can test this hypothesis by using the full model/restricted model approach. Here the restriction states that the functions join at $X = 250$, implying that $\beta_0 + \beta_1(250) = (\beta_0 + \beta_2) + (\beta_1 + \beta_3)(250)$, or $\beta_2 + \beta_3(250) = 0$, or $\beta_2 = -250\beta_3$. Incorporating this constraint gives the restricted model that forces join point at $X = 250$.

$$\begin{aligned}\text{Charity} &= \beta_0 + \beta_1\text{Income} + (-250\beta_3)\text{Ind} + \beta_3\text{Income} \times \text{Ind} + \varepsilon \\ &= \beta_0 + \beta_1\text{Income} + \beta_3(\text{Income} \times \text{Ind} - 250\text{Ind}) + \varepsilon\end{aligned}$$

You can fit this model by predicting Charity as a function of two X variables, $X_1 = \text{Income}$, and $X_2 = \text{Income} \times \text{Ind} - 250\text{Ind}$, then testing the unrestricted model against the restricted model using the anova function as described above.

A related application is called *regime analysis*, which researchers typically do with time-series data. The idea is that the system you are studying over time is separated into two distinct time regimes, one before an event that (perhaps) changed the system, and the other after. The event might be the passage of a new law, the implementation of trade sanctions, an announcement by the Federal Reserve to increase interest rates, etc. You can analyze such data by using the ANCOVA model, where the groups are defined by time regime. What follows is an example of such an analysis using real data.

10.7.2 Relationship Between Commodity Price and Commodity Stockpile

The following data set contains government-reported annual numbers for price (Price) and stockpiles (Stocks) of a particular agricultural commodity in an Asian country.

```
Comm = read.table("https://raw.githubusercontent.com/andrea2719/
URA-DataSets/master/Comm_Price.txt")
attach(Comm)
```

Figure 10.13 shows how the Stocks and Price have changed over time. Something happened in 2002 to the Stocks variable; perhaps a re-definition of the measurement in response to a policy change.

This abrupt shift in 2002 causes trouble in estimating the relationship between Price and Stocks, which would ordinarily be considered a negative one because of the laws of supply and demand. Figure 10.14 shows the (Stocks, Price) scatter, with data values before 2002 indicated by circles, as well as global and separate least-squares fits.

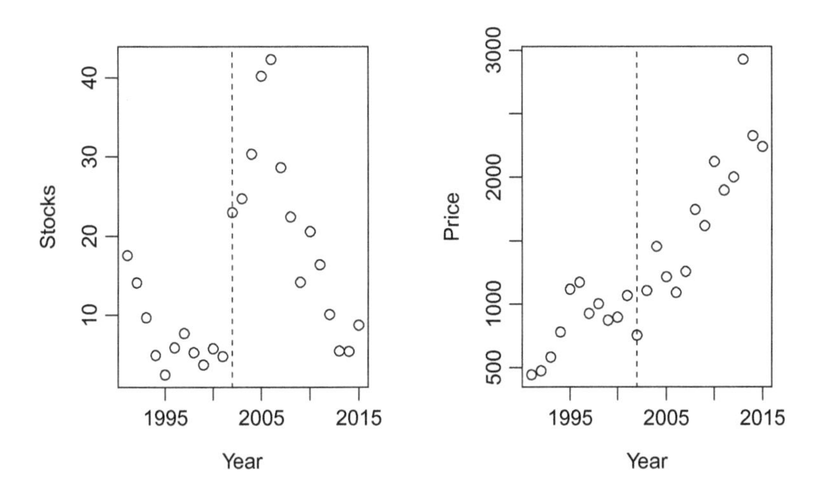

FIGURE 10.13
Plots of Stocks versus Year (left panel) and Price versus year (right panel). There appears to be a structural change in 2002 to the Stocks measurement.

R code for Figure 10.14

```
pch = ifelse(Year < 2002, 1, 2)
par(mfrow=c(1,2))
plot(Stocks, Price, pch=pch)
abline(lsfit(Stocks, Price))
plot(Stocks, Price, pch=pch)
abline(lsfit(Stocks[Year<2002], Price[Year<2002]), lty=1)
abline(lsfit(Stocks[Year>=2002], Price[Year>=2002]), lty=2)
```

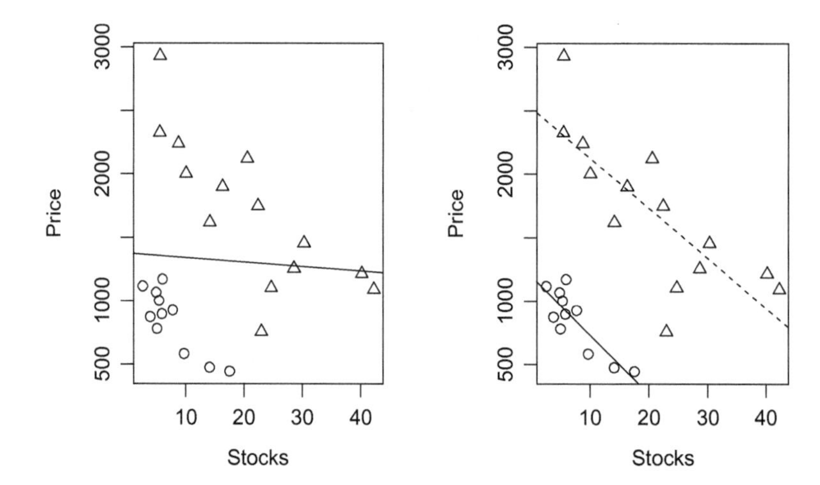

FIGURE 10.14
(Stocks, Price) scatterplots with pre-2002 data indicated by circles, 2002 and later indicated by triangles. Left panel shows ordinary least-squares fit using combined data. Right panel shows least-squares fits using pre-2002 (solid) and 2002 onward (dashed) data subsets.

As Figure 10.14 shows, there is clearly a need to separate the data into the two regimes. The expected negative relationship between Price and Stocks is barely visible in the left panel, and is not distinguishable from chance effects (T_{23} = –0.301, p = 0.766). But the expected negative trend is clearly visible in the right panel, where the data are analyzed separately by regime.

You can obtain the least-squares lines shown in the right panel of Figure 10.14 by using the indicator variable model with interaction, as follows:

```
Ind = (Year < 2002)
fit1 = lm(Price ~ Stocks + Ind + Stocks*Ind)
summary(fit1)
```

This gives you the following output:

```
Coefficients:
                 Estimate Std. Error t value Pr(>|t|)
(Intercept)      2522.551    172.267  14.643 1.71e-12 ***
Stocks            -39.440      7.242  -5.446 2.11e-05 ***
IndTRUE         -1325.090    251.508  -5.269 3.19e-05 ***
Stocks:IndTRUE     -7.029     22.336  -0.315    0.756
---
Signif. codes:  0 '***' 0.001 '**' 0.01 '*' 0.05 '.' 0.1 ' ' 1

Residual standard error: 307.7 on 21 degrees of freedom
Multiple R-squared:  0.7948,    Adjusted R-squared:  0.7654
F-statistic:  27.1 on 3 and 21 DF,  p-value: 2.051e-07
```

As always, you need to analyze the various categories separately to understand a model with indicator variables. Here, the categories are "Before 2002" and "2002 onward," and the theoretical models are as follows:

Group 1: Year < 2002

$$\begin{aligned}\text{Price} &= \beta_0 + \beta_1\text{Stocks} + \beta_2(1) + \beta_3\text{Stocks}(1) + \varepsilon \\ &= (\beta_0 + \beta_2) + (\beta_1 + \beta_3)\text{Stocks} + \varepsilon\end{aligned}$$

Group 2: Year ≥ 2002

$$\begin{aligned}\text{Price} &= \beta_0 + \beta_1\text{Stocks} + \beta_2(0) + \beta_3\text{Stocks}(0) + \varepsilon \\ &= \beta_0 + \beta_1\text{Stocks} + \varepsilon\end{aligned}$$

Thus, the first two coefficients in the output estimate the slope and intercept in the 2002 onward model, while the next two estimate differences from those coefficients in the pre-2002 model. In particular, note that the estimate of the difference of slopes differs from zero by an amount that is explainable by chance alone ($T_{21} = -0.315$, $p = 0.756$). If there is no difference in slope, then the structural relationship between Price and Stocks is consistent over time. The difference in regimes seems mainly due to a constant shift in the level of Stocks, again perhaps due to a re-definition of some sort in 2002, which affects the intercept but not the slope.

Is the difference between the left and right panel fits in Figure 10.14 explainable by chance? It seems unlikely, given the extreme separation of the data between the two groups. But if you need the test, you can again use the full-model restricted model test. The restricted model is just the global fit, which is a restricted version of the indicator variable model above, with the restriction $\beta_2 = \beta_3 = 0$. The code is as follows:

```
fit0 = lm(Price ~ Stocks)
anova(fit0, fit1)
```

The output is as follows:

```
Analysis of Variance Table

Model 1: Price ~ Stocks
Model 2: Price ~ Stocks + Ind + Stocks * Ind
  Res.Df     RSS Df Sum of Sq      F    Pr(>F)
1     23 9647976
2     21 1988060  2   7659916 40.456 6.265e-08 ***
---
Signif. codes:  0 '***' 0.001 '**' 0.01 '*' 0.05 '.' 0.1 ' ' 1
```

Thus, the separate lines shown in the right panel of Figure 10.14 are not easily explained by chance alone, assuming a model where there is no structural break in 2002 $\left(F_{2,21} = 40.456, p = 6.265 \times 10^{-8}\right)$. This test provides further evidence of a structural shift, but is unnecessary once you see the dramatic structural change effects in Figure 10.14. For reference, the test is called the "Chow test" (Chow 1960).

10.7.3 Using Indicator Variables to Represent an Ordinal *X* Variable

Ordinal variables are discrete variables, typically with only a few levels, where the levels have numerical meaning. A perfect example is the 1, 2, 3, 4, or 5 response you give to a survey item, such as "Rate your satisfaction with our customer service." Another example occurs in the Charity data set: The variable DEPS (number of claimed dependents) takes the values 0,1,2,...,6. An example where a discrete variable with few numerical levels is *not* an ordinal variable is the MAJOR variable of the GPA data set, which has levels 2, 66, 70, 114, 115, 118, and 203: Since these levels refer to distinct majors, they have no numeric meaning. In other words, 2 is *not less than* 66 when these numbers refer to MAJOR categories. (MS Accounting is not "less than" MS Finance.) Instead, MAJOR is a *nominal variable*, not an ordinal variable, despite having numerical values in the data set.

Linearity is nearly always wrong, as discussed in Chapter 1. When an X variable is ordinal, the cause of the nonlinearity is often due to *spacing* between the ordinal levels. For example, the difference between people with 0 versus 1 dependent might be much greater than the difference between people with 1 versus 2 dependents. Similarly, a survey taker might view the difference between a "1" versus "2" response as being much greater than the difference between a "2" versus "3" response.

Fortunately, with ordinal X data, there are few levels of the variable, which in turn generally implies much data per level (roughly n/g data values per level, assuming g levels). Thus, it is easy to construct accurate estimates of the within-level mean values without having to rely on the linearity assumption. You can do this using indicator variables.

Consider the analysis of the charitable contribution data, and model the mean charitable contributions (CHARITY) as a function of number dependents (DEPS). Clearly, Income will be a better explanatory variable, but number of dependents is also interesting, so let's expand the model to include income and number of dependents. Both INCOME and CHARITY are expressed in terms of natural logarithms of the actual dollar amounts in the code below.

Here is the `R` syntax for reading the data and fitting the model:

```
charity = read.csv("https://raw.githubusercontent.com/andrea2719/
URA-DataSets/master/charitytax.csv")
attach(charity)
fit1 = lm(CHARITY ~ INCOME + DEPS)
summary(fit1)
```

And here are the results:

```
            Estimate Std. Error t value Pr(>|t|)
(Intercept) -2.79955    1.21184  -2.310  0.02131 *
INCOME       0.86365    0.11485   7.520 2.83e-13 ***
DEPS         0.10984    0.03689   2.977  0.00306 **
---
Signif. codes:  0 '***' 0.001 '**' 0.01 '*' 0.05 '.' 0.1 ' ' 1

Residual standard error: 1.23 on 467 degrees of freedom
F-statistic: 33.96 on 2 and 467 DF,  p-value: 1.694e-14
```

Notice that the effect of DEPS is positive, which seems strange. The correct interpretation of its coefficient β_2 is as follows:

Consider two *potentially observable* sets of tax returns: (i) Those reporting 50,000 income and 3 dependents, and (ii) Those with 50,000 income and 4 dependents. Then β_2 is the mean charitable contributions (in natural log scale) in set (ii), minus the mean charitable contributions (in natural log scale) in set (i).

But according to the output above, β_2 is approximately in the range $0.10984 \pm 2(0.03698)$, or in the range (0.036,0.184), which indicates that β_2 is a positive number. This does seem odd: In the sets (i) and (ii) defined just above, people in set (ii) obviously will have less disposable income than those in set (i), so it is surprising that they nevertheless give more to charity.

This example shows the clear distinction between predictive models and causal models, described in Chapter 6. The positive coefficient of DEPS is correct from a predictive standpoint: If you want to predict the charitable contributions of people having the same income, but some of whom have more dependents, you should indeed predict that those with more dependents will give more. Predictive models are useful, even when the models do not make sense from a causal standpoint.

In this example, the predictive model is clearly not causal: Considering two types of families who are identical in all ways but for the fact that one type of family has more dependents, then families with more dependents will give less to charity, *on average*, simply because they will have less disposable income. What went wrong? Perhaps a failure of some major assumption? In this section, we investigate whether the incorrect sign might be due to a failure of the linearity assumption.

We will return to the causality issue in the next section. For now, let us consider the predictive model (despite its being wrong as a causal model). Again, there is nothing wrong with a predictive model that does not makes causal sense. Recall that, in the ice cream/drownings example, it is correct to say that in days with higher cream sales there will be more drownings. And in the current example, it is also correct to say that, among families with identical income, there will be higher charitable contributions in families having more dependents.

Let's explore whether the predictive model can be improved by accounting for curvature in the effect of DEPS on CHARITY.

The theoretical model is

$$\text{CHARITY} = \beta_0 + \beta_1\text{INCOME} + \beta_2\text{DEPS} + \varepsilon$$

Here, DEPS is a highly discrete ordinal variable, taking values 0, 1,..., 6 in the data set. As with the nominal variables discussed earlier in this chapter, you should summarize discrete variables (whether nominal or ordinal) by using a frequency table: `table(DEPS)` produces the output

```
DEPS
  0   1   2   3   4   5   6
 57  80 120 114  43  43  13
```

For example, 13 of the observed tax returns reported 6 dependents.

The theoretical model assumes that the DEPS effect is perfectly linear: For fixed INCOME it assumes that the mean of CHARITY is β_2 higher for DEPS = 1 than for DEPS = 0; and it assumes exactly the same for DEPS = 6 versus DEPS = 5. This seems a little weird: The groups having DEPS = 1 and DEPS = 0 are qualitatively much different, from a family finance perspective, than the groups where DEPS = 6 and DEPS = 5. So, following the logic above for nominal variables, it might make more sense to allow each DEPS category to have its own mean, rather than to force them onto a straight line. You can do this simply by representing DEPS with indicator variables using the `as.factor` function:

```
DEPS.F = as.factor(DEPS)
fit2 = lm(CHARITY ~ INCOME + DEPS.F)
summary(fit2)
```

This gives the following output:

```
Coefficients:
             Estimate Std. Error t value Pr(>|t|)
(Intercept) -2.88971    1.22481  -2.359   0.0187 *
INCOME        0.87755    0.11694   7.504 3.21e-13 ***
DEPS.F1     -0.03537    0.21581  -0.164   0.8699
DEPS.F2       0.17295    0.20011   0.864   0.3879
DEPS.F3       0.28580    0.20050   1.425   0.1547
DEPS.F4       0.55448    0.25066   2.212   0.0275 *
DEPS.F5       0.32147    0.25126   1.279   0.2014
DEPS.F6       0.69787    0.37924   1.840   0.0664 .
---
Signif. codes:  0 '***' 0.001 '**' 0.01 '*' 0.05 '.' 0.1 ' ' 1

Residual standard error: 1.234 on 462 degrees of freedom
Multiple R-squared:  0.1313,    Adjusted R-squared:  0.1181
F-statistic: 9.972 on 7 and 462 DF,  p-value: 1.274e-11
```

Notice that R left out the DEPS = 0 category, so all other indicator variable coefficients are differences from that category. In particular, among people having the same income, the estimated mean charitable contributions for people having DEPS = 1 is *lower* than the estimated mean charitable contributions for people having DEPS = 0 $\left(\hat{\beta}_2 = -0.3537\right)$. Already this is different from the model assuming a linear DEPS effect, which predicts that this difference is positive $\left(\hat{\beta}_2 = +0.10984\right)$.

Which model should you use? As before, to answer this question, you should first compare the two fitted models graphically in an attempt to answer the question. The following code shows how, and the result is shown in Figure 10.15.

R code for Figure 10.15

```
b = fit1$coefficients
b0 = b[1]; b1 = b[2]; b2 = b[3]
yhat.lin = b0 + b1*mean(INCOME) + b2*(0:6)

b = fit2$coefficients
b0 = b[1]; b1 = b[2]; b2 = b[3:8]
yhat.ord = b0 + b1*mean(INCOME) + c(0,b2*c(1,1,1,1,1,1))

plot(0:6, yhat.lin, ylim = c(6.2, 7.1), xlab="DEPS",
ylab = "CHARITY")
points(0:6, yhat.lin, type="l")
points(0:6, yhat.ord, pch = 2)
points(0:6, yhat.ord, type="l", lty=2)
legend(2.7, 6.45, c("Linear Model", "Indicator Variable Model"),
lty=c(1,2), pch = c(1,2))
```

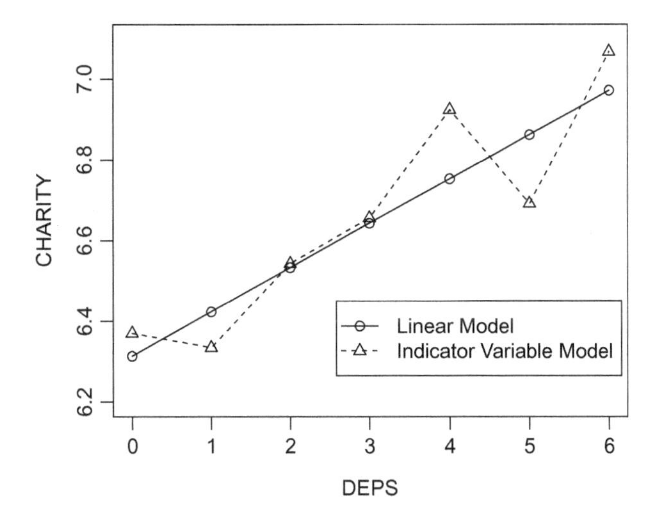

FIGURE 10.15
Estimates of mean ln(Charitable contributions) as a function of number of dependents, among people whose income is at the mean of ln(Income) (10.55208), using the linear model (solid) and indicator variable model (dashed). The linear model constrains the predicted means to fall exactly on a line, while the indicator variable model places no constraints on the predicted means.

For the most part, the indicator variable model and the linear model shown in Figure 10.15 show similar results. As mentioned above, there is a dip from DEPS = 0 to DEPS = 1 that may or may not be sensible from the subject matter of DEPS definition. There is also a dip from DEPS = 4 to DEPS = 5, then an increase from DEPS = 5 to DEPS = 6 that appears to be purely idiosyncratic. Thus, the strangely increasing effect of DEPS on CHARITY is not explained by failure of the linearity assumption.

The two fitted models shown in Figure 10.15 fall into the full model/restricted model framework: The indicator variable model makes no constraints on the mean as a function of DEPS, while the linear model constrains these means to fall on a straight line. You can easily test whether the deviations from a straight line for the indicator model are explainable by chance alone, assuming a model that is truly linear, by using `anova(fit1,fit2)`. The results are as follows:

```
Analysis of Variance Table

Model 1: CHARITY ~ INCOME + DEPS
Model 2: CHARITY ~ INCOME + DEPS.F
  Res.Df    RSS Df Sum of Sq      F Pr(>F)
1    467 706.58
2    462 703.11  5    3.4648 0.4553 0.8094
```

The deviations from linearity shown in Figure 10.15 are indeed explainable by a model that is linear $\left(F_{5,462} = 0.4553,\ p = 0.8094\right)$. As always, such a conclusion does not imply that the true mean function is linear; we already know that the true mean function is not linear. The test simply states that the deviations from linearity shown in Figure 10.15 are well within the size of the deviations that you would see if the data were from a process with a truly linear DEPS function.

Self-study question: What is the specific "null," or "chance-only" model that the *F* test above refers to?

Note that the indicator variable-based test above is yet another test for linearity. Earlier, we showed you how to do this by fitting a quadratic function, and by examining the "significance" of the quadratic term. An advantage of doing the test with indicator variables is that there is no presumption about the functional form of the curvature, whether quadratic or otherwise. A disadvantage is that the indicator variable test requires a few levels of the *X* variable. With a more continuous *X* variable, you will have too many indicator variables, and each estimated indicator variable coefficient adds randomness. Such additional randomness dilutes the power of the *F* test. So, if your *X* variable has few levels, you can use the indicator variable test, otherwise use the quadratic test.

But we are not going to give any ugly rule of thumb threshold here for how many levels of the *X* variable are needed, because the answer to "How many levels" depends on sample size and true nature of curvature. The best way you can answer such questions is to use simulation: Try the tests both ways with different simulation scenarios and compare the power. Then use the simulation scenarios that are most similar to your data to decide which test to use.

Tests are not the right way to assess linearity, anyway, so it's not worth splitting hairs about which test to use, and when. It is better to use graphs, as shown in Figure 10.15, aided by subject matter considerations, as well as by an understanding of the idiosyncratic (random) nature of data.

10.7.4 Repeated Measures, Fixed Effects, and Unobserved Confounding Variables

No matter whether the indicator variable model or the linear model was used, the charitable contributions example given above was strange, in that it predicts greater charitable contributions for people with more dependents, even when they have identical income. A possible explanation for this contradiction is that there are *unobserved confounding variables*. As noted in Chapter 6, confounding variables make causal claims dubious.

One possible unobserved confounding variable is *Religiosity*: Recall the two sets of potentially observable people identified above: (i) Those reporting 50,000 income and 3 dependents, and (ii) Those with 50,000 income and 4 dependents. It may be the case that in the group of people with 4 dependents you find a greater percentage of religious people, and that such a difference is even more pronounced when comparing the DEPS = 6 and DEPS = 0 groups. In general, one might surmise that, among the larger families, there is a greater percentage of religious people, or at least that the *degree* of religiosity among larger families is generally higher. If this is so, then the increasing effect of DEPS on CHARITY is not caused by DEPS at all; instead, it is caused by the fact that people with more DEPS tend to donate more money to their religious organizations. These donations are lumped into "Charitable contributions" on their tax returns.

Or it could be that people with more children simply tend to be more charitable because people with children tend to be more humanitarian. Thus, *Humanitarianism* is another possible unobserved confounding variable. There can be myriad other unobserved confounding variables as well.

Self-study question: What other unobserved confounding variables might explain the strange increasing relationship between DEPS and CHARITY among people with equal income?

One way that you can control for such confounding variables, *whether or not you know what those confounding variables are*, is to use an indicator variable model where each subject in the data set defines a category. To perform this kind of analysis, you need to have *repeated measurements* on each observational unit, sometimes called *longitudinal data*. (If you do not have such repeated measures, then your indicator variable model will have as many variables as observations, which leaves zero degrees of freedom for error.) In the charitable contributions data set, there are in fact 10 years of repeated observations for each of 47 taxpayers. So, the data set is comprised not of 470 *taxpayers*, but rather 470 *taxpayer-years*.

You can fit the indicator variable model easily in R; just make sure the "SUBJECT" variable (which identifies a particular taxpayer) has the R class "factor"; otherwise the results will be nonsense, just like treating the MAJOR variable as numeric in the Grade Point Average analysis was nonsense (see Figure 10.9). Here is the code:

```
sub.f  = as.factor(SUBJECT)
fit3  = lm(CHARITY ~ INCOME + DEPS + sub.f)
summary(fit3)
```

This code yields several lines of output because it estimates a different effect for each taxpayer:

```
Coefficients:
             Estimate Std. Error t value Pr(>|t|)
(Intercept) -1.97575    1.07401  -1.840 0.066531 .
INCOME       0.73063    0.10000   7.306 1.39e-12 ***
DEPS        -0.07133    0.05213  -1.368 0.171915
sub.f2       1.96613    0.30828   6.378 4.73e-10 ***
sub.f3       1.09687    0.31052   3.532 0.000457 ***
. . . (many more lines of output) . . .
sub.f46     -0.14376    0.31433  -0.457 0.647652
sub.f47      0.02113    0.32622   0.065 0.948377
---
Signif. codes:  0 '***' 0.001 '**' 0.01 '*' 0.05 '.' 0.1 ' ' 1

Residual standard error: 0.6822 on 421 degrees of freedom
Multiple R-squared:  0.7579,    Adjusted R-squared:  0.7303
F-statistic: 27.46 on 48 and 421 DF,  p-value: < 2.2e-16
```

The theoretical model is

$$\text{CHARITY} = \beta_0 + \beta_1\text{INCOME} + \beta_2\text{DEPS} + \beta_3\text{Subject}_2 + \beta_4\text{Subject}_3 + \ldots$$
$$+ \beta_{47}\text{Subject}_{46} + \beta_{48}\text{Subject}_{47} + \varepsilon$$

This model controls for unobserved confounders because it controls for subject effects: It is reasonable to assume that the unobserved confounders such as religiosity, charitability, or other person-specific characteristics are reasonably constant for a given person over the ten-year horizon. For a particular person, say subject s ($s = 2,\ldots,47$ taxpayers in the data set, with $s = 1$ being the "left-out" category) the model states that

$$\text{CHARITY} = \beta_0 + \beta_1\text{INCOME} + \beta_2\text{DEPS} + \beta_{s+1}\text{Subject}_s + \varepsilon$$

Thus, β_2 is interpreted as the difference between the mean of *potentially observable* charitable contributions *for particular subject s* in two different years, where INCOME stayed the same but DEPS differs by 1. This interpretation differs from the previous interpretation, where the two different subgroups (INCOME fixed, DEPS differ by 1.0) referred to potentially observable data on *different people*, and the group of people with higher DEPS were different in terms of unmeasured confounders like charitability or religiosity. Here, the different potentially observable subgroups refer to different life situations for the *same person*, so it is reasonable to assume that the unmeasured confounders are reasonably constant when comparing the two subgroups.

You have to imagine many different potentially observable past histories for subject *s* to make sense of the last paragraph, but that's nothing new. We have been doing that all along. That is the way we have always understood the regression model, $Y \mid X = x \sim p(y \mid x)$: The model describes all the *potentially observable* past data, not just the idiosyncratic data that you have in your data set.

So, now that the model refers only to subject *s*, all potential unmeasured confounders are controlled (again, presuming that these also do not change over the time horizon). The results now make sense: The estimated effect of DEPS is in the expected negative direction $\left(\hat{\beta}_2 = -0.07133\right)$. Thus, for a given hypothetical taxpayer, if DEPS changes by +1 and INCOME remains unchanged, the model predicts –0.07133 less charitable contributions, on average. (Here, "on average" refers to *different potentially observable data for the same given taxpayer.*) However, direction of the true effect cannot be ascertained: By using `confint(fit3)`, you will see in the output that $-0.17378748 < \beta_2 < 0.03112989$. Thus, the data are not precise as to whether there is an increase or a decrease in mean charitable contributions associated with more dependents. Since there are only 47 taxpayers, not 470, don't expect too much precision!

10.7.5 The Independence Assumption and Repeated Measurements

You know what? All the analyses we did on the charitable contributions prior to the subject/indicator variable model were grossly in error because the independence assumption was so badly violated. You may assume, nearly without question, that these 47 taxpayers are independent of one another. But you may not assume that the repeated observations on a given taxpayer are independent. Charitable behavior in different years is similar for given taxpayers; i.e., the observations are dependent rather than independent. It was wrong for us to assume that there were 470 independent observations in the data set. As you recall, the standard error formula has an "n" in the denominator, so it makes a big difference whether you use $n = 470$ or $n = 47$. In particular, all the standard errors for models prior to the analysis above were too small.

Sorry about that! We would have warned you that all those analyses were questionable earlier, but there were other points that we needed to make. Those were all valid points for cases where the observations are independent, so please do not forget what you learned.

But now that you know, please realize that you must consider the dependence issue carefully. You simply cannot, and must not, treat repeated observations as independent. All of the standard errors will be grossly incorrect when you assume independence; the easiest way to understand the issue is to recognize that $n = 470$ is quite a bit different from $n = 47$.

Confused? Simulation to the rescue! The following `R` code simulates and analyzes data where there are 3 subjects, with 100 replications on each, and with a strong correlation (similarity) of the data on each subject.

```
s = 3      # subjects
r = 100    # replications within subject
X = rnorm(s); X = rep(X, each=r) +rnorm(r*s,0,.001)
a = rnorm(s); a = rep(a, each=r)
```

```
e = rnorm(s*r,0,.001)
epsilon = a+e
Y = 0 + 0*X + rnorm(s*r) +epsilon    # Y unrelated to X
sub = rep(1:s, each = r)
summary(lm(Y ~ X))  # Highly significant X effect
summary(lm(Y ~ X + as.factor(sub))) # Insignificant X effect
```

Please run the code a few times. You will see that the model `Y ~ X` gives (usually) "highly significant" results, despite there being no relationship whatsoever between X and Y (see the "`0*X`" in the line in the code where Y is defined). The reason for these incorrectly "significant" results is that there are $n = 300$ observations, all assumed independent by the classical regression analysis. But in reality, there are only three independent subjects. Recall that there is an n in the denominator of the standard error formula: Here, n should be closer to 3 than 300, and the standard errors should be much larger. When you analyze the data using the subject indicator variable model (see the last line of the code), the relationship becomes (usually) correctly "insignificant."

Why is dependence not such a problem in the subject-indicator variable model? It is because the independence assumption refers to the residuals, which are deviations from the mean. In the taxpayer data set, even though all 10 years of data are on the same subject, it is more reasonable to assume that an individual year's charitable contribution deviation from the expected value that is *specific to that person* is independent of another year's deviation. But in the model without the subject indicators, the error deviation refers to deviation from expected value *specific to the entirety of the taxpayers.* In that case, a given taxpayer who is charitable will have residuals that may be all positive, if that person is charitable, or all negative if that person is not charitable. Thus, the error terms are clearly dependent when you use the model without subject indicators. Use of subject indicators does not completely guarantee independence, as the residuals in adjacent years are expected to be more similar than residuals in distant years; see Chapter 12 for a re-analysis of these data that allows such correlations. Still, the independence assumption is much more reasonable in the subject-indicator model than it is in the model without subject indicators.

A similar problem occurs with firm-level data. Firms are grouped by a particular industry in which the firm operates. These industry effects, like subject effects, can be removed by using an indicator variable model, which will reduce both unobserved confounding effects, and will reduce dependence effects.

These analyses are called *fixed effects* analyses because they assume that the subjects (people, firms, etc.) are fixed, and not representative of a larger set. This model is problematic because the scope of generalization of the model is limited to those specific subjects. In the case of the taxpayers, this is clearly undesirable, as you would rather state that your conclusions of your analysis generalize to a much larger set of taxpayers than just those 47. The same can be said of industry-level fixed effects analysis: Wouldn't you like your data analysis to support a theory about people or firms *in general*, rather than just the ones that happen to exist in a particular sample? General theories are usually much more interesting than narrow theories.

Thus, a problem with fixed effects indicator variable analysis is that the scope of inference is *narrow*, applying *only to* the given subjects (or firms, in the case of firm-level data). With fixed effects models, you cannot generalize beyond that narrow scope. If you want our data analysis to support a more general theory, you need to use *random effects* models. Such models allow a broad scope of generalization and also provide more accurate estimates of parameters (called *shrinkage estimates*), which properly account for the different sample sizes within different categories. These models also remove the effects of unobserved confounding variables, just like fixed-effects models. See Gelman and Hill 2006.

Some researchers address the issue of choosing fixed effects or random effects by using a hypothesis test (one such test is called the *Hausman test*). For reasons already given in this book, and as further supported by the ASA statement on *p*-values, you should not use tests to decide between models of any type. For example, a more important issue is scope of inference: Do you wish your conclusions to be limited to, for example, the given set of 47 taxpayers, or would you like your conclusions to refer to taxpayers in general? If you answered "in general," then you prefer the random effects model.

References

Bretz, F., Hothorn, T., & Westfall, P. (2010). *Multiple comparisons with R*. Boca Raton, FL: Chapman & Hall.

Chow, G. C. (1960). Tests of equality between sets of coefficients in two linear regressions. *Econometrica, 28*(3), 591–605.

Gelman, A., & Hill, J. (2006). *Data analysis using regression and multilevel/hierarchical models*. Cambridge, MA: Cambridge University Press.

Exercises

1. Read the data set

```
clinical = read.table("https://raw.githubusercontent.com/andrea2719/
URA-DataSets/master/clinical.txt")
```

The dependent variable is the variable T4 in the data set, which is a doctor's assessment of a patient asthma symptoms, on a 0 – 4 scale (4 = "healthier", 0 = "sicker.") Also use the variable "DRUG," which is either "A" (for the "active" asthma drug being investigated by the pharmaceutical company) or "P" (refers to "placebo," meaning no drug at all.) The goal is to compare patient health in those two groups. (The pharmaceutical company would be happy to see that patient health is better when they take the company's drug because then the company can market and make profits on the drug.)

a. You can model T4 (health) using an indicator variable model, where the indicator variable is X = 1 for the "DRUG = A" group, and X = 0 for the "DRUG = P" group. Interpret the three parameters of that model (β_0,β_1 and σ) in terms of potentially observable patients, and whether they took the drug or no drug. (This question does not involve data).

b. Fit the regression model of 1.a and estimate the difference in mean health for the "P" and "A" group *using the fitted regression model*.

c. Is the estimated difference in 1.b within the range of differences that are explained by chance alone? (Here, give for the logic underlying *p*-value here, not just the *p*-value).

d. As discussed in the book, you can obtain the two-sample t test by using the regression model you just fit. But just to verify, perform the two-sample t test using `t.test(T4 ~ DRUG, var.equal = T)` and verify that the *p*-values are identical. Use the *p*-values inside the fitted objects for comparison so you can compare more digits than shown in the default output.

e. Use "AGE" as a covariate and draw a graph as shown in Figure 10.5 to see whether there is interaction between AGE and DRUG. Use jitter(T4) on the graph (but don't use jitter(T4) in the lm fit; use T4) so that no discrete data values are hidden.

f. Does it appear from the graph of 1.e. that the drug works better for some AGE values than others?

g. Is the evidence of interaction between AGE and DRUG shown in 1.f explainable by chance alone? (Here, you can just use the *p*-value to answer.)

2. The following data set contains real data from TTU graduate students who filled out the survey upon graduation.

Run the following code:

```
grad = read.csv("https://raw.githubusercontent.com/andrea2719/
URA-DataSets/master/pgs.csv", header=T)
attach(grad)
Fac.Gen = FacKnowledge + FacTeaching + FacResInClass +
FacOutsideClass + FacIntSuccess
summary(lm(Fac.Gen ~ COL))
```

The variable "Fac.Gen" is an overall measure of perceived faculty performance (higher is better) and is the dependent variable of the model. The variable "COL" is a nominal variable indicating college within which the student graduated.

a. Interpret all the *theoretical* parameters (including σ) corresponding to the fitted model. (Your answer will not involve the actual data analysis WITH THE EXCEPTION OF THE LEFT-OUT CATEGORY that was chosen by R. (There are lots of parameters, so you may interpret enough so that the pattern is clear, and then say "the rest are interpreted similarly").

b. Interpret the actual numbers in the output now.

c. What does the R^2 tell you in this example?

d. What does the F test tell you in this example?

e. What is another name (other than regression) for the analysis you just performed?

3. Use the example of problem 2, but now include survey year (SurvYear) in the model, with no interaction terms.

a. Interpret all the theoretical parameters (including σ) corresponding to the fitted model. (Your answer will not involve the actual data analysis WITH THE EXCEPTION OF THE LEFT-OUT CATEGORY that was chosen by *R*. (There are lots of parameters, so you may interpret enough so that the pattern is clear, and then say "the rest are interpreted similarly").

b. Interpret the actual numbers in the output now.

c. What does the R^2 tell you in this example?

d. What does the F test tell you in this example?

e. What is another name (other than regression) for the analysis you just performed?
f. Draw a graph like Figure 10.6.
g. Fit the interaction model, and draw a graph like Figure 10.7. Based on this graph, does it appear that perceived improvements over time in faculty performance differ by college?

4. Read the data set

```
clinical = read.table("https://raw.githubusercontent.com/andrea2719/
URA-DataSets/master/clinical.txt"))
```

The dependent variable is the variable T4 in the data set, which is a doctor's assessment of a patient asthma symptoms, on a 0 – 4 scale. Consider first the two-way ANOVA model using GENDER and DRUG (A="Active", P="Placebo") as predictors, with no interaction.

a. Specify the theoretical model and interpret all its parameters.
b. Estimate the theoretical model and draw the graph as shown in Figure 10.10.
c. This model states that the effect of DRUG is the same for men and women. Why might the drug manufacturer like this type of effect to be true? (Hint: Money.)

5. See problem 4 above. Now consider first the two-way ANOVA model using GENDER and DRUG (A = "Active," P = "Placebo") as predictors, with interaction.

a. Specify the theoretical model and interpret all its parameters.
b. Estimate the theoretical model and draw the graph as shown in Figure 10.11.
c. See the graph in 5.b. Does the effect of DRUG appear to be different for Males and Females in the graph?

6. See problem 5 above. Now include "Age" as a covariate. Call this model2, call the model in problem 5. "model1," and the model in problem 4. "model0. "

a. Explain why this sequence of the models is "nested"; i.e., explain why each in the sequence is either full or restricted relative to the next.
b. Test all three models using the "anova" function in R and interpret the results. As part of your explanation, state the relevant null model for each F test, and use that model to clarify what is *"explained* by chance."

7. Consider the data set

```
firms = read.csv("https://raw.githubusercontent.com/andrea2719/
URA-DataSets/master/firms10.csv").
```

This is firm-level data, with a measure of Y = firm performance (perform_y) and a measure of X= the extent to which the firm has adopted a particular performance-enhancing strategy (strategy_x). There is also another variable called "code_3dig" which identifies the industry sector in which the firm operates.

a. Fit the model where Y is predicted as a function of X alone. Is the observed effect of X on Y explainable by chance alone?
b. Fit the model where Y is predicted as a function of X and the fixed effects of industry, as measured by "code_3dig." Is the observed effect of X on Y explainable by chance alone in this model?
c. Comparing your answers to 7.a and 7.b, was there any indication of unobserved confounding due to industry-specific unmeasured confounders?

11

Variable Selection

When you are contemplating which variables you should use to predict Y in your regression model, consider the following three questions:

1. **Global Variables**: Which variables V_1, V_2, ... are *possibly related to* Y? (This set of variables is likely to be infinite.)
2. **Measurable Variables**: Which subset of variables $\{W_1,\ldots,W_K\} \subseteq \{V_1,V_2,\ldots\}$ can you *actually measure*, and get into a data set amenable to estimation of regression models?
3. **Variables to Use in Your Estimated Model**: Which subset of measured variables $\{X_1,\ldots,X_k\} \subseteq \{W_1,\ldots,W_K\}$ should you *ultimately use* in your estimated model?

Question 1 is purely conceptual; it is a "think about it using your knowledge of the subject matter" kind of question. But such "brainstorming" about potential variables will be very useful to you when you frame your causality arguments; "brainstorming" will also be very useful to help you identify good predictive variables that you might decide to collect data upon.

Question 2 becomes more practical. When you are designing your study to collect measurements on $W_1,\ldots,W_K$, consider not only your answers to question 1, but also the following:

- Precedent. Which variables have been used in the past for similar prediction models?
- Availability. Which variables are actually available?

In addition, consider the goals of your analysis. Are you interested in predictive modeling, without regard for causality, or are you interested in establishing whether there is a causal link between a particular X variable and Y? In the predictive case, the candidate set $W_1,\ldots,W_K$ should include everything you can possibly lay your hands on. You are looking for the biggest, richest data set you can get.

On the other hand, if you wish to establish a causal connection, then you need to ask yourself, "What are the possible confounding variables?" Do some literature review, or simply think about it very hard. Use the two examples discussed previously in this book for guidance: (i) the drownings/ice cream sales case had a possible confounding variable, "Temperature," and (ii) the charitable contributions/number of dependents example had a possible confounding variable "Religiosity." You need to identify as many confounding variables as possible, or perhaps their surrogates, to include in the measured set $W_1,\ldots,W_K$, in order to have a better argument for causal effect of W_1 (say) on Y.

Now consider question 3 above, and notice the "K" versus "k" distinction. This distinction is not a "random vs. fixed" distinction as in the case of $\mathbf{X}$ vs. $\mathbf{x}$, but rather an indication that the set of variables that you ultimately use, $\{X_1,\ldots,X_k\}$, is a (perhaps much smaller) subset of the set of variables $\{W_1,\ldots,W_K\}$ that you can actually measure. Hence "k" can be much smaller than "K," but never larger.

One answer to question 3 above was already given in Chapter 6: Since additional variables can never make the predictions worse, assuming you know the parameter values, you should use all measured variables $\{W_1,\ldots,W_K\}$. Even if some are junk variables, they can make your predictions no worse, again, *assuming you know the parameter values*. However, the problem with this statement is that you *never know the parameter values* (outside of simulation studies); instead, you have to estimate them. Parameter estimation adds idiosyncratic error to the predictions, and negates the general theory that states "more variables are better." Variable selection, which is the subject of this chapter, concerns the effect of estimating parameters on prediction accuracy.

For the remainder of the chapter, we will assume that you have identified the measured set $W_1,\ldots,W_K$, and that those variables either will be, or are already in, your data set. The next question you should ask is, "Which subset of the variables $W_1,\ldots,W_K$ should I use?" We will re-frame your question as follows: "Which subset of the variables should I use in my estimated model to estimate $E(Y \mid W_1 = w_1,\ldots,W_K = w_K)$, as accurately as possible?"

Recall again, from Chapter 6, that if you knew the conditional means $E(Y \mid W_1 = w_1,\ldots,W_K = w_K)$, (i.e., if you knew all the model's parameters), then those conditional means are the optimal predictors of $Y \mid W_1 = w_1,\ldots,W_K = w_K$, so you would not need any subset—you would just use all the variables.

However, in practice you never know those conditional means, and you have to estimate them. Because there is error in the estimates, you can often estimate the conditional mean $E(Y \mid W_1 = w_1,\ldots,W_K = w_K)$ more precisely by using a regression model based on a subset $X_1,\ldots,X_k$ of the measured variables $W_1,\ldots,W_K$, than by using a regression model based on all of the variables $W_1,\ldots,W_K$. This result is surprising, so let us set it off so that you can see it clearly:

You can often estimate $E(Y \mid W_1 = w_1,\cdots,W_K = w_K)$ more precisely by using a regression model based on a subset $X_1,\ldots,X_k$ of the measured variables $W_1,\ldots,W_K$, rather than by using all the measured variables $W_1,\ldots,W_K$.

An example will make the general discussion given above more specific.

Suppose you wish to estimate a model to predict the graduating GPA (grade point average) of a college student, based on information the student presents at the time of admission to the college. The candidate set of variables, V_1, V_2 ..., is essentially infinite: Not only are the obvious choices in that set, such as High School GPA, entrance exam scores, family socioeconomic status, family college education, and extramural activities, but a whole host of other variables can be included as well, including nutritional intake, religiosity, cigarette smoking behavior, alcohol or drug use, motivation, humanitarianism, hours spent sleeping per night, commute time, height, weight, nearsightedness or farsightedness, depression, exercise, artistic ability, dating behavior, reading habits (not school related), number of vacations per year, existence of tattoos or piercings, ... Genetics are known to be important as well, and when you include DNA as a predictor, the number of possible variables increases dramatically, because there is so much information (so many potential V variables) in a single strand of DNA.

On the other hand, which variables $W_1, W_2, \ldots, W_K$ can you actually get your hands on? Genetic information may be unavailable for confidentiality reasons. If the data come from college applications, available variables may include information such as high school rank, high school GPA, entrance exam scores, and quantifiable indications of extramural activities and recommendation letters.

Perhaps there are $K = 30$ such measured (W) variables. You should ordinarily not use all of them in your estimated regression model; instead, you should use a subset of size k of these 30 variables for reasons we will explain. Perhaps you ultimately decide, based on statistical analysis and other considerations, to use just $k = 4$ out of the $K = 30$ measured variables: High school rank, High school GPA, Socioeconomic Status, and SAT score.

The above discussion was simply meant to help understand ideas in the first paragraphs of this chapter, and was not intended to give the "right answer." There are no "right answers" to the variable selection problem. But there are good statistical tools to help you select variables, including penalized likelihoods, cross-validation, and simulation. Hypothesis testing-based statistics (p-values) are low on this list. They are usually the wrong tools for the job, just like a blunt axe is the wrong tool for cutting lumber to build a custom home. Sure, you can use the blunt axe, but wouldn't a skill saw be a better tool for the job?

11.1 The Effect of Estimating Parameters on Prediction Accuracy

In Chapter 6, you learned that having more variables always improves prediction accuracy in theory, assuming that you know the conditional mean function. In practice, however, you have to estimate the conditional means, by first estimating the regression model parameters (the β's), and then by plugging in the X variables' values into the estimated regression function. Estimated parameters are always different from the true parameters; hence, there are additional errors not accounted for by the theory when you use estimated parameters. These additional errors can make predictions worse when including additional X variables, even when the additional X variables are in fact related to the response Y.

The following example explains this seeming contradiction. Consider the example of predicting GPA again.

11.1.1 Predicting Hans' Graduate GPA: Theory Versus Practice

Hans is applying for graduate school at Calisota Tech University (CTU). He sends CTU his quantitative score on the GRE entrance examination ($X_1 = 140$), his verbal score on the GRE ($X_2 = 160$), and his undergraduate GPA ($X_3 = 2.7$). What would be his final graduate GPA at CTU?

Of course, no one can say. But what we do know, from the Law of Total Variance discussed in Chapter 6, is that the variance of the conditional distribution of Y = final CTU GPA is smaller on average when you consider additional variables. Specifically,

$$\mathrm{E}\{\mathrm{Var}(Y \mid X_1, X_2, X_3)\} \le \mathrm{E}\{\mathrm{Var}(Y \mid X_1, X_2)\} \le \mathrm{E}\{\mathrm{Var}(Y \mid X_1)\}$$

Figure 11.1 shows how these inequalities might appear, as they relate to Hans. The variation in potentially observable GPAs among students who are like Hans in that they have GRE Math = 140 is shown in the top panel. Some of that variation is explained by different verbal abilities among students, and the second panel removes that source of variation by considering GPA variation among students who, like Hans, have GRE Math = 140, and GRE Verbal = 160. But some of that variation is explained by the general student diligence. Assuming undergraduate GPA is a reasonable measure of such "diligence," the final panel removes that source of variation by considering GPA variation among students who, like Hans, have GRE Math = 140, and GRE verbal = 160, and undergrad GPA = 2.7. Of course, this can go on and on if additional variables were available, with each additional variable removing a source of variation, leading to distributions with smaller and smaller variances.

The means of the distributions shown in Figure 11.1 are 3.365, 3.5, and 3.44, respectively. If you were to use one of the distributions to predict Hans, which one would you pick? Clearly, you should pick the one with the smallest variance. His ultimate GPA will be the same number

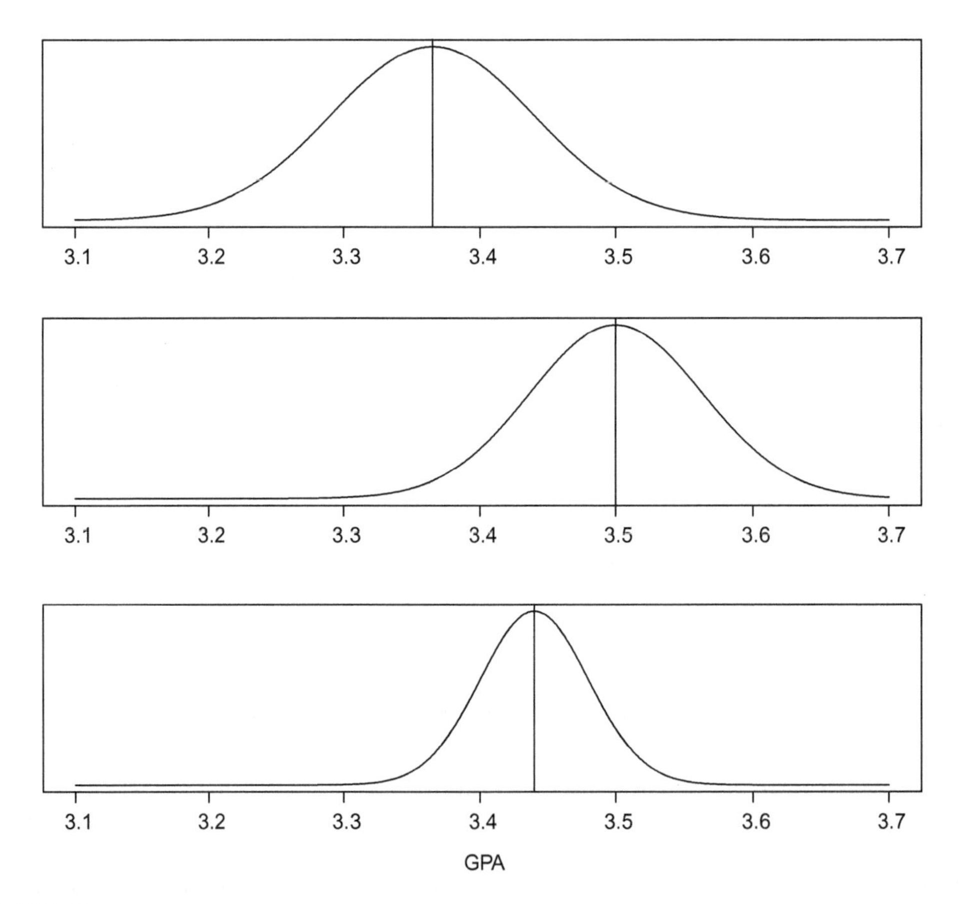

FIGURE 11.1
Conditional distributions of potentially observable GPA. Top panel: Distribution of GPA among students having GRE quant score = 140. Middle panel: Distribution of GPA among students having GRE quant score = 140 AND GRE verbal score = 160. Bottom panel: Distribution of GPA among students having GRE quant score = 140 AND GRE verbal score = 160 AND undergrad GPA = 2.7. Conditional means are 3.365, 3.5, and 3.44, respectively.

under all three distributions, and since the third distribution has the smallest variance, his GPA will likely be closer to its mean (3.44) than to the other distribution means (3.365 or 3.5).

While Figure 11.1 gives the right answer in theory, which is to use $E(GPA \mid X_1 = 140, X_2 = 160, X_3 = 2.7) = 3.44$ to predict Hans' GPA, the problem is that you do not know the number 3.44, since it is the mean of infinitely many potentially observable GPA values among students who have the combination $X_1 = 140$, $X_2 = 160$, and $X_3 = 2.7$. Instead, you have to estimate $E(GPA \mid X_1 = 140, X_2 = 160, X_3 = 2.7)$ using data. By the same logic, you do not know $E(GPA \mid X_1 = 140, X_2 = 160) = 3.5$ and $E(GPA \mid X_1 = 140) = 3.365$ either, and you would have to estimate them using data as well.

Suppose you have some historical GPA data at CTU. Refer back to the discussion of generalizability in Section 1.2 of Chapter 1, and suppose also that the processes that produced the historical GPA data are the same processes that will produce Hans' future GPA.

- You can estimate $E(GPA \mid X_1 = 140)$ by averaging the GPAs of all students who got 140 for GRE quant. Suppose there are 100 of these students.
- You can estimate $E(GPA \mid X_1 = 140, X_2 = 160)$ by averaging the GPAs of all students who got 140 for GRE quant *and* 160 for GRE verbal. Since this is a subset of the students in the first group, there has to be fewer students in it. Suppose there are 10 of these students.
- You can estimate $E(GPA \mid X_1 = 140, X_2 = 160, X_3 = 2.7)$ by averaging the GPAs of all students who got 140 for GRE quant *and* 160 for GRE verbal *and* 2.7 for undergrad GPA. Since this is a subset of the students in the second group, there has to be fewer students in it. Suppose there are 2 of these students.

Now, if you knew the theoretical mean for the infinite set of potentially observable GPAs among students that have 140 for GRE quant *and* 160 for GRE verbal *and* 2.7 for undergrad GPA (the bottom panel of Figure 11.1), then you should use that as a prediction of Hans' GPA. However, you only have an estimate of this mean, and it is based on *just two students rather than infinitely many.* We hope you can see that this is not a very good estimate of the mean of all such potentially observable GPAs. The estimate based on using only the GRE quant score could very easily be better estimates of Hans' future performance, since it is an average of 100 numbers, and will, therefore, have much less variability than an average of 2 numbers. (Recall that the variance of an average is equal to σ^2/n.)

Thus, even though it is better to use all the variables in theory, in practice your predictions may be better when you use a subset of the variables, simply because your estimates of the conditional means may be more accurate in the subsets where there is more data.

When using regression analysis to estimate these mean values, a similar phenomenon occurs. Consider three different fitted models using the same set of historical data:

```
fit1 = lm(GPA ~ GRE.quant)
fit2 = lm(GPA ~ GRE.quant + GRE.verbal)
fit3 = lm(GPA ~ GRE.quant + GRE.verbal + GPA.undergrad)
```

There are three different estimated models, leading to three different predictions:

Model 1: Hans' Predicted GPA $= \hat{\beta}_{01} + \hat{\beta}_{11}(140)$
Model 2: Hans' Predicted GPA $= \hat{\beta}_{02} + \hat{\beta}_{12}(140) + \hat{\beta}_{22}(160)$
Model 3: Hans' Predicted GPA $= \hat{\beta}_{03} + \hat{\beta}_{13}(140) + \hat{\beta}_{23}(160) + \hat{\beta}_{33}(2.7)$

Even if the linear, additive models were truly correct for all three conditional distributions, the estimates (the $\hat{\beta}$'s) differ randomly from the true parameters (the β's). Notice that there are four $\hat{\beta}$ estimates in Model 3, but only two $\hat{\beta}$ estimates in Model 1. Thus, there is more error due to estimation in Model 3 than in Model 1. This means that, despite the facts that the Model 3 estimate is an unbiased estimate of $E(GPA \mid X_1 = 140, X_2 = 160, X_3 = 2.7)$, and that the Model 1 estimate is a biased estimate of $E(GPA \mid X_1 = 140, X_2 = 160, X_3 = 2.7)$, the estimate from Model 1 can easily be closer to $E(GPA \mid X_1 = 140, X_2 = 160, X_3 = 2.7)$, despite its bias because there is less variability in terms of error due to $\hat{\beta}$ estimates.

To summarize: There are only two estimated $\hat{\beta}$'s in Model 1, but four estimated $\hat{\beta}$'s in Model 3; hence there is more error due to estimation of parameters in Model 3 than in Model 1.

11.2 The Bias-Variance Tradeoff

The previous example of predicting Hans' GPA suggests that you can predict Y better by using fewer X variables, even when the variables you exclude are in fact related to Y. The logic for this occurrence is known as the bias-variance tradeoff, which states that biased predictions are sometimes better because of a reduction in variance. What follows is an explanation of this phenomenon.

Suppose your measured variables are $W_1,\ldots,W_K$. In the example above with Hans, $K = 3$. However, you decide to use a subset of k of those K variables, say $X_1,\ldots,X_k$, to predict Y. Using regression or some other method, such as the simple averaging method described in the Hans GPA example, you get an estimate $\hat{\mu}(x_1,\ldots,x_k)$, where $(x_1,\ldots,x_k)$ are the observed values of $(X_1,\ldots,X_k)$.

Recall that the optimal prediction is $\mu(w_1,\ldots,w_K)$, but you cannot use this in practice because you do not have an infinite number of potentially observable Y at the given values $(w_1, \ldots, w_K)$ of the measured variables.

Let us define the optimal (but unusable) prediction, $\mu(w_1,\ldots,w_K) = \mu_K$ for simplicity of notation. Note that μ_K is *fixed*, not random.

Suppose you use an estimate of μ_K based on a subset of the variables, $\hat{\mu}(x_1,\ldots,x_k)$. Define $\hat{\mu}(x_1,\ldots,x_k) = \hat{\mu}_k$, again for simplicity of notation. Note that $\hat{\mu}_k$ is *random*, not fixed, because it is a function of your idiosyncratic data: Different data sets from the same GPA-generating process give different $\hat{\mu}_k$, but the μ_K is fixed because it is a characteristic of the data-generating process.

How accurate is $\hat{\mu}_k$ as a predictor of $Y \mid W_1 = w_1,\ldots,W_K = w_K$? Using expected squared deviation as a measure of accuracy, the following math shows how accurate it is:

$$E\{(Y - \hat{\mu}_k)^2\} = E\{(Y - \mu_K + \mu_K - \hat{\mu}_k)^2\}$$

$$= E\{(Y - \mu_K)^2\} + E\{(\mu_K - \hat{\mu}_k)^2\} + 2E\{(Y - \mu_K)\,(\mu_K - \hat{\mu}_k)\}$$

Now, the first summand in the last expression is just σ^2, the conditional variance of Y given the values $W_1 = w_1,\ldots,W_K = w_K$. Also, assuming the past data used to obtain $\hat{\mu}_k$ are independent of the "future" Y, the last summand is 0.

The middle summand can be expressed as follows:

$$\begin{aligned} E\{(\mu_K - \hat{\mu}_k)^2\} &= E\{(\mu_K - E(\hat{\mu}_k) + E(\hat{\mu}_k) - \hat{\mu}_k)^2\} \\ &= E\{(\mu_K - E(\hat{\mu}_k))^2\} + E\{(E(\hat{\mu}_k) - \hat{\mu}_k)^2\} + 2E\{(\mu_K - E(\hat{\mu}_k))(E(\hat{\mu}_k) - \hat{\mu}_k)\} \end{aligned}$$

The last summand in the above expression is 0, by the linearity property of expectation. The middle summand is, by definition, the variance of $\hat{\mu}_k$. The first summand is the squared bias of $\hat{\mu}_k$ as an estimate of μ_K.

Putting it all together, the average squared prediction error when using $\hat{\mu}_k$ as a predictor of $Y \mid W_1 = w_1, \ldots, W_K = w_K$ is given as follows:

The main equation that explains the bias-variance tradeoff

$$E\{(Y - \hat{\mu}_k)^2\} = \sigma^2 + \{\text{bias}(\hat{\mu}_k)\}^2 + \text{Var}(\hat{\mu}_k)$$

Refer to Figure 11.1 again. Suppose you decide to use $\hat{\mu}_2$ = average GPA of the ten students who got $X_1 = 140$ *and* $X_2 = 160$ to estimate Hans' GPA. The bias inherent in that estimate is, using the theoretical information given above, $3.5 - 3.44$, and the squared bias is $(3.5 - 3.44)^2$. The variance is $\tau^2/10$, where τ^2 is the variance of the middle distribution shown in Figure 11.1. Thus, the expected prediction error is

$$E\{(Y - \hat{\mu}_2)^2\} = \sigma^2 + (3.5 - 3.44)^2 + \tau^2/10$$

Suppose instead you decide to use $\hat{\mu}_3$ = average GPA of the two students who got $X_1 = 140$ *and* $X_2 = 160$ *and* $X_3 = 2.7$ to estimate Hans GPA. The bias inherent in that estimate is 0, and the variance is $\sigma^2/2$. Thus, the expected prediction error is

$$E\{(Y - \hat{\mu}_3)^2\} = \sigma^2 + (0)^2 + \sigma^2/2$$

It can easily happen that the unbiased prediction using $\hat{\mu}_3$ may be worse than the biased prediction using $\hat{\mu}_2$, because $\sigma^2/2$ can easily be larger than $(3.5 - 3.44)^2 + \tau^2/10$, depending of course, on the values of τ and σ. This is what is meant by the bias/variance trade-off: You *can* get more accurate predictions using a biased estimate, provided that the variance of the biased estimate is smaller than the variance of the unbiased estimate by an amount that is enough to compensate for the squared bias.

As always, to understand such complex arguments, it is wise to use simulation.

11.2.1 Simulation Study to Demonstrate the Bias-Variance Tradeoff

The following code is related to the study shown above in Figure 11.1. The numbers will not jibe exactly with those shown above, as the example is exaggerated slightly in the code below to make the points clearer. Also, rather than use averages of data in the subgroups to predict Y, the code uses regression models. The same idea holds, though: Estimated regression models using subsets of variables have lower variances of predicted values and parameter estimates, but give biased estimates relative to the complete set of measured variables.

Please run the following code, line-by-line, read the comments, and contemplate the results. Every time the comments say "stop," please stop, read the comments, look at the output, and understand the results. You should understand how predictions from *incorrect reduced* models can be more accurate than predictions from *correct full* models after doing so. If not, then do it again!

```
## With this code, you can simulate data from a model where a student's
## grad GPA truly depends on their GRE verbal (X1), GRE quant (X2),
## and Undergrad GPA (X3). Yet the estimates from the model using X1 only
## are better, because of the variance/bias trade-off.

nsimu = 10000    ## Number of simulated data sets
n = 20           ## Sample size of students within a particular data set

## The following code simulates the X's from a multivariate normal
## distribution, with variances that are sensible, and where the X's are
## correlated at .5 with one another.

Sig.X = diag(c(5,5,.3))%*% matrix(c(1,.5,.5,.5,1,.5,.5,.5,1),
nrow=3) %*%diag(c(5,5,.3))
Mean.X = c(150, 150, 3)
library("MASS")
X = mvrnorm(n, Mean.X, Sig.X)
X1 = round(X[,1],0)  # Simulated GRE verbal
X2 = round(X[,2],0)  # Simulated GRE quant
X3 = round(X[,3],2)  # simulated undergrad GPA
head(cbind(X1,X2,X3))  ## Stop and look at the simulated X data!

b0 = -.50            ## The true intercept
b1 =  .02            ## The true effect of verbal GRE
b2 =  .005           ## The true effect of quant GRE
b3 =  .02            ## The true effect of undergrad GPA
sg =  .20            ## The true conditional standard deviation of GPA
EY = b0 + b1*X1 + b2*X2 + b3*X3   ## The true conditional means of GPA

## Suppose Hans has X1=140, X2=160, X3=2.7. How well does the estimated
## regression model predict Hans?
X.H  = c(1, 140, 160, 2.7)   # Hans' X values, including intercept
X.Hr = c(1, 140)             # Hans' X values for the reduced model

## Here are nsimu *predictions* of Hans' GPA using two regression models,
## (i) the correct full model (p.F), and (ii) the incorrect reduced
## model (p.R).
p.F = replicate(nsimu, lm(EY +rnorm(n,0,sg) ~ X1+X2+X3)$coefficients%*%X.H)
p.R = replicate(nsimu, lm(EY +rnorm(n,0,sg) ~ X1)$coefficients%*%X.Hr)
head(p.F); head(p.R)
## Stop and look at the predictions of Hans' GPA!

## Hans' GPA comes from a distribution whose true mean is
## b0 + b1*140 + b2*160 + b3*2.7, and whose standard deviation is sg =.20.
Hans.M = b0 + b1*140 + b2*160 + b3*2.7; Hans.M
## Stop and notice the mean GPA for people like Hans is Hans.M = 3.154.
```

```
## Here are nsimu simulated *actual values* of Hans' ultimate GPA:
Hans = rnorm(nsimu, Hans.M, sg); head(Hans)
## stop and compare Hans' actual GPA with the predictions above!

## Here are the errors of prediction using the two different models.
Err.F = Hans - p.F ; Err.R = Hans - p.R

## Here is the error of prediction if you were mega-Hans, and knew the
## true conditional mean:
Err.T = Hans - Hans.M

head(Err.F); head(Err.R); head(Err.T)
# Stop and compare the errors of prediction using the three methods!

## Here are the root mean squared prediction errors for the estimated
## full model, the estimated reduced model, and the mega-Hans model:
sqrt(mean(Err.F^2)); sqrt(mean(Err.R^2)); sqrt(mean(Err.T^2))

## Stop and read this: The root mean square prediction errors measure
## how close the predictions using the various methods are to Hans' GPA.
## The mega-Hans model is best, but unusable in practice. You could use
## either the full or reduced models, as estimated from the data.
## Why is the full (correct) model worse? Because the predicted values have
## higher variance (less accuracy) when you use the full model, even though
## it is correct. This happens because all of the coefficient estimates are
## random, i.e., data-set specific: Different data sets imply different
## coefficient estimates. Although unbiased, the coefficient estimates from
## the full model are not equal to the true parameter values.

sd(p.F); sd(p.R)
## Stop and notice that there is more variability in the full model
## predictions.

mean(p.F) - Hans.M; mean(p.R) - Hans.M
## Stop and notice that there is no bias when the full model is used to
## predict Hans, but there is bias with the reduced model, which tends
## to give predictions that are too low:

par(mfrow=c(2,1))
hist(p.F, xlim = c(2.5,4), main="Predicted GPA Using Full Model")
abline(v=Hans.M, lwd=2, col="gray")
hist(p.R, xlim = c(2.5,4), main = "Predicted GPA Using Reduced Model")
abline(v=Hans.M, lwd=2, col="gray")
## Stop and look at the histograms to see (i) the unbiasedness of the full
## model predictions, (ii) the higher variability of the full model
## predictions, and (iii) the biasedness of the reduced model predictions,
## and (iv) the lower variability of the reduced model predictions.
```

The final graph produced by the code above is shown in Figure 11.2.

Self-study question: Did you run the R code above, stop where it said to stop, read the comments, and understand the lesson?

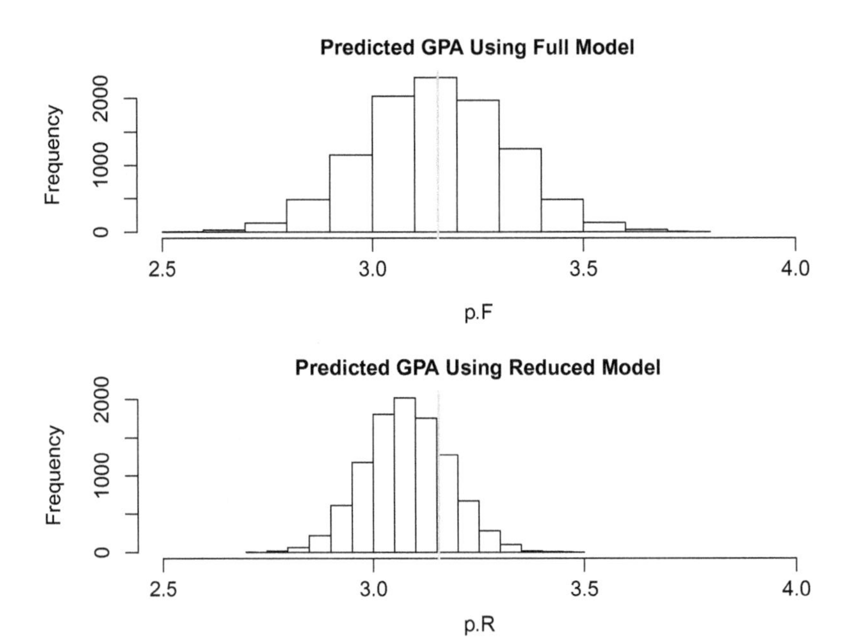

FIGURE 11.2
Predictions of Hans' GPA. Hans had verbal GRE = 140, quantitative GRE = 160, and undergrad GPA = 2.7. The full model uses a random sample of 20 students to estimate a regression model using all three variables; the reduced model uses the same 20 students to estimate a regression model using verbal GRE only. Predictions using the full model are $\hat{\beta}_{0F}+\hat{\beta}_{1F}(140)+\hat{\beta}_{2F}(160)+\hat{\beta}_{3F}(2.7)$ and are $\hat{\beta}_{0R}+\hat{\beta}_{1R}(140)$ using the reduced model. The mean GPA for students having predictors (140, 160, 2.7) is 3.154, the solid vertical line. Notice that the estimates using the reduced model are biased but have less variability.

The main equation, given above, that explains the variance-bias trade-off is $\mathrm{E}\{(Y-\hat{\mu}_k)^2\}=\sigma^2+\{\mathrm{bias}(\hat{\mu}_k)\}^2+\mathrm{Var}(\hat{\mu}_k)$. In the simulation study above, there were three different predictions $\hat{\mu}_k$: (i) the full model prediction, $\hat{\mu}=\hat{\beta}_{0F}+\hat{\beta}_{1F}(140)+\hat{\beta}_{2F}(160)+\hat{\beta}_{3F}(2.7)$, the reduced model prediction $\hat{\mu}=\hat{\beta}_{0R}+\hat{\beta}_{1R}(140)$; and the "Mega-Hans model" prediction, $\hat{\mu}_k=3.154$. The average squared difference between each of these predictions and Hans' ultimate GPA is what the term $\mathrm{E}\{(Y-\hat{\mu}_k)^2\}$ refers to. Figure 11.3 shows these differences, and you can see that the differences are generally largest for the full model, smaller for the restricted model, and smallest for the (unusable) Mega-Hans model.

R code for Figure 11.3

```
par(mfrow=c(3,1))
hist(Hans - p.F, xlim = c(-1,1), main= "Full Model
Prediction Errors")
hist(Hans - p.R, xlim = c(-1,1), main = "Reduced Model
Prediction Errors")
hist(Hans - Hans.M, xlim=c(-1,1), main = "Mega-
Hans Model Prediction Errors")
```

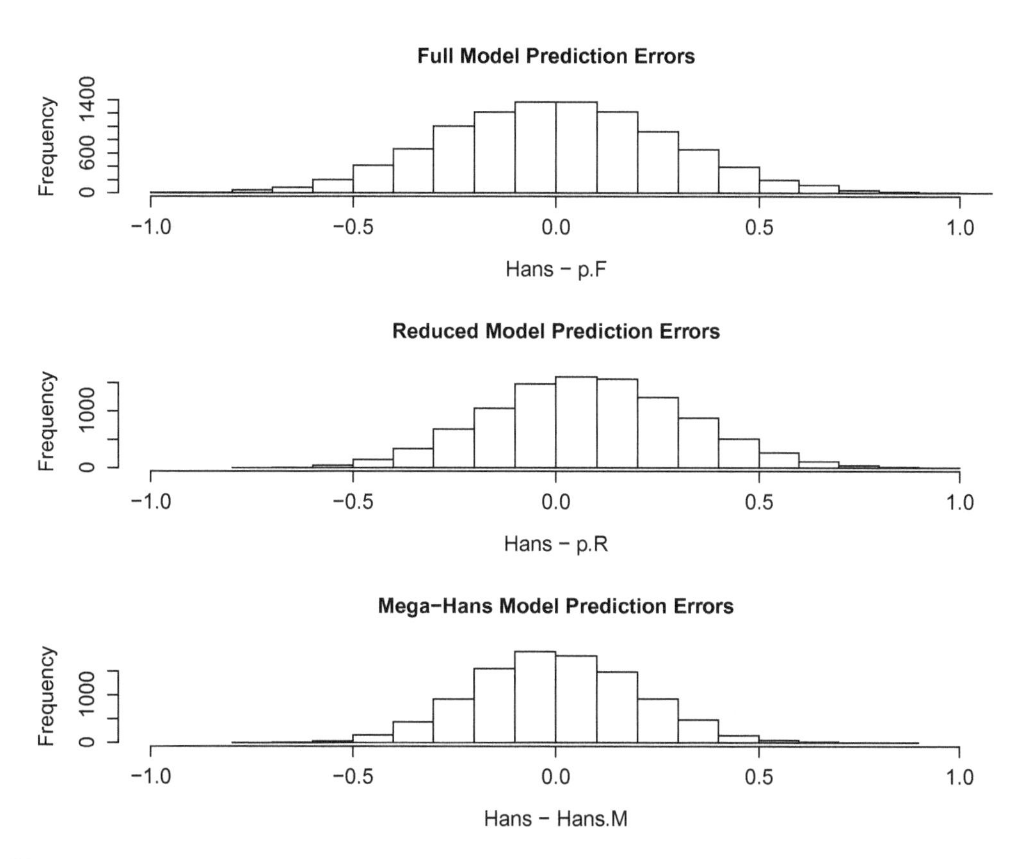

FIGURE 11.3
Prediction errors using three different models. The reduced model provides predictions that are biased low but are still generally more accurate (closer to Hans' GPA) than the unbiased full model. The "Mega-Hans model" uses the true conditional mean GPA of people like Hans to predict Hans, and it is the best model, but this model is unusable because this conditional mean is unknown in practice.

The term $\mathrm{E}\{(Y-\hat{\mu}_k)^2\}$ is the average squared deviation from zero of the prediction error, as shown in Figure 11.3. Table 11.1 identifies all terms in the variance-bias trade-off equation $\mathrm{E}\{(Y-\hat{\mu}_k)^2\}=\sigma^2+\{\mathrm{bias}(\hat{\mu}_k)\}^2+\mathrm{Var}(\hat{\mu}_k)$, as they refer specifically to the GPA prediction study. The final column of Table 11.1 shows the root mean square prediction errors (RMSEs) given by `sqrt(mean(Err.F^2))`, `sqrt(mean(Err.R^2))`, and `sqrt(mean(Err.T^2))` in the simulation code; any differences are due to simulation error and rounding error.

The essential lesson of this exercise is an extremely important, general, and foundational principle that applies to all statistical estimates, and in particular to regression model estimates:

You can trade more bias for less variance when you use a reduced model, sometimes giving you better predictions.

TABLE 11.1

Components (Some Estimated Via Simulation, Some Actual Values) of the Bias-Variance Tradeoff When Using Three Different Predictions of Hans' GPA. Explanations Are Given in Parentheses; Some Refer to the `R` Code of the Simulation Study

Prediction	σ^2	$\{\text{bias}(\hat{\mu}_k)\}^2$	$\text{Var}(\hat{\mu}_k)$	$E\{(Y-\hat{\mu}_k)^2\}$	RMSE
Full model ($\hat{\beta}_{0F}+\hat{\beta}_{1F}(140)+\hat{\beta}_{2F}(160)+\hat{\beta}_{3F}(2.7)$)	0.2^2 (`sg = 0.20`)	0 (Full model is unbiased)	0.0565 (`var(p. F)`)	0.0965(0.2^2 + 0.0 + 0.0565)	0.311 (Square root of 0.0965)
Reduced model ($\hat{\beta}_{0R}+\hat{\beta}_{1R}(140)$)	0.2^2 (`sg = 0.20`)	0.00521 (`(mean (p. R)-Hans. M)^2`)	0.00874 (`var(p. R)`)	0.05395 (0.2^2 + 0.00521 + 0.00874)	0.232 (Square root of 0.5395)
Mega-Hans model (3.154)	0.2^2 (`sg = 0.20`)	0.0 (Mega-Hans prediction = 3.154 = true mean)	0.0 (3.154 has 0 variance)	0.04 (0.2^2 + 0.0 + 0.0)	0.200 (Square root of 0.04)

11.3 Variable Selection Based on Penalized Fit

To recap the previous section, additional variables can hurt in practice, even though they can never hurt in theory. Thus, you often should use a regression model that is estimated using a subset of the measured variables. Despite the fact that a prediction using the subset of the measured X variables is a biased estimate of the conditional mean for the complete set of X variables, the prediction using the subset is often better. This happens when the reduction in variance of $\hat{Y}$ obtained by using the subset of X variables is enough to compensate for the bias incurred by using this subset.

The simulations above show one way to answer the question, "Which subset of variables should I use?" The answer is, you can develop a simulation model, try many different subsets, and pick the one that gives the most accurate predictions in terms of RMSE. However, this method requires you to develop a simulation model that reflects your reality very well, which can be difficult. The true parameters of the natural processes are never known, so your simulation model can never be a perfect representation of Nature.

In practice, it is more common to identify good prediction subsets by analyzing your data set rather than by using simulation. However, this is not an ideal solution either. There is a fundamental flaw in any analysis based on data, because your data set, no matter how large, is not the infinity of potentially observable data (unlike your simulation model). Instead, your data set is, as you have heard repeatedly throughout this book, *idiosyncratic*. So you cannot possibly get the absolute right answer using any statistic based on your idiosyncratic data set. Let's set that off so you can see it clearly:

No selection procedure based on your observed, idiosyncratic data can find the absolute best model.

Still, there are methods that you can use to identify the *better* models (if not the *best* model) using your idiosyncratic data.

There are two main strategies for selecting such models using your data:

- Use models with good penalized fit statistics
- Use models with good out-of-sample prediction accuracy

We discuss penalized fit statistics in this section. Statistics such as "Akaike's Information Criterion," or AIC, penalize you for including too many parameters.

Additional X variables will give you additional parameters because there is an additional β parameter for every X variable. But "additional parameters" does not necessarily mean "additional X variables." You can also have additional parameters without additional X variables. For example, suppose you have ANOVA data and you decide to estimate a separate variance parameter for each group. In this case, the number of X variables does not change, but the number of parameters increases; such models for different variances are described in the next chapter. For another example, you might consider a more flexible family of distributions for Y rather than the normal distribution such as the family of mixtures of normal distributions. Such distributions involve extra parameters in addition to the usual μ and σ, even though the number of X variables stays the same. Models for non-normal Y are described extensively in later chapters. For yet another example, models for dependence structure of the observations involve additional parameters of the *covariance matrix* that model dependencies between observations. In this case, there will also be additional parameters even when there are no additional X variables. Such models are described in the next chapter.

AIC is usually defined as

$$\text{AIC} = -2(\text{Log Likelihood}) + 2 \times (\text{\# of estimated parameters}).$$

Higher likelihood is better (remember, it is called *maximum* likelihood). Thus, *smaller* values of AIC indicate better models because of the "−2" that multiplies the Log Likelihood in the expression for AIC. If an additional parameter increases the likelihood only a little, it may make the AIC worse because of the penalty term 2 × (# of estimated parameters). Thus, even though the model with all measured variables (and many, many parameters) is best in theory, AIC will often select a model having only a subset of the X variables and fewer parameters. Such behavior is desirable from the standpoint of the bias-variance tradeoff.

There are many variants of penalized likelihood, including the "Bayesian Information Criterion," or BIC, defined as

$$\text{BIC} = -2(\text{Log Likelihood}) + \{\ln(n)\} \times (\text{\# of estimated parameters}).$$

Like AIC, the BIC statistic penalizes you for including too many parameters. With BIC, the penalty also becomes larger for larger sample size n.

11.3.1 Identifying Models with Low BIC for Predicting Crime Rate

The data set "`Boston`" in the "`MASS`" library of `R` has information on per capita crime rates in Boston suburbs, as well as numerous other possibly related variables.

In the following code, crime rate is predicted from various variables in the "Boston" data set, and all possible subsets of those variables are considered for use in the final model. Two R libraries, "leaps" and "car" are needed.

R code for Figure 11.4

```
library(MASS)
data("Boston")
library(leaps)
fits = regsubsets(crim ~ zn + indus + rm + age + dis + rad + tax +
lstat + medv, data=Boston, nbest=10)
library(car)
par(mfrow=c(1,2))
subsets(fits, statistic="bic")
subsets(fits, statistic="bic", ylim = c(-260,-250))
```

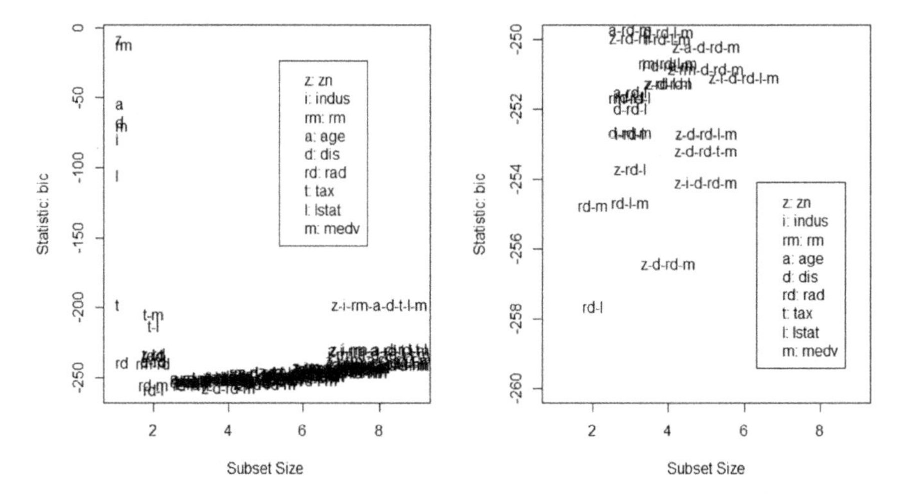

FIGURE 11.4
BIC statistics for various subset models. Right panel is a "zoomed-in" version of the left panel, where the model with lowest BIC is clearly identifiable. That model predicts Crime Rate as a function of "rad," an index of accessibility to major highways, and "lstat," a measure of socio-economic status.

The best model for predicting crime rate, according to the BIC statistic, includes the subset of $k = 2$ variables {rad, lstat} of the larger set of $K = 9$ variables. The fitted model corresponding to the chosen $k = 2$ variables is given by

```
Coefficients:
             Estimate Std. Error t value Pr(>|t|)
(Intercept) -4.38141    0.59872  -7.318 1.00e-12 ***
```

```
rad          0.52281    0.03842   13.607   < 2e-16 ***
lstat        0.23728    0.04685    5.065 5.75e-07 ***
---

Signif. codes:  0 '***' 0.001 '**' 0.01 '*' 0.05 '.' 0.1 ' ' 1

Residual standard error: 6.559 on 503 degrees of freedom
Multiple R-squared:  0.4208,    Adjusted R-squared:  0.4185
F-statistic: 182.7 on 2 and 503 DF,  p-value: < 2.2e-16
```

Even though this model is best based on the BIC statistics calculated from this particular idiosyncratic data set, you should not assume that it is the "best" or "correct" model. Other models are nearly equally viable according to the BIC statistic, as seen in Figure 11.4.

11.4 Variable Selection Based on Out-of-Sample Prediction Accuracy

Next, let's consider out-of-sample prediction accuracy as a criterion. Recall that additional variables always decrease SSE (or at least make it no larger). Thus, you can always make your fitted regression function's $\hat{y}$ values generally closer to the y values *in your idiosyncratic data set* by including additional variables in your fitted regression model. However, that should not be your goal—instead, you would like your model to generalize beyond your data. If a new observation came along after you estimated your regression model, with a fresh set of Y and X variable values that were not used in your estimated model, you would like your estimated model to predict that external Y value well. It is not good enough to know that the $\hat{y}$ values are generally close to the internal y values in your idiosyncratic data set.

You can estimate how well a model predicts such external Y data by estimating the model using a subset of your n observations, say s of them, and then predicting the remaining $r = n - s$ "external" observations. Then the sum of squared prediction errors is

$$\text{SSPE}^{(r)} = \sum_{i=1}^{r} \{y_i^{(r)} - (\hat{\beta}_0^{(s)} + \hat{\beta}_1^{(s)} x_{i1}^{(r)} + \cdots + \hat{\beta}_k^{(s)} x_{ik}^{(r)})\}^2.$$

In the expression above, the "(s)" superscript refers to estimates based on using a subset of $s < n$ of the observations (rows) in your data set. The "(r)" superscript refers to the remaining data. Thus, $\text{SSPE}^{(r)}$ is a measure of how well your model works to predict external data.

How to pick the "s" subset of your n observations? There are two main ways. The first is called "ten-fold cross-validation." With this method, you randomly divide the data set into ten non-overlapping subsets. So, if you have $n = 200$ observations, you will have the computer randomly pick 20 for group 1, then another 20 from the remaining 180 for group 2, etc. you will then leave out the first group of 20, estimate the model using the subset of $s = 180$ other observations, and get $\text{SSPE}^{(r)}$. You will then put the 20 back in the data set, leave out the next group of 20, estimate the model using the subset of $s = 180$ other observations, and get a different $\text{SSPE}^{(r)}$. Repeating for all subsets, you will have 10 $\text{SSPE}^{(r)}$ statistics; the global SSPE is the sum. Notice that there will be 200 numbers in this global sum, where each of the individual Y values is an "external" value that is predicted at some stage in the process.

The second main way is called "n-fold cross-validation," or simply "cross-validation." This method is similar to ten-fold cross-validation, except that there are n subsets rather than ten subsets. There is no need to randomly divide in this case because every single observation is a distinct subset. Thus, you leave out the first observation (x_1, y_1), fit the model on the remaining $n - 1$ observations, and predict y_1 using the regression model fit from the remaining $n - 1$ observations. This gives a $SSPE^{(r)}$ that is a single squared prediction error rather than a sum. Then the first observation is returned to the data set, the second observation (x_2, y_2) is removed, the model is estimated using $n - 1$ observations, giving another $SSPE^{(r)}$ that is the single squared prediction error for predicting y_2. This process is repeated for all n observations in the data set, and again the global SSPE is the sum of all the $SSPE^{(r)}$ values. In this way, each of the individual Y values that was predicted was an "external" value, because it was not used in the data that gave the estimated model.

Which to use, 10-fold cross-validation (CV) or n-fold CV? Hands down, the winner is n-fold, for two reasons. First, there is no randomness in the combined SSPE value when you use n-fold CV. With 10-fold CV, every time you run the computer code, you get a different number. Second, the goal of using CV is to assess the predictive ability of your estimated model based on *all* n observations, which is what you will ultimately use. With 10-fold CV, the models are all estimated using $(9/10)n$ observations, whereas with n-fold CV, the models are all estimated using $(n - 1)$ observations. Since $(n - 1)$ is much closer to n than $(9/10)n$ (e.g., 199 is closer to 200 than 180), the n-fold CV estimate is much closer to the true external prediction error that you would get using all n observations.

Self-study question: Then why not calculate SSPE using your estimated model with all n observations?

On the other hand, with large data sets, n-fold CV is too time-consuming. For example, if you have $n = 1{,}000{,}000$ observations, then you will have to fit 1,000,000 regression models. But ten-fold CV requires only 10 model fits, which is less than 1,000,000 model fits. Quite a bit less! So, if you can do it within an acceptable computer wait time, use n-fold CV. If not, use ten-fold CV.

The idea, then, is to use CV on a collection of different regression models; and like BIC, the models with the smallest SSPE are the better models. Again, notice the phrasing: Better *models*. Just because you have identified a model with smallest SSPE does not mean it is the best model, or that it is the correct model, because the smallest SSPE is calculated from your idiosyncratic data. A different data set produced by the exact same process will likely yield a different model, even when the same minimum SSPE criterion is used.

The following example is motivated by "Taylor's Theorem," a mathematical theorem stating that smooth functions $f(x)$ can be approximated with greater precision by using polynomial functions of higher order. Taylor's theorem suggests that you should use higher-order polynomials to estimate the true conditional mean function $f(x)$ because higher-order polynomial regression models will give less biased estimates of $f(x)$. However, you have learned in this chapter that such estimated models will also have higher variance, since they require more estimated parameters. Thus, even though higher-order polynomials are better in theory, when you know the true parameter values, in practice they may be worse due to the error incurred when estimating the parameter values. The following example illustrates this fact using the out-of-sample prediction error as a criterion to compare models.

11.4.1 Example Showing Decrease in SSE but Increase in SSPE

The fact the SSE always decreases with additional X variables applies to any kind of X variable. In particular, if you have a single X variable but include X^2, X^3, X^4, X^5, etc., to the model, then SSE will be smaller and smaller. You can show this graphically as follows.

R code for Figure 11.5

```
x = c(2.0, 3.2, 2.1, 2.00, 3.1, 8.0, 8.2, 7.0)
y = c(2.2, 2.1, 2.7, 2.36, 2.8, 3.6, 3.0, 3.2)

x1=x; x2 = x^2; x3 = x^3; x4 = x^4; x5 = x^5
fit1 = lm(y ~ x1); fit2 = lm(y ~ x1 + x2); fit3 = lm(y ~ x1 + x2 + x3)
fit4 = lm(y ~ x1 + x2 + x3 + x4); fit5 = lm(y ~ x1 + x2 +
x3 + x4 + x5)

## Now, graph the models and the data:
xg = seq(1.8, 8.4, .001)
pred = data.frame(xg, xg^2, xg^3, xg^4, xg^5)
colnames(pred) = c("x1","x2","x3","x4","x5")

plot(xg,predict(fit1, pred), type="l", ylim = c(0, 5),
xlab="x", ylab="y-hat");
points(x,y, cex=2, lwd=2); points(xg,predict(fit2, pred),
type="l", lty=2)
points(xg,predict(fit3, pred), type="l", lty=3)
points(xg,predict(fit4, pred), type="l", lty=1, lwd=2)
points(xg,predict(fit5, pred), type="l", lty=2, lwd=2)
```

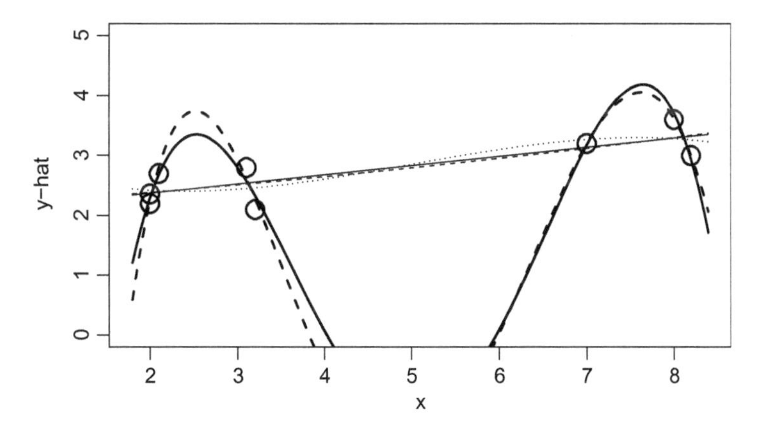

FIGURE 11.5
Scatterplot of data (n = 8 circles) with fitted linear (thin solid line), quadratic (thin dashed line), cubic (thin dotted line), quartic (thick solid line), and quintic (thick dashed line) functions.

The results are shown in Figure 11.5.

In Figure 11.5, you should be able to see that the SSE statistics are getting smaller and smaller with higher order polynomial functions, because the fitted functions are getting closer and closer to all the data points in the data set. However, suppose a new, external observation came along having $X = 5$. Where do you think the Y value will be? Probably somewhere around 2 or 3. For this external observation, the prediction error would be very high if you used the quartic or quintic models shown in Figure 11.5.

You can get SSPE for all the models above using n-fold CV using the `CVlm` (cross-validated `lm`) function in the `DAAG` library of `R`. In the code that follows, we have calculated both the SSE and SSPE, and have normalized them as RMSE = sqrt(SSE/n) and RMSPE = sqrt(SSPE/n) for easier connection to the graphs. (Note: The acronyms RMSE and RMSPE read "root mean square error" and "root mean square predictive error," respectively.) The statistics RMSE and RMSPE measure typical vertical deviation of actual observation from the fitted function, and typical vertical deviation of "future" observation from the fitted function, respectively.

Here is the R code:

```
library(DAAG)
x = c(2.0, 3.2, 2.1, 2.00, 3.1, 8.0, 8.2, 7.0)
y = c(2.2, 2.1, 2.7, 2.36, 2.8, 3.6, 3.0, 3.2)
x1=x; x2 = x^2; x3 = x^3; x4 = x^4; x5 = x^5

n = length(y)
data.f = data.frame(x1,x2,x3,x4,x5,y)

pred.1 = CVlm(data = data.f, form.lm=formula(y ~ x1), m=n)
pred.2 = CVlm(data = data.f, form.lm=formula(y ~ x1+x2), m=n)
pred.3 = CVlm(data = data.f, form.lm=formula(y ~ x1+x2+x3), m=n)
pred.4 = CVlm(data = data.f, form.lm=formula(y ~ x1+x2+x3+x4), m=n)
pred.5 = CVlm(data = data.f,
form.lm=formula(y ~ x1+x2+x3+x4+x5), m=n)

## In-sample root mean square errors (RMSE)
sqrt(mean((y-pred.1$Predicted)^2))
sqrt(mean((y-pred.2$Predicted)^2))
sqrt(mean((y-pred.3$Predicted)^2))
sqrt(mean((y-pred.4$Predicted)^2))
sqrt(mean((y-pred.5$Predicted)^2))

## Out-of-sample root mean square prediction errors (RMSPE)
sqrt(mean((y-pred.1$cvpred)^2))
sqrt(mean((y-pred.2$cvpred)^2))
sqrt(mean((y-pred.3$cvpred)^2))
sqrt(mean((y-pred.4$cvpred)^2))
sqrt(mean((y-pred.5$cvpred)^2))
```

The results are as follows:

Linear model shown in Figure 11.5: RMSE = 0.275, RMSPE = 0.373

Quadratic model shown in Figure 11.5: RMSE = 0.275, RMSPE = 0.455

Cubic model shown in Figure 11.5: RMSE = 0.265, RMSPE = 0.772

Quartic model shown in Figure 11.5: RMSE = 0.128, RMSPE = 0.723

Quintic model shown in Figure 11.5: RMSE = 0.108, RMSPE = 7.210

So, while the in-sample fit statistic RMSE always decreases with additional variables, in this example the out-of-sample fit statistic RMSPE is worse for all models except the simple linear model. The happy conclusion is that the simple linear fit is best using the RMSPE criterion; there is no need for additional complexity. Despite the fact that the more complex models fit the observed data better and better, our best guess is that the simple linear model will fit future data better than all of the other more complex models.

You can use `CVlm` to compare different models for predicting crime rate too. Refer to the `R` code that gives Figure 11.4, and add these lines:

```
results = cbind(summary(fits)$bic, summary(fits)$which)
results = results[order(summary(fits)$bic),]
head(results)
```

You then get the following output:

```
       (Intercept) zn indus rm age dis rad tax lstat medv
2 -258           1  0     0  0   0   0   1   0     1    0
4 -256           1  1     0  0   0   1   1   0     0    1
2 -255           1  0     0  0   0   0   1   0     0    1
3 -255           1  0     0  0   0   0   1   0     1    1
5 -254           1  1     1  0   0   1   1   0     0    1
3 -254           1  1     0  0   0   0   1   0     1    0
```

The 1's and 0's indicate whether the variable in question is in or out of the fitted model. Using these 6 models, you can calculate the RMSPE values as follows:

```
n = nrow(Boston)
library(DAAG)

pred1=CVlm(data=Boston,form.
lm=formula(crim ~ rad+lstat), m=n, printit=FALSE)

pred2=CVlm(data=Boston,form.
lm=formula(crim ~ zn+dis+rad+medv), m=n, printit=FALSE)

pred3=CVlm(data=Boston,form.
lm=formula(crim ~ rad+medv), m=n, printit=FALSE)

pred4=CVlm(data=Boston,form.
lm=formula(crim ~ rad+lstat+medv), m=n, printit=FALSE)

pred5=CVlm(data=Boston,form.
lm=formula(crim ~ zn+indus+dis+rad+medv), m=n, printit=FALSE)

pred6=CVlm(data=Boston,form.
lm=formula(crim ~ zn+rad+lstat), m=n, printit=FALSE)

y=Boston$crim
## In-sample root mean square prediction errors
sqrt(mean((y-pred1$Predicted)^2))
sqrt(mean((y-pred2$Predicted)^2))
```

```
sqrt(mean((y-pred3$Predicted)^2))
sqrt(mean((y-pred4$Predicted)^2))
sqrt(mean((y-pred5$Predicted)^2))
sqrt(mean((y-pred6$Predicted)^2))

## Out-of-sample root mean square prediction errors
sqrt(mean((y-pred1$cvpred)^2))
sqrt(mean((y-pred2$cvpred)^2))
sqrt(mean((y-pred3$cvpred)^2))
sqrt(mean((y-pred4$cvpred)^2))
sqrt(mean((y-pred5$cvpred)^2))
sqrt(mean((y-pred6$cvpred)^2))
```

The results are as follows:

$$\begin{array}{ll}
\text{Model 1, crim} \sim \text{rad} + \text{lstat:} & \text{RMSE} = 6.54, \text{RMSPE} = 6.60, \text{BIC} = -258 \\
\text{Model 2, crim} \sim \text{zn} + \text{dis} + \text{rad} + \text{medv:} & \text{RMSE} = 6.47, \text{RMSPE} = 6.53, \text{BIC} = -256 \\
\text{Model 3, crim} \sim \text{rad} + \text{medv:} & \text{RMSE} = 6.56, \text{RMSPE} = 6.62, \text{BIC} = -255 \\
\text{Model 4, crim} \sim \text{rad} + \text{lstat} + \text{medv:} & \text{RMSE} = 6.52, \text{RMSPE} = 6.60, \text{BIC} = -255 \\
\text{Model 5, crim} \sim \text{zn} + \text{indus} + \text{dis} + \text{rad} + \text{medv:} & \text{RMSE} = 6.44, \text{RMSPE} = 6.51, \text{BIC} = -254 \\
\text{Model 6, crim} \sim \text{zn} + \text{rad} + \text{lstat:} & \text{RMSE} = 6.53, \text{RMSPE} = 6.59, \text{BIC} = -254
\end{array}$$

According to the RMSPE criterion, model 5 is best.

When sifting through large numbers of potential subset models, automatic search strategies to narrow the collection of subsets can be useful. *Forward selection* is an example of such a strategy. For example, suppose you have identified $K = 100$ possible X variables, and you wish to select a subset of k of them. See Figure 11.4 again to understand the subset concept. With $K = 100$ variables, there are $2^{100} = 1.27 \times 10^{30}$ possible subsets, an impossibly large number for even by supercomputer standards. So instead of evaluating all possible subsets as indicated in Figure 11.4, you could narrow the collection of subsets by using *forward selection*.

Forward selection algorithm for identifying subsets of the X variables

Subset 1: Contains the best X variable among all K single X variable regressions of Y on X_j, $j = 1, 2, \ldots, K$.

Subset 2: Contains the X variable identified in Subset 1, as well as the best X variable among all $K - 1$ regressions involving two X variables, namely regression of Y on (X_j and the variable identified in subset 1).

Subset 3: Contains the X variables identified in Subset 2, as well as the best X variable among all $K - 2$ regressions involving three X variables, namely regression of Y on (X_j and the variables identified in subset 2).

...

Subset K: Contains all K of the X variables.

The term "best" that appears many times in the algorithm can be operationalized many ways; the `regsubsets` function uses BIC as its default measure of "best."

The beauty of the forward selection scheme, there are only K subset models to consider, rather than 2^K. You can then evaluate each subset model by using penalized fit or out-of-sample prediction measures to pick just one or more of the subsets.

There is a similar *backwards elimination* algorithm that starts with *all* X variables and eliminates the "worst" variables sequentially. The backwards elimination algorithm is better because it can identify variables that are predictive in combination but not individually (review the *predictive multicollinearity* concept discussed at the end of Section 8.4.). This method is best when there are not too many X variables, but when the number of possible X variables is too large it can be infeasible, and forward selection is needed instead.

No matter which statistic you use to evaluate "goodness" of a model or variable, BIC, AIC, RMSPE, etc., please realize that none can give the best model because they are all based on your idiosyncratic data. The BIC statistic has goals that we do not wish to discuss in detail, so you can just consider it to be a "penalized likelihood" statistic. That does not mean it is "best," even in a theoretical sense, let alone in an idiosyncratic data sense. By the same token, RMSPE re-uses the sample data to estimate average external prediction error, as measured by squared deviations. It is certainly not "best," either, even in a theoretical sense. For one thing, the data you have in your data set is certainly not "future" data, or even "hypothetical past" data that you want your model to represent. For another thing, the squared deviation measure is not necessarily the best measure of prediction error: As discussed in Chapter 2, Section 2.3, you should prefer absolute deviations over squared deviations when you have outlier-prone data.

Perhaps most importantly, none of these methods are designed to identify variables that are *causally* connected with Y. Instead, they are designed to find variables that are good for predicting external Y data values. The goal of prediction is to find $\hat{Y}$ values that are close to external Y values, regardless of whether the X variables used in your $\hat{Y}$ function are causally connected to Y.

You should note that, because of the random (data-specific, idiosyncratic) nature of the model selection, evidence concerning the importance of variables is distorted (usually, greatly biased toward greater importance) when you perform variable selection. One reason this happens is that, when you pick a model that fits your idiosyncratic data best, your estimate of residual variance will be biased downward, thus biasing standard errors downward. For another reason, in your idiosyncratic data, some of the X variables will incorrectly appear more important than others by chance alone, again giving inflated determinations of importance. Because of these reasons, you should not use a variable selection procedure as part of a data analysis plan to validate and publish a research hypothesis.

If your goal is simply to obtain predicted values, and not to evaluate research hypotheses, you can and should use such variable selection methods, aided of course, by penalized or out-of-sample measures of fit. But if your goal is to evaluate research hypotheses, you can safely use variable selection methods only in a pilot study. You can then legitimately use the selected model from your pilot study in a confirmatory way, suitable for publication, if you use the selected model on new data that was not part of the pilot study. In this way, the determinations of importance when using the confirmatory study data will not be biased by the variable selection that was performed on your pilot study data.

A final comment: Historical sources on regression analysis often recommended that variables be included based on "significance" (again, p-value < 0.05) or excluded based on "insignificance" (p-value > 0.05). As discussed repeatedly throughout this book, such a practice is fraught with numerous kinds of errors: It makes people think that "practical importance" and "statistical significance" are the same, it promotes basic misunderstanding of statistics by implying that an "insignificant" result means "no effect," and it is simply the wrong tool for the job of identifying good predictive models. Therefore, variable selection methods based solely on p-value thresholds should be avoided.

Exercises

1. See the mathematical logic at the beginning of Section 10.5, which proves that the SSE is always *smaller* (or at least can be no larger) in the full model as compared to the restricted model. Use the same logical to prove that the maximized log-likelihood of a full model *must be larger* (or at least can be no smaller) in the full model as compared to the restricted model.
2. In Chapter 11, Section 11.2, consider the simulation in "Simulation study to demonstrate the Bias-Variance tradeoff." The true value of β_1 is 0.02 in the study. While the simulation study demonstrated the bias-variance tradeoff for predicted values, you can also demonstrate the variance-bias trade-off for estimates of parameters. That is the purpose of this problem.
   a. Demonstrate via simulation that the estimate of β_1 in the full model is an unbiased estimate of $\beta_1 = 0.02$. To do so, draw the histogram of the estimates, with vertical lines indicating (i) the target 0.02, and (ii) the mean of the estimates.
   b. Demonstrate via simulation that the estimate of β_1 in the model with only X_1 is a biased estimate of the full model $\beta_1 = 0.02$. To do so, draw the histogram of the estimates, with vertical lines indicating (i) the target 0.02, and (ii) the mean of the estimates.
   c. Demonstrate using your simulation data that the estimate in 2.b has less variance than the estimate in 2.a. To do so, calculate the variances of the two estimates.
   d. Demonstrate by calculating the root mean square differences from the target $\beta_1 = 0.02$ that the estimate in 2.b tends to be closer to the target $\beta_1 = 0.02$, despite its bias. Relate that result to your graphs in 2.a and 2.b.
3. Consider the example in showing "Example showing decrease in SSE but increase in SSPE" in Section 11.4 of Chapter 11.
   a. Find the SSPE "manually" for the linear model by fitting the relevant 8 separate linear models and predicting the left-out Y value for each of those estimated models. Using that SSPE, find the RMSPE, and compare it to the value calculated using the `CVlm` fit. Show that your results are identical to many, many decimals, differing only at the round-off level of the computer, by calculating the difference between the two results. (Hint: A loop makes it easier, but you do not have to do it that way.)
   b. In that same example, find the RMSPE for the model that predicts Y just using the average of Y, rather than any X or any polynomial in X. (This is the intercept-only model, fit using `R` as `lm(Y ~ 1)`). You can do this using the `CVlm` function rather than "by hand." Is the intercept-only model better than any of the other models reported in the book, according to RMSPE?

12

Heteroscedasticity and Non-independence

In the previous chapter, we discussed methods for selecting *variables* $X_1, X_2, \ldots, X_k$ to use in your model $p(y \mid x_1, x_2, \ldots, x_k)$. But *which model* $p(y \mid x_1, x_2, \ldots, x_k)$ should you use? Recall that the model $p(y \mid x_1, x_2, \ldots, x_k)$ states that there is a distribution of possible values of Y when the X data are $X_1 = x_1, X_2 = x_2, \ldots, X_k = x_k$. The model $p(y \mid x_1, x_2, \ldots, x_k)$ does not state which distribution (normal or otherwise), nor how the mean function $\mathrm{E}(Y \mid X_1 = x_1, X_2 = x_2, \ldots, X_k = x_k)$ looks (planar, having quadratic curvature, having interaction curvature, etc.), nor how the variance function $\mathrm{Var}(Y \mid X_1 = x_1, X_2 = x_2, \ldots, X_k = x_k)$ looks (constant, a function of X_1 only, an exponential function of all the X's, etc.).

The classical regression model made all those choices for you: The distributions are assumed to be normal, the mean function is assumed to be planar, with

$$\mathrm{E}(Y \mid X_1 = x_1, X_2 = x_2, \ldots, X_k = x_k) = \beta_0 + \beta_1 x_1 + \beta_2 x_2 + \ldots + \beta_k x_k$$

and the variance function is assumed to be constant, or *homoscedastic*, not depending on the X data, with $\mathrm{Var}(Y \mid X_1 = x_1, X_2 = x_2, \ldots, X_k = x_k)$ assumed to equal σ^2, regardless of the X values.

Of course, the constant variance assumption is patently wrong. Why should all those variances be exactly the same number? For example, if the variance of GPA among people who score 150 on the GRE is $(0.4000000000000)^2$, does this automatically imply that the variance of GPA among people who score 120 on the GRE is also $(0.4000000000000)^2$? No, of course not. Nature does not work that way. There is an infinity of possible values for the latter variance, and $(0.40)^2$ is just one of those infinitely many values, so the chance that the latter variance is $(0.40)^2$ is $1/\infty = 0.0$.

You do not need, and usually should not even use, a hypothesis testing method to check the homoscedasticity assumption: If the test "passes" $(p > 0.05)$, it does not prove homoscedasticity; and if the test "fails" $(p < 0.05)$, it does not prove that the heteroscedasticity is so bad that you have to worry about it.

Instead, you can simply *model* the heteroscedasticity. You already learned how to model the conditional mean function as a function of the X values via

$$\mathrm{E}(Y \mid X_1 = x_1, X_2 = x_2, \ldots, X_k = x_k) = f(x_1, x_2, \ldots, x_k; \beta)$$

You can model the conditional variance function in similar fashion. Instead of assuming

$$\mathrm{Var}(Y \mid X_1 = x_1, X_2 = x_2, \ldots, X_k = x_k) = \sigma^2,$$

you can assume a non-constant variance function

$$\mathrm{Var}(Y \mid X_1 = x_1, X_2 = x_2, \ldots, X_k = x_k) = g(x_1, x_2, \ldots, x_k; \gamma).$$

Then you can estimate the variance function along with the mean function. In some cases, it is just as interesting to know the variance function as it is to know the mean function.

Take income inequality, for example. Higher variance of income reflects greater inequality. Related statistics such as the coefficient of the variation, also measure income inequality and also use the variance. If you want to study variables that affect income inequality, then you need to model the *variance* of income as a function of these variables.

For another example, consider stock market volatility, which is commonly measured using the variance of the market returns. What factors affect volatility? Under what conditions is the market more volatile and less volatile? If you want to study factors that affect market volatility, then you need to model the *variance* of market returns as a function of variables such as market trading volume, confidence in political leaders, confidence in the Federal Reserve Board, recent geo-political turmoil, etc.

But even if your ultimate goal is to estimate the mean function, you still would like to know the variance function, because the ML estimates of the mean function parameters (the β's) can be much better when you incorporate this variance function. The ML estimates, in this case, become *weighted least squares estimates* (WLS) rather than *ordinary least squares estimates* (OLS).

Further, if your goal is to predict Y, you also need to know the variance function to obtain the correct prediction intervals. The usual (approximate) prediction interval, $\hat{Y} \pm 2\hat{\sigma}$, presumes that the same conditional variance, σ^2, applies for all possible $X_1 = x_1, X_2 = x_2, \ldots, X_k = x_k$ combinations. But these intervals are too narrow for X data where $\text{Var}(Y \mid X_1 = x_1, X_2 = x_2, \ldots, X_k = x_k)$ is large, and are too wide for X data where $\text{Var}(Y \mid X_1 = x_1, X_2 = x_2, \ldots, X_k = x_k)$ is small. You can get a much more accurate prediction interval by using the variance function, i.e., by using $\hat{Y} \pm 2\sqrt{g(x_1, \ldots, x_k; \hat{\gamma})}$ rather than by using $\hat{Y} \pm 2\hat{\sigma}$.

12.1 Maximum Likelihood and Weighted Least Squares

Have a look at Figure 1.5 in Chapter 1, the "Personal Assets" data set. It appears that the vertical ranges of Y values increase as X increases. The visible vertical range that you can see in the graph is related to standard deviation, not variance, so let us suppose that the standard deviation of Y increases as a linear function of $X = x$, with unknown slope. Also, although we are usually loathe to exclude an intercept term in our models, let's go ahead and do that here for simplicity. Here is the model for the standard deviation in this case:

$$\sigma(Y \mid X = x) = \gamma x.$$

Recall that $\sigma(Y \mid X = x)$ is the standard deviation of the conditional distribution $p(y \mid x)$. Since $\sigma(Y \mid X = x) = \sqrt{\text{Var}(Y \mid X = x)}$, this model implies that the variance function is

$$\text{Var}(Y \mid X = x) = \gamma^2 x^2.$$

In this model, γ is the unknown slope of the standard deviation function, a parameter to be estimated using the data. In general, you can consider any non-negative function of x here, such as $g(x) = x^2$, $g(x) = e^x$, etc., giving a more general model

$$\sigma(Y \mid X = x) = \gamma g(x).$$

For individual observations $i = 1, \ldots, n$, this general form translates to

$$\sigma(Y_i \mid X_i = x_i) = \gamma d_i,$$

where $d_i = g(x_i)$ is a function of the X data that you assume to be proportional to the conditional standard deviation. Note that since d_i is a function of the X data, you can view the variable d_i as an additional column in your data set.

One example of such a d_i is $d_i = x_i$, implying that the standard deviation is proportional to x_i, as used above. Other possibilities are $d_i = x_i^{1/2}$, implying that the variance is proportional to x_i; $d_i = 1$, implying homoscedasticity; and $d_i = \exp(x_i)$, implying that the standard deviation grows exponentially as a function of x_i. In multiple regression, d_i can be any known, positive function of all k of the X variables.

There are infinitely many choices for a standard deviation function, just as there are infinitely many choices for a mean function. Fortunately, as with mean functions, there are graphical methods and likelihood-based statistics that allow you to estimate which standard deviation functions are best supported by your data. You have already seen a graphical method in Chapter 4, Section 4.6: Superimpose the LOESS fit over the $(x_i, |e_i|)$ data. The resulting function gives an estimate of a function that is proportional to the standard deviation function. Figure 12.1 shows this graph for the Personal Assets data set.

R code for Figure 12.1

```
Worth = read.table("https://raw.githubusercontent.com/andrea2719/
URA-DataSets/master/Pass.txt")
attach(Worth)
fit = lm(P.assets ~ Age)
abs.resid = abs(fit$residuals)
plot(Age, abs.resid); add.loess(Age, abs.resid)
add.loess(Age, abs.resid, span=2, lty=2)
```

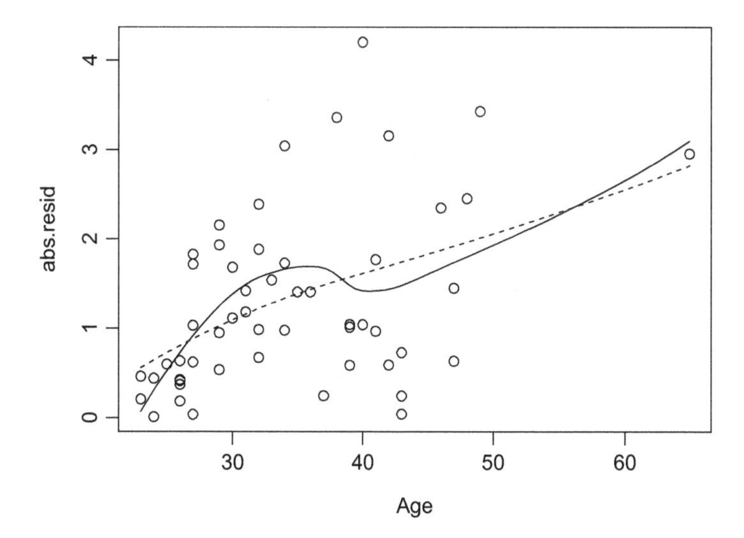

FIGURE 12.1
Plot of $(x_i, |e_i|)$ for the Personal Assets data, with LOESS fits superimposed: Solid line uses `span = 0.75`, and dashed line uses `span = 2.0`.

In Figure 12.1, the curves are estimates of a function that is proportional to the standard deviation function. The solid curve uses the default span of 0.75, and shows an up-then-down-then up appearance that is likely an idiosyncratic (random) characteristic of this particular data set. Choosing a larger span parameter reduces idiosyncratic noise, and shows a more sensible estimate that is always increasing, but with a slight concavity as might appear in a square root function. Thus, based on the graph, we can entertain standard deviation functions

$$\sigma(Y \mid X = x) = \gamma x,$$

which is suggested if the smooth LOESS curve in Figure 12.1 is estimating a straight line, and

$$\sigma(Y \mid X = x) = \gamma x^{1/2},$$

which is suggested if the smooth LOESS curve in Figure 12.1 is really estimating a square root function.

Fortunately, there are likelihood-based and additional graphical diagnostic tools that can help to decide between these (and other) possibilities for the standard deviation function. To describe the likelihood-based approach, consider the classical model, but change the homoscedasticity assumption to an assumption that the conditional standard deviation has the form

$$\sigma(Y_i \mid X_i = x_i) = \gamma d_i, \text{ for } i = 1, \ldots, n.$$

Here, γ is a single unknown parameter that you must estimate using the data. Using maximum likelihood, you can simultaneously estimate the three parameters β_0, β_1, and γ. Let the parameter vector be $\theta = (\beta_0, \beta_1, \gamma)$. Following Chapter 2, you view the Y data as produced by different distributions, depending on the X value, as shown in Table 12.1.

Assuming the probability distributions are the $N(\beta_0 + \beta_1 x_i, (\gamma d_i)^2)$ distributions, with

$$p(y_i \mid X = x_i, \theta) = \frac{1}{\sqrt{2\pi}\gamma d_i} \exp\left[-\frac{\{y_i - (\beta_0 + \beta_1 x_i)\}^2}{2(\gamma d_i)^2}\right],$$

TABLE 12.1

Contributions to the Likelihood Function in the Heteroscedastic Case. Note That There is An Additional Column, d_i, Which is Assumed to be Proportional to the Conditional Standard Deviation

Obs	X	d	Y	Contribution to the Likelihood Function
1	x_1	$g(x_1)$	y_1	$p(y_1 \mid X = x_1, \theta)$
2	x_2	$g(x_2)$	y_2	$p(y_2 \mid X = x_2, \theta)$
3	x_3	$g(x_3)$	y_3	$p(y_3 \mid X = x_3, \theta)$
...	...	...	...	...
n	x_n	$g(x_n)$	y_n	$p(y_n \mid X = x_n, \theta)$

the likelihood function in this case is

$$L(\theta \mid \text{data}) = p(y_1 \mid X = x_1, \theta) \times p(y_2 \mid X = x_2, \theta) \times \ldots \times p(y_n \mid X = x_n, \theta)$$

$$= \frac{1}{\sqrt{2\pi}\gamma d_1} \exp\left[-\frac{\{y_1 - (\beta_0 + \beta_1 x_1)\}^2}{2(\gamma d_1)^2}\right] \times \frac{1}{\sqrt{2\pi}\gamma d_2} \exp\left[-\frac{\{y_2 - (\beta_0 + \beta_1 x_2)\}^2}{2(\gamma d_2)^2}\right]$$

$$\times \cdots \times \frac{1}{\sqrt{2\pi}\gamma d_n} \exp\left[-\frac{\{y_n - (\beta_0 + \beta_1 x_n)\}^2}{2(\gamma d_n)^2}\right]$$

$$= (2\pi)^{-n/2} (\gamma^2)^{-n/2} \prod_{i=1}^{n} d_i^{-1} \exp\left[-\frac{1}{2\gamma^2} \times \frac{\sum_{i=1}^{n} \{y_i - (\beta_0 + \beta_1 x_i)\}^2}{d_i^2}\right]$$

Taking the natural logarithm, you get the log likelihood function.

Log likelihood function, assuming $\sigma(Y_i \mid X_i = x_i) = \gamma d_i$, for known d_i

$$LL(\theta \mid \text{data}) = -\frac{n}{2}\ln(2\pi) - \frac{n}{2}\ln(\gamma^2) - \sum_{i=1}^{n} \ln(d_i) - \frac{1}{2\gamma^2} \sum_{i=1}^{n} \frac{1}{d_i^2} \{y_i - (\beta_0 + \beta_1 x_i)\}^2$$

Notice that there is a "–" sign in front of the last term of the log likelihood. Thus, no matter what is the value of γ^2, the values of β_0 and β_1 that *maximize* the log likelihood function must *minimize* the *weighted* sum of squares:

$$\text{WSSE}(\beta_0, \beta_1) = \sum_{i=1}^{n} \frac{1}{d_i^2} \{y_i - (\beta_0 + \beta_1 x_i)\}^2$$

In words, the maximum likelihood estimates of the parameters β_0 and β_1 are the numbers $\hat{\beta}_0$ and $\hat{\beta}_1$ that give a line $\hat{f}(x) = \hat{\beta}_0 + \hat{\beta}_1 x$ such that the *weighted sum of squared vertical deviations from points y_i to the line $\hat{\beta}_0 + \hat{\beta}_1 x_i$ is the minimum.* These numbers $\hat{\beta}_0$ and $\hat{\beta}_1$ are called "weighted least squares," or WLS, estimates. The weights are $w_i = (1/d_i)^2$, and are observable numbers that are assumed to be inversely proportional to the conditional variances $\text{Var}(Y_i \mid X_i = x_i)$.

There is a nice intuition behind the WLS estimates. First, note that the number d_i is proportional to the standard deviation of Y_i. When the standard deviation is small, then d_i is small, and the observation y_i is close to the conditional mean (i.e., y_i is close to the regression function). The weight $w_i = (1/d_i)^2$ is large in this case, meaning that such an observation y_i will have greater impact on the estimated regression function. This is what you want, since such a y_i is known to be close to the regression function.

Conversely, when the standard deviation is large, then d_i is large, and the observation y_i tends to be far from the regression function. The weight $w_i = (1/d_i)^2$ is small in this case, meaning that such an observation y_i will have lesser impact on the estimated regression function. Again, this is also what you want, because such a y_i observation tends to be farther from the regression function.

By using the `lm` function of `R`, you can easily get the weighted least squares estimates, and you can compare different standard deviation functions in terms of maximized log likelihood. The following `R` code computes the ML (WLS) estimates of the β parameters and the maximized log likelihood using three different variance function models.

$$\text{Model 0: } Y \mid X = x \sim \text{N}(\beta_0 + \beta_1 x, (\gamma x^0)^2)$$
$$\text{Model 1: } Y \mid X = x \sim \text{N}(\beta_0 + \beta_1 x, (\gamma x^{1/2})^2)$$
$$\text{Model 2: } Y \mid X = x \sim \text{N}(\beta_0 + \beta_1 x, (\gamma x)^2)$$

Notice that Model 0 is identical to the classical regression model, where $\gamma^2 = \sigma^2$.

```
fit.0 = lm(P.assets ~ Age, weights=1/Age^0)
summary(fit.0); logLik(fit.0)
fit.1 = lm(P.assets ~ Age, weights=1/Age^1)
summary(fit.1); logLik(fit.1)
fit.2 = lm(P.assets ~ Age, weights=1/Age^2)
summary(fit.2); logLik(fit.2)
```

From the output, we have the following estimated models:

$$\text{Model 0: } \hat{\mu}(Y \mid X = x) = -6.25592 + 0.27928x; \hat{\sigma}(Y \mid X = x) = 1.651; LL = -100.7574$$
$$\text{Model 1: } \hat{\mu}(Y \mid X = x) = -5.87593 + 0.26825x; \hat{\sigma}(Y \mid X = x) = 0.2691x^{1/2}; LL = -97.66076$$
$$\text{Model 2: } \hat{\mu}(Y \mid X = x) = -5.65151 + 0.26137x; \hat{\sigma}(Y \mid X = x) = 0.0448x; LL = -95.6942$$

(Note: The "`Residual standard error`" reported in the output is an estimate of the parameter γ in the standard deviation function, but differs slightly from the ML estimate in that it is a "less biased" version, similar to what was described for the estimation of σ in Chapter 3.)

According to the log likelihoods, Model 2 fits the data best, while the classical model (Model 0) fits the data the worst.

For further validation, you can compare the three standard deviation functions in terms of how well they fit the graph of the raw data. The approximate prediction bands, which are given by

$$\hat{\mu}(Y \mid X = x) \pm 2\hat{\sigma}(Y \mid X = x)$$

should contain approximately 95% of the data for each $X = x$. Figure 12.2 shows the data set with each of these three prediction bands superimposed.

R code for Figure 12.2

```
par(mfrow=c(2,2))
par(mar=c(4, 4, 1, 4))
plot(Age, P.assets, xlab=""); abline(-6.25592, 0.27928, lty=1)
abline(-5.87593, 0.26825, lty=2); abline(-5.65151, 0.26137, lty=3)

xlist = seq(20, 70, .1)
Low0 = -6.2559+0.27928*xlist - 2*1.651
Upp0 = -6.2559+0.27928*xlist + 2*1.651
plot(Age, P.assets, ylab="", xlab="", ylim = c(-5, 20))
points(xlist, Low0, type="l", lty=1); points(xlist, Upp0, type="l",
lty=1)
```

(Continued)

```
Low1 =-5.87593+0.268255*xlist - 2*.2691*xlist^(1/2)
Upp1 =-5.87593+0.268255*xlist + 2*.2691*xlist^(1/2)
plot(Age, P.assets,ylim = c(-5, 20))
points(xlist, Low1, type="l"); points(xlist, Upp1, type="l")

Low2 = -5.65151+0.26137*xlist - 2*.0448*xlist
Upp2 = -5.65151+0.26137*xlist + 2*.0448*xlist
plot(Age, P.assets,ylab="",ylim = c(-5, 20))
points(xlist, Low2, type="l"); points(xlist, Upp2, type="l")
```

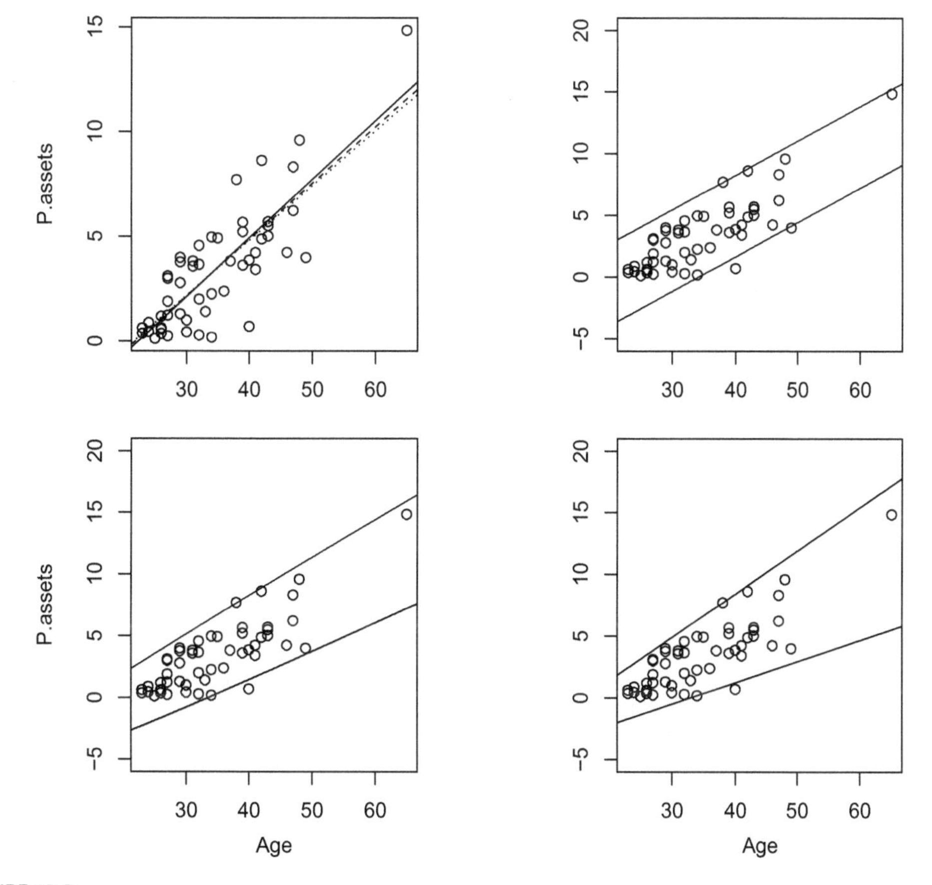

FIGURE 12.2
Upper left: Scatterplot with OLS (solid) and ML (WLS) estimates of mean functions. Upper right: Homoscedastic prediction limits, Lower left: Heteroscedastic prediction limits assuming $\hat{\sigma}(Y|X = x) = 0.2691x^{1/2}$. Lower right: Heteroscedastic prediction limits assuming $\hat{\sigma}(Y|X=x) = 0.0448x$.

Figure 12.2 shows little difference in estimated mean functions (upper left panel), but more noticeable differences in the prediction limits using difference standard deviation functions. The best-fitting (according to log likelihood) standard deviation function is used in the lower right panel, and appears better because it better matches the pattern of variability of Y for given $X = x$ values than do the other standard deviation functions. The fact that the limits extend to below zero is an indication that there are other problems (non-normality, specifically) with the model.

While this example shows a benefit to using a non-constant variance function in terms of prediction limits, the three estimated mean functions are similar. However, you can also show that the estimated mean functions are better when you use ML (WLS); the following code gives an example. Figure 12.3 shows that the OLS estimate misses the data badly where the variance of Y is small.

R code for Figure 12.3

```
par(mfrow=c(1,2))
n=500
set.seed(123)
x = 5*rexp(n)
y = 2*x + x*rnorm(n)
## The weight is inversely proportional to the conditional variance
w = 1/x^2
fit.ols = lm(y ~ x)
ols.pred = fit.ols$fitted.values
fit.wls = lm(y ~ x, weights = w)
wls.pred = fit.wls$fitted.values

## Scatterplot and fitted functions
plot(x, y, pch=".", cex=1.5, ylim=c(-5,120))
points(sort(x), sort(ols.pred), type="l", lty=2)
points(sort(x), sort(wls.pred), type="l", lty=1)
legend("topleft", c("WLS", "OLS"), lty = c(1,2))

## Fitted OLS and WLS lines and data where variance is small
plot(x[x<1], y[x<1], ylim = c(-1,4))
points(sort(x[x<1]), sort(ols.pred[x<1]), type="l", lty=2)
points(sort(x[x<1]), sort(wls.pred[x<1]), type="l", lty=1)
legend("topleft", c("WLS", "OLS"), lty = c(1,2))
```

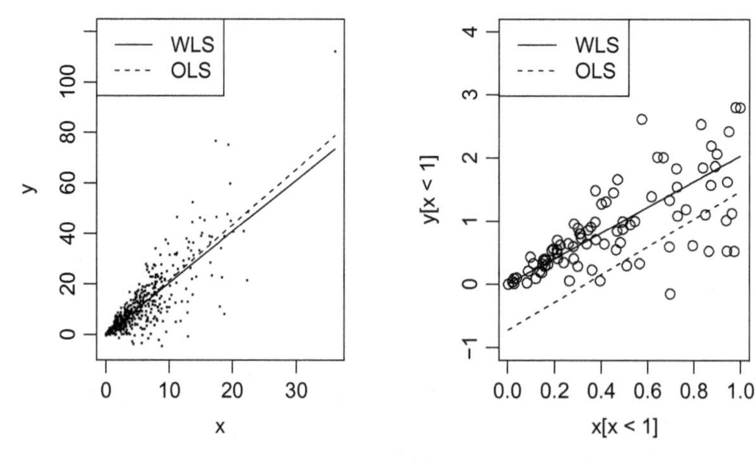

FIGURE 12.3
Left panel: OLS and WLS estimates of the mean function. Right panel: Same as left panel, but zoomed in on X data that are < 1. The OLS estimate misses the data badly in this range.

12.2 The Gauss-Markov Theorem, Revisited

Recall that, under the assumptions of the Gauss-Markov (G-M) model (which include all classical assumptions except normality, and notably which include homoscedasticity) the OLS estimates have the smallest variance in the class of estimates that are linear functions of the data. We say they are best linear unbiased estimators (BLUEs).

But, *if* the standard deviation function is the heteroscedastic function $\sigma(Y_i \mid X_i = x_i) = \gamma d_i$, for $i = 1,\ldots,n$, and all other assumptions of the G-M model hold, *then* the WLS estimates defined above are the BLUEs. In particular, the WLS estimates are more accurate than the OLS estimates in this case.

Thus, while the predicted values ($\hat{y}$'s) using the WLS estimates can be much more accurate than those from OLS estimates as shown in Figure 12.3, it is also the case that the WLS parameter estimates ($\hat{\beta}$'s) tend to be closer to the true β's than are the OLS parameter estimates, *if* $\sigma(Y_i \mid X_i = x_i) = \gamma d_i$ is correct. That's a big "if," because you can never specify the standard deviation function correctly. But what else is new? You can never specify the mean function correctly, either, as we have stated repeatedly throughout this book. In either case, whether specifying a mean function or a standard deviation function, you simply do your best, knowing that the result is always flawed.

To specify a reasonable standard deviation function, use the same steps that you use to specify a mean function: (i) think about the standard function in terms of your data-generating process, and ask yourself, which effects make sense? (ii) Evaluate different models, using graphs and (penalized) fit statistics. (iii) Perform simulation studies to resolve questions or ambiguities. As when specifying your conditional mean function, these steps allow you to make an informed decision about your choice of a conditional standard deviation function.

Here is a simulation study to illustrate the BLUE properties of WLS when you have specified the standard deviation function correctly. The parameters of the simulation model are suggested by the analysis of the "Personal Assets" data using a heteroscedastic model, specifically the model graphed in the lower right panel of Figure 12.2.

12.2.1 Simulation Study to Illustrate That WLS Is More Efficient than OLS

```
Worth = read.table("https://raw.githubusercontent.com/andrea2719/
URA-DataSets/master/Pass.txt")
attach(Worth); n = nrow(Worth); X = Age

## Conditional-x heteroscedastic simulation model parameters
gamma = 0.05; beta0 = -6; beta1 = 0.25

## Simulation study to understand efficiency of WLS versus OLS
NSIM = 20000
b1.ols = numeric(NSIM); b1.wls = numeric(NSIM)

for (i in 1:NSIM) {
    Y = beta0 + beta1*X + gamma*X*rnorm(n)
    b1.ols[i] = lm(Y ~ X)$coefficients[2]
    b1.wls[i] = lm(Y ~ X, weights=1/X^2)$coefficients[2]
}

par(mfrow=c(2,1))
hist(b1.ols, xlim = c(.10,.40), main = "Distribution of OLS Estimates")
hist(b1.wls, xlim = c(.10,.40), main = "Distribution of WLS Estimates")
sd(b1.ols); sd(b1.wls)
```

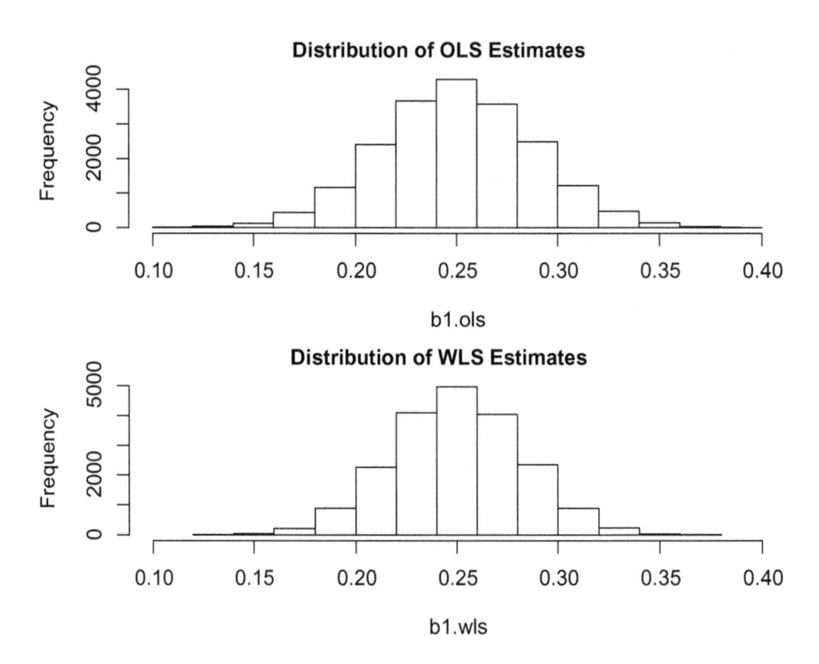

FIGURE 12.4
Histograms of OLS estimates and WLS estimates. Both are unbiased for $\beta_1 = 0.25$, but the WLS estimates tend to be closer to β_1.

Figure 12.4 suggests that the OLS slope estimates tend to be farther from the target $\beta_1 = 0.25$; this suggestion is verified by the standard deviations 0.037 and 0.032 for the OLS and WLS estimates, respectively.

If you specify a non-constant standard deviation function when the standard deviations are all the same (i.e., when there is homoscedasticity), then the situation shown in Figure 12.4 will be reversed. The reason is again the G-M theorem: Under homoscedasticity, the G-M theorem tells you that the OLS estimators will have smaller variance than any other linear unbiased estimators, which include the WLS estimates using any weight function.

In practice, you never know the true standard deviation function, just like you never know the true mean function. However, we hope you are getting some intuition about these kinds of issues: If you specify a standard deviation function that is close to the truth, then WLS will most likely be better than OLS. On the other hand, if the process is homoscedastic but you specify a grossly heteroscedastic standard deviation function, then OLS will probably be better than WLS. And if the previous two sentences are unclear, just perform a simulation study to clear up the confusion. Simulation makes it all concrete.

Self-study question: How can you use simulation to show that the WLS estimates using the model of Figure 12.3 are more efficient than the OLS estimates?

Self-study question: How can you modify the simulation leading to Figure 12.4 to show that the WLS estimates are less efficient than the OLS estimates when the data-generating process is homoscedastic?

12.3 More General Standard Deviation Functions

The assumption that $\sigma(Y_i \mid X_i = x_i) = \gamma d_i$ gives you the ML estimates (which are WLS estimates using weights $w_i = (1/d_i)^2$), and provides a case where the WLS estimates are BLUE. But this particular class of standard deviation function is rather limited. In multiple regression, you should allow the standard deviation of Y to depend on each of the X variables with different coefficients, just like you allow the conditional mean function of Y to depend on each of the X variables with different coefficients. One such possibility is, taking the cue from the usual mean function,

$$\sigma(Y_i \mid X_{i1} = x_{i1}, X_{i2} = x_{i2}, \ldots, X_{ik} = x_{ik}) = \gamma_0 + \gamma_1 x_{i1} + \gamma_2 x_{i2} + \ldots + \gamma_k x_{ik}$$

However, this function, like the mean function, can take negative values, and is therefore not recommended for the standard deviation function.

Instead, it makes more sense to model the logarithm of the standard deviation as a linear function:

$$\ln\{\sigma(Y_i \mid X_{i1} = x_{i1}, X_{i2} = x_{i2}, \ldots, X_{ik} = x_{ik})\} = \gamma_0 + \gamma_1 x_{i1} + \gamma_2 x_{i2} + \ldots + \gamma_k x_{ik}$$

Then the actual standard deviation cannot be negative:

$$\sigma(Y_i \mid X_{i1} = x_{i1}, X_{i2} = x_{i2}, \ldots, X_{ik} = x_{ik}) = \exp\{\gamma_0 + \gamma_1 x_{i1} + \gamma_2 x_{i2} + \ldots + \gamma_k x_{ik}\}$$

To estimate parameters of this model, you can use maximum likelihood. Consider "Grade Point Average" data set, with the conditional mean function

$$\mathrm{E}(\mathrm{GPA} \mid \mathrm{GMAT},\ \mathrm{PHD}) = \beta_0 + \beta_1 \mathrm{GMAT} + \beta_2 \mathrm{PHD},$$

as well as the conditional standard deviation function

$$\ln\{\sigma(\mathrm{GPA} \mid \mathrm{GMAT},\ \mathrm{PHD})\} = \gamma_0 + \gamma_1 \mathrm{GMAT} + \gamma_2 \mathrm{PHD}.$$

The classical model is a special case (or restricted form) of this model where $\gamma_1 = \gamma_2 = 0$ and $\sigma = \exp(\gamma_0)$, Note that there are 6 parameters in the heteroscedastic model, two more than the 4 parameters in the classical model.

The following code performs maximum likelihood estimation essentially "by hand," by providing the log likelihood function to the `maxLik` function within the `maxLik` library. There exist `R` functions to perform such calculations automatically, but it is useful and instructive to see the machinery at work. In particular, the `maxLik` function requires the log likelihood function, which, following the same method as shown above, is given as follows:

Log likelihood function, assuming $\sigma(Y_i \mid X_i = x_i) = \sigma_i(\gamma)$

$$LL(\gamma \mid \text{data}) = -\frac{n}{2}\ln(2\pi) - \frac{1}{2}\sum_{i=1}^{n} \ln\{\sigma_i^2(\gamma)\} - \frac{1}{2}\sum_{i=1}^{n} \frac{1}{\sigma_i^2(\gamma)}\{y_i - (\beta_0 + \beta_1 x_{i1} + \ldots + \beta_k x_{ik})\}^2$$

In the GPA example, the function $\sigma_i(\gamma)$ is given by $\sigma_i(\gamma) = \exp(\gamma_0 + \gamma_1 \mathrm{GMAT}_i + \gamma_2 \mathrm{PHD}_i)$, where $i = 1, 2, \ldots$ refers to an individual student.

R code to estimate the conditional variance function in the GPA prediction example

```
ba = read.table("https://raw.githubusercontent.com/andrea2719/
URA-DataSets/master/gpa_gmat.txt")
attach(ba)

y = gpa; x1 = gmat
x2 = ifelse(degree=="P", 1,0)

library(maxLik)
loglik <- function(param) {
  b0   = param[1]
  b1   = param[2]
  b2   = param[3]
  g0   = param[4]
  g1   = param[5]
  g2   = param[6]
mean = b0 + b1*x1 + b2*x2
ln.sd = g0 + g1*x1 + g2*x2 ; sd = exp(ln.sd)
    z  = (y - mean)/sd
   ll  = sum(dnorm(z,log=T) - ln.sd)
   ll
 }

fit = maxLik(loglik, start=c(lm(y~x1+x2)$coefficients,0,0,0))
summary(fit)
```

The results are as follows:

```
--------------------------------------------
Maximum Likelihood estimation
Newton-Raphson maximisation, 7 iterations
Return code 2: successive function values within tolerance limit
Log-Likelihood: -68.06105
6  free parameters
Estimates:
              Estimate Std. error t value  Pr(> t)
(Intercept)  2.7439803  0.1098789  24.973  < 2e-16 ***
x1           0.0013696  0.0001998   6.856 7.06e-12 ***
x2           0.1785019  0.0291449   6.125 9.09e-10 ***
            -1.5076116  0.2806901  -5.371 7.83e-08 ***
             0.0005050  0.0005078   0.994     0.32
            -0.7212940  0.1245030  -5.793 6.90e-09 ***
---
Signif. codes:  0 '***' 0.001 '**' 0.01 '*' 0.05 '.' 0.1 ' ' 1
--------------------------------------------
```

The standard errors shown in the `Std. error` column of the output are the *Wald standard errors* calculated from the log likelihood function; see Appendix B of this chapter for details. From the output, the estimated conditional standard deviation function is given by

$$\hat{\sigma}(\text{GPA} \mid \text{GMAT, PHD}) = \exp(-1.5076116 + 0.0005050\text{GMAT} - 0.7212940\text{PHD}).$$

Thus, the data suggest that the GPA distribution has more variability with larger GMAT scores, and smaller variability for PhD students. However, based on the approximate interval for the GMAT effect, which is $0.0005050 \pm 2 \times 0.0005078$, or (−0.00051 to 0.0015), the direction of the effect of GMAT on the GPA variance is uncertain.

The classical regression model, which assumes a constant variance function, can be estimated in a similar way as follows:

```
library(maxLik)
loglik0 <- function(param) {
  b0   = param[1]
  b1   = param[2]
  b2   = param[3]
  g0   = param[4]
mean = b0 + b1*x1 + b2*x2
ln.sd = g0; sd = exp(ln.sd)
   z  = (y - mean)/sd
  ll  = sum(dnorm(z,log=T) - ln.sd)
  ll
 }

fit0 = maxLik(loglik0, start=c(lm(y~x1+x2)$coefficients,0))
summary(fit0)
```

With results:

```
---------------------------------------------
Maximum Likelihood estimation
Newton-Raphson maximisation, 5 iterations
Return code 2: successive function values within tolerance limit
Log-Likelihood: -79.77668
4  free parameters
Estimates:
             Estimate Std. error t value  Pr(> t)
(Intercept)  2.750700   0.118820  23.150  < 2e-16 ***
x1           0.001357   0.000215   6.314 2.72e-10 ***
x2           0.179380   0.050154   3.577 0.000348 ***
            -1.257447   0.031817 -39.521  < 2e-16 ***
---
Signif. codes:  0 '***' 0.001 '**' 0.01 '*' 0.05 '.' 0.1 ' ' 1
---------------------------------------------
```

Using this model, the estimated variance function is constant, not depending on the X variables:

$$\hat{\sigma}(\text{GPA} \mid \text{GMAT, PHD}) = \exp(-1.257447) = 0.2843791.$$

You can understand these models better by using graph, as shown in the following code.

R code for Figure 12.5

```
g = fit$estimate; g0 = g[4]; g1 = g[5]; g2 = g[6]
gmat.g = seq(400,800,10)
s.phd = exp(g0 + g1*gmat.g + g2)
```

(Continued)

```
s.masters = exp(g0 + g1*gmat.g)
plot(gmat.g,s.phd, type="l", ylim = c(0,.4),
  ylab="Estimated Standard Dev. of GPA", xlab="GMAT Score")
points(gmat.g,s.masters, type="l", lty=2)
abline(h = exp(fit0$estimate[4]), lwd=2)
legend("bottomright", c("Ph.D.", "Masters"), lty = c(1,2))
```

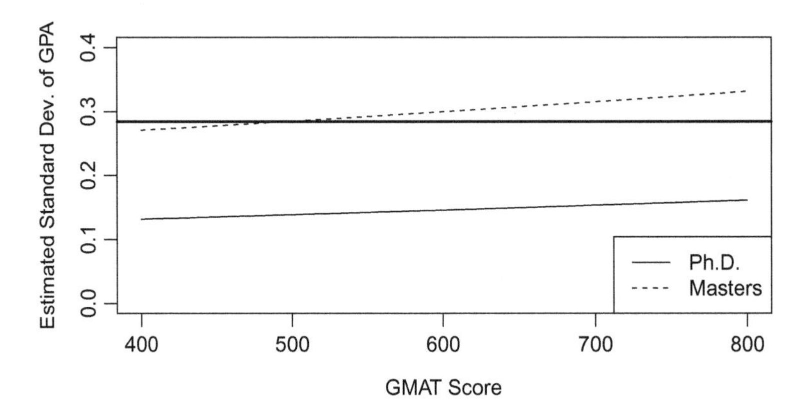

FIGURE 12.5
Estimated standard deviations of GPA as a function of GMAT score and degree program. The thicker flat line is the estimated standard deviation for the homoscedastic model, which can be thought as a kind of average standard deviation. The reason it is not closer to the PhD line is that there are relatively few PhD students in the sample.

We now give several comments about the heteroscedastic and homoscedastic estimated models.

1. The estimates of the β's in the heteroscedastic model are WLS estimates, using weights $1/\hat{\sigma}_i^2$ as indicated by Figure 12.5. They differ from the OLS estimates.
2. The estimates of the β's in the homoscedastic model are identical to the OLS estimates, which you can verify by using the command `lm(y ~ x1 + x2)$ coefficients`, giving you the following:
```
(Intercept)          x1          x2
2.750699944 0.001357218 0.179380537
```
3. The estimated σ in the homoscedastic model is $\hat{\sigma}(\text{GPA} | \text{GMAT, PHD}) = 0.2843791$, the maximum likelihood estimate SSE/n. This estimate differs slightly from the "unbiased" version printed by `lm` as "residual standard error": `summary(lm(y ~ x1 + x2))$sigma` which gives you 0.2852465.

   However, the difference is small and of no concern. Both are equally viable estimates of the conditional standard deviation of Y under homoscedasticity.
4. In Figure 12.5, the upward trend in the GPA standard deviation, as a function of GMAT, is explainable by a GPA-generating process where the GPA standard deviation is not affected by GMAT ($T = 0.994$, $p = 0.321$).

5. In Figure 12.5, the difference between the GPA standard deviation for PhD and Masters students is not easily explained by a GPA-generating process where GPA is identical for those two groups ($T = -5.793$, $p = 6.90 \times 10^{-9}$). It makes sense that the PhD standard deviation really is smaller because their grades are closer to the 4.0 boundary. Whenever data are closer to a boundary, the standard deviation has to be smaller.
6. A fantastic advantage of using likelihood-based methods over OLS-based methods is that you can compare models using likelihood-based statistics. You can get the AIC statistics for comparing these two models as follows: `AIC(fit0)` gives 167.5534, and `AIC(fit)` gives 148.1221. Thus, the heteroscedastic model is much better supported by the data than is the homoscedastic model, as determined by the comparison of the AIC statistics.

Self-study question: How can you calculate those AIC statistics manually, by using the log likelihood values and number of parameters given in the outputs above?

---

7. As shown in Appendix A of this chapter, you can test for a difference between the models by using the likelihood ratio test. Using the two fits, you can get the likelihood ratio chi-square as follows:

```
LL = fit$maximum
LL0 = fit0$maximum
Chi.sq = 2*(LL - LL0)
Chi.sq
pval = 1- pchisq(Chi.sq, 6-4)
pval
```

The results are $\chi_2^2 = 23.43127$, $p = 8.165146 \times 10^{-6}$. Thus, the difference between log likelihoods for the homoscedastic and heteroscedastic model is not easily explained by the homoscedastic model. Rather, it indicates that there is truly heteroscedasticity.

This test is (yet another) test for heteroscedasticity. As always, tests are not that useful for determining which models to use. Better tools are subject matter considerations (e.g., boundaries affect variance), graphs (see Figure 12.5 above), and simulation.

A final cautionary note about inferences (tests and intervals) for variances:

Inferences for the variance parameters (γ's) are considerably more sensitive to non-normality (particularly kurtosis) of the distributions $p(y|x)$ than are inferences for the mean parameters (β's).

Because of the cautionary note, use of simulation to validate the methods is even more important for variance parameters than it is for mean parameters.

8. As you might expect, `R` functions are available to estimate heteroscedastic models. Inside the "`gamlss`" library you can use the `gamlss` function like this: `mod= gamlss(y~x1+x2,sigma.fo=~x1+x2, data=ba); summary(mod)`

   The `sigma.fo=~x1+x2` formula assumes by default that the natural log of the standard deviation is a linear function of $X_1 = x_1$ and $X_2 = x_2$. The resulting ML estimates are the same and the log likelihoods are the same as what we got above "by hand," using the `maxLik` function.

   You can also estimate variance functions by using the `nlme` package of `R`.

12.4 The Effect of Estimating Parameters in Variance Functions

The previous sections might have you wondering, "Why not use the estimated variance function approach all the time?" After all, you get weighted least squares estimates, which are more efficient than OLS estimates when the variance is correctly estimated; you get more accurate prediction limits for Y, as shown in Figures 12.2 above; and you get additional, interesting analyses regarding the effect of the X variables on the variance of Y, as shown Figure 12.5. It seems all good!

Well, almost all good. We hope you have learned from real-life experience that there is never any "free lunch." We also hope you have also learned from this and other books in statistics that parameter estimates are always different from the true parameters, and that those differences have important consequences. Here, the consequence is that the weights used for WLS (which are the inverses of the estimated variances) are not the right weights, even if you have specified the general form of the variance function correctly, because estimates are never equal to their true values.

For example, suppose for the sake of argument that the variance is truly homoscedastic, but you fit a heteroscedastic model anyway. Then you will estimate non-constant variances, whose different values are explained by chance alone. But these non-constant variances will then be used as weights in the WLS estimates. And you know in this hypothetical set-up that OLS are BLUE; hence, you will lose efficiency by estimating a heteroscedastic variance model in this case.

In practice, of course, no process is truly homoscedastic. Suppose now that the variance is heteroscedastic, but only slightly. If you fit a heteroscedastic error variance model, it is likely that your estimates will show more heteroscedasticity than actually exists, by chance alone. Thus the weights again are not correct. It can easily happen in this case that OLS estimates are again more efficient than WLS, despite the fact that the process is truly heteroscedastic, simply because of the inaccuracy in estimating the weights.

When is OLS better than WLS with ML-estimated weights in the heteroscedastic case? When is it worse? Based on what you have learned in this book, what do you think is the best way to address these questions? Yes, by using simulation! You will find that if you have specified the variance function correctly, then the estimated parameters of the variance function will become more accurate with larger sample size. Thus, the WLS weights will be closer to correct with larger sample sizes than with smaller ones, and WLS will beat OLS with large sample sizes. On the other hand, if you have specified the variance function incorrectly, larger sample sizes will not necessarily give WLS estimates that are better than OLS; you must use simulation to be more specific.

12.5 The Blunt Axe Approach: Heteroscedasticity-Consistent Standard Errors

One response to the problem that the variance function is unknown, and to the problem that estimated parameters are (always, by definition) in error, is to avoid modeling the variance function entirely, and just use OLS-based methods, ignoring the problems caused by heteroscedasticity, which include:

1. inefficient OLS estimates of the β's
2. incorrect OLS prediction limits
3. inability to compare models using likelihood-based statistics
4. inability to investigate interesting relationships between the standard deviation of Y and X, and
5. incorrect OLS standard errors.

Researchers who promote the OLS approach simply accept the first four problems, but solve the last problem by correcting the standard errors for heteroscedasticity. We call this the "Blunt Axe" approach because, like cutting wood to build a fine cabinet wall by using a blunt axe (rather than by using an electric table saw), it is a rather crude approach, and arguably the wrong tool for the job.

To understand this method, we must first revisit the OLS standard error derivation given in Chapter 7. First, you need the covariance matrix of the estimated β vector:

$$\begin{aligned}\text{Cov}\left(\hat{\beta}|\mathbf{X}=\mathbf{x}\right) &= \text{Cov}\left(\left(\mathbf{X}^{\text{T}}\mathbf{X}\right)^{-1}\mathbf{X}^{\text{T}}\mathbf{Y}|\mathbf{X}=\mathbf{x}\right)\\ &= \left(\mathbf{x}^{\text{T}}\mathbf{x}\right)^{-1}\mathbf{x}^{\text{T}}\,\text{Cov}\left(\mathbf{Y}\mid\mathbf{X}=\mathbf{x}\right)\left(\left(\mathbf{x}^{\text{T}}\mathbf{x}\right)^{-1}\mathbf{x}^{\text{T}}\right)^{\text{T}}\\ &= \left(\mathbf{x}^{\text{T}}\mathbf{x}\right)^{-1}\mathbf{x}^{\text{T}}\,\text{Cov}\left(\mathbf{Y}\mid\mathbf{X}=\mathbf{x}\right)\mathbf{x}\left(\mathbf{x}^{\text{T}}\mathbf{x}\right)^{-1}\end{aligned}$$

Now, under homoscedasticity, $\text{Cov}(\mathbf{Y}\mid\mathbf{X}=\mathbf{x})=\sigma^2\mathbf{I}$, implying that $\text{Cov}\left(\hat{\beta}\mid\mathbf{X}=\mathbf{x}\right)=\sigma^2\left(\mathbf{x}^{\text{T}}\mathbf{x}\right)^{-1}$. The standard errors are then the square roots of the diagonal elements of $\hat{\sigma}^2\left(\mathbf{x}^{\text{T}}\mathbf{x}\right)^{-1}$.

However, in the heteroscedastic (but uncorrelated) case, $\text{Cov}\left(\mathbf{Y}\mid\mathbf{X}=\mathbf{x}\right)\neq\sigma^2\mathbf{I}$. Instead,

$$\text{Cov}\left(\mathbf{Y}\mid\mathbf{X}=\mathbf{x}\right)=\begin{bmatrix}\sigma_1^2 & 0 & \cdots & 0\\ 0 & \sigma_2^2 & \cdots & 0\\ \vdots & \vdots & \ddots & \vdots\\ 0 & 0 & \cdots & \sigma_n^2\end{bmatrix},$$

with every observation Y_i having a different conditional variance, depending on the particular values of the X observations. See Figure 12.5 again to understand how this can happen: Each student's GPA comes from a different distribution, and each of those distributions has a different standard deviation, depending on the student's GMAT score and PhD status.

As we have seen so far, we can estimate all the σ_i values using maximum likelihood. It makes sense to replace the σ_i in the covariance matrix with these estimates, getting an estimated covariance matrix

$$\hat{\Omega}_{ML} = \begin{bmatrix} \hat{\sigma}_1^2 & 0 & \cdots & 0 \\ 0 & \hat{\sigma}_2^2 & \cdots & 0 \\ \vdots & \vdots & \ddots & \vdots \\ 0 & 0 & \cdots & \hat{\sigma}_n^2 \end{bmatrix},$$

and then using this estimated covariance matrix to get an estimated covariance matrix of the OLS estimates:

$$\hat{\mathrm{Cov}}(\hat{\beta}_{\mathrm{OLS}}) = \left(\mathbf{x}^{\mathrm{T}}\mathbf{x}\right)^{-1}\mathbf{x}^{\mathrm{T}}\hat{\Omega}_{ML}\mathbf{x}\left(\mathbf{x}^{\mathrm{T}}\mathbf{x}\right)^{-1}.$$

Then, you could obtain valid standard errors of the OLS estimates by taking the square roots of the diagonal elements of the matrix above. But if you were going to use the estimates $\hat{\Omega}_{ML}$ obtained from the variance function model, you would go ahead and use the WLS/ML estimates of the β's rather than the OLS estimates, so this approach is not generally used.

Instead, researchers that love OLS so much that they are willing to ignore all of its problems (1, 2, 3 and 4 listed above) tend to avoid ML-based estimation. Their reasoning is that the form of the variance function is always unknown, and even if the variance function is correctly specified, then the randomness of parameter estimates degrades the efficiency. So instead they estimate $\mathrm{Cov}\left(Y \mid \mathbf{X} = \mathbf{x}\right)$ in a way that does not assume any particular variance function, and use that estimate to get the standard errors. Here is how they do it. They use

$$\hat{\Omega} = \begin{bmatrix} e_1^2 & 0 & \cdots & 0 \\ 0 & e_2^2 & \cdots & 0 \\ \vdots & \vdots & \ddots & \vdots \\ 0 & 0 & \cdots & e_n^2 \end{bmatrix}$$

as an estimate of

$$\mathrm{Cov}\left(Y \mid \mathbf{X} = \mathbf{x}\right) = \begin{bmatrix} \sigma_1^2 & 0 & \cdots & 0 \\ 0 & \sigma_2^2 & \cdots & 0 \\ \vdots & \vdots & \ddots & \vdots \\ 0 & 0 & \cdots & \sigma_n^2 \end{bmatrix}.$$

Then, they calculate *heteroscedasticity-consistent (HC) robust standard errors* as the square roots of the diagonal elements of $\left(\mathbf{x}^{\mathrm{T}}\mathbf{x}\right)^{-1}\mathbf{x}^{\mathrm{T}}\hat{\Omega}\,\mathbf{x}\left(\mathbf{x}^{\mathrm{T}}\mathbf{x}\right)^{-1}$.

The logic for this approach rests on the idea that e_i^2 is an estimate of σ_i^2. To understand this logic, first note that ε_i^2 is in fact an unbiased estimator of σ_i^2, since $\mathrm{E}(\varepsilon_i^2) = \mathrm{E}\{(\varepsilon_i - 0)^2\} = \mathrm{Var}(\varepsilon_i) = \sigma_i^2$. But there is a problem with this estimate: Even if you knew the value of ε_i, this is an estimate based on only $n = 1$ observation, and statistical estimates based on just $n = 1$ observation are usually quite inaccurate.

For example, suppose you want to estimate mean US Income, and you decide to sample just one person, and use that person's income as an estimate of the mean US income. Your estimate will be unbiased but potentially highly inaccurate, depending on the person you select. Estimating σ_i^2 using just one observation, ε_i^2, has precisely the same problem.

But it's even worse than that. You do not know ε_i^2, so you have to use e_i^2 instead. Not only is this estimate still based on just $n = 1$ observation, and therefore can be quite far from the target, but it is also biased, meaning that it misses the target in a systematic fashion as well.

How do you think we could investigate whether e_i^2 is a reasonable estimate of σ_i^2? You guessed it: Simulation!

12.5.1 Simulation to Investigate Whether e_i^2 Is a Reasonable Estimate of σ_i^2

We have done quite a few simulations so far; any one of them will work. Earlier in this chapter, in the code that led to Figure 12.4, we simulated Y = Personal Assets data conditional on X = Age, using a heteroscedastic model for which the conditional standard deviation of Y was $0.05X$. These estimates e_i^2 are supposed to estimate σ_i^2 for all $i = 1, 2, \ldots, n$; let's just pick $i = 2$ for the sake of concreteness. Here is `R` code to study the accuracy of e_2^2 as an estimate of σ_2^2. Since person $i = 2$ was 43 years old, the true variance is $\sigma_2^2 = (0.05 \times 43)^2 = 4.6225$.

```
Worth = read.table("https://raw.githubusercontent.com/andrea2719/
URA-DataSets/master/Pass.txt")
attach(Worth); n = nrow(Worth); X = Age;

## Conditional-x heteroscedastic simulation model parameters:
gamma = 0.05; beta0 = -6; beta1 = 0.25

## Simulation study to understand how well e2^2 estimates sigma2^2
NSIM = 20000
e2  = numeric(NSIM)
for (i in 1:NSIM) {
    eps = gamma*X*rnorm(n)
    Y = beta0 + beta1*X + eps
    e2[i] = lm(Y ~ X)$residuals[2]
}
hist(e2^2, main="", breaks=50, xlab="Squared Second Residual")
abline(v=(0.05*43)^2, lty = 1, lwd=2)
abline(v=mean(e2^2), col="gray", lty=2, lwd=2)
```

Results shown in Figure 12.6 show how e_2^2 estimates 4.6225, so we can try to answer the question, "Does this make any sense at all?" posed above.

As you can see in Figure 12.6, the squared sample residual is not a very good estimate of the conditional variance. Why should it be? Again, it's an estimate based on just one observation. So you can see that

$$\hat{\Omega} = \begin{bmatrix} e_1^2 & 0 & \cdots & 0 \\ 0 & e_2^2 & \cdots & 0 \\ \vdots & \vdots & \ddots & \vdots \\ 0 & 0 & \cdots & e_n^2 \end{bmatrix},$$

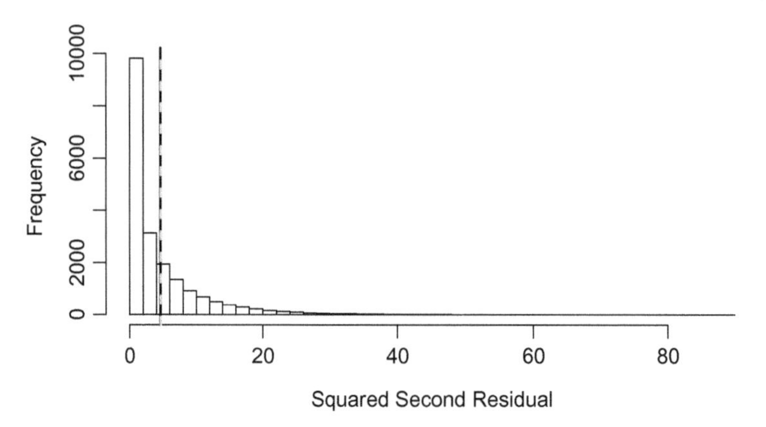

FIGURE 12.6
Histogram of distribution of e_2^2, an estimate of $\sigma_2^2 = (0.5 \times 43)^2 = 4.6225$, shown as the solid line. The mean of the simulated e_2^2 values is shown as the dashed line. The estimate e_2^2 is quite inaccurate, giving values higher than 100.

which is used by the heteroscedasticity-consistent standard errors, is not a very good estimate of

$$\text{Cov}(Y \mid \mathbf{X} = \mathbf{x}) = \begin{bmatrix} \sigma_1^2 & 0 & \cdots & 0 \\ 0 & \sigma_2^2 & \cdots & 0 \\ \vdots & \vdots & \ddots & \vdots \\ 0 & 0 & \cdots & \sigma_n^2 \end{bmatrix}.$$

However, the story is not all bad. Using $\hat{\Omega}$, the estimate of $\text{Cov}(\hat{\beta} \mid \mathbf{X} = \mathbf{x})$ is $(\mathbf{x}^T\mathbf{x})^{-1}\mathbf{x}^T\hat{\Omega}\,\mathbf{x}(\mathbf{x}^T\mathbf{x})^{-1}$. As shown in Figure 12.6, there are large errors in the estimates inside of $\hat{\Omega}$, but these errors "average out" when plugged into the matrix function $(\mathbf{x}^T\mathbf{x})^{-1}\mathbf{x}^T\hat{\Omega}\,\mathbf{x}(\mathbf{x}^T\mathbf{x})^{-1}$, which performs a kind of averaging of the terms in $\hat{\Omega}$. Because of this averaging, the gross errors shown in Figure 12.6 are reduced; and as a result, the estimated covariance matrix

$$\hat{\text{Cov}}(\hat{\beta} \mid \mathbf{X} = \mathbf{x}) = (\mathbf{x}^T\mathbf{x})^{-1}\mathbf{x}^T\hat{\Omega}\,\mathbf{x}(\mathbf{x}^T\mathbf{x})^{-1}$$

can be a reasonably accurate estimate of the true covariance matrix $\text{Cov}(\hat{\beta} \mid \mathbf{X} = \mathbf{x}) = (\mathbf{x}^T\mathbf{x})^{-1}\mathbf{x}^T\Omega\,\mathbf{x}(\mathbf{x}^T\mathbf{x})^{-1}$, even though $\hat{\Omega}$ is a horribly inaccurate estimate of Ω. In fact, under certain conditions on the nature of the heteroscedasticy pattern, inferences (tests and intervals) for the β's that are based on the HC covariance matrix become closer to correct with larger sample sizes. The term "consistency" in statistics often refers to such large-sample convergences, and is the reason that these standard errors are called "consistent."

Sometimes, the covariance matrix estimate $\hat{\text{Cov}}(\hat{\beta} \mid \mathbf{X} = \mathbf{x}) = (\mathbf{x}^T\mathbf{x})^{-1}\mathbf{x}^T\hat{\Omega}\,\mathbf{x}(\mathbf{x}^T\mathbf{x})^{-1}$ is called a "sandwich" estimate, because it places the "meat," $\hat{\Omega}$, inside a "sandwich" with "bread slices" $(\mathbf{x}^T\mathbf{x})^{-1}\mathbf{x}^T$ and $\mathbf{x}(\mathbf{x}^T\mathbf{x})^{-1}$.

Notice from Figure 12.6 that the estimates inside of $\hat{\Omega}$ are not only highly variable (hence inaccurate), but also slightly biased. There are various modifications to the HC form $(\mathbf{x}^T\mathbf{x})^{-1}\mathbf{x}^T\hat{\Omega}\,\mathbf{x}(\mathbf{x}^T\mathbf{x})^{-1}$ that correct for sources of bias that we will not discuss here. The main theory is all we really wanted to get across. You can see further details on these bias corrections in Long and Ervin (2000).

In R, you can access these HC standard errors using the "hccm" (short for heteroscedasticity-consistent covariance matrix) function in the "car" library. Here is an example using the GPA prediction study.

```
ba = read.table("https://raw.githubusercontent.com/andrea2719/
URA-DataSets/master/gpa_gmat.txt")
attach(ba)

y = gpa
x1 = gmat
x2 = ifelse(degree=="P", 1,0)

library(car)
fit = lm(y ~ x1 + x2)

## Homoscedastic analysis
summary(fit)$coefficients[,1:3]

## Heteroscedasticity-consistent analysis using method HC3
se.hc3 = sqrt(diag(hccm(fit, type = "hc3")))
b = fit$coefficients
t = b/se.hc3
cbind(b, se.hc3, t)
```

Results are as follows for the usual OLS analysis:

```
              Estimate   Std. Error   t value
(Intercept) 2.750699944 0.1191639363 23.083326
x1          0.001357218 0.0002155794  6.295674
x2          0.179380537 0.0503072073  3.565703
```

The analysis with heteroscedasticity-consistent standard errors is given as follows:

```
                      b       se.hc3         t
(Intercept) 2.750699944 0.1285724433 21.394164
x1          0.001357218 0.0002343912  5.790396
x2          0.179380537 0.0306261686  5.857100
```

Notice that the estimated β's are the same in either case—they are both the OLS estimates. However, the standard errors differ. There really is non-constant variance, so the standard errors from the "usual analysis" are in error because they assume homoscedasticity.

Self-study question: What are the consequences of having incorrect standard errors?

What follows is a simulation study to investigate the performance of the usual standard errors and the HC standard errors within a nominal (or "supposed") 95% confidence interval.

We will evaluate the intervals obtained using ±2 standard errors rather than taking the multiplier from the T distribution. As it turns out, the T ratio is not T-distributed under heteroscedasticity, so there is no "exact" critical value; hence we may as well use the approximate value 2.

```
Worth = read.table("https://raw.githubusercontent.com/andrea2719/
URA-DataSets/master/Pass.txt")
attach(Worth); n = nrow(Worth); X = Age;

## Conditional-x heteroscedastic simulation model parameters:
gamma = 0.05; beta0 = -6; beta1 = 0.25

## Conditional-x heteroscedastic simulation model

## Simulation study to estimate the true confidence level of ordinary
## and HC intervals
library(car)

NSIM = 10000
b1            = numeric(NSIM)
se.b1         = numeric(NSIM)
se.b1.robust  = numeric(NSIM)

for (i in 1:NSIM) {
    eps = gamma*X*rnorm(n)
    Y = beta0 + beta1*X + eps
    fit = lm(Y~X)
    fit1 = summary(lm(Y~X))
    b1[i] = fit1$coefficients[2,1]
    se.b1[i] = fit1$coefficients[2,2]
    se.b1.robust[i] = sqrt(diag(hccm(fit, type = "hc3")))[2]
}

chk1 = (b1 - 2*se.b1 < beta1) *(b1 + 2*se.b1 > beta1)
mean(chk1)
chk2 = (b1 - 2*se.b1.robust < beta1) *(b1 + 2*se.b1.robust > beta1)
mean(chk2)
```

The results are as follows (your numbers will differ slightly due to randomness):

```
> mean(chk1)
[1] 0.8739
> mean(chk2)
[1] 0.9318
```

The output shows that the usual homoscedasticity-assuming 95% confidence intervals for β_1 only cover the true $\beta_1 = 0.250$ in 87.39% of the samples. Thus, the homoscedasticity-assuming standard errors are too small. On the other hand, the intervals using the HC standard errors are correct in 93.18% of the samples. Thus, while the HC standard errors are only approximations, (the true confidence level is smaller than the target 95%), they give better approximations than the usual (classical) regression method.

Self-study question: Options for HC standard errors other than "hc3" are "hc0," "hc1," "hc2," and "hc4." Is the true coverage rate closer to 95% than the 93.18% of "hc3" for any of these other options?

12.6 Generalized Least Squares for Non-independent Observations

A problem that is closely related to non-constant variance is the problem of correlated observations. To introduce the problem, suppose your colleague Hans has collected the following small data set, having only $n = 3$ observations, shown in Table 12.2.

Hans runs the regression on his data set as follows.

```
X = c(3.1, 4.3, 0.7); Y = c(1.2, 2.0, 1.3)
# Analysis based on n=3
XY.1 = data.frame(X,Y); summary(lm(Y~X, data=XY.1))
```

Hans gets the following results and is not pleased.

```
Coefficients:
            Estimate Std. Error t value Pr(>|t|)
(Intercept)   1.0661     0.5411   1.970    0.299
X             0.1607     0.1753   0.917    0.528

Residual standard error: 0.4543 on 1 degrees of freedom
Multiple R-squared:  0.4568,    Adjusted R-squared:  -0.08647
F-statistic: 0.8408 on 1 and 1 DF,  p-value: 0.5276
```

Hans is specifically unhappy that his X variable is "insignificant" ($p = 0.528$). But he also knows that a larger sample size will decrease the standard errors, thereby giving a larger T statistic and a chance for "significance." But rather than collect more data, Hans just decides to duplicate his existing data set, getting a "new" data set with $n = 6$ observations, shown in Table 12.3.

Using the "new" data set of Table 12.3, here is Hans' code for the analysis:

```
# Analysis based on n=6
XY.2 = rbind(XY.1,XY.1); summary(lm(Y~X, data=XY.2))
```

TABLE 12.2

A Data Set Collected by Hans

Obs	X	Y
1	3.1	1.2
2	4.3	2.0
3	0.7	1.3

TABLE 12.3

Hans' Second Data Set: The Last Three Observations Simply Duplicate the First Three Observations of Table 12.2

Obs	X	Y
1	3.1	1.2
2	4.3	2.0
3	0.7	1.3
4	3.1	1.2
5	4.3	2.0
6	0.7	1.3

And here are his results:

```
Coefficients:
            Estimate Std. Error t value Pr(>|t|)
(Intercept)  1.06607    0.27053   3.941    0.017 *
X            0.16071    0.08763   1.834    0.141
---
Signif. codes:  0 '***' 0.001 '**' 0.01 '*' 0.05 '.' 0.1 ' ' 1

Residual standard error: 0.3213 on 4 degrees of freedom
Multiple R-squared:  0.4568,    Adjusted R-squared:  0.321
F-statistic: 3.363 on 1 and 4 DF,  p-value: 0.1406
```

Hans is getting excited! His estimated regression model is the same as before, and his R^2 statistic is also the same as before, but now the T statistic for his X variable is much higher, and his p-value is approaching "significance" (p = 0.141). So Hans decides to continue in this way, by simply duplicating the observations. The following code shows his analyses based on further duplications of the same n = 3 observations.

```
# Analysis based on n=12
XY.3 = rbind(XY.2,XY.2); summary(lm(Y~X, data=XY.3))

# Analysis based on n=24
XY.4 = rbind(XY.3,XY.3); summary(lm(Y~X, data=XY.4))

# Analysis based on n=48
XY.5 = rbind(XY.4,XY.4); summary(lm(Y~X, data=XY.5))
```

The results of his final analysis, based on n = 48 (his original n = 3 data 16 times) are as follows:

```
Coefficients:
            Estimate Std. Error t value Pr(>|t|)
(Intercept)  1.06607    0.07978  13.363  < 2e-16 ***
X            0.16071    0.02584   6.219 1.36e-07 ***
---
Signif. codes:  0 '***' 0.001 '**' 0.01 '*' 0.05 '.' 0.1 ' ' 1

Residual standard error: 0.268 on 46 degrees of freedom
Multiple R-squared:  0.4568,    Adjusted R-squared:  0.445
F-statistic: 38.68 on 1 and 46 DF,  p-value: 1.358e-07
```

Again, his estimated regression model and his R^2 are the same as before, but now his p-value is "highly significant" $\left(p = 1.36 \times 10^{-7}\right)$. Hans is thrilled, so he writes up the analysis based on n = 48 observations, and prepares to submit the results to a high-quality research journal.

Of course, this discussion of Hans' method is completely "tongue in cheek." In all seriousness, we hope you understand that Hans' method is seriously flawed, and that Hans would be guilty of "scientific misconduct" for fabricating data. Still, there are important lessons to be learned from Hans' method. Specifically, what is wrong with using the n = 48 observations? After all, the estimated regression model was exactly the same as the model with n = 3, and the R^2 statistics were exactly the same as well. What is the problem?

The problem is that the standard errors reported by the software assume that the data are uncorrelated. But the $n = 48$ observations in Hans' final analysis cannot possibly be uncorrelated, because they are simple duplicates. Hence, there is a serious violation of the uncorrelatedness assumption in Hans' final analysis, and this leads to standard errors that are too small.

Rather than $n = 48$ uncorrelated observations in Hans' final analysis, as assumed by the software, there are really only (at most) $n = 3$ uncorrelated observations. The standard error reported by Hans' first analysis (0.1753) is valid, but the ones reported in his other analyses, especially his final analysis (0.02584), are too small because of a violation of the uncorrelatedness assumption.

Recall that the standard errors come from the covariance matrix of the estimated β vector. From Section 12.5 you know that, for the OLS estimates, this covariance matrix is

$$\mathrm{Cov}\left(\hat{\beta}_{\mathrm{OLS}} \middle| \mathbf{X} = \mathbf{x}\right) = \left(\mathbf{x}^{\mathrm{T}}\mathbf{x}\right)^{-1} \mathbf{x}^{\mathrm{T}} \left\{\mathrm{Cov}\left(\mathbf{Y} \middle| \mathbf{X} = \mathbf{x}\right)\right\} \mathbf{x} \left(\mathbf{x}^{\mathrm{T}}\mathbf{x}\right)^{-1}$$

When the observations are correlated, the covariance of the Y data will have non-zero elements off of the diagonal, like this:

$$\mathrm{Cov}\left(\mathbf{Y} \middle| \mathbf{X} = \mathbf{x}\right) = \begin{bmatrix} \sigma_1^2 & \sigma_{12} & \sigma_{13} & \cdots & \sigma_{1n} \\ \sigma_{21} & \sigma_2^2 & \sigma_{23} & \cdots & \sigma_{2n} \\ \sigma_{31} & \sigma_{32} & \sigma_3^2 & \cdots & \sigma_{3n} \\ \vdots & \vdots & \vdots & \ddots & \vdots \\ \sigma_{n1} & \sigma_{n2} & \sigma_{n3} & \cdots & \sigma_n^2 \end{bmatrix}.$$

The correct standard errors must, therefore, account for all dependencies between the observations, which are shown as non-zero covariances σ_{ij} of the matrix above.

Typically, it is impossible to estimate all the covariances σ_{ij} using regression data because there are no repeats. For example, σ_{12} is the covariance between the errors ε_1 and ε_2. Even if you had observed ε_1 and ε_2, you still would have only one pair with which to estimate their covariance, because you only have one observation 1 and one observation 2 in your data set. Usually, you need many pairs (u_i, v_i) to estimate the covariance between any variables (U,V).

Thus, to estimate the covariances inside $\mathrm{Cov}(\mathbf{Y} \mid \mathbf{X} = \mathbf{x})$, you usually need to assume a special structure for the covariance matrix. Two notable types of such covariance structures are given as follows:

Autoregressive covariance structure

$$\mathrm{Cov}(\mathbf{Y} \mid \mathbf{X} = \mathbf{x}) = \sigma^2 \begin{bmatrix} 1 & \phi & \phi^2 & \cdots & \phi^{n-1} \\ \phi & 1 & \phi & \cdots & \phi^{n-2} \\ \phi^2 & \phi & 1 & \cdots & \phi^{n-3} \\ \vdots & \vdots & \vdots & \ddots & \vdots \\ \phi^{n-1} & \phi^{n-2} & \phi^{n-3} & \cdots & 1 \end{bmatrix}$$

The parameter ϕ is called the *autocorrelation coefficient*. Since ϕ is a correlation, it is a number between −1 and 1.

The autoregressive covariance structure is commonly used in time series data, where the adjacent observations (e.g., one day apart if the data are collected daily) have correlation equal to ϕ, but observations two apart are less correlated, with correlation ϕ^2. Observations three apart are even less correlated, with correlation ϕ^3, and so on.

Notice that there are only two parameters in this covariance matrix; namely, ϕ and σ^2. This parameterization allows you to estimate the parameters via maximum likelihood, unlike in the unstructured case described above. In particular, there are $n-1$ residual pairs $(e_2,e_1),(e_3,e_2),\ldots,(e_n,e_{n-1})$ available to estimate the correlation coefficient ϕ, unlike the unstructured case above, where only the pair (e_2,e_1) was available to estimate σ_{12}.

Other types of covariance structures have the same parameterization issues; namely, the number of parameters to estimate generally must be far less than the number of terms in the covariance matrix. Another such covariance structure that is commonly used is called the block-diagonal *covariance structure.*

Block-diagonal covariance structure

$$\text{Cov}(Y \mid \mathbf{X}=\mathbf{x}) = \begin{bmatrix} \boldsymbol{\Phi} & 0 & 0 & \cdots & 0 \\ 0 & \boldsymbol{\Phi} & 0 & \cdots & 0 \\ 0 & 0 & \boldsymbol{\Phi} & \cdots & 0 \\ \vdots & \vdots & \vdots & \ddots & \vdots \\ 0 & 0 & 0 & \cdots & \boldsymbol{\Phi} \end{bmatrix}$$

This term *block-diagonal* refers to the fact that each of the elements in the matrix is actually a matrix (a "block"). The term $\boldsymbol{\Phi}$ is an unstructured covariance matrix that repeats, and the "0's" are actually matrices indicating groups of observations that are uncorrelated.

The block-diagonal covariance structure is commonly used in repeated measures and longitudinal data, where the first set of observations are repeated data collected on an observational unit (e.g., a person, a company, and animal, etc.), and the matrix $\boldsymbol{\Phi}$ refers to the covariance matrix of those repeated observations. The "0's" off the diagonal indicate that observations taken from one observational unit (person, company, etc.) are uncorrelated with observations taken from another observational unit (person, company, etc.). When there are many observational units, there is sufficient data to estimate the common covariance matrix $\boldsymbol{\Phi}$. For example, the (1,2) element of $\boldsymbol{\Phi}$ (ϕ_{12}) is the covariance between the 1st and 2nd measurements on an observational unit. If there are 20 such units, then there are 20 pairs with which to estimate ϕ_{12}.

Whatever covariance structure is assumed, whether block-diagonal, autoregressive, or other, maximum likelihood is the most commonly used method of estimation. Let that assumed covariance structure be denoted by $\Sigma(\gamma)$; the parameters of the covariance matrix (such as ϕ and σ^2 in the autoregressive covariance structure) are collectively γ. The likelihood function for the regression model with covariance matrix $\Sigma(\gamma)$ uses the n-dimensional multivariate normal distribution:

$$L(\beta,\gamma \mid \text{data}) = \frac{\exp[-0.5(Y - \mathbf{x}\beta)^{\mathrm{T}}\{\Sigma(\gamma)\}^{-1}(Y - \mathbf{x}\beta)]}{(2\pi)^{n/2} \mid \Sigma(\gamma) \mid^{1/2}}$$

Thus, in this most general case (which includes all of the likelihood functions for normal distributions considered so far in the book as special cases), the log likelihood function for the parameters is given as follows:

Log likelihood function for the general normal model

$$LL(\beta,\gamma \mid \text{data}) = -\frac{n}{2}\ln(2\pi) - \frac{1}{2}\ln\{\mid \Sigma(\gamma) \mid\} - \frac{1}{2}(Y - \mathbf{x}\beta)^{\mathrm{T}}\{\Sigma(\gamma)\}^{-1}(Y - \mathbf{x}\beta)$$

The vectors β and γ that maximize the likelihood are the maximum likelihood estimates, denoted by $\hat{\beta}_{ML}$ and $\hat{\gamma}$. The maximum likelihood estimates $\hat{\beta}_{ML}$ are also called *generalized least squares (GLS) estimates*, because they can be mathematically proven to have the following form, which is reminiscent of the matrix form of the OLS estimates:

$$\hat{\beta}_{ML} = \hat{\beta}_{GLS} = (\mathbf{x}^{\mathrm{T}}\{\Sigma(\hat{\gamma})\}^{-1}\mathbf{x})^{-1}\mathbf{x}^{\mathrm{T}}\{\Sigma(\hat{\gamma})\}^{-1}Y$$

The GLS estimates are more efficient than OLS and WLS estimates when the covariance structure ($\Sigma(\gamma)$) is correctly specified, and when its parameters are well estimated (e.g., when the sample size is large enough).

These estimates are called *generalized least squares* because ordinary least squares and weighted least squares are special cases.

When $\Sigma(\hat{\gamma})$ is a diagonal matrix with common diagonal value, then the generalized least squares estimates are equal to the OLS estimates: $\hat{\beta}_{GLS} = \hat{\beta}_{OLS}$.

When $\Sigma(\hat{\gamma})$ is a diagonal matrix with different diagonal values, then the generalized least squares estimates are equal to the WLS estimates: $\hat{\beta}_{GLS} = \hat{\beta}_{WLS}$. In this case, the weights used in the WLS estimates are the inverses of the diagonal elements of $\Sigma(\hat{\gamma})$.

Often, the estimates themselves, whether OLS, WLS, or GLS, do not differ that much, but the standard errors differ greatly. For ML/GLS estimates, the standard errors are the square roots of the diagonal elements of

$$\widehat{\mathrm{Cov}}(\hat{\beta}_{\mathrm{GLS}} \mid \mathbf{X} = \mathbf{x}) = (\mathbf{x}^{\mathrm{T}}\{\Sigma(\hat{\gamma})\}^{-1}\mathbf{x})^{-1},$$

which incorporate the dependencies between the observations through $\Sigma(\hat{\gamma})$. Often, these standard errors are much larger than the OLS or WLS standard errors, which assume uncorrelated observations. As shown in the farcical example involving Hans' data fabrication at the beginning of this section, ignoring dependencies can easily give you standard errors that are too small.

On the other hand, it can occasionally happen that by using a model that ignores correlations, your standard errors for certain parameter estimates will be too high. For example, when estimating the difference between positively correlated data values (U, V), if you assume no correlation between U and V you get $\mathrm{Var}(U - V) = \mathrm{Var}(U) + \mathrm{Var}(V)$. This is larger than the correct variance, which is $\mathrm{Var}(U - V) = \mathrm{Var}(U) + \mathrm{Var}(V) - 2\mathrm{Cov}(U,V)$. So in cases where your estimated β is essentially a difference between positively correlated data values, the computer-reported standard error of your estimate will be (incorrect and) too large when you assume uncorrelated observations.

Generalized least squares estimates are used in many regression applications where observations are dependent, including time series, repeated measures, panel data (cross-sectional time series). They are also used with spatial data, where observations that are geographically near to each other are more highly correlated than observations that are geographically far apart.

There are many other possible covariance structures than the two listed above. As with mean functions and standard deviation functions, no restriction on covariance structure is correct, although some restrictions are more reasonable than others. Also, as with mean functions and standard deviation functions, you simply need to do the best that you can when specifying the covariance structure: (1) Think about it clearly in terms of your data-generating process as is appropriate for your study data (e.g., time-series versus repeated measures), (2) evaluate different models in terms of parameter estimates and fit statistics, and (3) perform simulation studies to understand the effects of parameter estimation and covariance structure misspecification. Then will you be in a position to make a more informed decision about your choice of a covariance structure.

12.6.1 Generalized Least Squares Estimates and Standard Errors for the Charitable Contributions Study

In Chapter 10, Section 10.7, we gave an example where charitable contributions were predicted in terms of income, number of dependents, and subject indicators. We discussed that the use of indicator variables reduced the dependence between residuals for the same subject, but also noted that "use of subject indicators does not completely guarantee independence, as the residuals in adjacent years are expected to be more similar than residuals in distant years."

In the following analysis, we model this type of dependence using the autoregressive covariance structure indicated above, in block-diagonal form, with one block per taxpayer. Here is the `R` code to fit the model, and also to fit the classical model that was estimated in Section 10.7.

```
charity = read.csv("https://raw.githubusercontent.com/andrea2719/
URA-DataSets/master/charitytax.csv")
attach(charity)
sub.f = as.factor(SUBJECT)

library(nlme)

# ML/GLS fit assuming classical sig^2*I covariance structure
fit.ols = gls(CHARITY ~ INCOME + DEPS + sub.f, method="ML")
```

```
# ML/GLS fit assuming block-diagonal AR1 covariances
fit.gls = gls(CHARITY ~ INCOME + DEPS + sub.f,
     corr=corAR1(form = ~1 | SUBJECT), method="ML")
```

Since there is an enormous amount of output in `summary(fit.ols)` and `summary(fit.gls)`, we will select output more judiciously. First, by using `summary(fit.ols)$tTable`, you get

```
                  Value  Std.Error      t-value      p-value
(Intercept) -1.97574797 1.07401072 -1.83959800 6.653117e-02
INCOME       0.73062559 0.10000008  7.30624994 1.388961e-12
DEPS        -0.07132880 0.05212552 -1.36840457 1.719153e-01
sub.f2       1.96613452 0.30827791  6.37779881 4.734889e-10
sub.f3       1.09686954 0.31051738  3.53239341 4.573710e-04
. . . (many more lines of output) . . .
sub.f46     -0.14375937 0.31432794 -0.45735473 6.476520e-01
sub.f47      0.02113385 0.32622349  0.06478336 9.483772e-01
```

Notice that these values correspond exactly with the analysis shown in Section 10.7 for the same data, so it is clear that the same model is being fit. On the other hand, the estimate of σ differs: Using the ML fit above, you can find the estimate of σ by using `summary(fit.ols)$sigma`, which gives $\hat{\sigma} = 0.6456665$. The analysis using the `lm` function, as shown in the output reported Section 10.7, returned the value $\hat{\sigma} = 0.6822$. The difference is that the latter uses the "unbiased" version of the estimate, $\sqrt{\text{SSE}/(470-29)}$, whereas the former uses the ML form, $\sqrt{\text{SSE}/470}$. Still, the main point is that the same model is estimated, but two different estimates of the same σ are reported.

By using `summary(fit.gls)$tTable`, you are showing the results of fitting a model that allows correlated observations, and you get the following results:

```
                 Value  Std.Error    t-value      p-value
(Intercept) -0.4204662 1.23409895 -0.3407071 7.334941e-01
INCOME       0.5674705 0.11333874  5.0068536 8.151023e-07
DEPS        -0.0564384 0.05666402 -0.9960183 3.198134e-01
sub.f2       2.1071850 0.44821047  4.7013292 3.508182e-06
sub.f3       1.2692770 0.45020947  2.8193032 5.039981e-03
. . . (many more lines of output) . . .
sub.f46     -0.2147991 0.45237788 -0.4748223 6.351599e-01
sub.f47      0.2611733 0.46324680  0.5637887 5.731981e-01
```

Now, while the essential conclusion regarding the effect of DEPS is unchanged in that the estimated coefficient is still negative, you will notice that the estimates and standard errors are different. This happened because both the GLS estimates and the standard errors of the GLS estimates incorporate correlations between the Y observations, conditional on the X variables. The OLS estimates, on the other hand, assumed these were all zero.

Recall that there are 47 subjects (taxpayers) in this data set, and that each subject is followed over 10 years. In the analysis above, the covariance matrix that is assumed for the 470 Y observations is a 470×470 block-diagonal matrix, with 10×10 blocks along the diagonal, each having the autoregressive covariance structure shown above. You can obtain the estimate of σ using `summary(fit.gls)$sigma`, giving $\hat{\sigma} = 0.6558013$, which is close to the ML estimate of σ in the classical model, shown above as $\hat{\sigma} = 0.6456665$.

You can obtain an estimate of ϕ using `fit.gls$modelStruct$corStruct`, which gives you the following output:

```
Correlation structure of class corAR1 representing
      Phi
0.4018739
```

Using $\hat{\phi} = 0.4018739$ and $\hat{\sigma} = 0.6558013$, you can estimate the covariance matrix for the repeated measures on a particular taxpayer as

$$\hat{\text{Cov}}(\mathbf{Y} \mid \mathbf{X} = \mathbf{x}) = 0.656^2 \begin{bmatrix} 1 & 0.402 & 0.402^2 & \cdots & 0.402^9 \\ 0.402 & 1 & 0.402 & \cdots & 0.402^8 \\ 0.402^2 & 0.402 & 1 & \cdots & 0.402^7 \\ \vdots & \vdots & \vdots & \ddots & \vdots \\ 0.402^9 & 0.402^8 & 0.402^7 & \cdots & 1 \end{bmatrix}$$

This matrix shows that excess charitable contributions in a given year are correlated with those in the previous year (and the next year) with an estimated correlation 0.402. It also shows that excess charitable contributions in a given year are correlated with those two years previous (and two years hence) with estimated correlation $0.402^2 = 0.162$, and so on.

On the other hand, the classical regression model assumes that these data are all uncorrelated. Using $\hat{\sigma} = 0.6456665$, you can estimate the covariance matrix for the repeated measures data on a particular taxpayer as

$$\hat{\text{Cov}}(\mathbf{Y} \mid \mathbf{X} = \mathbf{x}) = 0.646^2 \begin{bmatrix} 1 & 0 & 0 & \cdots & 0 \\ 0 & 1 & 0 & \cdots & 0 \\ 0 & 0 & 1 & \cdots & 0 \\ \vdots & \vdots & \vdots & \ddots & \vdots \\ 0 & 0 & 0 & \cdots & 1 \end{bmatrix}$$

Since you have fit both of these models using maximum likelihood, you have likelihood-based fit measures for comparing model. By using `AIC(fit.ols)` and `AIC(fit.gls)`, you see that the classical regression model has $AIC = 1{,}022.578$ while the model that allows correlated observations has $AIC = 964.7098$. Thus the model that allows correlated observations fits the data much better than does the classical model, which assumes zero correlations.

Appendix A: Likelihood Ratio Tests

In the case of the classical regression model, you can use F tests to compare the β parameters in "full" versus "restricted" models, as shown in Chapter 10, Section 10.5. However, in this chapter and throughout the rest of the book, we are interested in models other than the classical regression model, and we are interested in parameters other than the β parameters. The *likelihood ratio test* (LRT) is used to compare full versus restricted models more generally. In particular, the F statistic arises from the LRT procedure.

To construct the LRT test, let LL_F denote the maximized log likelihood when you use the full model, and let LL_R denote the maximized log likelihood when you use the restricted model. By the same logic that implies $SSE_F \leq SSE_R$, as shown in Section 10.5, it is a mathematical fact that

$$LL_F \geq LL_R$$

If the restricted model is, in fact, the true data-generating process, then the apparent "improvement in fit" of the full model, $LL_F - LL_R$, is *explained* by chance alone. And, just like the case of the classical regression model, this improvement in fit is entirely predictable.

The likelihood ratio χ^2 statistic for comparing a full versus restricted (null) model

$$\chi^2 = 2(LL_F - LL_R)$$

The reason for calling this a "likelihood ratio" is that the statistic is equal to twice the logarithm of the likelihood ratio: $\chi^2 = 2\ln(L_F/L_R) = 2(LL_F - LL_R)$.

As always, the chance-only (null) model is the restricted model. It is a mathematical theorem that, *if* the data are in fact produced by the restricted model, with restricted model's assumptions true, *then* distribution of the LRT χ^2 statistic is given as follows:

Null distribution of the χ^2 statistic

$$\chi^2 \stackrel{\cdot}{\sim} \chi^2_{p_F - p_R},$$

where p_F is the number of parameters in the full model, and p_R is the number of parameters in the restricted model.

Notice the "$\stackrel{\cdot}{\sim}$" symbol, it indicates that the null distribution is only *approximately* chi-squared. Thus, this result is similar to the Central Limit Theorem, which likewise asserts *approximate* normality of the distribution of the sample average.

If the restricted (null) model is false and the full model is true, then you will see larger values of $LL_F - LL_R$, making the distribution of the LRT χ^2 statistic shifted to the right of the $\chi^2_{p_F - p_R}$ distribution. Thus, like the F test, the p-value for the LRT test is the area under the $\chi^2_{p_F - p_R}$ distribution *to the right* of the observed χ^2 statistic.

Self-study question: How can you check that the null distribution of the χ^2 statistic is in fact approximately the $\chi^2_{p_F - p_R}$ distribution by using simulation? And how can you check, also using simulation, that its non-null distribution is shifted to the right?

Appendix B: Wald Standard Errors

The Wald standard errors are calculated by approximating the likelihood function $L(\theta)$ using a multivariate normal distribution. Since the vector of maximum likelihood estimates, $\hat{\theta}$, gives the maximum of $L(\theta)$, the approximating multivariate normal distribution uses $\hat{\theta}$ for its mean vector. The Wald standard errors are then obtained as the square roots of the diagonal elements of the covariance matrix that provides the approximation of the multivariate normal distribution to the likelihood function.

To start, suppose that the likelihood is approximately proportional to a multivariate normal distribution with mean vector $\hat{\theta}$ (the maximum likelihood estimates) and covariance matrix $\hat{\Sigma}$ (to be determined). Then

$$L(\theta \mid \text{data}) \cong c \times \exp\left\{-0.5(\theta - \hat{\theta})^{\mathrm{T}} \hat{\Sigma}^{-1} (\theta - \hat{\theta})\right\}$$

Suppose for a minute that the likelihood is exactly proportional to a multivariate normal pdf, so that

$$LL(\theta \mid \text{data}) = \ln(c) - 0.5(\theta - \hat{\theta})^{\mathrm{T}} \hat{\Sigma}^{-1} (\theta - \hat{\theta})$$

The *Hessian* matrix is defined as

$$\mathbf{H}(\theta) = \left\{ \frac{\partial^2 LL(\theta \mid \text{data})}{\partial \theta_i \partial \theta_j} \right\}.$$

Using results from matrix differentiation you will get $\mathbf{H} = -\hat{\Sigma}^{-1}$, which does not depend on the value of θ, by the exact proportionality assumption. Thus, in such a case, $\hat{\Sigma} = -\mathbf{H}^{-1}$.

However, the likelihood function is not exactly proportional to a multivariate normal distribution, and hence second derivative (Hessian) matrix is not constant for all θ. Instead, you must pick a value of θ, evaluate the Hessian at that θ, and solve for $\hat{\Sigma}$. If you pick $\theta = \hat{\theta}$, then the multivariate curvature of the approximating multivariate normal distribution function will match the curvature of the likelihood function at the MLE. This gives you the Wald covariance matrix as follows: Let

$$\mathbf{H}(\hat{\theta}) = \left\{ \frac{\partial^2 LL(\theta \mid \text{data})}{\partial \theta_i \partial \theta_j} \right\} \Bigg|_{\theta = \hat{\theta}}$$

Then

$$\hat{\Sigma} = -\mathbf{H}^{-1}(\hat{\theta})$$

is the *Wald Covariance Matrix*. The *Wald Standard Error* of $\hat{\theta}_j$ (abbreviated W.S.E $(\hat{\theta}_j)$) is then the square root of the jth diagonal element of $\hat{\Sigma}$. These standard errors are used in the usual way, with $\hat{\theta}_j \pm 2\text{W.S.E}(\hat{\theta}_j)$ giving an approximate 95% confidence interval for the parameter θ_j.

Reference

Long, J. S., & Ervin, L. H. (2000). Using heteroscedasticity-consistent standard errors in the linear regression model. *The American Statistician*, 54, 217–224.

Exercises

1. In Chapter 12, under Figure 12.4, the second self-study question asks "How can you modify the simulation leading to Figure 12.4 to show that the WLS estimates are less efficient than the OLS estimates when the data-generating process is homoscedastic?" Go ahead and perform that simulation study.
2. As discussed in the book, Chapter 12, when you specify "`weights=`" in the "`lm`" function, then apply the "`logLik`" function to the fitted object, `R` returns the maximized log likelihood from the heteroscedastic model. That means it is easy to compare different weights by using the log likelihoods. Try two other weight functions other than the ones used in Section 12.1, and see if your two new models fit any better than the three discussed in the book.
3. Why does it not matter whether you use log likelihood, AIC, or BIC to make the comparisons between all the models in problem 2? (Hint: Consider the number of parameters in all models, and use simple algebra with the definitions of AIC and BIC.)
4. Use the data

```
ge = read.csv("https://raw.githubusercontent.com/andrea2719/
URA-DataSets/master/ge_ret_vol.csv").
```

Inside are variables "return" and "volume", which are the daily return on GE stock (Y), and the associated trading volume (X), respectively.

a. See point 8 just prior to Section 12.4 in Chapter 12. Fit the indicated heteroscedastic model using the `gamlss` function, and report the estimated mean function and the estimated standard deviation function.
b. Using the result of 4.a, graph the estimated standard deviation function (as a function of trading volume, of course).
c. Using the results of 4.a., graph the (X,Y) scatterplot with the estimated mean function and the approximate 95% prediction intervals overlaid. Does the estimated standard deviation function look reasonable, based on this plot?

d. Compare the fit of the heteroscedastic model with the homoscedastic model using the likelihood ratio test and draw a conclusion.
e. Why is the homoscedastic model a "restricted model" relative to the heteroscedastic model? What is the specific parameter restriction? (Comment: The test in 4.d requires such full/restricted form, otherwise, you could get a negative chi-square statistic.)
f. Is the variance function better modeled as a function of log(X)? Perform the analysis and compare AIC statistics with what you got in 4.a to answer.

5. Just before Section 12.6 is a self-study question regarding HC standard errors. Please perform the necessary simulations to answer the question. Based on your simulations, make a recommendation for choice of HC standard error.

13

Models for Binary, Nominal, and Ordinal Response Variables

Did you ever hear anyone tell you that the regression model is really $p(y \mid x)$, not $\beta_0 + \beta_1 x$? We hope the logic for that definition has become clearer by now: The regression model $p(y \mid x)$ covers all aspects of the potentially observable $Y \mid X = x$ data, from the mean function, to the variance function, to the covariance function, to the actual distributions themselves.

So far, we have discussed mainly normal distributions $p(y \mid x)$, with occasional digressions to the lognormal distribution, the Laplace distribution, and the generic distributions of the Gauss-Markov model. In this chapter, we must discard the normal distribution completely because we are considering Y data that are grossly non-normal. Binary data, such as $Y = 1$ if a customer repays a loan, or $Y = 0$ if not, are as discrete as they can possibly be; they are also quite obviously non-normal because the normal distribution is a continuous distribution. Ordinal data, such as a No/Maybe/Yes response to a question on a questionnaire, can be coded as 1, 2, 3, and are also highly discrete, implying that they are also quite non-normal. Nominal data, such as a customer's choice of a brand of product (e.g., toothpaste brand A, B, C, or other), are not only discrete, but they are also non-numeric. Unlike the binary and ordinal cases, where you might squint your eyes very, very hard, and envision some sort of normal or rounded-off normal distribution, nominal data are not even numbers, so you cannot even consider using the normal distribution.

In all the cases given above, there is a distribution $p(y)$ that you can display in list form as follows:

Generic form of a discrete distribution

y	$p(y)$
y_1	π_1
y_2	π_2
...	...
y_c	π_c
Total	1.0

The generic form of the discrete distribution covers all the binary, ordinal and nominal cases that we consider in this chapter. In the binary case, $c = 2$; in the other cases, c is the number of possible values of Y (we chose the letter "c" to represent "categories"). The possible values $y_1, y_2, \ldots, y_c$ can be ordinal or nominal.

When you have X data, you need to string these $p(y)$ distributions along the values $X = x$ to construct your model $p(y \mid x)$. As models for Nature, it makes sense to have these discrete distributions "morph" continuously as a function of x: Usually, Nature favors continuity over discontinuity.

We could have gotten by without using maximum likelihood prior to this chapter, but it would have been more difficult. We would not have had as much flexibility in the types of models we have considered, and we would not have had the ability to compare models based on penalized likelihoods. Estimation of the variance parameters as in the previous chapter also would have been very awkward without maximum likelihood. But in this chapter, you will see that likelihood is even more important, as it is the default estimation and analysis method employed by statistical software for the analysis of binary, ordinal, and nominal Y data.

Recall that you can obtain the likelihood function for the unknown parameters $\boldsymbol{\theta}$ from the data in the usual way as shown in Table 2.3 of Chapter 2. Assuming independence, your likelihood function is given as in the following display. Notice that, unlike the likelihood for continuous distributions (like the normal distribution), the likelihood function is a product of probabilities rather than densities.

Likelihood and log likelihood function for discrete *Y*

Likelihood

$$L(\boldsymbol{\theta} \mid \text{data}) = p(y_1 \mid X = x_1, \boldsymbol{\theta}) \times p(y_2 \mid X = x_2, \boldsymbol{\theta}) \times \ldots \times p(y_n \mid X = x_n, \boldsymbol{\theta})$$
$$= \Pi_i p(y_i \mid X = x_i, \boldsymbol{\theta})$$

Log likelihood

$$LL(\boldsymbol{\theta} \mid \text{data}) = \ln\{p(y_1 \mid X = x_1, \boldsymbol{\theta})\} + \ln\{p(y_2 \mid X = x_2, \boldsymbol{\theta})\} + \ldots + \ln\{p(y_n \mid X = x_n, \boldsymbol{\theta})\}$$
$$= \sum_i \ln\{p(y_i \mid X = x_i, \boldsymbol{\theta})\}$$

Recall also that you can use the likelihood function to do the following:

- Obtain maximum likelihood estimates of the parameters $\boldsymbol{\theta}$, and use those estimates to make predictions and/or attempt to make causal connections
- Compare alternative models using likelihood-based statistics such as AIC and BIC.
- Perform inferences (likelihood ratio tests, Wald intervals) for the parameters $\boldsymbol{\theta}$.

The remainder of the chapter is devoted to the use of different probability models $p(y \mid X = x, \boldsymbol{\theta})$ for Y as appropriate for binary, ordinal and nominal Y data.

13.1 The Logistic Regression Model for Binary Y

Your choice of a distribution $p(y)$ in the case of binary Y, which takes the values 0 or 1, is easy: Here, Y has the Bernoulli distribution. In the generic form given above, $y_1 = 0$, $y_2 = 1$, $c = 2$, $\pi_1 = \Pr(Y = 0)$, and $\pi_2 = \Pr(Y = 1)$. Since $\pi_1 + \pi_2 = 1$, it is common to give the probabilities as $\Pr(Y = 1) = \pi$ and $\Pr(Y = 0) = 1 - \pi$. Specifically, $p(y)$ is given by:

The Bernoulli distribution

y	$p(y)$
0	$1-\pi$
1	π
Total	1.0

To construct your model $p(y \mid x)$, all you need to do is to create a function $\Pr(Y = 1 \mid X = x) = \pi(x)$ for different values $X = x$. The probabilities $\Pr(Y = 0 \mid X = x)$ are then automatically $1 - \pi(x)$. Here is the resulting distribution:

The regression model $p(y \mid x)$ for binary Y

y	$p(y \mid x)$
0	$1-\pi(x)$
1	$\pi(x)$
Total	1.0

The emphasis on the "1" category here makes sense because it ties directly to our usual interest in the mean function $E(Y \mid X = x)$, which in this case is given by:

$$E(Y \mid X = x) = 0 \times \{1 - \pi(x)\} + 1 \times \pi(x) = \pi(x).$$

So $\pi(x)$ is both the probability of the "1" response (often called a "success" in the binary data parlance), and it is also the conditional mean function that we have come to know and love.

What conditional mean function $\pi(x)$ should you use? As with all conditional mean functions, there are infinitely many possible choices. But whatever function you pick, you should ensure that $\pi(x)$ always lies between 0 and 1, because it is a probability. Thus, the linear model $\pi(x) = \beta_0 + \beta_1 x$ that you would estimate by using OLS is *not* a viable candidate, because whenever $\beta_1 \neq 0$, the linear model will give probabilities greater than 1 and less than 0 for some values of x.

The logistic regression model assumes a particular form of a probability function. Notice the word "assumes." We hope you realize by now that any time anyone says "assume," it means they are assuming something that is patently false. But, as always, assumptions are made for convenience and efficiency. And as always, assumptions are reasonable to make as long as they are not violated too badly. Just like the linear function is not Nature's function, the particular function assumed by the logistic regression model is also not Nature's function. There are many other choices for probability functions that you can use, and you can check each of their conformance with the data by using penalized likelihood statistics such as AIC and BIC. Just realize that none of them (not even the one with the best data-based fit) will be Nature's function.

Still, logistic regression is the most commonly used function in models for binary Y, just like the linear function is the most commonly used mean function in classical regression. Here is the probability function that you assume to be equal to Nature's function when you use logistic regression:

The logistic regression function

$$\pi(x) = \frac{\exp(\beta_0 + \beta_1 x)}{1 + \exp(\beta_0 + \beta_1 x)}$$

It is instructive to draw graphs of this function for different β_0 and β_1 to help understand it. That's what you see in Figure 13.1.

R code for Figure 13.1

```
x = seq(0,20,.1)
p1 = 1/(1 + exp(-(-1 + .2*x))) # Mathematically equivalent form
p2 = 1/(1 + exp(-(-4 + .4*x)))
p3 = 1/(1 + exp(-(3  - .4*x)))
plot(x,p1,type="l", ylim=c(0,1),ylab="Pr( Success | x )",yaxs="i")
points(x,p2,type="l",lty=2)
points(x,p3,type="l",lty=3)
legend(14.5,.5, c("b0= -1, b1=  .2", "b0= -4, b1=  .4", "b0=  3,
b1= -.4"), lty=c(1,2,3))
```

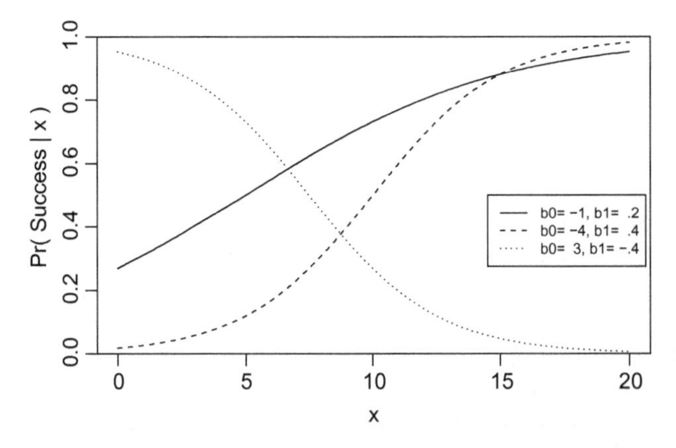

FIGURE 13.1
Graphs of the logistic regression probability function for different β_0, β_1 combinations.

Figure 13.1 highlights several facts about the logistic regression probability function $\pi(x)$:

- The probabilities assigned by this model are always between 0 and 1.
- If the parameter β_1 is >0 then the probability function increases; the higher the value of β_1, the more rapid the increase. The converses apply for $\beta_1 < 0$.
- When $x = 0$, the probability of a "1", and the conditional mean of Y, is the intercept, $\exp(\beta_0)/\{1+\exp(\beta_0)\}$. If $\beta_0 = 0$, the intercept is 0.5. If $\beta_0 > 0$, the intercept is > 0.5. If $\beta_0 < 0$, then the intercept is < 0.5.

There is a simple way to interpret the logistic regression model using the following mathematical identity, which, to some extent, explains the form of the logistic regression probability function. Define the "logit" function (also called the "log odds" function) as follows:

$$\text{logit}(x) = \ln\{x/(1-x)\}$$

Then for the $\pi(x)$ of the logistic regression model,

$$\text{logit}\{\pi(x)\} = \beta_0 + \beta_1 x$$

Self-study question: Why is the equation given above for $\text{logit}\{\pi(x)\}$ true?

Hence, the coefficients β_0 and β_1 can be interpreted in the ordinary linear way, but in terms of the *logit* of the response probability (which is the logit of the mean of Y), rather than in terms of the ordinary mean of Y as in the case of ordinary regression. This linear form of the logistic regression model also implies, using the same argument that was given in Chapter 1, Section 1.7, that the logistic regression model cannot be precisely true when there are three or more levels of the X variable.

Figure 13.2 shows how the logit function looks. Notice that the logit function transforms the (0,1) probability interval to the $(-\infty,\infty)$ range, thus it makes sense to model $\text{logit}\{\pi(x)\}$ as a linear function $\beta_0 + \beta_1 x$ that can also go from $-\infty$ to ∞.

R code for Figure 13.2

```
pi.seq = seq(0.0005, .9995, .0005)
logit = log(pi.seq/(1-pi.seq))
plot(pi.seq, logit, type= "l", xaxs="i",   xlim = c(0,1), xlab="Success
Probability, p", ylab="logit(p)")
abline(v=.5, lty=2)
abline(h=0, lty=3)
```

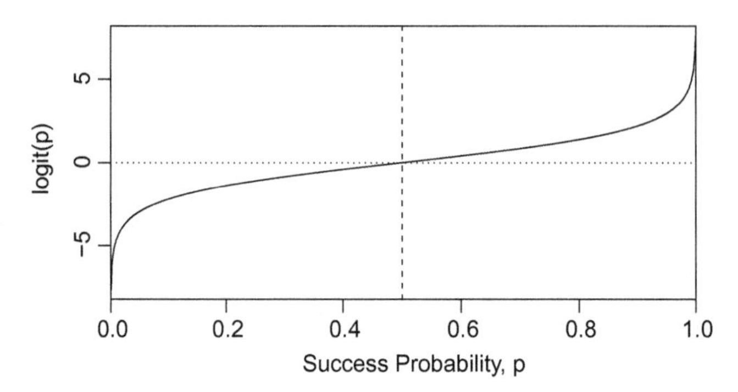

FIGURE 13.2
Graph of the function $\text{logit}(p) = \ln\{p/(1-p)\}$ as a function of p.

13.1.1 Estimating the Probability of Successfully Throwing a Piece of Wadded-up Paper into a Trash Can

Have you ever tossed a piece of crumpled-up paper towards a trash can and missed? What affects the chance that you will miss? Clearly, your distance from the trash can matters. The size of the trash can, the tightness of the wadding of the paper, and your aim will also affect your probability of success.

Table 13.1 displays data on distance versus success. In the experiment, a student was asked to stand 3 feet, 6 feet, 12 feet, …, and 21 feet away, and take three shots at the trash can from each distance.

You can enter and graph these data using R as follows:

```
dist    = rep(c(3,6,9,12,15,18,21), each=3)
success = c("Y", "Y", "Y",  "N", "Y", "Y",  "Y", "Y", "Y",  "N", "N", "N",
            "Y", "N", "N",  "N", "N", "N",  "N", "Y", "N")

Y = ifelse(success=="Y",1,0)
par(mar=c(4,4,2,2))
par(mfrow=c(2,1))
plot(dist,Y)
set.seed(12345)
plot(jitter(dist,.3), jitter(Y,.2))
```

TABLE 13.1
Data Set Indicating Success (Y or N) at Tossing a Wadded Piece of Paper into a Trash Can

Distance	Success
3	Y, Y, Y
6	N, Y, Y
9	Y, Y, Y
12	N, N, N
15	Y, N, N
18	N, N, N
21	N, Y, N

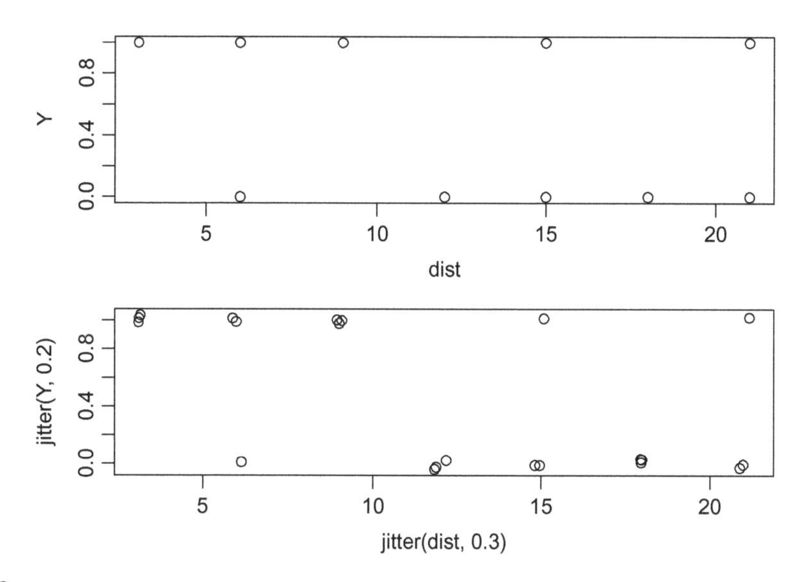

FIGURE 13.3
Plot of success as a function of distance. Top panel: Non-jittered data. Bottom panel: Jittered data. The "jitter" function is essential to view all distinct $n = 21$ data points.

There is a clear and logical trend shown in Figure 13.3: As distance increases, successes (1's) become less frequent. But the raw data are nearly useless for estimating the probability of success: For example, when distance = 12, the raw data predict 0/3 = 0% chance of success. This is clearly wrong, as there is certainly a chance to be successful at 12 feet. After all, there were two successes at longer distances.

Because regression models $p(y \mid x)$ "morph" continuously as a function of $X = x$, the regression-based estimates (whether from ordinary regression from previous chapters, or from logistic regression that we are discussing now) "borrow strength" from nearby observations to estimate means for a particular x value. So, instead of just using the three observations at distance = 12, the regression-based estimates use all 21 observations to estimate the probability of success, not only at $X = 12$ feet, and also at all other $X = x$ distances.

And, as in other regression methods discussed in this book, such "borrowing strength" occurs via the following steps:

1. Specify a theoretical model $p(y \mid x, \boldsymbol{\theta})$ that continuously morphs as a function of x.
2. Estimate the parameter(s) $\boldsymbol{\theta}$ via maximum likelihood.
3. Use the resulting estimated parameters ($\hat{\boldsymbol{\theta}}$) inside the theoretical model $p(y \mid x, \boldsymbol{\theta})$ to make predictions.

In the case of a binary Y, such as with the "trashball" experiment, these three steps become:

1. The theoretical model $p(y \mid x, \boldsymbol{\theta})$ is the Bernoulli distribution shown above, where the success probability is defined by the logistic regression function (note that the parameter vector is $\boldsymbol{\theta} = (\beta_0, \beta_1)$).
2. The parameters $\boldsymbol{\theta} = (\beta_0, \beta_1)$ are chosen by maximizing the likelihood function, obtained using the specific "trashball" data set as shown in Table 13.2 below.
3. Use the estimated parameters to predict success at different values of x, e.g., when distance = 12 feet.

TABLE 13.2

Contributions to the Likelihood Function in Terms of the Parameters of the Logistic Regression Model

OBS	X	Y	Contribution to Likelihood Function
1	3	1	$\exp\{\beta_0+\beta_1(3)\}/\left[1+\exp\{\beta_0+\beta_1(3)\}\right]$
2	3	1	$\exp\{\beta_0+\beta_1(3)\}/\left[1+\exp\{\beta_0+\beta_1(3)\}\right]$
3	3	1	$\exp\{\beta_0+\beta_1(3)\}/\left[1+\exp\{\beta_0+\beta_1(3)\}\right]$
4	6	0	$1/\left[1+\exp\{\beta_0+\beta_1(6)\}\right]$
…	…	…	…
21	21	0	$1/\left[1+\exp\{\beta_0+\beta_1(21)\}\right]$

Notice in Table 13.2 that when there is a success $(Y=1)$, then contribution to the likelihood is the logistic regression probability of success, and when there is a failure $(Y=0)$, then the contribution to the likelihood function is the logistic regression probability of a failure.

Self-study question: In Table 13.2, why is $1/\left[1+\exp\{\beta_0+\beta_1(6)\}\right]$ the probability of a failure? (Hint: Find and simplify the quantity [one minus the probability of a success].)

Thus, the likelihood function is

$$\begin{aligned} L(\beta_0,\beta_1\mid\text{data}) = {} & \exp\{\beta_0+\beta_1(3)\}/\left[1+\exp\{\beta_0+\beta_1(3)\}\right]\times \\ & \exp\{\beta_0+\beta_1(3)\}/\left[1+\exp\{\beta_0+\beta_1(3)\}\right]\times \\ & \exp\{\beta_0+\beta_1(3)\}/\left[1+\exp\{\beta_0+\beta_1(3)\}\right]\times \\ & 1/\left[1+\exp\{\beta_0+\beta_1(6)\}\right]\times\ldots\times \\ & 1/\left[1+\exp\{\beta_0+\beta_1(21)\}\right] \end{aligned}$$

The ML estimates of (β_0,β_1) are the values $(\hat{\beta}_0,\hat{\beta}_1)$ that maximize this function.

It is easy to maximize the likelihood function $L(\beta_0,\beta_1\mid\text{data})$ "by hand" using the `maxLik` function of `R` shown in Chapter 12, but it is even easier to do it using the "`glm`" function of `R`. Here is the code:

```
fit <- glm(Y ~ dist, family = "binomial")
summary(fit)
```

The following is the output:

```
Call:
glm(formula = Y ~ dist, family = "binomial")
```

```
Deviance Residuals:
    Min       1Q    Median       3Q      Max
-1.7683  -0.8338  -0.4271   0.6855   2.2091

Coefficients:
            Estimate Std. Error z value Pr(>|z|)
(Intercept)   2.7995     1.3528   2.069   0.0385 *
dist         -0.2452     0.1067  -2.297   0.0216 *
---
Signif. codes:  0 '***' 0.001 '**' 0.01 '*' 0.05 '.' 0.1 ' ' 1

(Dispersion parameter for binomial family taken to be 1)

    Null deviance: 29.065  on 20  degrees of freedom
Residual deviance: 21.347  on 19  degrees of freedom
AIC: 25.347

Number of Fisher Scoring iterations: 4
```

Thus, an example of the third step of the "recipe" outlined above, "Use the resulting estimated parameters ($\hat{\theta}$) inside the theoretical model $p(y \mid x, \theta)$ to make predictions," is given as follows.

The estimated success probability is

$$\hat{\pi}(x) = \frac{\exp(\hat{\beta}_0 + \hat{\beta}_1 x)}{1 + \exp(\hat{\beta}_0 + \hat{\beta}_1 x)} = \frac{\exp(2.7995 - 0.2452x)}{1 + \exp(2.7995 - 0.2452x)}$$

In particular, when $X = 12$, the estimated success probability is

$$\hat{\pi}(12) = \frac{\exp\{2.7995 - 0.2452(12)\}}{1 + \exp\{2.7995 - 0.2452(12)\}} = 0.464 = 46.4\%$$

rather than $0/3 = 0.0\%$, which is what you would get if you just used the raw data at $X = 12$ to estimate the probability of success at $X = 12$. Again, regression (of any type, whether ordinary or logistic) is beneficial because of the way it "borrows strength" from all the data to estimate conditional means and conditional probabilities regarding your Y variable at a particular value $X = x$.

Self-study question: Using the estimates given above, what is the estimate of the probability distribution $p(y \mid X = 12)$? How does this distribution look when graphed? How does this distribution differ from the distribution $p(y \mid X = 12)$ that is postulated by the classical regression model?

The jittered scatter of the data, along with the overlaid logistic regression curve, is shown in Figure 13.4.

R code for Figure 13.4

```
par(mfrow=c(1,1))
b = fit$coefficients; b0 = b[1]; b1 = b[2]
dist.plot = seq(0,24,.01)
prob.est = exp(b0+b1*dist.plot)/(1+exp(b0+b1*dist.plot))
plot(dist.plot, prob.est, type="l", ylim=c(0,1), xlab="Distance",
ylab="Probability")
set.seed(12345)
points(jitter(dist, .3), jitter(Y,.2))
abline(v=12, lty=2); abline(h=0.464, lty=2)
```

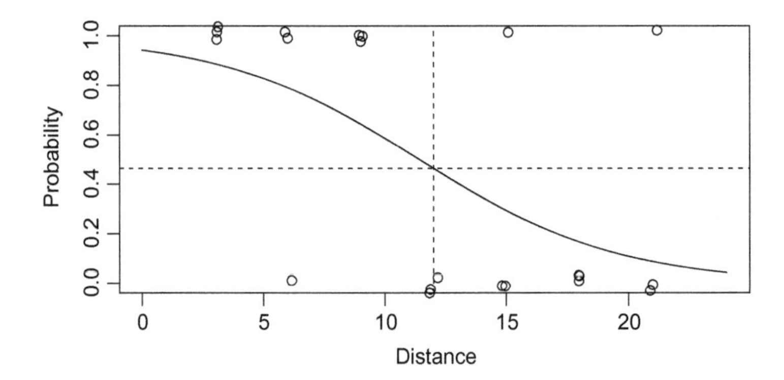

FIGURE 13.4
Jittered scatterplot of data with fitted probability function overlaid. The dashed lines show the estimated success probability, 0.464, at Distance = 12.

You can interpret the parameter estimates, $\hat{\beta}_0 = 2.7995$ and $\hat{\beta}_1 = -0.2452$, conveniently in terms of the logit probability function:

$$\text{logit}\left\{\hat{\pi}(x)\right\} = 2.7995 - 0.2452x$$

Thus,

- The estimated logit of the probability of success when $X = 0$ is 2.7995. From Figure 13.2, this corresponds to a large probability of success in this case, which is reasonable because you should be able to throw the paper into the trash successfully when you are only 0 feet away from the trash can.
- The increase in the estimated logit probability of success per one foot of additional distance is (−0.2452). To understand this number more clearly in terms of probability, let's consider two cases: (i) an increase from 0 feet to 1 foot, in which case the logit decreases by 0.2452 from 2.7995 to 2.5543, and (ii) an increase from 12 feet to 13 feet, in which case the logit decreases again by 0.2452, but from −0.1429 to −0.3881. Again, look at Figure 13.2. Case (i) corresponds to a very small reduction

in the probability (plugging in, we get a decrease from 0.9426 to 0.9279, a 0.0148 difference), while case (ii) corresponds to a larger reduction in the probability (plugging in, we get a decrease from 0.4643 to 0.4042, a 0.0601 difference).

The second bullet brings up an important issue: The logit function "dilates" small differences in probabilities when the probability is near 0 or 1. This "dilation" is sensible. For example, consider the probability of winning the lottery. If the probability were 1 in 100 instead of 1 in 100 million, you would think that is a big difference, right? On the logit scale, that difference is $\ln(0.01/0.99) - \ln(0.00000001/0.99999999) = 13.$. See Figure 13.2 again: The difference 13.8 on the logit scale is huge, despite the fact that the probabilities themselves only differ by about 0.01. The logit function naturally performs an intuitively beneficial service by amplifying small differences in the probabilities when they are close to 0.00 or 1.00.

Gamblers often express probabilities in terms of odds: If the bet has equal probability of winning and losing, then the odds are 1:1, indicating "1 way to win, 1 way to lose." If the bet has a 25% chance of winning and a 75% chance of losing, then the odds are 1:3, indicating "1 way to win, 3 ways to lose." If the bet has a 75% chance of winning and a 25% chance of losing, then the odds are 3:1.

Expressing the a:b odds as the ratio a / b, you can relate the probability to odds via

$$\text{Odds} = \text{Probability} / (1 - \text{Probability})$$

Thus, if the probability is 0.5, then $\text{Odds} = 0.5/(1-0.5) = 1.0$; for probabilities 0.25 and 0.75, the odds are 0.3333 and 3.000, respectively. Clearly, odds is *not* a probability because the numbers can be greater than 1.0.

The logistic regression model is a model for log odds: Recall that the model assumes

$$\text{logit}\{\pi(x)\} = \beta_0 + \beta_1 x$$

Hence, by definition of the logit function, the model assumes

$$\ln\left[\pi(x)/\{1-\pi(x)\}\right] = \beta_0 + \beta_1 x$$

or that

$$\ln\{\text{odds}(x)\} = \beta_0 + \beta_1 x$$

An equivalent representation of the logistic regression model is thus

$$\text{odds}(x) = \exp(\beta_0 + \beta_1 x)$$

This re-expression of the logistic regression model gives you another way to interpret the parameter β_1. Consider the potentially observable data in two groups:

Group 1: $X = x$. Here, $\text{odds}(x) = \exp(\beta_0 + \beta_1 x)$.

Group 2: $X = x + 1$. Here,

$$\begin{aligned} \text{odds}(x+1) &= \exp(\beta_0 + \beta_1(x+1)) = \exp(\beta_0 + \beta_1 x + \beta_1) = \exp(\beta_0 + \beta_1 x)\exp(\beta_1) \\ &= \text{odds}(x)\exp(\beta_1) \end{aligned}$$

Thus, $\exp(\beta_1)$ is the *proportional increase* in the odds of success associated with a one unit increase in X. In the "trashball" example, the estimate of this proportional increase in the odds of success at trashball associated with an additional foot of distance is

$$\exp(-0.2452) = 0.7825$$

In other words, the odds of success for any $X = x+1$ is estimated to be 0.7825 times the odds of success at $X = x$. Since 0.7825 is less than 1.0, the odds of success are estimated to decrease for larger X.

You can convert this number to a percentage via

$$\text{Percent increase in odds} = 100\{\exp(\beta_1) - 1\}$$

Hence, for each one additional foot of distance in the trashball example, the odds of success are estimated to increase by $100(0.7825 - 1) = -21.75\%$. Since a negative increase is a decrease, a more direct interpretation is that the odds of success are estimated to decrease by 21.75% for each additional foot.

A major assumption of the logistic regression model is that it constrains the logit to be a linear function of the $X = x$ value. So, all of the previous caveats about linearity (and more general, correct functional specification) apply. You can assess curvature effects in the same way as before: (i) fit a quadratic model and compare, or (if X is ordinal) (ii) fit the indicator model in terms of the X variable and compare. While the graphic comparison is the best way to compare the linear and curved fits, you can also supplement the graph with the full model/restricted model likelihood ratio test as discussed in Chapter 12 to assess whether deviations from linearity shown in the graph are explainable by chance alone.

Using the "trashball" data above, you can assess the degree of curvature in the true logit function in exactly the same way as or ordinary regression by using a quadratic model.

```
dist2 = dist^2
fit2 <- glm(Y ~ dist + dist2, family = "binomial")
summary(fit2)

c = fit2$coefficients; c0 = c[1]; c1 = c[2]; c2 = c[3]
dist.plot = seq(0,24,.01)
logit.lin  = b0 + b1*dist.plot
logit.quad = c0 + c1*dist.plot + c2*dist.plot^2

par(mfrow=c(1,2))
plot(dist.plot, logit.lin, type="l", ylab="logit", xlab="Distance")
points(dist.plot, logit.quad, type="l", lty=2)
legend(15,2.5, c("Linear", "Quad"), lty = c(1,2))

prob1.est = exp(logit.lin)/(1+exp(logit.lin))
prob2.est = exp(logit.quad)/(1+exp(logit.quad))
plot(dist.plot, prob1.est, type="l", ylim = c(0,1), xlab="Distance",
ylab="Probability")

points(dist.plot, prob2.est, type="l", lty=2)
legend(15,.8, c("Linear", "Quad"), lty = c(1,2))
points(jitter(dist), Y)
```

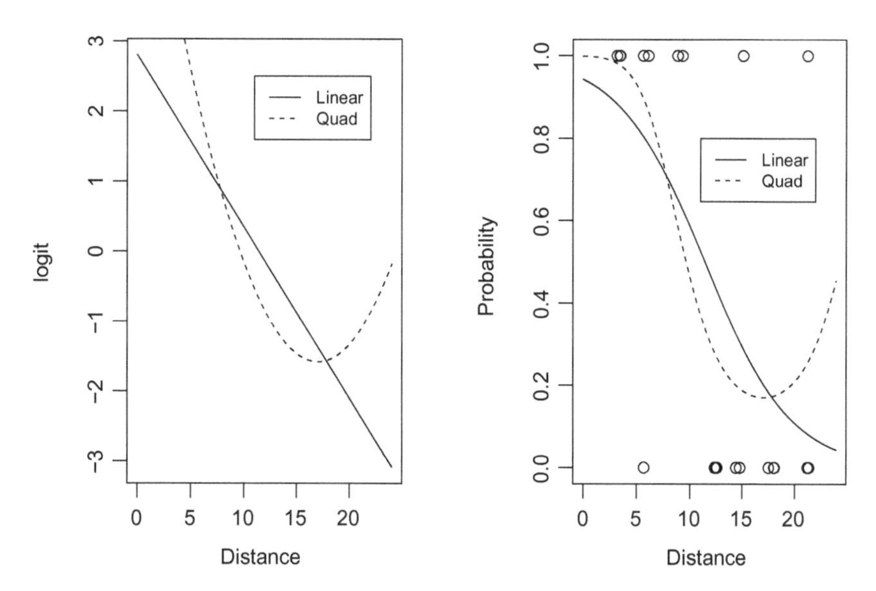

FIGURE 13.5
Left panel: Linear and quadratic fits to the logit probability of success. Right panel: Left panel fits converted to probability estimates.

Notice that the quadratic fit in Figure 13.5 has good and bad properties: On the good side, it gives much higher probability estimates for distances close to zero. This makes sense. When you are 1 foot away it is very hard to miss. And when you are zero feet away, the only way you can miss is if you suddenly drop dead, fall over backward, and drop the crumpled-up paper onto the ground, missing the trash can! Don't you hate it when that happens?

On the bad side, the quadratic model predicts an increasing probability of success for larger distances. That is a negative feature of quadratic functions. You can use different functions to assess curvature other than the quadratic, but as long as you understand that the quadratic function is just a tool, and that all tools have limitations, then you can use it to assess curvilinearity. As discussed in Chapter 4, the quadratic model is a useful tool to diagnose curvature, but if curvature is *detected* using a quadratic model, you still do not necessarily wish to *use* the quadratic model. For example, in the model used to predict car sales as a function of interest rates, curvature was detected using the quadratic model, but ultimately, we suggested modeling the curvature using the inverse of interest rates rather than a quadratic function of interest rates, because of the same extrapolation concerns that we have here with the trashball example.

A main benefit (here and elsewhere) of using the quadratic in this way as a tool to diagnose curvature is that the linear model is just a special case of quadratic. In other words, if you assume quadratic, then you are also assuming, *a fortiori*, that the model might be linear because linear models are in the family of quadratic models. Because of this nesting of linear functions within the family of quadratic functions, you can test for curvature using the full model/restricted model approach. To test whether the deviation from linearity shown in the graphs above is explainable by the linear model, you can use the likelihood ratio test (LRT) introduced in Chapter 12. In the case of `glm` fits in R, you can still use the anova function in R as anova(fit,fit2, test="LRT"). This will give you the following output:

```
Analysis of Deviance Table

Model 1: Y ~ dist
Model 2: Y ~ dist + dist2
      Resid. Df Resid. Dev Df Deviance Pr(>Chi)
      1        19      21.347
      2        18      19.547  1    1.8001    0.1797
```

The "Resid.Dev" column contains the "deviance" values, defined as −2×(maximized log likelihood). Thus, the difference between the deviances is equal to the likelihood ratio chi-square statistic. The difference between the log-likelihoods for the linear and quadratic logit models is explainable by chance alone, assuming a data-generating process where the logit is truly a linear function of distance ($\chi_1^2 = 1.8001$, $p = 0.1797$). Of course, it is also explainable by a data-generating process where the logit is a curved function of distance, which is the reason that linearity is not proven by this (or any other) test. No test, LRT or otherwise, can prove that the linear model is valid. In fact, you know, *a priori*, that regression functions are almost always curved when there are more than two values of the X variable. With larger sample sizes, you are better able to estimate the true, curved functional form of the relationship using flexible models such as LOESS functions and neural network functions (see Chapter 17).

Much of what you have learned in previous chapters, including LOESS, indicator variables, confounding, multicollinearity, etc., apply equally to logistic regression. Just view the logistic regression model like you do an ordinary regression model, except with a transformed (logit transformed, specifically) mean function. Also, realize that model fit is now expressed in terms of likelihood, rather than error sums of squares.

There are many R^2 statistics for logistic regression, but which one to pick in the case of logistic and other non-standard regressions is debatable. In Chapter 8, we defined a "population" R^2 by $\Omega^2 = \mathrm{Var}\{E(Y \mid X)\} / \mathrm{Var}(Y)$. In logistic regression, the numbers $E(Y \mid X)$ are the numbers on the probability curve, whose estimates are called "fitted.values" in the glm fit object. So, you can estimate Ω^2 by using var(fit$fitted.values)/var(Y). The linear "trashball" logistic regression model gives the estimate 0.3305 and the quadratic model gives 0.3916. One problem with R^2 statistics when you have discrete data is that, from certain points of view, the discreteness biases the estimated R^2 towards zero. Alternative definitions of the R^2 statistic in logistic regression (and in discrete regression models generally) attempt to remove such bias.

As far as assumptions go, other than linearity of the logits, the only other assumption needed for logistic regression is conditional independence. The independence assumption is needed to construct the likelihood function as the product of likelihoods as shown above; it is also needed for obtaining the resulting standard errors and likelihood-based fit statistics. Heteroscedasticity is automatically allowed, correctly, because the variance of the Bernoulli distribution is $\pi(1-\pi)$. Thus, when π is close to 0 or 1, the conditional variance of Y is smaller than when π is near 0.5. Obviously, non-normality is automatically allowed as well, since the distribution of Y is assumed to be Bernoulli, not normal.

13.2 The Multinomial Regression for Nominal *Y*

When your *Y* data are nominal, the choice of distribution $p(y)$ is easy: It is the multinomial distribution.

The multinomial distribution

y	$p(y)$
C_1	π_1
C_2	π_2
…	…
C_{c-1}	π_{c-1}
C_c	π_c
Total	1.0

We use the symbol "C" to underscore the fact that the possible values of your *Y* variable are "categories," often "choices," rather than ordinal numbers.

Examples where *Y* has the multinomial distribution include the following:

- Y = choice of brand of toothpaste by a shopper. Then C_1 = "Crest", C_2 = "Colgate," etc. There are lots of brands, so it would make sense to pick those with sizeable market share and lump the rest into an "other" category.
- Y = species of fish caught by a person fishing at a Texas lake. Then C_1 = "catfish," C_2 = "largemouth bass," C_3 = "crappie," C_4 = "striped bass," C_5 = "perch," etc.
- Y = disease of a patient: C_1 = "acute myeloid lymphoma," C_2 = "acute lymphoid lymphoma," or C_3 = "other lymphoma."
- Y = laundry duty for a married couple: C_1 = "mainly wife," C_2 = "mainly husband," or C_3 = "shared."

Self-study question: What "X" variables for each of the above cases will modify the distribution of *Y*? How will π_1 change, depending on an *X* value? How will π_2 change?

To make the multinomial model a regression model, you need to specify that the probabilities π_j are functions of $X = x$. Since the c probabilities add to 1.0, you only need functions $\Pr\left(Y = C_j \mid X = x\right) = \pi_j\left(x\right)$ for $\pi_1(x), \pi_2(x), \ldots, \pi_{c-1}(x)$, since $\pi_c(x) = 1 - \{\pi_1(x) + \pi_2(x) + \ldots + \pi_{c-1}(x)\}$. Then the general multinomial regression model has the following form:

The general multinomial regression model

y	$p(y\|x)$
C_1	$\pi_1(x)$
C_2	$\pi_2(x)$
…	…
C_{c-1}	$\pi_{c-1}(x)$
C_c	$1-\{\pi_1(x)+\pi_2(x)+\ldots+\pi_{c-1}(x)\}$
Total	1.0

When there are only $c = 2$ categories, multinomial regression is the same as binary regression, where $C_1 = 1$ and $C_2 = 0$. A big difference with multinomial regression is that, with more than $c = 2$ categories, there is no mean function $\text{E}(Y \mid X = x)$. What does it mean to average a perch, a catfish, and a bass? They are categories, not numbers. So, a major distinguishing characteristic of multinomial regression, different from every other regression model we have discussed so far, is that there is no mean function. Everything is now concerned with estimating and making predictions from $p(y \mid x)$, the distribution of possible values of Y for a given $X = x$.

Now is it becoming even clearer that regression is a model for $p(y \mid x)$? There is simply no other way to view regression for multinomial data. There is no mean function, linear or otherwise. There is only the collection of conditional distributions, $p(y \mid x)$.

So that you can use the data to "borrow strength" in the estimates of the specific distributions, you need a function form to string the multinomial distributions $p(y \mid x)$ together as a function of $X = x$. As with other regression models, it makes sense to require these distributions to "morph" continuously as a function of $X = x$, because Nature generally favors continuity over discontinuity.

The multinomial logistic regression (MLR) model assumes a particular probability function $p(y \mid x)$, one that generalizes the logistic regression function from $c = 2$ categories to $c > 2$ categories. Notice again the word "assume," which means you are assuming something that is patently false. But, as always, assumptions are made for convenience and efficiency, and as always, the assumptions are reasonable to make as long as they are not violated too badly. There are many other choices for a probability function that you can use other than the default MLR function, and you can check such choices for conformance with your data by using penalized likelihoods. But none of these choices will be Nature's function. The standard MLR model is the most commonly used regression model for nominal data, and is the only one we will discuss. Be aware that there are many other possible regression models for nominal data, and some may fit the data better than the classic MLR model. In the same way, you have seen that alternative regression models (such as heteroscedastic) often fit the data better than the classical regression model.

To get the multinomial function $p(y \mid x)$, recall the logit (log odds) form of the logistic regression model:

$$\text{logit}\{\pi(x)\} = \ln\left\{\frac{\Pr(Y = 1 \mid X = x)}{\Pr(Y = 0 \mid X = x)}\right\} = \beta_0 + \beta_1 x$$

With MLR, this form is generalized by choosing a "baseline" category, and then modeling the logits for all remaining categories as follows:

$$\ln\left\{\frac{\Pr(Y = C_1 \mid X = x)}{\Pr(Y = C_c \mid X = x)}\right\} = \beta_{01} + \beta_{11}x$$

$$\ln\left\{\frac{\Pr(Y = C_2 \mid X = x)}{\Pr(Y = C_c \mid X = x)}\right\} = \beta_{02} + \beta_{12}x$$

...

$$\ln\left\{\frac{\Pr(Y = C_{c-1} \mid X = x)}{\Pr(Y = C_c \mid X = x)}\right\} = \beta_{0,c-1} + \beta_{1,c-1}x$$

Notice that there are $c-1$ slope β's for *just one* X variable. If there are k "X" variables, then there are $k(c-1)$ slope β's in the MLR model.

These equations imply that the individual probabilities $\Pr(Y = C_j \mid X = x), j = 1,\ldots,c-1$, are functions of the baseline probability $\Pr(Y = C_c \mid X = x)$. Applying the constraint that these c probabilities add to 1.0 allows us to solve for all the probabilities as follows:

$$\Pr(Y = C_1 \mid X = x) = \frac{\exp(\beta_{01} + \beta_{11}x)}{1 + \exp(\beta_{01} + \beta_{11}x) + \exp(\beta_{02} + \beta_{12}x) + \cdots + \exp(\beta_{0,c-1} + \beta_{1,c-1}x)}$$

$$\Pr(Y = C_2 \mid X = x) = \frac{\exp(\beta_{02} + \beta_{12}x)}{1 + \exp(\beta_{01} + \beta_{11}x) + \exp(\beta_{02} + \beta_{12}x) + \cdots + \exp(\beta_{0,c-1} + \beta_{1,c-1}x)}$$

...

$$\Pr(Y = C_{c-1} \mid X = x) = \frac{\exp(\beta_{0,c-1} + \beta_{1,c-1}x)}{1 + \exp(\beta_{01} + \beta_{11}x) + \exp(\beta_{02} + \beta_{12}x) + \cdots + \exp(\beta_{0,c-1} + \beta_{1,c-1}x)}$$

$$\Pr(Y = C_c \mid X = x) = \frac{1}{1 + \exp(\beta_{01} + \beta_{11}x) + \exp(\beta_{02} + \beta_{12}x) + \cdots + \exp(\beta_{0,c-1} + \beta_{1,c-1}x)}$$

Self-study question: How do the equations above follow from the linear equations for the logits?

Self-study question: Why is the MLR model with $c = 2$ identical to the logistic regression model?

Notice that the MLR model probabilities sum to 1.0 because the denominator is the same for each of the c probabilities, and because the numerators sum to that denominator. Notice also that these probabilities are all continuous functions of x, so that these probabilities all continuously "morph" as x changes, as is appropriate for a model that describes how Nature works.

13.2.1 Who Does the Laundry?

In a (fictitious) survey of $n = 1{,}643$ married couples, the couples were asked (1) Who usually does the laundry in your household? The possible responses were "Husband," "Wife," or "Alternating," and this is the Y variable for the study. Clearly, Y is nominal.

One predictor variable is X_1 = husband's annual income, recorded as either 1, 2, 3, 4, or 5, where 1 = "less than \$10,000," 2 = "\$10,001 to \$20,000," 3 = "\$20,001 to \$35,000," 4 = "\$35,001 to \$60,000," and 5 = "more than \$60,000." The other predictor variable is X_2 = wife's annual income, recorded in the same way. The goal is to assess how the couple's income profile affects choice of laundry duties.

To estimate the parameters of the MLR model, you can use (guess what?) Maximum Likelihood. You should view these data in the standard way, as produced by the regression model $p(y \mid x)$, with probability distributions defined above, as shown in Table 13.3.

For example, in the first line, assuming the categories are C_1 = "Husband," C_2 = "Wife," and C_3 = "Alternating," we have that

$$\Pr(Y = \text{Wife} \mid X_1 = 3, X_2 = 1, \beta) = \frac{\exp\{\beta_{02} + \beta_{12}(3) + \beta_{22}(1)\}}{1 + \exp\{\beta_{01} + \beta_{11}(3) + \beta_{21}(1)\} + \exp\{\beta_{02} + \beta_{12}(3) + \beta_{22}(1)\}}.$$

Further, assuming conditional independence of the potentially observable entries in the Y column, you get the likelihood function for the parameter vector $\boldsymbol{\beta}$ by multiplying all entries in the rightmost column. From this likelihood function, you get the estimated β's, the Wald standard errors, the likelihood ratio tests, and the likelihood-based measures of fit such as AIC and BIC.

Just like when the X data are nominal (Chapter 10), when the Y data are nominal, you should first display the table of possible Y values to understand your data better. The results are as follows:

```
Laundry
    Alt Husband    Wife
    323      418     902
```

From the output of the "`table`" function, notice that there are 323 couples with "Alternating" laundry duties, 418 couples where the Husband primarily does the laundry, and in the remaining 902 couples, the wife has primary laundry duties. This gives you the estimate of $p(y)$, which is the marginal distribution of Y, shown in Table 13.4.

TABLE 13.3

Contributions to the Likelihood Function in Terms of the Parameters of the Multinomial Regression Model

Obs	Y	X_1	X_2	Contribution to Likelihood Function
1	Wife	3	1	$\Pr(Y = \text{Wife} \mid X_1 = 3, X_2 = 1, \beta)$
2	Wife	5	1	$\Pr(Y = \text{Wife} \mid X_1 = 5, X_2 = 1, \beta)$
3	Husband	1	1	$\Pr(Y = \text{Husband} \mid X_1 = 1, X_2 = 1, \beta)$
4	Husband	5	2	$\Pr(Y = \text{Husband} \mid X_1 = 5, X_2 = 2, \beta)$
…	…	…	…	…
1,643	Wife	3	1	$\Pr(Y = \text{Wife} \mid X_1 = 3, X_2 = 1, \beta)$

TABLE 13.4

Estimated Marginal Distribution of Y = Laundry Person

y	$\hat{p}(y)$
Alternating	323 / 1643 = 19.7%
Husband	418 / 1643 = 25.4%
Wife	902 / 1643 = 54.9%
	Total: 100.0%

As we have illustrated repeatedly throughout this book, regression is a model that shows how the distribution of Y changes for different given values of X. The same is true here: The goal of the multinomial logistic regression is to estimate how the distribution of laundry duty changes for different given values of husband's income and wife's income. What do you think will happen to the percentages listed in Table 13.4 when the husband's income is high? When the wife's income is high? Since you have not looked at any analysis yet, your answers to these questions are merely hypotheses. Data analysis via the multinomial logistic regression model, as shown by the following `R` code will allow you to answer such questions more definitively.

```
library(nnet)
fit.mnlr = multinom(Laundry ~ H.Inc + W.Inc)
summary(fit.mnlr)
```

The results are given as follows:

```
Call:
multinom(formula = Laundry ~ H.Inc + W.Inc)

Coefficients:
        (Intercept)       H.Inc         W.Inc
Husband    1.986346 -0.55833030  0.0005732999
Wife       3.232370  0.05599298 -0.7869644576

Std. Errors:
        (Intercept)      H.Inc      W.Inc
Husband   0.3008948 0.06306935 0.06241707
Wife      0.2754433 0.05851732 0.05597366

Residual Deviance: 2748.182
AIC: 2760.182
```

Notice in the output of the `multinom` function that there is no mention of the "Alt," or "Alternating Laundry Duty" category. Thus, the software automatically chose the "Alt" category as the "baseline" category, and you can write down the estimated model equations, using equations above for the log probability ratios, as follows:

$$\ln\left\{\frac{\hat{\Pr}(Y = \text{Husband} \mid X_1 = x_1, X_2 = x_2)}{\hat{\Pr}(Y = \text{Alt} \mid X_1 = x_1, X_2 = x_2)}\right\} = 1.986 - 0.558x_1 + 0.00057x_2$$

and

$$\ln\left\{\frac{\hat{P}r(Y = \text{Wife} \mid X_1 = x_1, X_2 = x_2)}{\hat{P}r(Y = \text{Alt} \mid X_1 = x_1, X_2 = x_2)}\right\} = 3.232 + 0.0560x_1 - 0.787x_2$$

Interpreting the first equation, larger husband income corresponds to smaller probability of husband doing laundry, relative to probability of alternating laundry duty, for fixed wife income. The wife's income seems to affect the relative probability ratio very little.

Interpreting the second equation, larger wife income corresponds to smaller probability of wife doing laundry, relative to probability of alternating laundry duty, for fixed husband income. The husband's income seems to affect that relative ratio very little.

Of course, the interpretations above are based on estimates from idiosyncratic data. The true coefficients of the data-generating process can be assumed to lie within the ±2(standard error) range. For example, using the "`Std. Errors:`" in the output from the `multinom` function, the true β corresponding to the coefficient 0.0560 in the second equation is most likely in the range $0.0560 \pm 2(0.0585)$, or in the range from −0.061 to 0.173. Thus, it is not clear from the data whether the husband effect in the second equation is negative or positive.

These model equations are tricky to interpret because they refer to effects of X variables on probability ratios, relative to a baseline category that is chosen by the software. To analyze the data in a more direct and relevant way, we suggest that you estimate and display, either in tabular or graphical form, the conditional distributions $p(y \mid x)$ for particular $X = x$ values, and then base your interpretations on those graphs and tables.

First, recall that the estimated marginal distribution, from the raw data only, was presented above in Table 13.4. As shown above, for particular values x_1 and x_2 of Husband's and Wife's income, respectively, we have the following equations for the *actual* probabilities:

$$\Pr(Y = \text{"Alt"} \mid x_1, x_2) = \frac{1}{1 + \exp(\beta_{01} + \beta_{11}x_1 + \beta_{21}x_2) + \exp(\beta_{02} + \beta_{12}x_1 + \beta_{22}x_2)}$$

$$\Pr(Y = \text{"Husband"} \mid x_1, x_2) = \frac{\exp(\beta_{01} + \beta_{11}x_1 + \beta_{21}x_2)}{1 + \exp(\beta_{01} + \beta_{11}x_1 + \beta_{21}x_2) + \exp(\beta_{02} + \beta_{12}x_1 + \beta_{22}x_2)}$$

and

$$\Pr(Y = \text{"Wife"} \mid x_1, x_2) = \frac{\exp(\beta_{02} + \beta_{12}x_1 + \beta_{22}x_2)}{1 + \exp(\beta_{01} + \beta_{11}x_1 + \beta_{21}x_2) + \exp(\beta_{02} + \beta_{12}x_1 + \beta_{22}x_2)}$$

The marginal estimated distribution shown in Table 13.4 is kind of an average across all possible Husband's and Wife's incomes, so it should be reasonably close (but not equal) to the conditional probabilities as estimated using the averages of the X values. The following code shows how you can get such predicted values.

```
mean(H.Inc); mean(W.Inc)
X.new = data.frame(mean(H.Inc), mean(W.Inc))
colnames(X.new) = c("H.Inc", "W.Inc")
predict(fit.mnlr, X.new, "probs")
```

TABLE 13.5

Estimated Conditional Distribution of Y = Laundry Person, Given Husband's Income = 3.30 and Wife's Income = 2.96

y	$\hat{p}(y \mid X_1 = 3.30, X_2 = 2.96)$
Alternating	19.5%
Husband	22.6%
Wife	58.0%
	Total: 100.1%

Note: Total adds to 100.1% because of rounding error.

The results are

```
      Alt   Husband       Wife
0.1947578 0.2256471 0.5795951
```

The means of the Husband and Wife's incomes are 3.30 and 2.96, so you can arrange these estimated probabilities in the table to show the conditional distribution in Table 13.5.

It is comforting that this conditional distribution is similar to the marginal distribution shown in Table 13.4. To see the effects of different Husband's and Wife's incomes (X_1 and X_2) on laundry duty, you can consider four cases: (i) X_1 low, X_2 low, (ii) X_1 low, X_2 high, (iii) X_1 high, X_2 low, and (iv) X_1 high, X_2 high. While you should usually not take "low" and "high" values to be the mins and maxs because such values are usually atypical outliers, in this example it makes sense because a large percentage of the incomes are 1's and 5's. The following code gives you these four estimated distributions:

```
X1.new = c(1,1,5,5)
X2.new = c(1,5,1,5)
X.newer = cbind(X1.new, X2.new)
colnames(X.newer) = c("H.Inc", "W.Inc")
predict(fit.mnlr, X.newer, "probs")
```

The results are as follows:

```
         Alt    Husband       Wife
1 0.05756273 0.24019815 0.70223912
2 0.17524539 0.73294398 0.09181062
3 0.05984700 0.02676403 0.91338897
4 0.47536449 0.21307468 0.31156083
```

You can arrange these results into four tables similar to Table 13.6.

TABLE 13.6

Estimated Conditional Distribution of Y = Laundry Person, Given Husband's Income = 1.0 and Wife's Income = 1.0

y	$\hat{p}(y \mid X_1 = 1.0, X_2 = 1.0)$
Alternating	5.8%
Husband	24.0%
Wife	70.2%
	Total: 100.0%

You can also display the effects of the X variables on the distribution of Y using graphs, as shown by the following code.

R code for Figure 13.6

```
Husb.eff.data = data.frame(cbind(1:5, rep(mean(W.Inc),5)))
Wife.eff.data = data.frame(cbind(rep(mean(W.Inc),5), 1:5))
colnames(Husb.eff.data) = c("H.Inc", "W.Inc")
colnames(Wife.eff.data) = c("H.Inc", "W.Inc")
Husb.eff.probs = predict(fit.mnlr, Husb.eff.data, "probs")
Wife.eff.probs = predict(fit.mnlr, Wife.eff.data, "probs")
par(mfrow=c(1,2))
plot(1:5, Husb.eff.probs[,1], type="o", ylim = c(0,1),
ylab = "Probability", xlab="Husband Income")
points(1:5, Husb.eff.probs[,2], type="o", ylim = c(0,1), pch=2, lty=2)
points(1:5, Husb.eff.probs[,3], type="o", ylim = c(0,1), pch=3, lty=3)
legend("topleft", c("Alt", "Husb", "Wife"), lty=1:3, pch=1:3)
plot(1:5, Wife.eff.probs[,1], type="o", ylim = c(0,1), ylab="",
xlab="Wife Income")
points(1:5, Wife.eff.probs[,2], type="o", ylim = c(0,1), pch=2, lty=2)
points(1:5, Wife.eff.probs[,3], type="o", ylim = c(0,1), pch=3, lty=3)
legend("topright", c("Alt", "Husb", "Wife"), lty=1:3, pch=1:3)
```

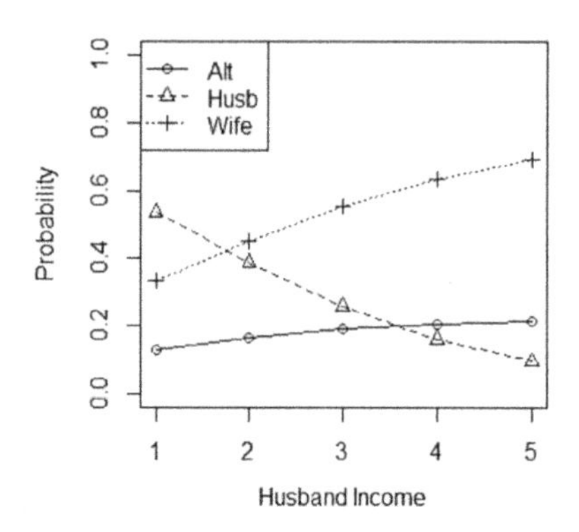

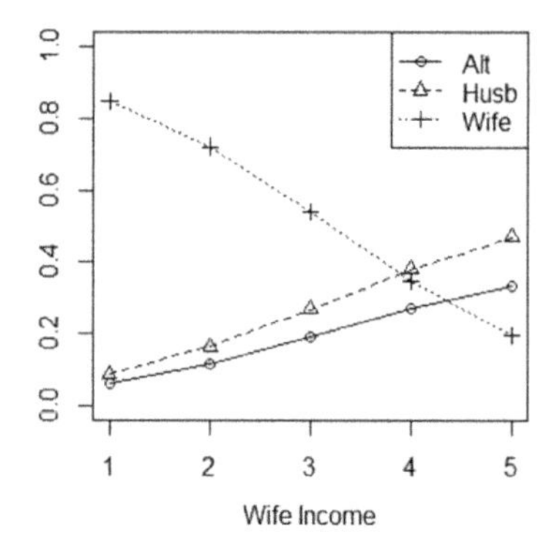

FIGURE 13.6
Estimated probabilities of laundry duty as a function of husband income, with wife income fixed at the mean (left panel); and as a function of wife income, with husband income fixed at the mean (right panel). Probabilities in vertical strips add to 1.0 and are the distributions $p(y \mid x_1, x_2)$ as estimated using the MLR model.

Notice in Figure 13.6 that the spouse with lower income tends to do laundry more often. When both spouses have high income, the laundry duty is more often shared than when the spouses have low income.

The choice of "Alt" as a baseline category may seem questionable. However, like in the case of indicator variable models presented in Chapter 10, the choice of baseline category affects parameter estimates but not predicted models.

There are many parameter estimates associated with a single X variable in the MLR model. Hence, the effect of a single X variable in MLR regression involves many β

parameters rather than just one β parameter. To test that all β's for a particular X variable are equal to 0, you can use a likelihood ratio test. For example, to assess whether the Husband's Income has any effect on the distribution of laundry duty, you can fit the restricted (null) model which excludes the husband income, and therefore restricts all of the husband income β parameters to zero, as follows:

```
fit.null = multinom(Laundry ~ W.Inc)
```

Then you can compare it to the "full" model `fit.mnlr` that you already fit by using `anova(fit.null, fit.mnlr)`, which gives you the likelihood ratio test. The results are:

```
          Model Resid. df Resid. Dev    Test    Df LR stat. Pr(Chi)
1         W.Inc      3282   2899.736            NA       NA      NA
2 H.Inc + W.Inc      3280   2748.182 1 vs 2      2 151.5541       0
```

You can write the interpretation as follows: In the test of whether Husband Income affects laundry duty distribution, $\chi_2^2 = 151.5541$ and $p \cong 0$. Hence, the difference between log likelihoods in the models with and without Husband Income is not easily explained by a data-generating process in which Husband Income has no effect on laundry duty. We conclude that Husband Income truly affects the laundry duty distributions; e.g., that the differences in the estimated differences seen in Figure 13.6, left panel, are outside the realm of chance differences.

Notice that there is no R^2 statistic in the output of the `multinom` function. The reason is that $\text{Var}(Y)$ has no meaning for nominal Y, so you cannot partition the variance in the usual way. There are a variety of "pseudo-R^2" statistics available for this case; many of these statistics compare the log-likelihood of the model with the X variables with the log-likelihood of the model with no X variables (the intercept-only model), but we will not explore them further.

13.3 Models for Ordinal Y

The example above with categories "Wife," "Husband," and "Alternating" of the Y variable have no order: Wife laundry duty is neither "less than" nor "more than" husband laundry duty, and neither category is "more than" nor "less than" alternating laundry duty.

On the other hand, consider the case where the Y variable is an employee's rating of the boss, either low $(Y = 1)$, medium $(Y = 2)$ or high $(Y = 3)$. You can analyze such data by using the MLR model, but you can often construct a better model by accounting for the fact that the categories (1, 2, 3) are ordinal. In this example, 1 is indeed less than 2, and 2 is indeed less than 3 when it comes to rating of the boss. The ordinal regression model specifically uses this ordering to create a different model, one that has fewer parameters than the MLR model. Because there are fewer parameters to estimate, there is less variance in the estimated model using ordinal regression than in MLR. However, yes, you guessed it: Because the model has fewer parameters, the ordinal model suffers from greater bias than the MLR model, and the variance-bias trade-off appears yet again.

Here are some data that will help to set ideas.

```
Rating = c( 1,  1,  2,  2,  1,  3,  3,  1,  1,  2,  3,  2,  1,  2)
Salary = c(15, 30, 20, 30, 40, 45, 49, 16, 20, 40, 55, 56, 24, 31)
plot(Salary, Rating)
```

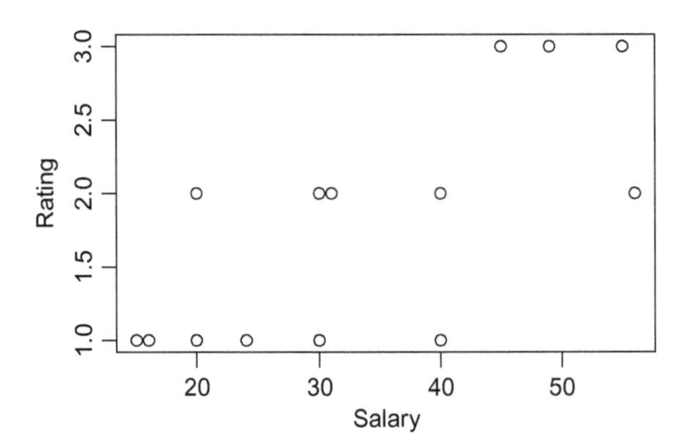

FIGURE 13.7
Scatterplot of Rating (1, 2 or 3) versus Salary.

You can easily visualize the distributions $p(y\,|\,x)$ in Figure 13.7. They are all 3-point distributions on the numbers 1, 2, and 3, and they "morph" towards higher probabilities on the "2" and "3" ratings as salary increases. The trick is to string these 3-point distributions together along the X variable, getting the $p(y\,|\,x)$ model, then estimating the unknown parameters of that model using maximum likelihood.

The standard way of doing this with ordinal Y data involves introducing a *latent variable* as a device to construct these probability functions. To understand this latent variable, consider the people who rated the boss a "1." It is not hard to imagine that, within this group of people, someone really hated the boss, and did not hesitate to enter "1" for their response. It is also not hard to imagine that someone else in this group was more ambivalent, because she kind of liked the boss in some ways, so she was debating in her mind whether to enter "1" or "2," but she finally decided to enter "1." The second person has a higher "boss liking" than the first, even though both persons ultimately entered "1" on the survey.

This "boss liking" variable is an example of a *latent variable,* which, by definition, is a variable that is not directly observed. The observed 1,2,3 responses are only indirect measures of this latent variable: Presumably, the latent "boss liking" would be much higher for people who rated the boss "3" than in the group that rated the boss a "1," but there is still unobserved variation in the latent "boss liking" variable among people *within* each of the three groups.

Let's define the latent variable as Z = "boss liking." We cannot measure Z directly, but we can easily assume that it is a continuous measure, so that each person has their own unique "boss liking" that is different from everyone else's "boss liking." Here is a familiar-looking model for Z:

$$Z = \gamma_1 X + \delta$$

Or, in more words,

$$(\text{Latent "boss liking" for a person}) = \gamma_1(\text{that person's Salary}) +$$
$$(\text{that person's deviation from mean boss liking among people with the same Salary})$$

Since Z, boss liking, cannot be observed directly, we cannot possibly estimate the standard deviation of Z. Nor can we estimate its overall mean level, which explains why there is no intercept in the model. The model assumes that $\delta \sim \mathrm{N}(0,1)$, which is not a typical kind of assumption because usually the error variance is a parameter to estimate, rather than the number 1.0. Along with the assumption of no intercept in the latent variable model, this assumption is needed to ensure that the remaining parameters can be estimated (i.e., so that the parameters are identifiable).

Figure 13.8 shows how these hypothetical latent "boss liking" data values might look as a function of Salary for 200 employees. As Figure 13.8 shows, the model for the latent Z = "boss liking" translates to a model for the observable Y = "actual boss rating" according to where Z is with regard to the horizontal thresholds, called "zeta" (ζ) parameters in the "`polr`" function that is used to fit these models:

$$\text{If } Z < \zeta_{1|2}, \qquad \text{then } Y = 1$$
$$\text{If } Z \geq \zeta_{1|2} \text{ but } Z < \zeta_{2|3}, \qquad \text{then } Y = 2$$
$$\text{If } Z \geq \zeta_{2|3}, \qquad \text{then } Y = 3$$

While the latent variable is a convenient device, you do not really have to believe that such a latent "boss liking" variable actually exists. Rather, the latent variable simply gives you a convenient way to specify models $p(y \mid x)$ for the observable (not latent) ordinal data Y. Specifically,

$$\begin{aligned}
&\Pr(Y = 1 \mid X = x) \\
&= \Pr(Z < \zeta_{1|2} \mid X = x) && (\text{by assumption relating the latent } Z \text{ to the observable } Y) \\
&= \Pr(\gamma_1 X + \delta < \zeta_{1|2} \mid X = x) && (\text{by substitution}) \\
&= \Pr(\delta < \zeta_{1|2} - \gamma_1 x) && (\text{by algebra, and incorporating the condition } X = x) \\
&= \Phi(\zeta_{1|2} - \gamma_1 x) && \left(\begin{array}{l}\text{using the assumption } \delta \sim \mathrm{N}(0,1)\text{, and introducing the symbol } \Phi \\ \text{to denote the cumulative standard normal distribution function}\end{array}\right)
\end{aligned}$$

Similar calculations yield

$$\Pr(Y = 2 \mid X = x) = \Phi(\zeta_{2|3} - \gamma_1 x) - \Phi(\zeta_{1|2} - \gamma_1 x)$$

and

$$\Pr(Y = 3 \mid X = x) = 1 - \Phi(\zeta_{2|3} - \gamma_1 x)$$

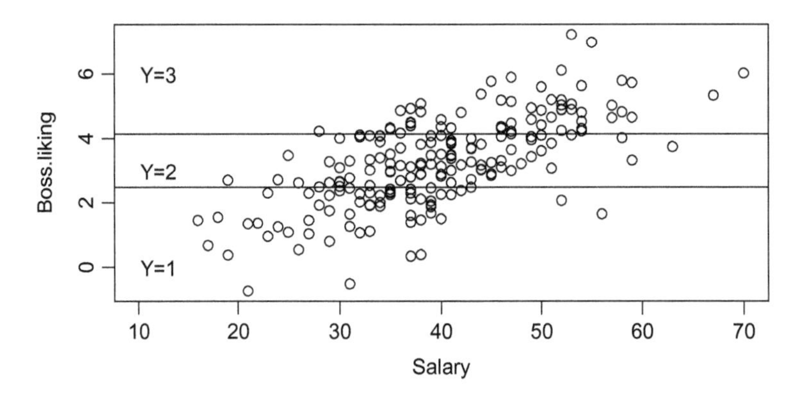

FIGURE 13.8
Latent "boss liking" measures as related to salary. The horizontal lines indicate estimated thresholds ("zeta" parameters) that determine whether the employee rates the boss as $Y=1$, $Y=2$, or $Y=3$. The estimated $\zeta_{1|2}$ threshold is 2.47, and the estimated $\zeta_{2|3}$ threshold is 4.14.

Hence, the ordinal regression model $p(y \mid x)$ is given as follows. Notice that there is no latent variable in the model.

Ordinal regression model assuming normal latent response

y	$p(y\|x)$
1	$\Phi(\zeta_{1\|2} - \gamma_1 x)$
2	$\Phi(\zeta_{2\|3} - \gamma_1 x) - \Phi(\zeta_{1\|2} - \gamma_1 x)$
3	$1 - \Phi(\zeta_{2\|3} - \gamma_1 x)$
Total	1.0

Notice also that there is a negative coefficient "$-\gamma_1$" in the probabilities. This makes sense because when the X variable is positively related to Y, you will have higher probabilities on the highest response (here, $Y=3$), and thus, lower probabilities on the lower responses (here, $Y=1$ and $Y=2$).

Notice also that the model depends only on the observable (x,y) data, as well as three unknown parameters, γ_1, $\zeta_{1|2}$, and $\zeta_{2|3}$, which are estimated via maximum likelihood. Using the "Boss Rating" data set, having observation pairs $(y_i, x_i) = \{(1,15),(1,30),(2,20),\ldots,(2,31)\}$, the likelihood function is given as follows:

$$L(\gamma_1,\zeta_{1|2},\zeta_{2|3} \mid \text{data}) = \Phi(\zeta_{1|2}-\gamma_1(15)) \times \Phi(\zeta_{1|2}-\gamma_1(30)) \times \{\Phi(\zeta_{2|3}-\gamma_1(20)) - \Phi(\zeta_{1|2}-\gamma_1(20))\}$$
$$\times \ldots \times \{\Phi(\zeta_{2|3}-\gamma_1(31)) - \Phi(\zeta_{1|2}-\gamma_1(31))\}$$

An `R` function that chooses γ_1, $\zeta_{1|2}$, and $\zeta_{2|3}$ to maximize this likelihood function is "`polr`," which you can use as follows:

```
library(MASS);
fit = polr(as.factor(Rating) ~ Salary, method="probit")
summary(fit)
```

TABLE 13.7

Estimated Conditional Distribution of Y = Boss Rating, Given that X = Salary = x

y	$\hat{p}(y \mid X = x)$
1	$\Phi\{2.4745 - 0.08423(x)\}$
2	$\Phi\{4.1352 - 0.08423(x)\} - \Phi\{2.4745 - 0.08423(x)\}$
3	$1 - \Phi\{4.1352 - 0.08423(x)\}$
	Total: 1.0

The results are:

```
Call:
polr(formula = as.factor(Rating) ~ Salary, method = "probit")

Coefficients:
        Value Std. Error t value
Salary 0.08423    0.03187   2.643

Intercepts:
    Value  Std. Error t value
1|2 2.4745 1.0537     2.3484
2|3 4.1352 1.4477     2.8563

Residual Deviance: 20.55558
AIC: 26.55558
```

Using this output, the estimated probability distribution $p(y \mid x)$ is given in Table 13.7.

For example, when $X = 20$, we have $\hat{p}(1 \mid X = 20) = \Phi\{2.4745 - 0.08423(20)\} = 0.7852$; $\hat{p}(2 \mid X = 20) = \Phi\{4.1352 - 0.08423(20)\} - \Phi\{2.4745 - 0.08423(20)\} = 0.2077$; and $\hat{p}(3 \mid X = 20) = 1 - \Phi(4.1352 - 0.08423(20)) = 0.0071$. Thus, 78.52% of workers having Salary = 20 are predicted to rate the boss as a "1"; 20.77% as a "2"; and 0.71% as a "3."

These distributions are automatically produced by the `polr` fit as `fit$fitted.values`. You can append these fitted values to your X data via `cbind(Salary, fit$fitted.values)` to see the various estimated $p(y \mid x)$ distributions for the X values in your data set. The results are as follows:

```
   Salary          1         2           3
1      15 0.88708313 0.1108761 0.002040815
2      30 0.47917597 0.4669503 0.053873759
3      20 0.78524585 0.2076243 0.007129817
4      30 0.47917597 0.4669503 0.053873759
5      40 0.18553336 0.5926717 0.221794952
6      45 0.09415340 0.5408060 0.365040560
7      49 0.04921565 0.4540219 0.496762397
8      16 0.87011510 0.1272295 0.002655350
9      20 0.78524585 0.2076243 0.007129817
```

```
10       40 0.18553336 0.5926717 0.221794952
11       55 0.01546937 0.2940414 0.690489226
12       56 0.01247793 0.2679863 0.719535797
13       24 0.67477289 0.3079589 0.017268164
14       31 0.44573382 0.4905337 0.063732518
```

The third row is a prediction when $X = 20$: `3    0.78524585 0.2076243 0.007129817`. Notice that the "prediction" is actually the entire estimated distribution rather than a single number.

Self-study question: You also get a prediction of the distribution of Y when you use the classical regression model. Using the "Boss Rating" data, what is the predicted distribution of Y when $X = 20$, when you use the classical regression model?

Notice also that in Figure 13.7, there are only two employees with Salary = 20. One of these rated the boss with a "1," and the other with a "2," hence the empirical distribution of Boss Rating using these raw data, and no model, gives probabilities 50%, 50%, and 0%, rather than 78.5%, 20.8%, and 0.7%. Again, the benefit of using regression models is clear: By "borrowing strength" across the entire data set, you can get better estimates in local neighborhoods.

You can construct a graph of the estimated conditional distributions similar to what is shown in Figure 13.6 as the following code shows. The code also shows a more standard regression-type plot of the estimated regression function $\mathrm{E}(Y \mid X = x)$. Unlike the case of nominal data, where the average of Y makes no sense, it does make sense to calculate the average of ordinal data values, and this graph is shown in the right panel of Figure 13.9.

R code for Figure 13.9

```
Plot.data = data.frame(10:60)
colnames(Plot.data) = c("Salary")
pred = predict(fit, Plot.data, "probs")

par(mfrow=c(1,2))
plot(10:60, pred[,1], type="l", ylim = c(0,1.0), ylab =
"Probability", xlab="Salary")
points(10:60, pred[,2], type="l", ylim = c(0,1), lty=2)
points(10:60, pred[,3], type="l", ylim = c(0,1), lty=3)
legend(22,1.05, c("1=Lowest", "2=Middle", "3=Highest"), lty=1:3)

EY = 1*pred[,1] + 2*pred[,2] + 3*pred[,3]
plot(10:60, EY, type="l", ylim = c(1,3), ylab="Expected Rating",
xlab = "Salary")
```

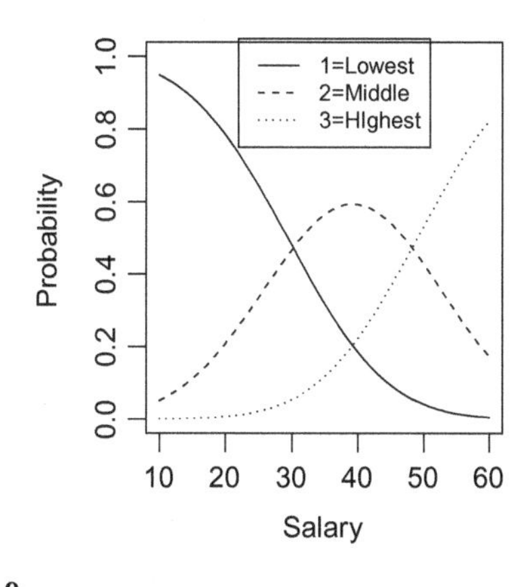

FIGURE 13.9
Left panel: Estimated probabilities of Boss Rating (1, 2, or 3) as a function of Salary. Probabilities in vertical strips add to 1.0 and are the distributions $p(y\,|\,x)$ as estimated using the ordinal probit model. Right panel: Estimated expected rating ($= 1\times p_1 + 2\times p_2 + 3\times p_3$, where the p_j are as shown in the left panel) as a function of Salary.

If the coefficient γ_1 is positive, then the distributions morph to higher probabilities on the higher ratings as $X = x$ increases; if γ_1 is negative then the distributions morph to lower probabilities on the higher rating as $X = x$ increases. As shown in Figure 13.9, the case $\gamma_1 > 0$ corresponds to higher $\mathrm{E}(Y\,|\,X = x)$ for larger x. The trend shown in Figure 13.9 is not easily explained by chance alone because the coefficient $\hat{\gamma}_1 = 0.08423$ is $T = 2.643$ standard errors from zero. This T-statistic does not have a precise T distribution, so you can get a valid asymptotic p-value from the standard normal distribution using $p =$ `2*pnorm(-2.643)` $= 0.0082$.

An assumption inherent in this model is one of parallel slopes: Note that the model implies that $\Pr(Y = 1\,|\,X = x) = \Phi(\zeta_{1|2} - \gamma_1 x)$, and that $\Pr(Y \le 2\,|\,X = x) = \Phi(\zeta_{2|3} - \gamma_1 x)$. Like the logit function, there is a function called the "probit" function that converts these expressions to linear functions. The "probit" function is the inverse cumulative standard normal distribution function (the quantile function), defined as Φ^{-1}:

$$\text{Probit}\{\Pr(Y = 1\,|\,X = x)\} = \Phi^{-1}\{\Pr(Y = 1\,|\,X = x)\} = \Phi^{-1}\{\Phi(\zeta_{1|2} - \gamma_1 x)\} = \zeta_{1|2} - \gamma_1 x,$$
$$\text{and Probit}\{\Pr(Y \le 2\,|\,X = x)\} = \Phi^{-1}\{\Pr(Y \le 2\,|\,X = x)\} = \Phi^{-1}\{\Phi(\zeta_{2|3} - \gamma_1 x)\} = \zeta_{2|3} - \gamma_1 x$$

Notice that these transformed probability functions have the same slope, $-\gamma_1$. Of course, Nature does not decide to make slopes equal, any more than Nature decides to make variances constant or mean functions linear. But still, the common slope model is good if the slopes do not differ greatly. The multinomial logistic regression (MLR) model described in the previous section makes no such common slope assumption (that's why there are so many slope parameters!), so one way to evaluate this common slope assumption is to fit both the ordinal regression model and the MLR model, and compare the results using tables, graphs, and penalized likelihoods. When you fit the MLR model to these data, you get AIC = 28.4, worse than that of the ordinal probit fit shown above, where AIC = 26.6.

The ordinal regression model also can be used when there are just c = 2 categories. In that case, it becomes a competitor to the logistic regression model, called the "probit regression model." Both models usually give very similar probability estimates, but they are different models. You can compare the models by fitting both and graphing their estimated probability functions, and you can also check which model fits the data better by comparing their log-likelihoods.

13.3.1 A Note on Comparing Classical, Normally Distributed Models with Ordinal Regression Models

In many cases, data can be analyzed using either ordinal regression models or classical, normality-assuming regression models. Clearly, with ordinal data, the normality assumption is violated due to discreteness. However, that is not necessarily a problem, because the normality assumption is always violated.

When data are more discrete, their distributions are farther from normal distributions. Hence, as a general principle, you should use ordinal regression rather than classical regression when there are few levels of the Y variable (in the boss rating example, there are only three levels). Where to draw the line in terms of number of levels? Should we use the ordinal regression model, rather than classical regression, if the number of levels is 5 or less? 7 or less? 10 or less? Unfortunately, there is no rule of thumb that can be universally applied.

Often, we like to compare likelihood-related statistics like AIC and BIC to resolve such conundrums. But since the normal model is continuous and the ordinal model is discrete, it is not valid to compare their likelihoods. However, you can count parameters, and notice that there are fewer parameters in the classical, normally distributed model, hence less variability due to parameter estimation in that model. So, despite being obviously wrong due to non-normality, the classical model could be better because of the variance/bias trade-off: With more and more levels of the ordinal Y variable, the number of parameters in the ordinal regression model (the number of ζ's and the γ's) increases, while the number of parameters in the classical model (the number of β's and σ) stays the same. Hence, you tend to prefer the classical, normally distributed model more when the number of levels of the Y variable is large.

And, as always, further guidance or rules of thumb should be determined based on simulation: Simulate data similar to the data that you plan to analyze, and use both ordinal logistic regression and classical regression on your simulated data. Perform the simulation many times to answer the question: Which method tends to work better? The answer will depend strongly on the nature of your planned data collection, especially on its sample size. The answer also tells you which method you should choose for your planned data analysis.

Exercises

1. Do the self-study problem right below Figure 13.1. (It's an exercise in algebra.)
2. Use the data set `pron = read.csv("https://raw.githubusercontent.com/andrea2719/URA-DataSets/master/pron.csv")`

   The first few lines of code you need are

```
pron = read.csv("https://raw.githubusercontent.com/andrea2719/
URA-DataSets/master/pron.csv")
attach(pron)
X = Age
Y = ifelse(Q4 ==1, 1,0)
```

This Y variable is 1 if the survey respondent pronounces "data" as "day – tuh," and is 0 if the survey respondent pronounces "data" as "daa – tuh." The X variable is the age of the survey respondent.

a. Fit the logistic regression to model $\Pr(Y = 1|X = x)$, and report the estimate of this probability function.
b. Graph the function of 2.a over the relevant jittered scatterplot (see the book for examples).
c. Is the trend in the fitted function of 2.b within the range of trends that you would see (by chance alone) if Age really had no effect whatsoever on pronunciation of "data"?

3. See problem 2. In that data set, there is a variable Q8, with values 1, 2 and 3. The question was "have you noticed a shift in the pronunciation of "data" during your life? The responses are

Q8 = 1 means "no shift noticed"

Q8 = 2 means "shift toward day-tuh"

Q8 = 3 means "shift toward daa-tuh"

Let Y = Q8, and let X = Age.

a. You might consider using (i) classical regression, (ii) logistic regression, (iii) multinomial logistic regression, and (iv) ordinal regression for this example. Explain why each of the options (i), (ii), and (iv) is wrong for these data.
b. Perform the multinomial logistic regression and interpret the parameter estimates.
c. Report (i) the estimated marginal distribution of Y (no model), (ii) the model-based estimate of the conditional distribution of Y for a "younger" person (pick an age value yourself), and (iii) the model-based estimate of the conditional distribution of Y for an "older" person (pick an age value yourself).
d. Use the fitted model to draw a graph to show the effect of Age on Perceived Pronunciation shift. (See Figure 13.6 for examples).
e. Are the trends shown in 3.c and 3.d within the range of trends that you would see (by chance alone) if Age really had no effect whatsoever on Perceived Pronunciation Shift? Answer by using a single likelihood ratio chi-square test. Report the chi-square statistic, its degrees of freedom, and its p-value. Use the p-value to answer the stated question.

14

Models for Poisson and Negative Binomial Response

This chapter provides yet another reminder that the regression model is really $p(y \mid x)$, not $\beta_0 + \beta_1 x$. In this chapter, we describe models you can use when your Y variable is discrete, as in Chapter 13. However, in this chapter we consider cases where the potentially observable Y values lie in the set $\{0,1,2,3,4,\ldots\}$, with no precisely defined maximum value. Such data commonly occur when Y is a *count* of something, such as Y = number of children of a person (0,1,2,3,4, or ...), when Y = number of lawsuits filed against a company in a given year (0,1,2,3,4, or ...), when Y = number of bottles of wine in a grocery basket (0,1,2,3,4, or ...), and so on.

You could use the ordinal probit model described in the previous chapter to analyze such Y data. However, the ordinal probit model requires that you estimate an additional parameter for every additional possible Y value, adding variability to estimated model. Because the number of possible values $\{0,1,2,3,4,\ldots\}$ of Y is large, the ordinal probit model can be quite inefficient relative to the standard models for such data, which are the Poisson and negative binomial regression models that are described in this chapter. These standard models do not require an additional parameter for each possible Y value. In addition, the ordinal probit model requires that the potentially observable Y data can only be values in a finite list with a precisely defined maximum value (e.g., a boss rating of 1, 2 or 3); while potentially observable count data typically have no precisely defined maximum value (e.g., it is hard to define an absolute maximum possible number of children that a person might have).

14.1 The Poisson Regression Model

In general, you should use a particular distribution $p(y)$ (normal, Bernoulli, Laplace, etc.) if your Y data appear as if produced by that distribution. Thus, you should use the Poisson distribution if your data appear as if produced by a Poisson distribution. Figure 14.1 shows different Poisson distributions. If your Y data have a frequency distribution that looks somewhat like what is shown in Figure 14.1, then the Poisson distribution might be a good model for your data.

R code for Figure 14.1

```
par(mfrow=c(2,2), mar=c(4,4,1,1))
Y = 0:10
d1=dpois(Y, .1); d2=dpois(Y, .6); d3=dpois(Y, 1.1); d4=dpois(Y, 2.0)
plot(Y, d1, type="h", yaxs="i",ylim=c(0,1), lwd=2, ylab="Poisson
Probability", xlab="")
legend(6,.8, expression(paste(lambda," = 0.1")))
plot(Y, d2, type="h", yaxs="i",ylim=c(0,1), lwd=2, ylab="", xlab="")
legend(6,.8, expression(paste(lambda," = 0.6")))
plot(Y, d3, type="h", yaxs="i",ylim=c(0,1), lwd=2, ylab="Poisson
Probability", xlab="y")
legend(6,.8, expression(paste(lambda," = 1.1")))
plot(Y, d4, type="h", yaxs="i",ylim=c(0,1), lwd=2, ylab="",
xlab="y")
legend(6,.8, expression(paste(lambda," = 2.0")))
```

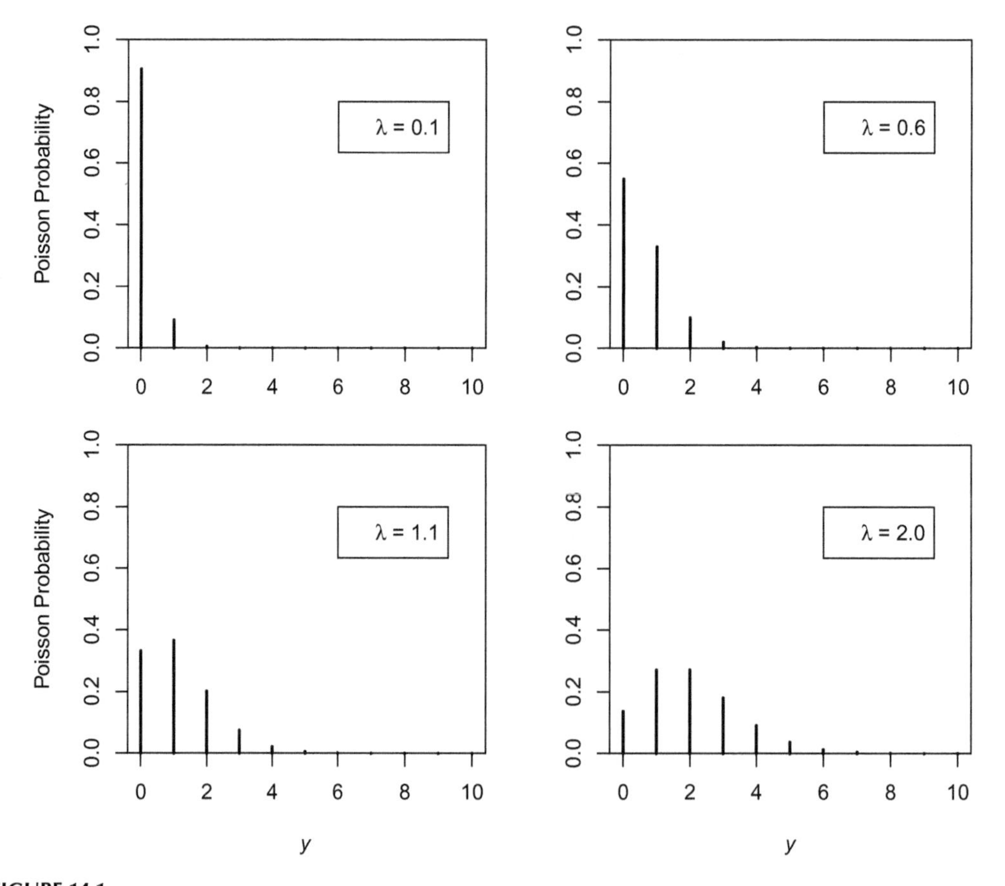

FIGURE 14.1
Poisson distributions having mean values $\lambda = 0.1$ (upper left panel), $\lambda = 0.6$ (upper right panel), $\lambda = 1.1$ (lower left panel), and $\lambda = 2.0$ (lower right panel).

Notice in Figure 14.1 the following:

- In all graphs, there is a substantial probability of seeing $Y = 0$. This is a hallmark of Poisson regression. If your Y data have no zeros, you ordinarily will not use Poisson regression.
- There are only discrete Y values that are integers. Thus, you should not use Poisson regression if your Y values can have decimals.
- There are no gaps between, e.g., 0 and 5: All integers in between are represented by substantial probabilities. That means, for example, if your Y data can only be multiples of 5, like 0, 5, 10, 15, etc., then you should not use Poisson regression.
- There is no distinct upper bound on Y. That means if your Y data are bounded above, as in a survey response that can be 0, 1, 2, 3, or 4, with nothing higher, then you should not use Poisson regression.
- The distributions are all right-skewed. That means if your Y data are left-skewed, you should not use Poisson regression.
- The distributions shown are typical Poisson distributions for which you would apply Poisson regression. That is to say, you should ordinarily only use Poisson regression when the mean of $Y(\lambda)$ is small, say 5 or less. The logic is that for large λ, the Poisson distribution is approximately a normal distribution. So, even if your data really do come from a Poisson distribution, the classical model which assumes normally distributed Y will be adequate in this case. Figure 14.2 shows the Poisson distribution when $\lambda = 10$. Notice that the Poisson distribution is well approximated by the superimposed normal curve. So, the normal model is adequate in this case. With higher mean values, the normal model is even better because (a) the range of possible Y values looks closer to "continuous" because there are so many of them, and (b) the skewness of the Poisson distribution that is shown in Figure 14.2 lessens.

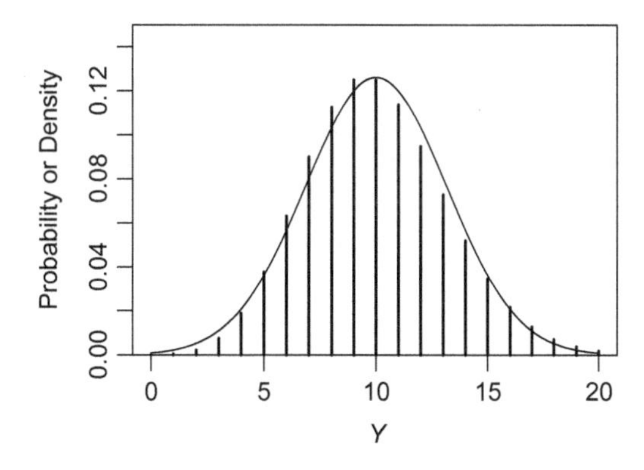

FIGURE 14.2
Poisson probability distribution (solid vertical lines) with mean $\lambda = 10$, along with approximating normal density distribution (continuous curve).

Additional examples where Poisson regression (or its close relative, negative binomial regression, described later in this chapter) might be appropriate are as follows

1. Y = number of accidents on the shop floor in a given day
2. Y = number of sexual partners that a 21-year-old person has had
3. Y = number of fish caught on a trip to the lake
4. Y = number of financial planners a person has used

Self-study question: Underneath the statement above "Notice in Figure 14.1 the following:" there are several bullet points regarding when *not* to use Poisson regression. For each of the four examples above, is there any obvious reason why you should *not* use Poisson regression?

Self-study question: For each of the four examples above, what X variable might affect the probability distribution of Y? And in what way will the probability distribution of Y be different, for different values of that X variable? (Figure 14.1 shows how different Poisson distributions will appear.)

The Poisson distribution $p(y)$ is given by:

$$p(y) = e^{-\lambda}\lambda^{y}/y!, \text{ for } y = 0,1,2,3,\ldots$$

Examples of graphs of these functions are shown in Figures 14.1 and 14.2.

To make this distribution a regression model $p(y \mid x)$, you need to make this distribution "morph" continuously as a function of $X = x$. Such continuous morphing gives a realistic model for Nature, because (again) Nature favors continuity over discontinuity. And (again), as with all regression models, such "morphing" in the model allows the estimates to "borrow strength" from nearby X data when estimating the distribution of Y at a particular $X = x$.

The number λ in the Poisson distribution is the mean of the distribution of Y. Thus, you can make the distribution morph continuously as a function of $X = x$ by making the mean (λ) a function of $X = x$. With classical regression, the default model for the mean is $\mathrm{E}(Y \mid X = x) = \beta_0 + \beta_1 x$. However, like the variance function models presented in Chapter 12, such a linear model can give you negative numbers, because a line will cross the horizontal axis somewhere (if it is not a flat line). The mean of the Poisson distribution must be a positive number; hence, like the variance function models, the default in Poisson regression is to model the logarithm of the mean as a linear function:

$$\ln\{\lambda(x)\} = \ln\{\mathrm{E}(Y \mid X = x)\} = \beta_0 + \beta_1 x$$

This gives you the Poisson regression model:

$$p(y \mid x;\boldsymbol{\theta}) = e^{-\lambda(x;\boldsymbol{\theta})}\{\lambda(x;\boldsymbol{\theta})\}^{y} / y!$$

where

$$\lambda(x;\boldsymbol{\theta}) = \exp(\beta_0 + \beta_1 x)$$

is the mean of the distribution of $Y \mid X = x$, and $\theta = (\beta_0, \beta_1)$ is the parameter vector. Unlike classical regression models, there is no separate variance parameter σ^2 in the Poisson regression model. The Poisson distribution is peculiar in that its mean and variance are the same number:

$$\mathrm{Var}(Y \mid X = x) = \mathrm{E}(Y \mid X = x) = \exp(\beta_0 + \beta_1 x).$$

In particular, the Poisson regression model is a heteroscedastic model, which makes sense when you look at Figures 14.1 and 14.2: With larger mean values, there is more variance in the Poisson distribution.

The constraint imposed by the Poisson model that the mean is the same as the variance limits its applicability, and is the reason that the negative binomial model (discussed below), which allows the mean and variance to differ, is often preferred.

The way that you estimate the parameters θ based on the data should be second nature by now. Of course, you can do this easily by using maximum likelihood. Table 14.1 shows how you can construct the likelihood function.

Assuming conditional independence of the potentially observable Y data, the product of all the terms in the last column of Table 14.1 is your likelihood function $L(\theta \mid \text{data})$, a function of the parameters $\theta = (\beta_0, \beta_1)$. If you plug in different $\theta = (\beta_0, \beta_1)$ into $L(\theta \mid \text{data})$, you will get different likelihood values. The combination $\hat{\theta} = (\hat{\beta}_0, \hat{\beta}_1)$ that gives the maximum of all such possible likelihood values gives you the maximum likelihood estimates $\hat{\beta}_0$ and $\hat{\beta}_1$.

14.1.1 Predicting Number of Financial Planners Used by a Person as a Function of Gender and Age

You can read the data set used in this example and see its first three lines as follows:

```
planners = read.csv("https://raw.githubusercontent.com/andrea2719/
URA-DataSets/master/planners.csv", header=T); head(planners, n=3)
```

This gives you the following output:

```
  nplanners age gender
1         0  55      0
2         2  55      1
3         0  40      1
```

TABLE 14.1

Contributions to the Likelihood Function in Terms of the Parameters θ of the Poisson Regression Model

Obs	X	Y	Contribution to Likelihood Function
1	x_1	y_1	$e^{-\lambda(x_1;\theta)}\{\lambda(x_1;\theta)\}^{y_1}/y_1!$
2	x_2	y_2	$e^{-\lambda(x_2;\theta)}\{\lambda(x_2;\theta)\}^{y_2}/y_2!$
3	x_3	y_3	$e^{-\lambda(x_3;\theta)}\{\lambda(x_3;\theta)\}^{y_3}/y_3!$
...	...	...	...
n	x_n	y_n	$e^{-\lambda(x_n;\theta)}\{\lambda(x_n;\theta)\}^{y_n}/y_n!$

The variables are "Nplanners," the number of financial planners used by the surveyed person in their lifetime; "Age," the survey respondent's age at time of the survey; and "Gender," the gender claimed by the survey respondent, coded as "1" for "Female" and "0" for "Male."

Recall that regression is a model for the conditional distribution of Y given $X = x$. If you had many repeats of Y values at specific $X = x$ values, you could estimate those distributions using either histograms (for continuous Y) or needle/bar plots (for discrete Y), and compare them to known forms, such as normal, Poisson, lognormal, etc. However, if you do not have enough repeats of Y values for specific X values, a good starting point is to estimate of the *marginal* distribution $p(y)$ of Y to get at least a rough idea about how to model the conditional distributions $p(y \mid x)$.

Using the "financial planners" data set, you can graph the estimate of the marginal distribution $p(y)$ as follows:

```
attach(planners)

plot(prop.table(table(nplanners)), ylab="p(y)", xlab="Number of Planners,
y", yaxs="i", ylim = c(0,1), lwd=3)
```

Comparing the distribution shown in Figure 14.3 with those shown in Figure 14.1, the dependent variable, Y = Nplanners, has a distribution that looks Poisson-like. However, there seems to be excess variance in the marginal Y data that makes the Poisson model questionable, since `mean(nplanners)` gives 0.564 and `var(nplanners)` gives 0.964. If the marginal distribution is truly Poisson, these numbers should be estimating the same quantity. On the other hand, recall that in the classical regression model, you assume that the conditional distributions $p(y \mid x)$ are normal distributions, and not that the marginal distribution $p(y)$ is a normal distribution. In the same way, when you use the Poisson regression model, you assume that the conditional distributions $p(y \mid x)$ are Poisson distributions, and not that the marginal distribution $p(y)$ is a Poisson distribution. It can easily happen that the marginal distribution $p(y)$ has extra-Poisson variance as suggested by the raw data shown in Figure 14.3, even when the conditional distributions $p(y \mid x)$ are all precisely Poisson.

Despite the fact that it is a marginal distribution and not a conditional distribution, Figure 14.3 is a useful first step to identify the nature of the data. For now, let us simply assume that Poisson regression is a reasonable model. Your first goal in using Poisson

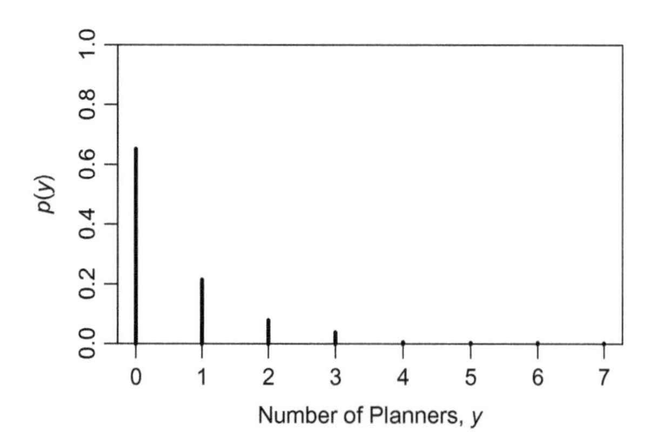

FIGURE 14.3
Estimate of the distribution $p(y)$, which is the marginal distribution of number of financial planners used.

regression is, as with all regression analyses, to understand how (and whether) the distribution of Y changes for different possible values $X = x$.

You can fit the Poisson regression model using the "glm" function of R as follows:

```
fit.pois = glm(nplanners~age+gender, family = poisson)
summary(fit.pois)
```

The results are as follows:

```
Call:
glm(formula = nplanners ~ age + gender, family = poisson)

Deviance Residuals:
    Min        1Q    Median       3Q       Max
-1.4519   -1.1274   -0.7785   0.3139   4.1149

Coefficients:
             Estimate Std. Error z value Pr(>|z|)
(Intercept) -1.168691   0.293516  -3.982 6.84e-05 ***
age          0.017448   0.005142   3.393 0.000691 ***
gender      -0.915015   0.097100  -9.423  < 2e-16 ***
---
Signif. codes:  0 '***' 0.001 '**' 0.01 '*' 0.05 '.' 0.1 ' ' 1

(Dispersion parameter for poisson family taken to be 1)

    Null deviance: 1220.0  on 891  degrees of freedom
Residual deviance: 1111.6  on 889  degrees of freedom
AIC: 1834.3

Number of Fisher Scoring iterations: 6
```

Here are some notes about the output:

- The first line, "Deviance residuals," shows likelihood-based quantities that correspond to the residuals in ordinary regression. Like ordinary regression, these quantities measure discrepancy between data and fitted model, and are very useful to detect observations that are not well explained by the model. Such observations could be outliers (discussed in detail in Chapter 16), or they could be indications of a poor model.
- The coefficients give you the linear model for the logarithm of the estimated mean function: Letting $\mu(x_1, x_2) = \mathrm{E}(\mathrm{Nplanners} \mid \mathrm{Age} = x_1, \mathrm{Gender} = x_2)$, the estimates give you

$$\ln\{\hat{\mu}(x_1, x_2)\} = -1.168691 + 0.017448x_1 - 0.915015x_2.$$

  Equivalently,

$$\hat{\mu}(x_1, x_2) = \exp\{-1.168691 + 0.017448x_1 - 0.915015x_2\}.$$

Unlike the case of the log-transformation of Y, this back-transformation is truly an estimate of the mean of the distribution of Y, rather than the median. The reason for the difference is that in the Poisson regression model, the log-transformation is applied to the mean of the distribution of Y, and not to Y itself. When you log transform Y itself, Jensen's inequality tells you that $\exp[\mathrm{E}\{\ln(Y)\}]$ is different from E(Y), hence, you cannot back-transform to estimate E(Y). But Poisson regression directly models $\ln\{\mathrm{E}(Y)\}$, so you can back-transform to estimate E(Y).

Self-study question: How do you interpret 0.017448 in terms of (i) effect on the logarithm of the mean number of financial planners, and in terms of (ii) effect on the mean number of financial planners?

The standard errors are the Wald standard errors described in Appendix B of Chapter 12. You can interpret these standard errors in the ordinary way: The true parameter of the data-generating process (DGP) lies in the range $\{\text{estimate} \pm 2(\text{Wald standard error})\}$ in approximately 95% of samples from the same DGP. Here, $0.017448 \pm 2(0.005142)$, or (0.007164, 0.027732) is an approximate 95% confidence interval for β_1.

- If, hypothetically, the data were produced by a DGP where $\beta_1 = 0$, then the estimate $\hat{\beta}_1$ would differ from 0 by chance alone. Letting $Z = \hat{\beta}_1 / \{\text{s.e.}(\hat{\beta}_1)\}$, the potentially observable Z-values have an approximately standard normal distribution when $\beta_1 = 0$. Thus, if the observed Z statistic is inside the central 95% range of the standard normal distribution (from –1.96 to +1.96), then the difference from $\hat{\beta}_1$ to 0 is explainable by a DGP where $\beta_1 = 0$. (As discussed previously, such a result is also explainable by a DGP where $\beta_1 \neq 0$, which is the reason for the word *explainable* rather than *explained*.) If the observed Z statistic is outside the central 95% range of the standard normal distribution (from –1.96 to +1.96), then difference from $\hat{\beta}_1$ to 0 cannot be easily explained by a DGP where $\beta_1 = 0$.
- The p-value is twice the probability in the tail region of the standard normal distribution beyond the observed Z. If this p-value is greater than 0.05, then the observed Z statistic is inside the central 95% range (from –1.96 to +1.96) of the standard normal distribution. If this p-value is less than 0.05, then the observed Z statistic is outside that range.
- The comment "`(Dispersion parameter for poisson family taken to be 1)`" in the `glm` output is a reminder that, unlike the classical regression model, there is no "extra" parameter in the Poisson distribution to estimate the variance, since the variance is identical to the mean of the Poisson distribution. (The term "dispersion" generally refers to any quantification of variability, not necessarily the variance. For example, mean absolute deviation can also be called "dispersion.")

- The deviance statistics are often $-2\times(\text{maximized log likelihood})$, but as we noted with the Box-Cox transformations, non-essential constants might be omitted. According to the `glm` documentation, deviance is "up to a constant, minus twice the maximized log-likelihood."
- The AIC statistic, on the other hand, is the usual $-2\times(\text{maximized log likelihood})+2k$, where k is the number of estimated parameters. Here, the β's are the only estimated parameters, because there is no extra variance parameter, and there are $k = 3$ of them. So, the "by hand" calculation of AIC gives the identical result as reported: The command `-2*logLik(fit.pois)[1] + 2*3` gives you

```
[1] 1834.256
```

To understand any estimated regression model discussed in this book, whether classical, heteroscedastic, logistic, multinomial, ordinal, etc., you should draw graphs of the estimated conditional distributions $p(y \mid x)$. These graphs will

1. Help you understand the possible problems with your model, allowing you to pick another model if necessary;
2. Allow you to understand the processes you are modeling better (if the model is a good model); and
3. Allow you to make predictions and decisions.

Figure 14.4 displays four such estimated distributions. Notice how the graph facilitates both the understanding of the model and its potential uses.

R code for Figure 14.4

```
X.pred = data.frame(c(45,65,45,65), c(0,0,1,1))
colnames(X.pred) = c("age", "gender")
pred.mu = exp(predict(fit.pois, X.pred))

par(mfrow=c(2,2), mar=c(4,4,1,1))
plot(0:6, dpois(0:6, pred.mu[1]), type="h", lwd=3, ylim = c(0,1),
  yaxs="i", ylab="Poisson Probability", xlab="")
legend(1.8,.9, c("45-Year old Males"))
plot(0:6, dpois(0:6, pred.mu[2]), type="h", lwd=3,ylim = c(0,1),
  yaxs="i", ylab="", xlab="")
legend(1.8,.9, c("65-Year old Males"))
plot(0:6, dpois(0:6, pred.mu[3]), type="h", lwd=3, ylim = c(0,1),
  yaxs="i", ylab="Poisson Probability", xlab="Financial Planners
  Used")
legend(1.4,.9, c("45-Year old Females"))
plot(0:6, dpois(0:6, pred.mu[4]), type="h", lwd=3, ylim = c(0,1),
  yaxs="i", ylab="", xlab="Financial Planners Used")
legend(1.4,.9, c("65-Year old Females"))
```

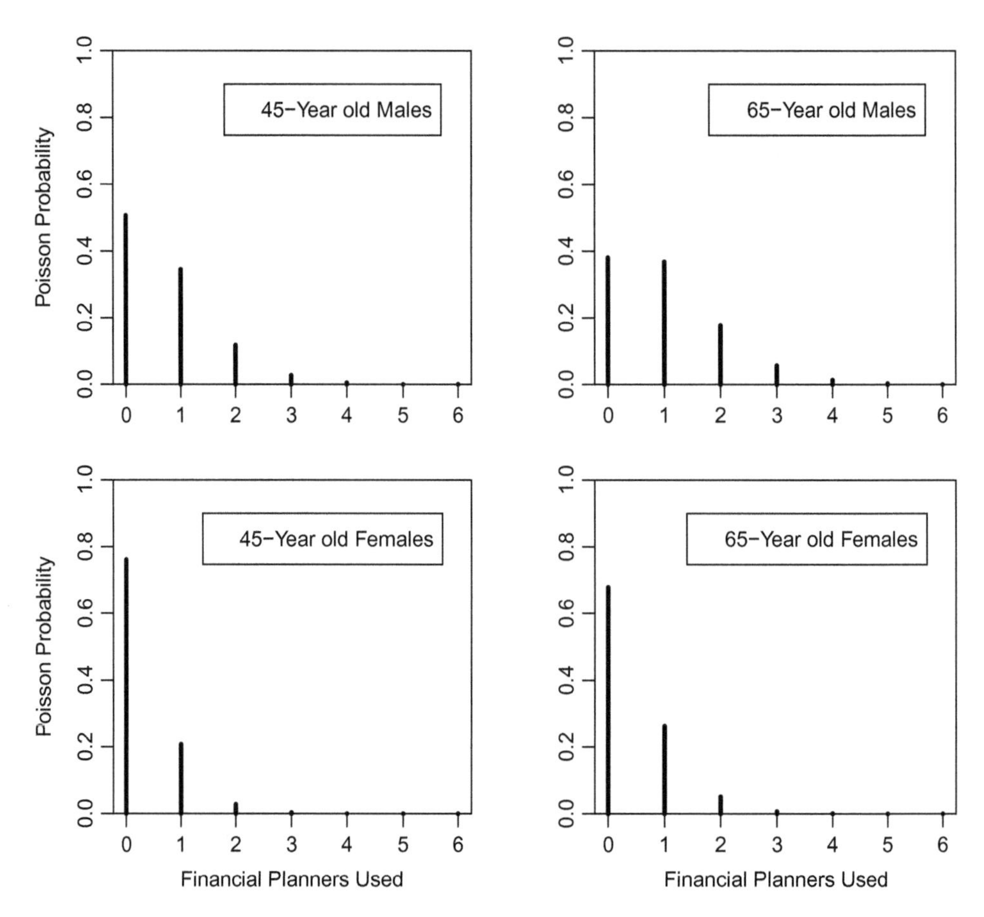

FIGURE 14.4
Estimated distributions of number of financial planners used, for different Age/Gender combinations. All distributions are estimated using the Poisson regression model.

Self-study question: What do the graphs of Figure 14.4 tell you about people's use of financial planners?

14.2 Negative Binomial Regression

Unlike the normal distribution, which is determined by two parameters, μ and σ the Poisson distribution $p(y)$ depends only on a single parameter, λ, which is both the mean and the variance of the distribution. But often, the data are more dispersed than this constraint dictates; i.e., often the variance is larger than the mean. When this happens, you have *extra-Poisson variation*.

To model extra-Poisson variation, you need a distribution with an extra parameter, similar to the normal distribution which also has the extra parameter, σ^2. Then you can estimate the degree of extra-Poisson variation using the data, just like you can estimate σ^2 of the normal distribution using the data.

The negative binomial (NB) distribution is the most commonly-used model for extra-Poisson variation. There are many other distributions that serve this purpose as well, notably the "zipf" (zero-inflated Poisson) distribution, but the NB distribution is the only one that we will discuss.

Unlike the Poisson distribution $p(y)$, which has the relatively simple mathematical form given above, the mathematical form of the NB distribution is considerably more complicated and not particularly illuminating, so we will not present it explicitly. Instead, we will discuss its mean and variance explicitly, its relation to the Poisson distribution, and we will display this information using graphs.

The negative binomial distribution has two parameters, μ and τ; the parameter τ is a measure of the excess Poisson variation. As implemented by the `R` software, the NB distribution has

$$\text{Var}(Y) = \mu + \mu^2/\tau.$$

Other software may specify that $\text{Var}(Y) = \mu + \tau\mu^2$, which is the same model, but with a different meaning of τ. Assuming the `R` implementation, when $\tau = \infty$, the NB distribution is identical to the Poisson distribution. For large τ, the NB and Poisson distributions are similar; for small τ, the NB distribution has excess variation compared to the Poisson distribution. You can compare how well each model fits the data by using penalized likelihood statistics such as AIC and BIC.

Figure 14.5 shows how the parameter τ affects the distribution when $\mu = 1$. Notice that the parameter τ is called "size" in the `dnbinom` probability function.

`R` code for Figure 14.5

```
par(mfrow=c(2,2), mar=c(4,4,1,1))
y = 0:15
plot(y, dnbinom(y, mu=1, size=.1), type="h", ylim=c(0,.8),
  ylab="Probability", xlab="", yaxs="i", lwd=2)
plot(y, dnbinom(y, mu=1, size=1), type="h", ylim=c(0,.8), ylab="",
  xlab="", yaxs="i", lwd=2)
plot(y, dnbinom(y, mu=1, size=2), type="h", ylim=c(0,.8),
  ylab="Probability", xlab="y", yaxs="i", lwd=2)
plot(y, dnbinom(y, mu=1, size=10000), type="h", ylim=c(0,.8),ylab="",
  xlab="y", yaxs="i", lwd=2)
```

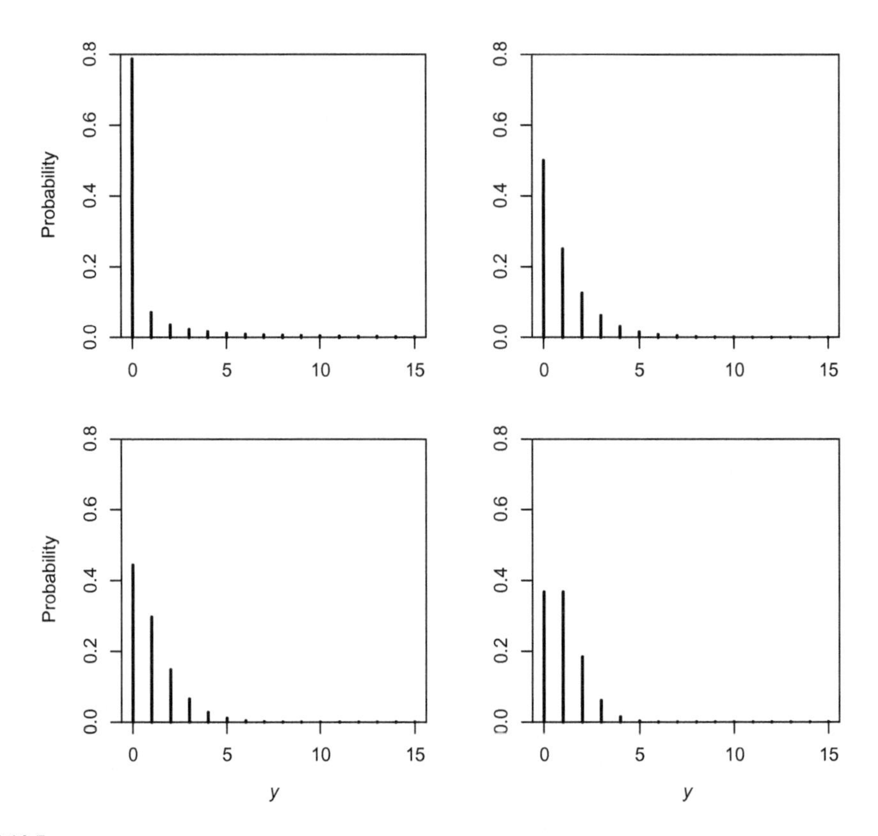

FIGURE 14.5
Left portions of negative binomial (NB) distributions, all having mean $\mu = 1$. Upper left panel: $\tau = 0.1$ ($\text{Var}(Y) = 1 + 1^2/0.1 = 11.0$). Upper right panel: $\tau = 1$ ($\text{Var}(Y) = 1 + 1^2/1 = 2.0$). Lower left panel: $\tau = 2$ ($\text{Var}(Y) = 1 + 1^2/2 = 1.5$). Lower right panel: $\tau = 10{,}000$ ($\text{Var}(Y) = 1 + 1^2/10{,}000 = 1.0001$). The lower right panel shows a distribution that is nearly identical to the Poisson distribution with $\lambda = 1$.

14.2.1 Predicting Number of Financial Planners Used by a Person as a Function of Gender and Age, Using Negative Binomial Regression

You can fit the NB regression model using the "`glm.nb`" function in the `MASS` library as follows:

```
library(MASS)
fit.nb = glm.nb(nplanners~age+gender)
summary(fit.nb)
```

The results are as follows:

```
Call:
glm.nb(formula = nplanners ~ age + gender, init.theta = 1.122254999,
    link = log)

Deviance Residuals:
    Min       1Q   Median       3Q      Max
-1.2253  -0.9999  -0.7303   0.2422   2.6358

Coefficients:
             Estimate Std. Error z value Pr(>|z|)
(Intercept) -1.209900   0.364851  -3.316 0.000913 ***
age          0.018230   0.006437   2.832 0.004625 **
gender      -0.920311   0.116245  -7.917 2.43e-15 ***
---
Signif. codes:  0 '***' 0.001 '**' 0.01 '*' 0.05 '.' 0.1 ' ' 1

(Dispersion parameter for Negative Binomial(1.1223) family taken to be 1)

    Null deviance: 823.69  on 891  degrees of freedom
Residual deviance: 750.45  on 889  degrees of freedom
AIC: 1761

Number of Fisher Scoring iterations: 1

              Theta:  1.122
          Std. Err.:  0.200

 2 x log-likelihood:  -1753.030
```

Notice that the output looks quite similar to the Poisson regression output, but also that there is an additional parameter, "`Theta`" that is estimated. We have called this same parameter "τ" above; we did not use "θ" (the Greek letter "theta") because we have reserved the "θ" symbol to denote generic parameters (β's, σ's, etc.) throughout this book. As in Poisson regression, the coefficients as estimated by NB regression give you the linear model for the logarithm of the estimated mean function:

$$\ln\{\hat{\mu}(x_1, x_2)\} = -1.209900 + 0.018230x_1 - 0.920311x_2.$$

Equivalently,

$$\hat{\mu}(x_1, x_2) = \exp\{-1.209900 + 0.018230x_1 - 0.920311x_2\}.$$

The estimated NB model for the mean is similar to the estimated Poisson regression model for the mean in this example.

The standard errors from the NB regression fit are again the Wald standard errors. These values are somewhat larger than those of the Poisson regression model, as expected, because NB allows extra variation. The extra variation estimate is $\hat{\tau} = 1.122$, implying that the conditional variance of the number of planners (Y) is estimated to be equal to $(\text{conditional mean of } Y) + (\text{conditional mean of } Y)^2 / 1.122$. Even though marginal and conditional differences are quite different in general, it is comforting that the estimated marginal variance, 0.964, is in the neighborhood of what the model predicts from the marginal mean, which is $0.564 + 0.564^2 / 1.122 = 0.848$. The law of total variance predicts that the conditional variance is smaller than the marginal variance, so the direction of the inequality $(0.848 < 0.964)$ also makes sense.

The AIC statistic is the usual $-2(\text{maximized log likelihood}) + 2k$, where k is the number of estimated parameters. Here, the β's and τ are the estimated parameters, hence there are $k = 4$ parameters. The "by hand" calculation of AIC gives `-2*logLik(fit.nb)[1] + 2*4` = `1761.03`, which is also displayed in the output. According to the AIC statistic, the NB regression model fits the data much better than the Poisson regression, which has AIC = 1,834.26. Thus, the data suggest that there is indeed extra-Poisson variation in the conditional distributions of financial planners.

Figure 14.6 shows four of the conditional distributions, as estimated using the NB regression model. It is subtle, but you can see the extra-Poisson variation in Figure 14.6 when you compare it with Figure 14.4, in that the NB model shown in Figure 14.6 predicts that there will be occasional, extreme values, unlike the Poisson model shown in Figure 14.4.

R code for Figure 14.6

```
X.pred = data.frame(c(45,65,45,65), c(0,0,1,1))
colnames(X.pred) = c("age", "gender")
pred.mu = exp(predict(fit.nb, X.pred))
theta = fit.nb$theta
par(mfrow=c(2,2), mar=c(4,4,1,1))
plot(0:6, dnbinom(0:6, mu=pred.mu[1], size=theta), type="h", lwd=3,
  ylim = c(0,1), yaxs="i", ylab="NB Probability", xlab="")
legend(1.6,.9, c("45-Year old Males"))
plot(0:6, dnbinom(0:6, mu=pred.mu[2], size=theta), type="h",
  lwd=3,ylim = c(0,1), yaxs="i", ylab="", xlab="")
legend(1.6,.9, c("65-Year old Males"))
plot(0:6, dnbinom(0:6, mu=pred.mu[3], size=theta), type="h", lwd=3,
  ylim = c(0,1), yaxs="i", ylab="NB Probability", xlab="Financial
  Planners Used")
legend(1.2,.9, c("45-Year old Females"))
plot(0:6, dnbinom(0:6, mu=pred.mu[4], size=theta), type="h", lwd=3,
  ylim = c(0,1), yaxs="i", ylab="", xlab="Financial Planners Used")
legend(1.2,.9, c("65-Year old Females"))
```

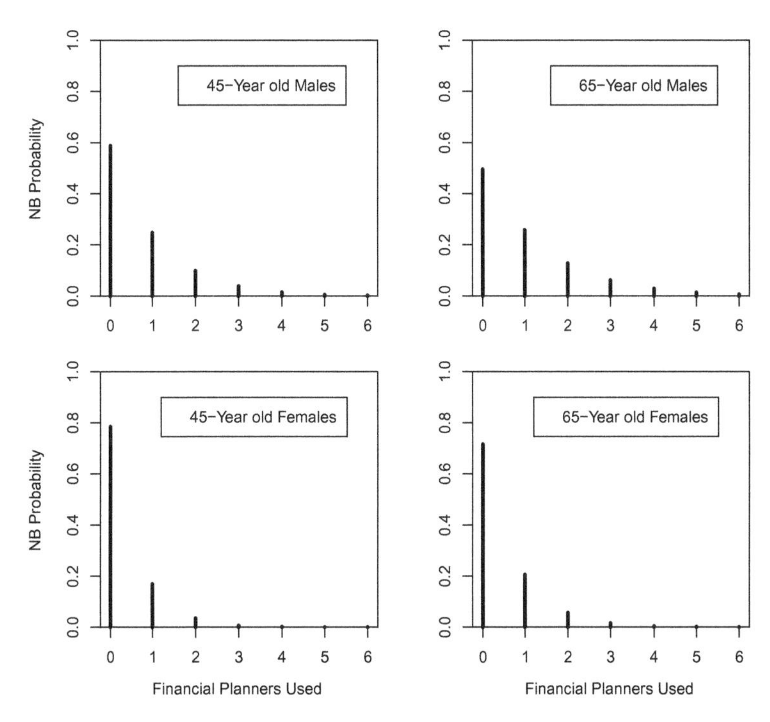

FIGURE 14.6
Estimated distributions of number of financial planners used, for different Age/Gender combinations. All distributions are estimated using the negative binomial (NB) regression model.

You can also estimate and graph the conditional mean functions that are implied by the fitted NB and Poisson regression as follows.

R code for Figure 14.7

```
X.male = data.frame(40:70, rep(0,length(40:70)))
colnames(X.male) = c("age", "gender")
pred.male.NB = exp(predict(fit.nb, X.male))

X.female = data.frame(40:70, rep(1,length(40:70)))
colnames(X.female) = c("age", "gender")
pred.female.NB = exp(predict(fit.nb, X.female))

pred.male.pois = exp(predict(fit.pois, X.male))
pred.female.pois = exp(predict(fit.pois, X.female))

par(mfrow=c(1,1))
plot(40:70, pred.male.NB, type="l", ylim = c(0,1.5), ylab="Mean # of
  Financial Planners", xlab = "Age", main="Estimated Mean Functions")
points(40:70, pred.female.NB, type="l", lty=2)
points(40:70, pred.male.pois, type="l", col="gray")
points(40:70, pred.female.pois, type="l", lty=2, col="gray")
legend(40,1.3, c("Males, Poisson", "Males, NB", "Females, Poisson",
  "Females, NB"), lty=c(1,1,2,2), col =c("gray", "black", "gray",
  "black" ))
```

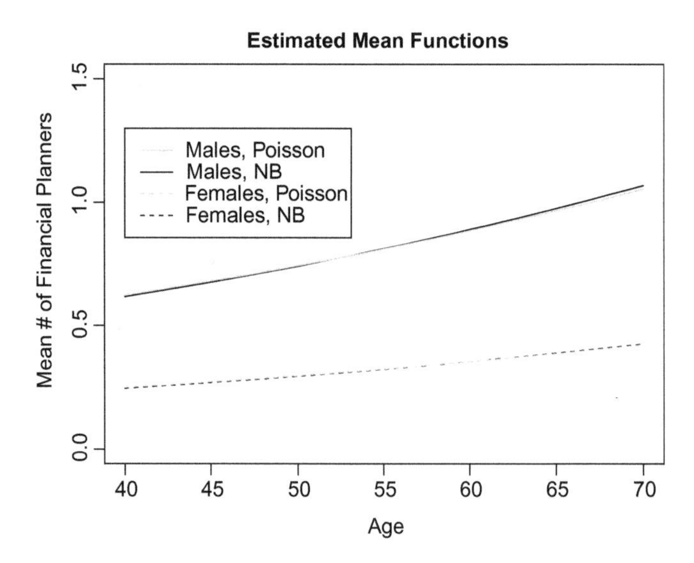

FIGURE 14.7
Estimated mean number financial planners as a function of Age, for males and females, using both Poisson and negative binomial (NB) regression models.

It is comforting that the estimated mean functions as shown in Figure 14.7 are nearly identical for the Poisson and NB models. Because of this happy occurrence, your conclusions about the effects of Age on mean use of financial planners are fairly robust to your choice of distribution, whether Poisson or negative binomial. On the other hand, since the NB model fits the data so much better, you should use NB regression to estimate the distributions themselves, as shown in Figure 14.6, and you should use NB regression to obtain confidence intervals and p-values.

14.2.2 A Note on Replicability and Preregistration

As a general note about likelihoods and AIC (or BIC or other penalized fit statistic) that covers all models, not just those in this chapter, we are *not* suggesting a general "recipe" of the form "use whichever model has the lowest AIC." Ideally, you should make the choice prior to your data analysis based on *a priori* theory, since any data-based choice method will affect the properties of inferences. In particular, models that have the best data-based measures of fit tend to be too optimistic, or "overfit," due to the idiosyncratic nature of the particular data. Such over-optimism gives not only fit measures that are biased toward "good" values, but also standard errors that are biased low. These facts can be demonstrated easily by using simulation. A consequence of standard errors that are biased low is that "significance" (p-value < 0.05) is too easily obtained, leading to "significant" published results that do not replicate in further studies.

This problem of non-replication has even been referred to as the *replication crisis*, and it undermines the entire scientific enterprise, which is built on the dissemination of replicable scientific truths. A solution to this problem is *preregistration*, where studies are published only after their designs, measurement methods, and proposed methods of analysis are *preregistered* by the publishing journal, prior to data collection. Methods based on data snooping, such as "pick the model with the best AIC," would ordinarily not be acceptable for such a preregistered study.

Instead, the "pick the model based on lowest AIC" analysis could be performed on a data set from a *pilot study*; such an analysis could then inform the proposed methods of the main, preregistered study, which will be performed on data that are completely separate from the pilot study data. Based on the analysis of the pilot study, which may include minimum AIC and other data exploration techniques, the actual choice of distribution, the set of X variables, transformations, etc., all will be completely determined for the preregistered study. This will ensure better replicability of the results of the preregistered study, since they will not depend so greatly on the idiosyncrasies of the particular data set. In the preregistration paradigm, "insignificant" results also would be published, removing researcher incentives to manipulate data to achieve "significance," which also contributes to the problem of non-replicability.

Exercises

1. Although the normality assumption is obviously badly violated, you can use OLS to analyze these data. Find and graph the estimated mean functions shown in Figure 14.7 using OLS, and compare the results with the Poisson and NB estimated mean functions.
2. Using the output from `summary(fit.pois)`, calculate the means of the four distributions shown in Figure 14.4 "by hand."
3. Interpret the 0.018230 of the NB regression in terms of (i) the logarithm of the mean number of financial planners, and in terms of (ii) the mean number of financial planners.
4. Using the output from `summary(fit.nb)`, calculate the means of the four distributions shown in Figure 14.6 "by hand."
5. Compute and compare the actual probabilities graphed in Figure 14.6 with those in Figure 14.4 to show that the NB model gives larger probabilities in the tails of the distributions.
6. Regression is a model for the conditional distribution of Y given particular values of the X variable(s). Using the Financial planners' data of Chapter 14, fit the classical regression model, and graph the resulting estimate of the conditional distribution of number of financial planners used by 45-year-old females. Compare that distribution to the corresponding distributions that were estimated using the Poisson and negative binomial models, shown in the book. Using only these graphs, as well as what you know about the number of financial planners that a person might use, critique the conditional distribution as estimated by the classical regression model.
7. Compute and compare the actual probabilities graphed in Figure 14.6 (bottom right panel) with those in Figure 14.4 (bottom right panel) to show that the NB model gives larger probabilities in the tails of the distributions. The numbers will be small in either case, seemingly close to each other because they are both close to zero. So, compare the logits of the probabilities, in addition to the probabilities themselves. Give a brief comment as to why the logit comparison is more meaningful than the comparison of the actual probabilities; see the discussion of lotteries in Chapter 13 for guidance.

15

Censored Data Models

Suppose you have a data set from a call center on how long the customers stayed on the phone, until either hanging up or being answered. The data set might look as shown in Table 15.1.

Here is the data set in R code.

```
time    = c(4.5, 3.9, 1.0, 2.3, 4.8, 0.1, 0.3, 2.0, 0.8, 3.0, 0.3, 2.1)
answer  = c( 0,   1,   1,   1,   0,   1,   0,   0,   1,   0, 1,   0)
```

Notice that the variable called "`answer`" has "Hung up the phone" coded as "0" and "Was answered" coded as 1. This variable is an indicator (0/1) variable that indicates whether the observation was directly observed (`answer` = 1), or *censored* (`answer` = 0). A *censored* observation is one that is not directly observed—instead, you only know that the observation lies in some known *range* of possible values. In the call center data above, the actual time to answer is not observed for those customers who hung up the phone: All you know is that the time to answer had to be *greater than* the time at which they hung up the phone. Such data are called *right censored.*

Now, suppose you wish to estimate the *mean* (or *median*) waiting time until being answered by customer support. How should you treat the censored cases, where the customer hung up before getting answered? If you ignore the censoring and use all the data in the "Time Waiting" column to estimate the average, you will get 2.091667 minutes (`mean(time)`) as an estimate. But this is clearly biased low because the time it *would have taken* customer 1 to be served is not 4.5 minutes, it is some unknown number *greater than 4.5 minutes.* Similarly, the time it *would have taken* customer 7 to be served is some unknown number *greater than 0.3 minutes*, and so on.

TABLE 15.1

Call Center Data Set

Customer	Time Waiting (Minutes)	Hung Up Before Answer?
1	4.5	Yes
2	3.9	No
3	1.0	No
4	2.3	No
5	4.8	Yes
6	0.1	No
7	0.3	Yes
8	2.0	Yes
9	0.8	No
10	3.0	Yes
11	0.3	No
12	2.1	Yes

You might consider simply deleting all the censored cases, getting an average of 1.4 minutes (`mean(time[answer==1])`) for the 6 remaining cases where the customer was answered. But this may give an even more biased result because typically, people hang up when they have been on the phone for too long. Not always, of course–someone may have to run to the bathroom or something, perhaps like customer 7. But as long as there is a general tendency for people who have been on the phone too long with no answer to hang up, which seems reasonable here, then deleting cases where people hung up will bias the resulting estimate of the mean. For the people who hung up, you need a reasonable way to use their data. What should you do in this case? Once again, maximum likelihood provides an ideal solution to the problem.

To introduce the ideas, see Figure 15.1, which shows the exponential distribution with mean $\mu = 4.0$.

R code for Figure 15.1

```
time1 = seq(0,15,.01)
pdf.time = dexp(time1, 1/4)    # The parameter is lambda = 1/mu
plot(time1, pdf.time, type="l", yaxs="i", ylim = c(0, 1.2*max (pdf.
time)), xaxs="i", xlim = c(0,15.5), xlab="time")
points(c(4.5,4.5), c(0, dexp(4.5,1/4)), type="l")
points(c(3.9,3.9), c(0, dexp(3.9,1/4)), type="l", lty=2)
x = seq(4.5, 15,.1); y = dexp(x,1/4)
col1 <- rgb(.2,.2,.2,.5)
polygon(c(x,rev(x)), c(rep(0,length(y)), rev(y)),
col=col1, border=NA)
```

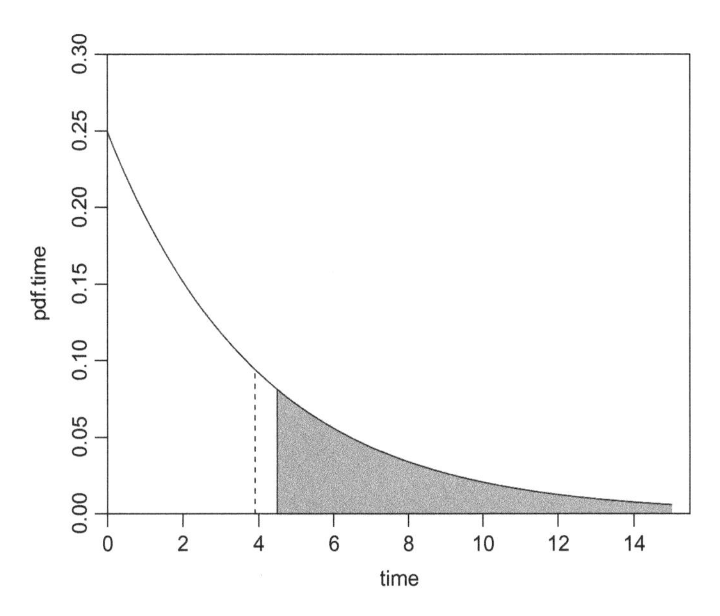

FIGURE 15.1
The exponential distribution with mean 4.0. Customer 1 hung up after waiting 4.5 minutes, so his or her true time is some number under the shaded area; the *area under the curve* is the contribution to the likelihood function. Customer 2 was answered at 3.9 minutes, and that customer's contribution to the likelihood function is the *height of the curve* atop the dashed line.

If Figure 15.1 were the true distribution of call times until actual answer, then you can assume that the actual time for customer 1 to be answered will be some number that is randomly sampled from the curve above 4.5. Hence, the probability associated with this customer's true response is the area under the curve to the right of 4.5. For customers who were served, the actual density, rather than the area under the density, is the contribution to the likelihood function.

In this case, the likelihood function is constructed as shown in Table 15.2.

TABLE 15.2
Contributions to the Likelihood Function When Data Values are Censored

Obs	Time (Y)	Answer (Censoring Indicator)	Contribution to Likelihood Function
1	4.5	0	$\Pr(Y > 4.5 \mid \theta)$
2	3.9	1	$p(3.9 \mid \theta)$
3	1.0	1	$p(1.0 \mid \theta)$
4	2.3	1	$p(2.3 \mid \theta)$
5	4.8	0	$\Pr(Y > 4.8 \mid \theta)$
…	…	…	…
12	2.1	0	$\Pr(Y > 2.1 \mid \theta)$

The likelihood function is thus given as follows:

$$L(\theta \mid \text{data}) = \Pr(Y > 4.5 \mid \theta) \times p(3.9 \mid \theta) \times p(1.0 \mid \theta) \times p(2.3 \mid \theta) \times \Pr(Y > 4.8 \mid \theta) \times \ldots \times \Pr(Y > 2.1 \mid \theta)$$

There are infinitely many choices for $p(y \mid \theta)$, but since "Time until event" is always greater than zero, distributions that only produce positive numbers are preferred. Popular choices of distribution include Weibull, exponential, logistic, lognormal, and loglogistic. The normal distribution can be used, but it is obviously wrong in that it can produce negative numbers, and thus, it is only appropriate for data values that are several standard deviations from zero.

The exponential distribution shown in Figure 15.1 has density

$$p(y \mid \lambda) = \lambda e^{-\lambda y}, \text{ for } y > 0,$$

where the parameter λ is the reciprocal of the mean $(\lambda = 1/\mu)$. The contribution to the likelihood of an observation censored at y is

$$\Pr(Y > y \mid \lambda) = \int_y^\infty \lambda e^{-\lambda t} dt = e^{-\lambda y}$$

In general, the function $\Pr(Y > y)$ is called the "survivor function" (or "survival function"), and is given the symbol $S(y)$. The reason that $S(y)$ is called the "survivor function" is that if the Y variable is years from birth until death, $S(y)$ is the probability that a person survives at least y years. For example, $S(70)$ is the proportion of people who live to be at least 70 years old when Y = years from birth to death.

Using the survival function $S(y) = e^{-\lambda y}$ for the censored observations and the density function $p(y) = \lambda e^{-\lambda y}$ for the non-censored observations, the likelihood function for the call center data when assuming an exponential distribution is given by

$$L(\lambda \mid \text{data}) = e^{-\lambda(4.5)} \times \lambda e^{-\lambda(3.9)} \times \lambda e^{-\lambda(1.0)} \times \lambda e^{-\lambda(2.3)} \times e^{-\lambda(4.5)} \times \ldots \times e^{-\lambda(2.1)}.$$

To estimate the parameter λ you can use the `maxLik` function:

```
loglik <- function(lambda) {
  answer*log(dexp(time, lambda)) + (1-answer)*log(1-pexp(time, lambda))
  }

library(maxLik)

## The starting value, 0.5 here, is a "guess" of lambda.
results <- maxLik(loglik, start=c(0.5))
summary(results)
```

The results are:

```
--------------------------------------------
Maximum Likelihood estimation
Newton-Raphson maximisation, 4 iterations
Return code 1: gradient close to zero
Log-Likelihood: -14.58665
```

```
1  free parameters
Estimates:
     Estimate Std. error t value Pr(> t)
[1,]  0.23904    0.09759   2.449  0.0143 *
---
Signif. codes:  0 '***' 0.001 '**' 0.01 '*' 0.05 '.' 0.1 ' ' 1
---------------------------------------------
```

The maximum likelihood estimate of the parameter λ of the exponential distribution is $\hat{\lambda} = 0.23904$. Thus, the maximum likelihood estimate of the mean time to be answered by a customer support representative is $\hat{\mu} = 1/0.23904 = 4.18$ minutes (the ML-estimated median is $(1/0.23904)\ln(2) = 2.90$ minutes). Compared to the other estimates of the mean, 2.09 and 1.40 minutes, which ignored or incorrectly handled the censoring, the ML estimate is sensible in that it incorporates the fact that the true times for the censored observations are greater than the censoring times.

As you can see from this example, you cannot analyze censored data using any "data-centered" method. Instead, you absolutely must consider the *data-generating process*, which involves the censoring mechanism. But that is nothing new—all estimation procedures that we have presented in this book have involved the data-generating process.

And as you might guess, "there is an app for that." Actually, there are many—censored data is a popular topic! So, you do not have to program the likelihood-based analysis yourself. Instead, you can use the `survreg` function, which is located in the "`survival`" library, as follows.

```
library(survival)
cens.fit = survreg(Surv(time, answer) ~ 1, dist = "exponential")
summary(cens.fit)
```

The results are:

```
Call:
survreg(formula = Surv(time, answer) ~ 1, dist = "exponential")
             Value Std. Error    z        p
(Intercept)  1.43      0.408 3.51 0.000456

Scale fixed at 1

Exponential distribution
Loglik(model)= -14.6   Loglik(intercept only)= -14.6
Number of Newton-Raphson Iterations: 5
n= 12
```

This model uses the logarithmic function to link the mean to the parameters, and there are no predictor variables, just the intercept term:

$$\ln(\mu) = \beta_0$$

Thus, the estimated mean of the distribution is `exp(cens.fit$coefficients)`, or $\exp(1.43)$, giving 4.18, exactly as we got above doing the likelihood calculations "by hand." Also, note that the log-likelihood, –14.6, is also identical to what we got "by hand."

TABLE 15.3

Fits of Different Distributions in the Call Center Data with Censored Observations

Distribution	AIC
Exponential	31.173
Lognormal	32.540
Loglogistic	32.775
Weibull	32.791
Gaussian	38.942
Logistic	39.610

The maximum likelihood method of handling censored data is highly dependent on the choice of distribution. Fortunately, you can compare the fits of different distributions using penalized log-likelihoods, as shown in Table 15.3. You can see that the exponential distribution fits best, but there is little practical difference in the degree of fit for all of the top four in the list. And, keep in mind that the AIC statistic cannot prove which distribution is right, only which ones are more consistent with your idiosyncratic data.

15.1 Regression Analysis with Censored Data

The analysis developed in the introductory section above extends easily to the regression case. The regression model is, as always, the distribution of Y considered as a function of $X = x$. Examples of distributions that you might use include the exponential distribution considered above, as well as Weibull, Gaussian (aka normal), logistic, lognormal, and loglogistic. As always, to get the regression model $p(y \mid x)$, you must allow the distribution of Y to depend on the particular value of X $(X = x)$. And, similar to most regression models, you can do this by allowing the mean (or more generally, location) parameter of the distribution of Y to depend on $X = x$ in a sensible way.

To start simply, suppose we have no censored data. Then we are right back to where we were before this chapter, except now we have a software tool that can fit a variety of distributions $p(y \mid x)$.

Consider the lognormal regression model introduced in Chapter 5, Section 5.2, under the heading, "Comparing log-likelihoods with the Charity data set," repeated here as follows:

```
charity = read.csv("https://raw.githubusercontent.com/andrea2719/
URA-DataSets/master/charitytax.csv")
attach(charity)
plot(DEPS, Income.AGI)
ln.char = log(Income.AGI)

fit.orig  = lm(Income.AGI ~ DEPS)
fit.trans = lm(ln.char ~ DEPS)

LL.orig = logLik(fit.orig)
LL.trans = logLik(fit.trans) - sum(log(Income.AGI))

LL.orig; LL.trans; LL.trans - LL.orig
```

The results were:

Normal (Gaussian) model log-likelihood: –5,381.359

Lognormal model log-likelihood: –5,294.951

The lognormal model fits the data much better, with log-likelihood that is 86.4 points higher than for the Gaussian model. You may have been a little uncomfortable with the "by hand" nature of the log-likelihood calculation for the lognormal model shown in the above code. But you can get those log-likelihoods easily using the `survreg` function with no special "by hand" calculations. Since all data are observed and none are censored, you can create a "censoring" variable called "`observed`" with all 1's to indicate that there are no censored data values.

```
n = nrow(charity)
observed = rep(1,n)
fit.normal    =survreg(Surv(Income.AGI, observed) ~ DEPS,
dist = "gaussian")
fit.lognormal =survreg(Surv(Income.AGI, observed) ~ DEPS,
dist = "lognormal")
```

These fits give you precisely the same log-likelihood statistics as shown above and in Chapter 5. And, if you are curious about other distributions, you can give them a try. For example, the exponential distribution fits the data relatively much worse than either normal or lognormal, with a log-likelihood of –5,486.103. The Weibull model fits better than the normal distribution model, with log-likelihood –5,329.956, but still not as good as the lognormal model.

15.1.1 Survival of Marriage as a Function of Education

The "divorce" data set (Lillard and Panis 2000) contains information on $n = 3{,}371$ married couples. You can read this data set as follows:

```
divorce = read.table("https://raw.githubusercontent.com/andrea2719/
URA-DataSets/master/divorce.txt")
```

The censoring variable is 0 if one of the couple died during the study period, it is also 0 if the couple was still married at the time the study ended. The dependent variable, Y = years until divorce, is equal to the value of the `years` variable when `div = 1`, and is known to be greater than the value of the `years` variable when `div = 0`.

This brings up an interesting point: Why should Y = (years until divorce) be considered as necessarily having a finite result? The usual probability models $p(y)$ for Y presume that people *will* get divorced eventually because the usual models all assume $\Pr(Y < \infty) = 1.0$. It might make more sense to model Y using a distribution $p(y)$ that puts *positive* probability on the outcome $Y = \infty$, which would allow that, in the extremely hypothetical case of couples living forever, some might never divorce. Such distributions exist, but we will use the standard types of distributions, which assume $\Pr(Y < \infty) = 1.0$, while acknowledging that these distributions may be restrictive in that $Y = \infty$ is disallowed.

Consider a model to estimate the effect of the husband's education on the distribution of time until divorce. The education variable is ordinal, but there are only three levels, so we

TABLE 15.4

First 6 Rows of the "Divorce" Data Set

	Id	Heduc	Heblack	Mixed	Years	Div
1	9	12–15 years	No	No	10.546	No
2	11	<12 years	No	No	34.943	No
3	13	<12 years	No	No	2.834	Yes
4	15	<12 years	No	No	17.532	Yes
5	33	12–15 years	No	No	1.418	No
6	36	<12 years	No	No	48.033	No

can use an indicator variable model rather than a linear model without much loss of parsimony (there is only one more parameter). The indicator variable also allows non-linear effects of education on the divorce distributions.

As a first step in any data analysis, you should become acquainted with your data.

```
divorce = read.table("https://raw.githubusercontent.com/andrea2719/
URA-DataSets/master/divorce.txt")
attach(divorce)
head(divorce)
table(div)
```

The data set looks as shown in Table 15.4.

The divorce indicator is summarized as follows by using the `R` command `table(div)`:

```
div
  No  Yes
2339 1032
```

Thus, there are 1,032 divorces (uncensored observations) and 2,339 non-divorces, which are actually censored observations, where one member of the couple died, or the couple was still married at the end of the study. The ultimate time until divorce for these 2,339 is unknown, except that it must bc *greater than* the value of the "`years`" variable for that couple. For example, the first row in Table 15.4 shows a couple who was not divorced after 10.546 years in the study. Thus, their time until divorce is *greater than* 10.546 years. On the other hand, the third couple shown in Table 15.4 was divorced at 2.834 years after marriage.

Let's explore the data more to understand the time until divorce and the censoring times.

R code for Figure 15.2

```
par(mfrow=c(2,1)); par(mar=c(4, 4, 2, 1))
hist(years[div=="Yes"], xlim = c(0,80), breaks=seq(0,80,2),
xlab = "Years Until Divorce", main = "", ylim = c(0,150))
hist(years[div=="No"], xlim = c(0,80), breaks=seq(0,80,2),
xlab = "Years Until Censoring", main = "")
```

FIGURE 15.2
Histograms of time until divorce for the 1,032 couples who were observed to divorce (top panel), and time until censoring for the 2,339 couples who were not observed to divorce during the study period (bottom panel).

The distribution shown in the top panel of Figure 15.2 shows time until divorce for couples who were actually observed to divorce in the study. The bottom panel shows time until the study ended (or a partner died) for couples who were not observed to divorce in the study. Thus, for couples in the study who had not divorced during the study, their times until divorce are greater than, or *to the right of* the times shown in the bottom panel of Figure 15.2. Thus, for the entire group of 3,371 couples, the actual times to divorce is a mixing of the 1,032 observed data values in the top panel of Figure 15.2 with 2,339 unobserved data values that are to the right of the data shown in the bottom panel of Figure 15.2. It is the latter group of unobserved data that causes all the problems: If all the 3,371 times until divorce were actually observed, we would have no problem estimating their distribution, e.g., via a histogram.

Notice that the "time until divorce" distribution must be a distribution on positive numbers, and one that it is right-skewed, so you want to use a distribution that shares these characteristics, such as the lognormal distribution. Fortunately, there is no need to assume a distribution form (lognormal or otherwise) for the censoring time distribution that corresponds to the bottom panel of Figure 15.2.

Among the probability distributions available in `survreg`, the lognormal distribution for marriage survival time fits best by the AIC statistic. Here is the `R` code to fit the lognormal $p(y \mid x)$, where X = husband's education coded with indicator variables.

```
library(survival)
div.ind = ifelse(div=="Yes",1,0)
educ = as.factor(heduc)
fit.lognormal =survreg(Surv(years, div.ind) ~ educ, dist = "lognormal")
summary(fit.lognormal)
```

The results are as follows:

```
Call:
survreg(formula = Surv(years, div.ind) ~ educ, dist = "lognormal")
                    Value Std. Error       z         p
(Intercept)         3.996     0.0688 58.096  0.00e+00
educ12-15 years    -0.274     0.0811 -3.371  7.49e-04
educ16+ years       0.109     0.1266  0.865  3.87e-01
Log(scale)          0.553     0.0241 22.916 3.22e-116

Scale= 1.74

Log Normal distribution
Loglik(model)= -5170.7   Loglik(intercept only)= -5178.9
        Chisq= 16.31 on 2 degrees of freedom, p= 0.00029
Number of Newton-Raphson Iterations: 4
n= 3371
```

To help understand the output, let's first understand the "educ" variable a little better:

`table(educ)` gives you

```
 < 12 years   12-15 years   16+ years
       1288          1655         428
```

The left-out category in the `survreg` output is "< 12 years" education. As shown in Chapter 10, the best way to interpret models involving indicator X variables is to separate the model into the various groups represented by the indicator variables. The same is true with survival models. Here, the model estimates a different lognormal distribution of time until divorce for each of the three education groups, with estimated parameters as follows:

Group 1, Education <12 years (n = 1,288): $\hat{\mu} = 3.996, \hat{\sigma} = 1.74$.

Group 2, Education 12–15 years (n = 1,655): $\hat{\mu} = 3.996 - 0.274 = 3.722, \hat{\sigma} = 1.74$.

Group 3, Education 16+ years (n = 428): $\hat{\mu} = 3.996 + 0.109 = 4.105, \hat{\sigma} = 1.74$.

Hence, you can interpret the remaining parameters and their p-values as follows:

- The mean of the logarithm of potentially observable time until divorce (regardless of censoring) is estimated to be 0.274 less for the "12–15" years category than for the "< 12 years" category. This difference is not easily explained by chance alone $(p = 7.49 \times 10^{-4})$; here "chance alone" refers to a marriage longevity process where there is no difference between marriage survival distributions for the middle and low education groups.
- The mean of the logarithm of potentially observable time until divorce is estimated to be 0.109 more for the "16+" years category than for the "< 12 years"

category. This difference is explainable by a marriage longevity process where there is no difference between marriage survival distributions for the high and low education groups $(p = 0.387)$. Of course, and as always, the word "explainable" rather the "explained" conveys the understanding that the 0.109 difference is also explainable by a marriage longevity process where there *is* a difference between marriage survival distributions for the high and low education groups.

In the interpretations above, the phrase "potentially observable" refers not only to the other potentially observable couples *not in* the sample (as in previous uses of the phrase), but also to couples *in* the sample whose times until divorce were censored.

When using survival analysis, you typically will present the conditional distributions graphically using their *survivor functions*, rather than their density functions. The conditional survivor function is defined as

$$S(y \mid X = x) = \Pr(Y > y \mid X = x) = 1 - \Pr(Y \le y \mid X = x) = 1 - P(y \mid x)$$

where $P(y \mid x)$ is the conditional cumulative distribution function (cdf) of Y that corresponds to our usual conditional probability distribution function $p(y \mid x)$. In the present example, the conditional survivor function has an intuitively appealing interpretation: It is the probability that a marriage will survive for y years, given characteristic(s) $(X = x)$ of the couple.

Figure 15.3 shows the estimated survivor distributions for data in the three education groups.

R code for Figure 15.3

```
b = fit.lognormal$coefficients
b0 = b[1]; b1 = b[2]; b2 = b[3]
m1 = b0
m2 = b0+b1
m3 = b0+b2
s = fit.lognormal$scale

years.graph = seq(0,70,.1)
sdf.y1 = 1 - plnorm(years.graph, m1, s)
sdf.y2 = 1 - plnorm(years.graph, m2, s)
sdf.y3 = 1 - plnorm(years.graph, m3, s)

plot(years.graph, sdf.y1, type="l", ylim = c(0,1),
xlab="Years Married", ylab="Marriage Survival Probability")
points(years.graph, sdf.y2, type="l", lty=2)
points(years.graph, sdf.y3, type="l", lty=3)
abline(v=20, col="gray"); legend("bottomright", c("< 12 Yrs Educ",
"12-15 Yrs Educ", ">15 Yrs Educ"), lty=c(1,2,3))
```

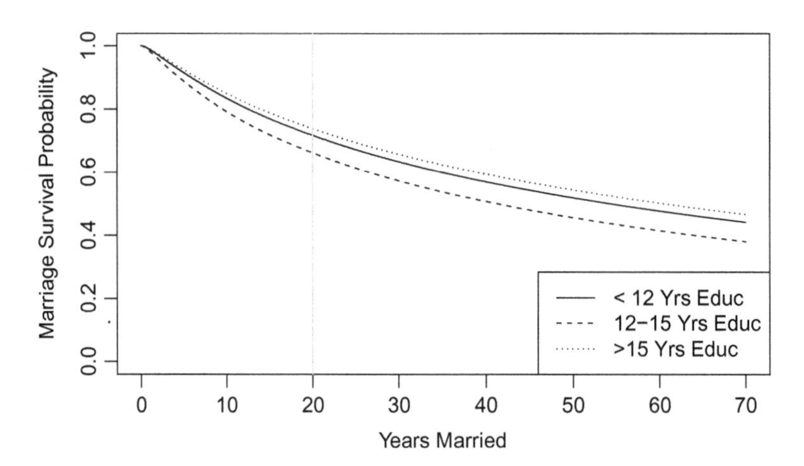

FIGURE 15.3
Estimated marriage survival functions for three different education groups.

From Figure 15.3 we see that the highest marriage survival rates are in the highest husband education group. The middle education group has the lowest marriage survival rate, thus the effect of education on marriage survival is non-linear. Note also, as in all regression models, this regression model does not prove that education *causes* marriage survival because there may be unobserved confounders.

Self-study question: Which unobserved confounding variables might explain the apparent relationship between marriage survival and education?

The 20-year marriage survival threshold is indicated by the vertical line in Figure 15.3. The estimated marriage survival probability at 20 years for the low education group is calculated from the lognormal distribution as `1 - plnorm(20, m1, s)`, giving 0.7175. In the other two groups, these probabilities are 0.6621 and 0.7384. Thus, we estimate that 71.75% of marriages in the low education group survive at least 20 years, that 66.21% of the marriages in the middle education group survive at least 20 years, and that 73.84% of the marriages in the high education group survive at least 20 years. You can locate these numbers on the vertical axis of Figure 15.3.

These predicted percentages of 20-year marriage survival, 71.75%, 66.21%, and 73.84%, all use the combined data from both the non-censored (divorced) people and the censored (died or study ended) people. As indicated in the introduction to this chapter, any data-based estimates that do not handle the censoring correctly will give biased results. For example, consider the lower education group, whose data are exhibited in Table 15.5, obtained using the R commands

```
div.Low.Ed = div[heduc=="< 12 years"]
GT.20.Low.Ed = years[heduc=="< 12 years"] > 20
table(div.Low.Ed, GT.20.Low.Ed)
```

If you consider only the couples who got divorced in the low education group, the 20-year survival rate is very low: There were 393 people in this group, and 75 were divorced after 20 years, leading to a 20-year marriage survival rate of 19.1%. But this result is obviously biased low because it ignores couples who were never divorced at all in the time

TABLE 15.5

Twenty-Year Marriage Results for the 1,288 Couples in the Low Education Group

	Divorce Observed	No Divorce Observed	Total
Time > 20 years	75	509	584
Time ≤ 20 years	318	386	704
Total	393	895	1,288

of the study. If you look at all cases where Time > 20 years, you get 45.3% (584/1,288), but this is also biased low because the time until divorce is always greater than the censoring time. A third data-based calculation combines the 509 couples who were never observed to divorce after 20 years with the 75 couples who were observed to divorce after twenty years, giving $509+75=584$ couples whose time until divorce is greater than 20 years. For the denominator, you could use $584+318=902$, the number of couples whose divorce status by 20 years can be ascertained with certainty. This leads to an estimate $584/902=64.75\%$, much closer to the model-based estimate of 71.75%. But still, this estimate is inferior in that it ignores the 386 couples whose divorce status at 20 years is unknown. So again, any data-based estimates using censored data will be biased, in some cases very badly so. You need a model to infer what might happen with the censored data.

Parametric survival analysis assumes that the Y variable (observed or not) follows a particular distribution. We assumed the lognormal distribution in the divorce example, but that assumption (as is the assumption any particular distribution) is always questionable.

On the other hand, it is a happy fact that you need not assume anything whatsoever about the distribution of censoring time, whether lognormal, or Weibull, or Gaussian, or anything else. There is an assumption about the censoring time that is important however: Censoring must be *non-informative*. That is, the distribution of Y should be the same, regardless of whether the observation is censored. This is a difficult assumption to check because you do not have the actual Y data from the censored cases. Consider the following scenario in the marriage survival case: Suppose that some of the censored cases due to death of a spouse are cases where the surviving spouse would have initiated a divorce earlier, but decided to stay in the marriage simply to care for an ailing partner. In such a scenario, the censored cases include couples where the time until divorce would have happened earlier. This means that the censoring is informative about the possible value of Y, and hence the model-based estimates will be biased towards higher marriage survival rates.

Another example of informative censoring occurs in clinical trials that study survival after treatment for a life-threatening medical condition. Many patients are censored because they stopped coming to the clinic for follow-up evaluation, so their survival times are unknown except for being greater than the time of their last clinic visit. But this group of drop-outs tends to include people who are not tolerating the treatment very well, and typically die sooner, so the censoring is informative. The estimated survival rates will again be biased on the high side.

In the case of the marriage study, perhaps it is the case that the couples who dropped out of the study did so because they were experienced marital troubles. If true, then the censoring is informative in that time until divorce tends to be lower for the censored couples.

To deal with informative censoring correctly, you need a good model for how the censoring mechanism relates to the distribution of Y. There is much published research on this topic, but we will not discuss it here.

15.2 The Proportional Hazards Regression Model

A popular alternative to the parametric survival analysis model presented in the previous section is the Cox proportional hazards regression (CPHR) model, named after Sir David Cox, who was made a knight of the British Empire (hence, the "Sir") for developing this methodology. The CPHR model is a standard tool in clinical trials where time until death (or other event, such as next heart attack) is analyzed. It is not hyperbole to assert that the CPHR model has contributed to more lives saved than any other statistical method.

This model is called "semi-parametric" because it does not require you to specify any distribution of Y, whether lognormal, Weibull, exponential, etc. But it is partially parametric because it assumes a parametric linear model relationship between parameters of the distribution of Y and the X variables. The Gauss-Markov model presented in early chapters is a similarly semi-parametric model, in that it assumes a parametric mean function (typically linear), but makes no assumption about the distribution of Y.

We will present the model in a slightly different way from other sources. The reason for doing this is in the title of the book, "... A Conditional Distribution Approach." Throughout this book, our persistent view of the subject of regression is how to model $p(y \mid x)$, the distribution of Y as it relates to $X = x$, and that is how we will present the CPHR model as well.

The CPHR model can be expressed in terms of the conditional survival functions $S(y \mid x)$, which are directly related to the conditional probability distribution functions $p(y \mid x)$ since

$$(\partial/\partial y)S(y|x) = (\partial/\partial y)\{1 - P(y|x)\} = -(\partial/\partial y)P(y|x) = -p(y|x)$$

Also,

$$S(y \mid x) = 1 - P(y \mid x) = 1 - \int_0^y p(t \mid x)dt$$

Hence, by specifying the model in terms of $S(y \mid x)$, you are also specifying the model in terms of $p(y \mid x)$: As the equations above show you, if you know one of $S(y \mid x)$ or $p(y \mid x)$, then you know the other.

While typical sources express the CPHR model in terms of "log hazards," we will present it in an equivalent way by using the survival function.

The Cox Proportional Hazards Regression (CPHR) model, in terms of survivor functions

$$S(y \mid X_1 = x_1, \ldots, X_k = x_k) = \{S_0(y)\}^{\exp(\beta_1 x_1 + \ldots + \beta_k x_k)}$$

In the CPHR model, $S_0(y) = S(y \mid X_1 = 0, \ldots, X_k = 0)$ is the "baseline" survival function, which is the survival distribution of Y when all X variables are 0.

The model is non-parametric as regards the baseline distribution $S_0(y)$, which can be anything, not restricted to lognormal, exponential, or other parametric forms. But the model is parametric in that it assumes the survival distributions for other values of the X variables are tied to the baseline survival function via the given parametric form involving the β parameters.

The estimation procedure used to obtain the $\hat{\beta}$ values is also semi-parametric, using a pseudo-maximum likelihood technique. This method provides excellent estimates, but unfortunately cannot be compared with the classical likelihood functions that we have discussed previously, so there is no way to tell whether the semi-parametric CPHR model fits the data better than the parametric survival models presented above. Instead, you can compare the results graphically, you can use subject matter considerations, and you can use simulation analysis to decide which is better.

The estimate of the CPHR model in the marriage survival study is found as follows.

```
divorce = read.table("https://raw.githubusercontent.com/andrea2719/
URA-DataSets/master/divorce.txt")
library(survival)
divorce$SurvObj <- with(divorce, Surv(years, div == "Yes"))
cox.phrm = coxph(SurvObj ~ heduc, data = divorce)
cox.phrm
```

The results are as follows:

```
Call:
coxph(formula = SurvObj ~ heduc, data = divorce)


                    coef exp(coef) se(coef)      z       p
heduc12-15 years  0.2388    1.2697   0.0669   3.57 0.00036
heduc16+ years   -0.0778    0.9251   0.1080  -0.72 0.47106


Likelihood ratio test=17.4  on 2 df, p=0.00017
n= 3371, number of events= 1032
```

Hence, the estimated model is

$$\hat{S}(y \mid \text{Educ} = x) = \left\{\hat{S}_0(y)\right\}^{\exp\{0.2388 I(x=\text{"12–15 years"}) - 0.0778 I(x=\text{">15 years"})\}}$$

where $\hat{S}_0(y)$ is the estimated survival function where both indicator variables are zero, i.e., $\hat{S}_0(y)$ is the estimated survival function in the low education group.

As is always the case with indicator variables, it is best to separate the model by categories to understand it, as shown in Table 15.6.

Now, the survivor function $S(y)$ is a probability function, and therefore always a number between 0 and 1. If you raise a number between 0 and 1 to an exponent greater than 1,

TABLE 15.6

Estimated Conditional Survival Functions for the Divorce Data, Given Education Category, Using the CPHR Model

Education	Estimated Survival Function
<12 years	$\hat{S}_0(y)$
12–15 years	$\left\{\hat{S}_0(y)\right\}^{\exp(0.2388)} = \left\{\hat{S}_0(y)\right\}^{1.270}$
>15 years	$\left\{\hat{S}_0(y)\right\}^{\exp(-0.0778)} = \left\{\hat{S}_0(y)\right\}^{0.925}$

you get a *smaller* number, but still one between 0 and 1. And if you raise a number between 0 and 1 to an exponent that is between 0 and 1, you get a *larger* number, but still one between 0 and 1.

Self-study question: What happens if you raise 0.7 to the 1.2 power? What happens if you raise 0.7 to the 0.9 power? How do the answers to those questions help you to understand the preceding paragraph?

Hence, you know that the CPHR estimate of survival function for the 12–15 education category is *below* that of the "<12" category, and that the CPHR estimate of survival function for the ">15" education category is *above* that of the "<12" category.

The discussion above explains a seeming contradiction between the computer output for CPHR and that of the lognormal parametric survival model: The coefficients have opposite signs. A positive coefficient in the CPHR means *less* survival and a negative coefficient indicates *more* survival. The Cox model estimating survival functions that look like those in Figure 15.3, in the sense that the "12–15" category has the lowest survival and the ">15" category has the highest survival.

The results are best understood graphically, as shown in Figure 15.4.

R code for Figure 15.4

```
divorce = read.table("https://raw.githubusercontent.com/andrea2719/
URA-DataSets/master/divorce.txt")
library(survival)
divorce$SurvObj <- with(divorce, Surv(years, div == "Yes"))
cox.phrm = coxph(SurvObj ~ heduc, data = divorce)
plot(survfit(cox.phrm, newdata=data.frame(heduc="< 12 years"),
  conf.int=0),main="Cox Survival Model", xlab = "Years Married",
  ylab="Survival Probability",ylim=c(0,1.0))
lines(survfit(cox.phrm, newdata=data.frame(heduc="12-15 years"),
  conf.int=0), lty=2)
lines(survfit(cox.phrm, newdata=data.frame(heduc="16+ years"),
  conf.int=0), lty=3)
legend("bottomright", c("< 12 Yrs Educ", "12-15 Yrs Educ",
  ">15 Yrs Educ"), lty=1:3)
```

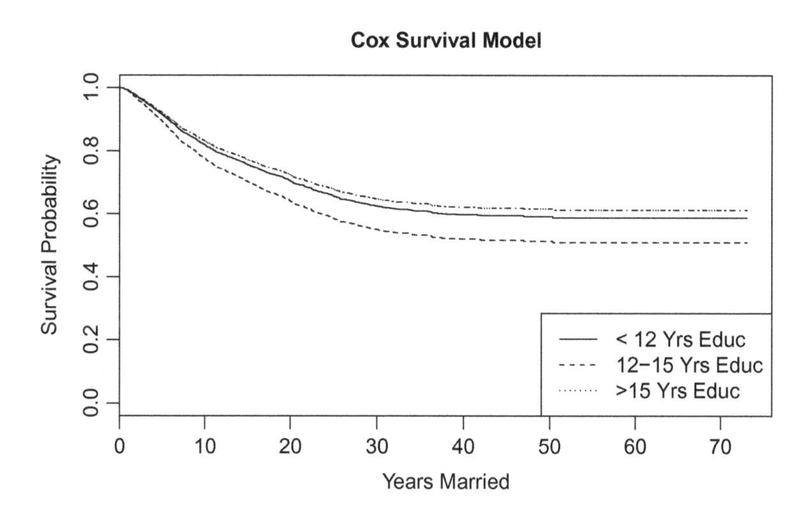

FIGURE 15.4
Estimated survival curves using the CPHR with the divorce data.

You can see that Figure 15.4 (from the semi-parametric survival model) and Figure 15.3 (from the parametric survival model) give similar results. Unfortunately, you cannot tell which one is better based on log-likelihood.

This is just a brief introduction, and not a definitive treatise on survival analysis using CPHR. To learn more about survival analysis, you will need to learn about the "hazard function," which is in the very name "Cox proportional hazards regression," but which we have not included here because it is somewhat of a digression from the main theme of this book. Mainly, we just want you to realize that (i) censored data needs special handling, and (ii) the framework within which to view censored data is the same framework within which to view all of regression; namely, as a model for $p(y \mid x)$, and (iii) while maximum likelihood provides a good method for estimating all regression models, it is essential that you use a likelihood-based method (rather than any purely data-centric technique) in the presence of censored data.

15.3 The Tobit Model

In Chapter 14, you learned models for when there are many 0's in your observable Y data; namely, Poisson and negative binomial regression. For those models, the Y data must be integers, and the values of Y should not be "too large" (say, not more than 100 if you need an ugly rule of thumb), because such large values are extremely unlikely with the Poisson and negative binomial distributions when there are also 0's.

The Tobit model is used for cases where (i) there are zeros in the Y data, and (ii) the Y data can be (but do not have to be) non-integer, and (iii) the Y data can be numbers that are large, even in the millions or billions. Such cases abound in financial and accounting reporting data. A reported financial or accounting number might be zero for many or even most firms, but can be some large positive amount for others.

For example, consider the variable Y = "Court-mandated payout due to successful lawsuit" in a given year. Most firms report nothing ($Y = 0$) in this category, but others report varying amounts, in the millions or even billions of dollars, depending on the severity of their infraction. For another example, consider Y = jail time assigned for a convicted defendant. Many convicts are assigned no jail time whatsoever $(Y = 0)$; others are assigned jail time with varying lengths.

Self-study question: What X variables might you use to predict Y in the two examples given above?

Your main goal in using regression (as stated persistently in this book) is to model $p(y \mid x)$, the conditional distribution of Y given $X = x$. The Tobit model is a model $p(y \mid x)$ that puts a certain probability on the discrete outcome $Y = 0$, but allows a continuous distribution of Y for $Y > 0$. This is an unusual case—a probability distribution that has both discrete and continuous components! You might have assumed that probability distributions could be entirely classified as either discrete or continuous, but now you know that there is another possibility: The Tobit specification of $p(y \mid x)$ is neither entirely discrete nor entirely continuous. Figure 15.5 shows examples of Tobit distributions $p(y \mid x)$. Since the total probability is 1.0 (that fact is *always* true), the probability of the 0 $(\Pr(Y = 0))$ plus the area under the curve to the right of zero $(\Pr(Y > 0))$ must always add to 1.0. Thus, as indicated in Figure 15.5, as the Tobit distributions $p(y \mid x)$ morph to the right, the probability on the discrete $(Y = 0)$ event must decrease. (And if such behavior is not appropriate for your data-generating process, then you should not use the classic Tobit model.)

R code for Figure 15.5

```
Y.list = seq(0,10, .01)
d1 = dnorm(Y.list, -2, 3)
d2 = dnorm(Y.list,  1, 3)
par(mfrow=c(1,2))
plot(Y.list, d1, type="l", yaxs="i", ylim = c(0,.8),
  xlab="Lawsuit Payout (in Millions of $),y",
  ylab = "p(y | X = Low)", xlim = c(-.5,10))
points(c(0,0), c(0, pnorm(0, -2,3)), type="l", lwd=2, col="gray")
plot(Y.list, d2, type="l", yaxs="i", ylim = c(0,.8),
  xlab="Lawsuit Payout (in Millions of $),y",
  ylab = "p(y | X = High)", xlim = c(-.5,10))
points(c(0,0), c(0, pnorm(0, 1,3)), type="l", lwd=2, col="gray")
```

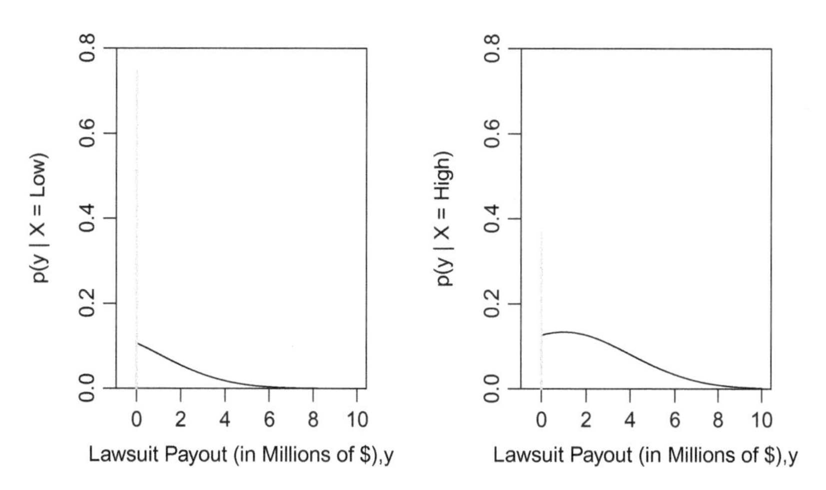

FIGURE 15.5
Tobit distributions $p(y \mid x)$ for different $X = x$.

You might have a hard time distinguishing the Tobit model from the Poisson/negative binomial models. After all, both models are suggested by Y data that have both zeros and positive values. To make this distinction, you should first understand the distributions $p(y \mid x)$ before deciding which model is more appropriate. Compare Figure 15.5 with Figure 14.1 and with Figure 14.5: Which one is most appropriate for your Y data? That is the one you should pick.

Censoring is used to define the Tobit distributions shown in Figure 15.5, although in a different way than it is used with survival analysis. Specifically, observations having $Y = 0$ are considered "left censored," as if the actual observation could be negative. Like the latent variables used in ordinal regression, this construction is just a convenient fiction that is used to define the model $p(y \mid x)$. Usually, there are no potentially observable negative Y values when you use Tobit models.

To define the models $p(y \mid x)$ as shown in Figure 15.5, first define a fictitious latent random variable Z as follows:

$$Z = \beta_0 + \beta_1 X + \sigma\varepsilon, \text{ where } \varepsilon \sim \mathrm{N}(0,1)$$

The Tobit distributions $p(y \mid x)$, as shown in Figure 15.5, are the distributions of $Y \mid X = x$, where

$$Y = Z, \text{ if } Z > 0, \text{ and}$$

$$Y = 0, \text{ if } Z \leq 0$$

For example, suppose $Z = -3 + 1.0X + 3\varepsilon$. Suppose also that $X = 1$ is a "low" value of X, and $X = 4$ is a "high" value of X. Then the Tobit distributions $p(y \mid x)$ are exactly as shown in Figure 15.5. To understand this better, see Figure 15.6, which shows simulated Z data, X data, and the associated Y data, according to this model.

R code for Figure 15.6

```
X = rnorm(2000, 2.5,1)
Z = -3 + 1*X + rnorm(1000, 0, 3)
plot(X,Z, col="gray", xlab = "Net Profits",
  ylab="Litigation Award (Millions of $)")
Y = ifelse(Z<0,0,Z)
points(X,Y)
abline(v=c(1,4), lwd=2)
abline(-3,1, lty=2)
```

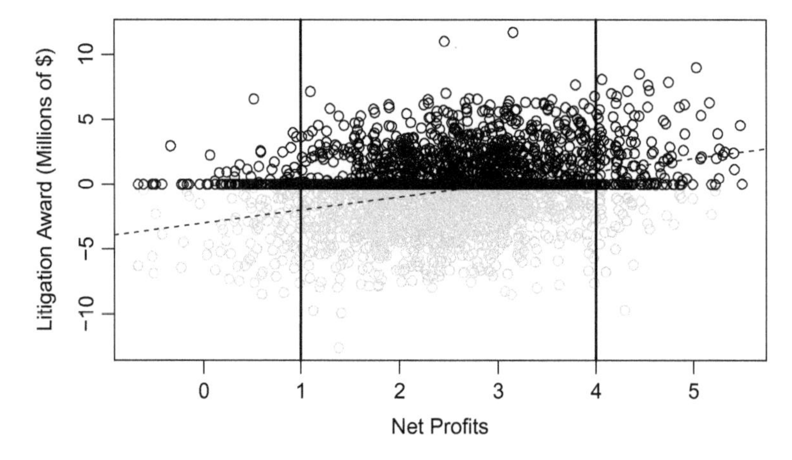

FIGURE 15.6
Scatterplot illustrating Tobit model. Dark points are observed data; note that there are many 0's. Light points are fictitious "latent" data values; these do not really exist but are used to define the model $p(y \mid x)$: If the light point is less than zero, then the observed Y is zero. The distributions $p(y \mid X = 1)$ and $p(y \mid X = 4)$ are shown by the locations of the Y data (black points) along the vertical lines; these distributions are exactly the ones shown in Figure 15.5. The upward sloping line determines the overall level and the rate of morphing of the distributions $p(y \mid x)$ shown in Figure 15.5.

You can estimate the Tobit model using the "`survreg`" function presented above, but you have to specify left censoring rather than the default of right censoring.

15.3.1 Predicting Number of Days Lost to Back Injury

Worker compensation is a huge expense. In manufacturing and warehousing, much of this expense is due to lower back injuries. An Industrial Engineering department at a university developed an index of job-induced stress (X = Stress Index) on the lower back, and used it to predict Y = number of days lost to back injury, for a collection of warehouse workers. The regression model is $p(y \mid X = x, \boldsymbol{\theta})$, which is determined from

$$Z = \beta_0 + \beta_1 X + \sigma\varepsilon$$

$$Y = Z, \text{ if } Z > 0, \text{ and}$$

$$Y = 0, \text{ if } Z \leq 0$$

While the classical Tobit model assumes a normal distribution for ε, this assumption can easily be relaxed to allow non-normal distributions using the "survreg" function. The parameters $\theta = (\beta_0, \beta_1, \sigma)$ are estimated using maximum likelihood; as indicated above, "censored" observations $(y_i = 0)$ contribute the probability $\Pr(Z_i \leq 0 \mid x_i, \theta)$ to the likelihood function, while "non-censored" observations $(y_i > 0)$ contribute the density $p(z_i \mid x_i, \theta)$ to the likelihood function.

First (always first!), you should get to know your data.

```
back = read.csv ("https://raw.githubusercontent.com/andrea2719/
URA-DataSets/master/liles.csv")
Y = back$dayslost
X = back$alr
plot(jitter(X), Y, xlab="Index of Lower Back Stress", ylab="Days Lost")
```

Notice in Figure 15.7 that there seems to be more workers losing time to back injury for larger back stress, but there is also one outlier in the upper left that severely goes against this trend.

To fit the Tobit model, use the following code.

```
fit.normal <- survreg(Surv(Y, Y>0, type='left') ~ X, dist='gaussian')
summary(fit.normal)
```

The output is:

```
Call:
survreg(formula = Surv(Y, Y > 0, type = "left") ~ X, dist = "gaussian")
              Value Std. Error     z        p
(Intercept) -112.25     30.655 -3.66 2.50e-04
X             14.71      8.759  1.68 9.31e-02
Log(scale)     4.08      0.207 19.68 3.45e-86

Scale= 59

Gaussian distribution
Loglik(model)= -121.1   Loglik(intercept only)= -122.8
        Chisq= 3.47 on 1 degrees of freedom, p= 0.063
Number of Newton-Raphson Iterations: 5
n= 206
```

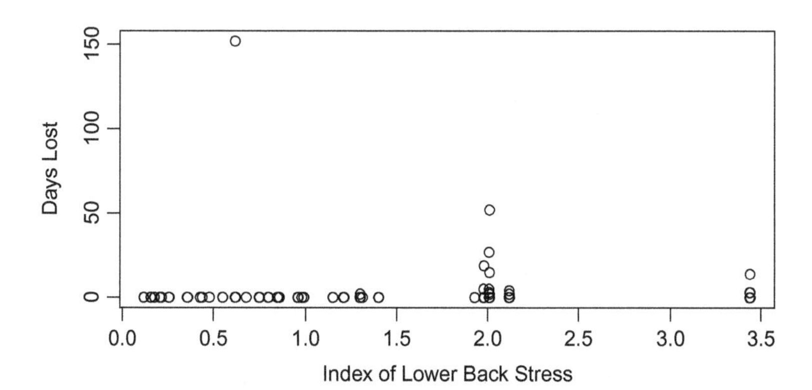

FIGURE 15.7
Work days lost to back injury as a function of index of stress on the lower back.

The Tobit model estimate of $p(y \mid x)$ is the $N(-112.25+14.71x, 59^2)$ density for $y > 0$, with $\Pr(Y = 0 \mid x)$ estimated to be the probability that a $N(-112.25+14.71x, 59^2)$ random variable is less than 0. Figure 15.8 shows the estimates of these distributions for $X = 0.5$ and $X = 3.0$; the shaded area corresponds to the probability that $Y = 0$.

R code for Figure 15.8

```
ylist = seq(-300, 200, .1)
d1 = dnorm(ylist, -112.25 + 14.71*.5, 59)
d2 = dnorm(ylist, -112.25 + 14.71*3, 59)
par(mfrow=c(1,2))
plot(ylist, d1, type="l", yaxs="i", ylim = c(0,.008), xaxt="n",
   xlab = "Days Lost, y", ylab = "p(y | Stress Index = 0.5)")
axis(1, at = c(0,50,100,150,200), lab=c(0,50,100,150,200))

x1 = seq(-300, 0,.1); y1 = dnorm(x1,-112.25 + 14.71*.5, 59)
col1 <- rgb(.2,.2,.2,.5)
polygon(c(x1,rev(x1)), c(rep(0,length(y1)), rev(y1)), col=col1,
border=NA)
plot(ylist, d2, type="l",yaxs="i", ylim = c(0,.008), xaxt="n",
    xlab = "Days Lost, y", ylab = "p(y | Stress Index = 3.0)")
axis(1, at = c(0,50,100,150,200), lab=c(0,50,100,150,200))
x2 = seq(-300, 0,.1); y2 = dnorm(x1,-112.25 + 14.71*3, 59)
col1 <- rgb(.2,.2,.2,.5)
polygon(c(x2,rev(x2)), c(rep(0,length(y2)), rev(y2)), col=col1,
border=NA)
```

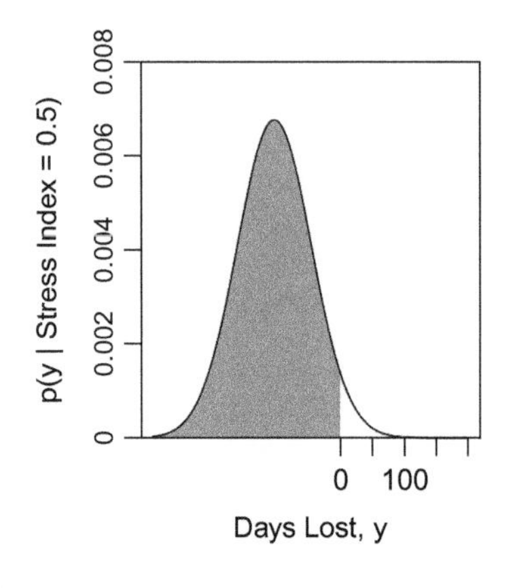

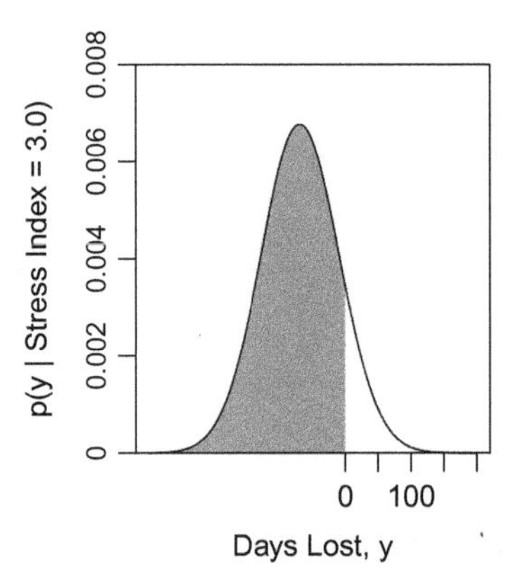

FIGURE 15.8
Estimated distributions $p(y \mid X = x)$ for Y = days lost to back injury as related to X = lower back stress index. Only the part of the curve for $y \geq 0$ is part of this distribution. For $y > 0$, $p(y \mid x)$ is the actual density. For $y = 0$, the probability is given by the area of the shaded region. Left panel shows estimate of $p(y \mid X = 0.5)$; right panel shows estimate of $p(y \mid X = 3.0)$.

Notice that in the output from survreg, the trend towards higher days lost for larger stress index (represented by the morphing to the right shown in the graphs of Figure 15.8) is explainable by a model where days lost is unrelated to stress index, since

TABLE 15.7

Fits of Tobit Models Using Different Distributions for the Y = Days Lost Variable

Distribution	AIC
Gaussian	248.1868
T(4)	233.4808
Logistic	240.6989

$p = 9.31 \times 10^{-2} = 0.0931$ does not meet the standard 0.05 threshold. But note also the outlier in the upper left of Figure 15.7, where $y = 152$ days lost, for a person whose X is near 0.5. As seen in the left panel of Figure 15.8, the value $y = 152$ has very low likelihood under the normal model. Since the technique is *maximum likelihood*, this point has had inordinate influence in pulling the mean of the distribution upwards, toward the outlier, for smaller X. A better model would be one that accommodates such outliers. In particular, a more heavy-tailed distribution would allow more extreme outliers to exist without pulling the mean function inappropriately toward the outlier.

The `survreg` function allows two such distributions, the T distribution (the software defaults to df = 4, but you can pick other df) and the logistic distribution. These distributions are symmetric and range from $-\infty$ to ∞, like the normal distribution, but have heavier tails than the normal distribution, and therefore their likelihood functions are not so influenced by outliers. Changing `dist = "gaussian"` to `dist = "t"` and `dist = "logistic"` gives the results shown in Table 15.7.

Thus, the T distribution provides the best fit among the three. And, this distribution gives results not so heavily influenced by the outlier, where the apparent trend in the distributions is not so easily explained by a model where days lost is unrelated to stress index $\left(p = 4.22 \times 10^{-2} = 0.0422\right)$, as shown in the following output:

```
Call:
survreg(formula = Surv(Y, Y > 0, type = "left") ~ X, dist = "t")
              Value Std. Error     z        p
(Intercept) -37.17      12.984 -2.86 4.20e-03
X             6.78       3.335  2.03 4.22e-02
Log(scale)    2.62       0.314  8.34 7.22e-17

Scale= 13.8

Student-t distribution: parmameters= 4
Loglik(model)= -113.7   Loglik(intercept only)= -117
        Chisq= 6.48 on 1 degrees of freedom, p= 0.011
Number of Newton-Raphson Iterations: 8
n= 206
```

While it is sensible that non-normal distributions should be considered in this example, the particular strategy suggested in the analysis above, which is "pick model with lowest AIC and use that one" tends to give p-values that are biased low. Hence, we could be accused of "p hacking" in the analysis above to get a p-value (0.0422) that met the standard $p < 0.05$ threshold. Review the section "A note on replicability" at the end of Chapter 14 for further discussion of this issue.

15.4 Interval Censored Data

Have you ever taken a survey that asked your income, and you were supposed to select a category range such as the following?

Income less than 10,000

Income between 10,000 and 20,000

Income between 20,000 and 50,000

Income between 50,000 and 120,000

Income greater than 120,000

These are called *interval censored data*, because the actual Y value is unknown, except that it lies in a known interval. Did you know that you could estimate the distribution of the *actual* (*unobserved*) Y, $p(y)$ using such data? The lognormal model would fit well. You can also model how this distribution morphs with $X = x$, i.e., you can model the distribution $p(y \mid x)$, using such data. And you can compare different distributions' fit to the data using penalized log-likelihoods.

For example, consider Table 15.8, which displays such data based on n = 145 survey respondents.

Let $P(y;\theta)$ denote the *cumulative distribution function* corresponding to whichever probability density function $p(y;\theta)$ you consider. Assuming that the n = 145 potentially observable responses corresponding to the observed data in Table 15.8 are independent, the likelihood function for the sample is

$$L(\theta \mid \text{data}) = \{P(10{,}000;\theta)\}^{11} \times \{P(20{,}000;\theta) - P(10{,}000;\theta)\}^{32} \times \{P(50{,}000;\theta) - P(20{,}000;\theta)\}^{78}$$

$$\times \{P(120{,}000;\theta) - P(50{,}000;\theta)\}^{20} \times \{1 - P(120{,}000;\theta)\}^{4}$$

You can estimate θ by choosing $\hat{\theta}$ to maximize $L(\theta \mid \text{data})$, and you can compare different distributions via their AIC or other penalized likelihood statistics. You can also extend this methodology to regression, where each individual observation i contributes a different amount to the likelihood, depending on the x_i value. See the `R` package "`icenReg`" for further details.

TABLE 15.8

Interval Censored Income Data for n = 145 Survey Respondents

Category	Count
Income less than or equal to 10,000	11
Income from 10,001 to 20,000	32
Income from 20,001 to 50,000	78
Income from 50,001 to 120,000	20
Income greater than 120,000	4
Total	145

Reference

Lillard, L. A., & Panis, C. W. (2000). *aML multilevel multiprocess statistical software, release 1.0.* California: EconWare.

Exercises

1. Verify the numbers in Table 15.3 by fitting the appropriate models.
2. Consider the "Time until divorce" example in Chapter 15. Include the "mixed" variable in the model, which is an indicator of whether the marriage is a mixed-race marriage, along with the "educ" variable that is used in the book.

   Some comments on the "mixed-race" variable: We are not sure what the intent of the study was, but it seems to us to be singling out the Black race for some reason, and we are not particularly comfortable with that emphasis. There are many other mixed-race possibilities. In addition, it appears from the data analysis that marriage survival is less in such mixed-race couples, but what about other types of mixed-race couples? At the time this particular data set was collected, this reduction in marriage survival may have been caused by additional familial and societal stresses for mixed-race couples, due to societal norms. However, we suspect that more current data would show less of an effect, because times and attitudes have changed such that a "mixed-race" marriage is hardly worth a mention, let alone be a cause of marital stress due to societal pressures. Or at least, that is the way we would hope things are by now. In any event, it is an interesting exercise in data analysis, so please take it that way.

   a. Is the estimated effect of the "mixed-race" variable explainable by a data generating process where "mixed-race" has absolutely no effect on marriage survival? (See our comment above on the "mixed-race" variable. One might wish that such a data generating process was, in fact, the real data-generating process.)
   b. Draw graphs of the estimated survival functions to compare mixed-race with non-mixed-race couples, in all three education groups. (Three separate graphs, each with two overlaid curves.)
   c. Estimate the 20-year marriage survival rates in the graphs of 2.b (six numbers). Then write three sentences, one for each education group, that compares the 20-year marriage survival rates for mixed and non-mixed couples. Use plain language, suitable for presentation to a non-statistician audience, in these sentences. Pretend your audience is interested in marriage longevity. Do not delve into causal issues, just report the survival rates, and what they mean. Since these are all conditional distributions, the word "among" can be particularly helpful in your answers.

16

Outliers: Identification, Problems, and Remedies (Good and Bad)

What is an outlier? Here is a definition:

Definition of an outlier

An outlier is a data point that is (i) far from the rest, and (ii) rare.

Let's parse that definition for a bit.

First, "data point" indicates either an individual number, or a combination of numbers, such as a point on a scatterplot which indicates two numbers (x, y). Second, "far," in the English language, refers to distance. "Far" also refers to distance in statistics, but in statistics, "distance" is defined *relative* to the typical range (or standard deviation) of the data. Third, "rare" means "unusual," meaning that only a small percentage of the data are near the outlier.

At this point, we should note that, in some sources on statistics, "outlier" is defined as a "mistake." And sometimes, a rare, extreme data point is indeed a mistake. But just because a data value is rare and extreme does not make it "wrong": Jeff Bezos's income is rare and extremely large, but Jeff Bezos is not a "mistake." Further, if there are mistakes in the data, they can occur within the middle of the data as well as at the extremes. So, to reiterate, we define an outlier only as a rare, extreme data point. It may or it may not be a mistake.

Later in this chapter, we will discuss the differing effects of outliers in the X and Y variables. For now, consider the following univariate data from a variable "V," which could be either an X or a Y variable.

$$v_1 = 2.31, v_2 = 1.28, v_3 = 2.15, v_4 = 23.02, v_5 = 3.02, v_6 = 1.10$$

Here, $v_4 = 23.02$ is an obvious outlier, because it is "far" from the rest, and there is only one such unusual value. If there were several other values near 23.02, such as 20.1, 39.4, 23.5, and 30.0, you could say that the number 23.02 is indeed "far" from the lower data values (2.31, 1.28, 2.15, 3.02, and 1.10), but it would not be "rare," and would therefore not be an outlier in that case.

In terms of ordinary distance, $v_4 = 23.02$ is "far" from the rest because it is 20 units away from the next nearest observation, $v_5 = 3.02$.

Now consider another variable, say "W":

$$w_1 = 231, w_2 = 128, w_3 = 215, w_4 = 322, w_5 = 302, w_6 = 110$$

Here, $w_4 = 322$ is also 20 units away from the next nearest observation $(w_5 = 302)$, but it is no longer an outlier. See Figure 16.1.

R code for Figure 16.1

```
par(mfrow=c(1,2))
v = c(2.31, 1.28, 2.15, 23.02, 3.02, 1.10)
plot(v, rep(0,6), ylab="", yaxt="n")
points(23.02, 0, pch=16)
w = c(231, 128, 215, 322, 302, 110)
plot(w, rep(0,6), ylab="", yaxt="n")
points(322, 0, pch=16)
```

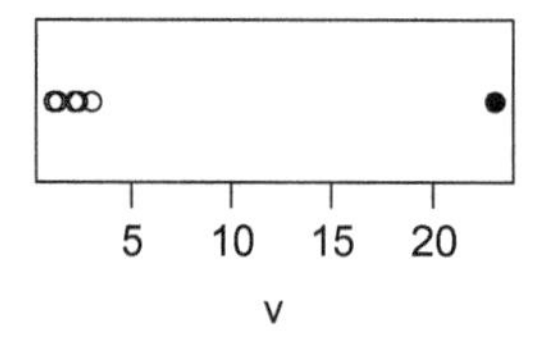

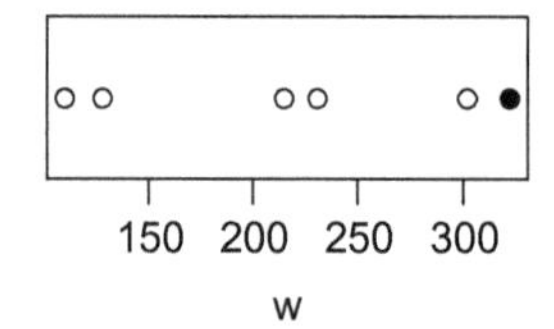

FIGURE 16.1
Left panel: The solid point is an outlier. Right panel: The solid point is not an outlier. In both data sets, the solid point is 20 units distant from the next nearest value.

As seen in Figure 16.1, the definition of "far," as it relates to statistical outliers, must use some other metric than ordinary straight-line (also called *Euclidean*) distance. The problem indicated by Figure 16.1 is that the standard deviations of V and W differ greatly; they are 8.62 and 87.0, respectively. By measuring distance in terms of standard deviations, you can more easily identify outliers. In the left panel of Figure 16.1, the outlier is $(23.02-3.02)/8.62=2.32$ standard deviations from the next closest value; in the right panel the red point is $(322-320)/87.0=0.023$ standard deviations from the next closest value.

The z-value measures distance from the *average* of the data, in terms of standard deviations.
(The z-value is technically called a "signed distance" because it can be positive or negative.

Definition of z-value

$$z\text{-value} = \{\text{data value} - (\text{average of data values})\} / (\text{standard deviation of data values})$$

True distances can never be negative.)

You can easily identify an outlier in a particular variable (i.e., in a column of your data frame) by calculating its z-value. The "`scale`" function in R does this for you automatically. Using `scale(v)` and `scale(w)` on the data above gives the z-value for v_4 to be 2.034, indicating that $v_4 = 23.02$ is 2.034 standard deviations above the average of the V data. The z-value for w_4 is 1.196, indicating that $w_4 = 322$ is 1.196 standard deviations above the average of the W data.

Ugly rule of thumb for identifying outliers based on z-values

Observations with $|z| > 3$ can be considered "outliers."

The z-value-based rule of thumb, like most rules of thumb, is "ugly" because you should not take it too seriously. An observation having $z = 2.98$ is as much of an outlier as one having $z = 3.01$. Further, there are many statistical methods to identify outliers other than z-values.

As the word "ugly" in the "ugly rule of thumb" given above indicates, there is no hard boundary, or threshold, by which a point is classified as "outlier" or "not an outlier." There are just degrees of extremity; degrees of "*outlierness*." The z-value is one statistic that measures *outlierness*, with larger z-values indicating greater "*outlierness*." Below, we give other measures of *outlierness* that have special relevance in regression analysis.

More important than any "ugly rule of thumb" is the logic behind it. In the case of the "large z indicates outlier" rule of thumb, the logic is based on Chebychev's inequality, which states that at most $(100/k^2)\%$ of the data will have $|z|$ values greater than k. For examples: (i) at most 11.1% of the $|z|$ values can be greater than 3; (ii) at most 1.0% of the $|z|$ values can greater than 10. Thus, larger $|z|$ indicates a data value that is both more distant and more rare; i.e., more of an outlier.

Self-study question: How can you verify that Chebychev's inequality works for $k = 3$ and $k = 10$ by using simulated data from a normal distribution? By using simulated data from a Cauchy distribution?

Suppose you saw a data set where about $n = 10{,}000$ data values were between 0 and 1, but there was one data value around 1,000. To see what kind of an outlier this is, let's simulate some data.

```
set.seed(12345)
Most.data = runif(10000)
Outlier.data.value = 1000
z.all = scale(c(Most.data, Outlier.data.value))
z.all[10001]
```

The z-value is 99.95 for the unusual data value, 1,000. So that data value is about 100 standard deviations from the mean—Wow! This is a very extreme outlier!

In the case where the data points are bivariate or multivariate, "outlierness" takes on another dimension (pun intended!) Consider the scatterplot of Figure 16.2.

R code for Figure 16.2

```
set.seed(12345)
W = rnorm(100, 70,10)
U = .05*W + rnorm(100,0,.1)
W = c(W,80); U = c(U,2.8); plot(W,U)
points(80,2.8, pch=19)
scale(W)[101]; scale(U)[101]
```

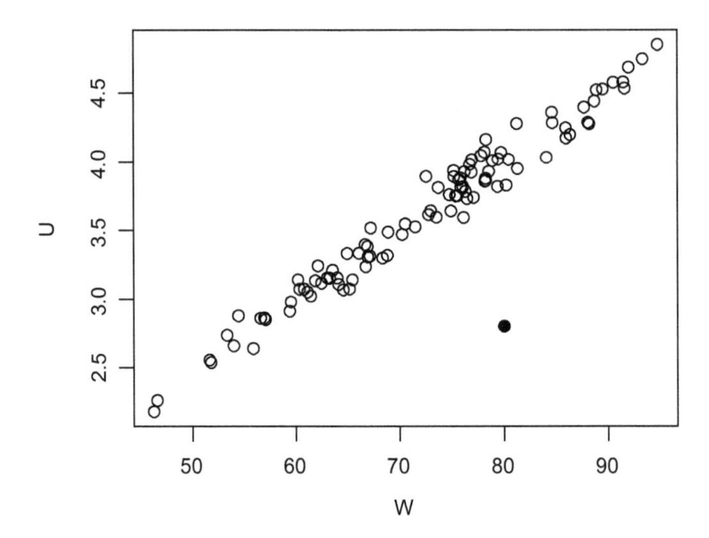

FIGURE 16.2
Scatterplot of (W,U) data. The solid point $(W,U)=(80,2.8)$ is an outlier, even though neither value is an outlier in the univariate sense. The value $W = 80$ is only $z = 0.67$ standard deviations above the mean of the W data. Likewise, the value $U = 2.8$ has z-value $z = -1.41$, and is thus only 1.41 standard deviations below the mean of the U data.

As seen in Figure 16.2, you cannot always identify outliers in multiple dimensions by using univariate z-values alone. You must also consider the joint distribution of the data values, because, depending on the correlation structure, certain *combinations* of values (such as (80, 2.8) shown in Figure 16.2) may be unusual, even when the *individual* values, by themselves, are not unusual.

16.1 What Is the Problem with Outliers?

The problem with outliers is that they can have an inordinate effect on your estimated model, particularly if you use ordinary least squares (OLS). Recall that your regression model $p(y \mid x)$ is one you assume to have produced your data. More importantly, $p(y \mid x)$ is also what you assume to produce potentially observable data that you have not observed. If your model states that the $p(y \mid x)$ are normal distributions (which would imply that the OLS estimates are optimal), then an observed Y data value that is far (e.g., 5 standard deviations) from the rest clearly indicates that your model is wrong. If your data really are produced by normal distributions, you will rarely see an observation that is 5.0 standard deviations from the regression mean line, let alone 99.95 standard deviations as shown above. If an observation is 99.95 standard deviations from a possible value of the mean, its contribution to the normal distribution likelihood function is

$$(1/\sigma)(2\pi)^{-1/2}\exp(-0.5\times 99.95^2)$$

Notice that this a *very small* number. So, if you have such a data point in your data set, the estimate of the mean value that maximizes the likelihood will have to be chosen to make this number not so small. Specifically, both the estimated mean (line) and the standard deviation will be chosen by the maximum likelihood algorithm to make the z-value smaller, so that the data point does not "drag down" the likelihood so much. Hence, this one outlier data point will greatly force the mean function and standard deviation towards the outlier, when you assume a normal distribution.

This problem, where an outlier has an inordinate effect on the fitted model, is more of a problem with the normal distribution than it is with other distributions. Distributions that produce extreme values (called heavy-tailed distributions), such as T distributions with small degrees of freedom and the Laplace distribution, have higher density values at the extremes than does the normal distribution (which is why they are called "heavy-tailed"). Thus, with heavy-tailed distributions, the likelihood of outliers is not as small as it is with the normal distribution. When you use heavy-tailed distributions $p(y \mid x)$, outliers have less influence on the likelihood function, and thus the model parameter estimates (mean, scale) are not so highly influenced by outliers. With such heavy-tailed distributions, outliers are expected.

We introduced this issue in Chapter 2, Section 2.3. See Figure 2.3 for a vivid illustration of how maximum likelihood with a heavy-tailed distribution (the Laplace distribution, in the case of Figure 2.3) gives estimates that are not influenced inordinately by outlier(s).

16.2 Why Outliers Are Important

Outliers provide valuable insights, and often they are much more important than the "common" data. In production and manufacturing applications, outliers identify conditions where the production process is "out of control," indicating need for immediate investigation and/or remedial action. They also might indicate good things: A famous engineer/statistician claimed that many of his patents were the results of locating outliers that showed unexpectedly good results. So he uncovered their mechanisms through further study, replicated the effects, and then got the patents.

In health care, the "1 in a 10,000" type of person will be allergic to certain drugs, foods, or therapies, resulting in outlier data values. These people need to be identified so that you don't make them sick or kill them inadvertently. A statistician in the pharmaceutical industry was heard to say, "All this talk about deleting outliers is backward. In my work, I ignore the main body of the data and just study the outliers."

Outliers are also extremely important in finance. Nassim Nicholas Taleb, in his book *The Black Swan* (Taleb, 2007), notes that most of the money that changes hands in financial markets is caused by outliers. The famous "80–20" rule states that 80% of the activity is caused by 20% of the potential causes, but Taleb argues for a "99–1" rule rather than an "80–20" rule, because around 99% of the change of money may be caused by 1% of the financial instances (these 1% are the outliers).

Outliers have an influence on society in general. Author Malcolm Gladwell has a book about those rare people that have "made it big" and have had great influence. The title of his book: *Outliers.*

As the examples above show, outlier identification is one of the most important things you can do with statistical data, because the most interesting facets of a scientific process are often its outliers. Thus, at best, automatic outlier deletion (called *truncation*), and its relative, *Winsorization* (discussed later in this chapter) are silly because you are likely throwing away the most interesting features of your data. At worst, truncation and Winsorization are bad scientific practices, verging on academic and scientific misconduct.

Here is an amusing true story. A non-statistician consultant for a medical device start-up company was seeking advice. His client company was making a device that purports to measure the same thing as electroencephalograph (EEG) data on the human heart, but one that can be used easily at home to give early warning signals of heart problems in people with heart conditions. The gold standard is the EEG, but the patient has to be in the medical office hooked up to electrodes to get such data. The medical device company has lots of data comparing the EEG with the device, and found that most of the time, the device works very well, giving measurements within 1% of the gold standard EEG. However, in the rare occasions where the heart is unusually stressed, the device misses the EEG target badly. The non-statistician consultant had heard that in some academic disciplines it is considered "OK" to delete or Winsorize the outliers. So he deleted the rare occasions where the device badly missed the EEG target. After deletion, he found there were no cases where the device performed badly. Based on that analysis, he was thinking of recommending to his client company that they continue with product development, "as is."

Self-study question: What do you think about the non-statistician consultant's recommendation?

You should always identify and consider outliers carefully. You should never ignore them, or "sweep them under the rug." Of course, if an outlier is truly an incorrect data value, then you should do some research and replace it with the correct data value. If you cannot do that, the second-best option is to delete the erroneous data value, but realize, as you learned in the previous chapter on censoring, that such deletion can cause bias. Also, while you are finding errors in the data, do not just look at the outliers. If some of the outliers are mistakes, it is likely that the mechanism that caused these mistakes (e.g., data entry error) also caused mistakes that are not outliers. Thus, you should scour your entire data set for mistakes and do the needed research to correct them all. Mistakes are garbage, and if you put garbage into your estimated model, then your estimated model will likewise be garbage (also known as "garbage in, garbage out," or GIGO). It does not matter whether the garbage are outliers; they are still garbage, and your estimated model will likewise be garbage.

If your outliers are not mistakes, then you should model them. You want your regression model $p(y \mid x)$ to produce data that look like what you will actually observe, right? Therefore, if you are modeling a process that produces outliers every once in a while, then your model $p(y \mid x)$ should also produce outliers once in a while. Remember, a "good" model is one that produces realistic potentially observable data.

We will not recommend automatic outlier deletion (truncation or Winsorizing), as it is usually just bad statistical and scientific practice. After reading this chapter, you will understand why these methods are so bad. You will also be able to (i) understand what an outlier is, (ii) know how to recognize outliers in your data, (iii) understand when one observation is more of an outlier than another, (iv) understand what effects outliers have on your regression analysis, and (v) understand how to model outlier-prone data-generating processes.

16.3 Identifying Outliers in Regression Data: Overview

Here and everywhere else in this book, "regression data" refers to data where you have a single "Y" variable and one or more "X" variables. Each variable is indexed by "i," the observation number, where $i = 1,2,\ldots,n$. For such data, the following two classifications of outliers are useful.

1. The combination of data values $(x_{i1}, x_{i2}, \ldots, x_{ik})$ may indicate that observation i is an *outlier in X space*.
2. The combination of data values $(y_i, x_{i1}, x_{i2}, \ldots, x_{ik})$ may indicate that observation i is an *outlier in Y | X space*.

We can also consider whether an observation y_i is an *outlier in Y space*, but this case is not nearly as much of a concern as the above two cases.

To understand "outlier in X space," see Figure 16.2. If both W and U are "X" variables, then the solid point indicates an outlier in X space. Of course, any other point well outside the main scatter would also be an outlier in X space as well; for example, $(W,U) = (100,5)$ would also be an outlier in X space.

Self-study question: In Figure 16.2, why would $(W,U) = (100,5)$ be an outlier in X space if both W and U were X variables?

To understand "outlier in Y | X space," see Figure 16.2 again, but now suppose U is the "Y" variable and W is the "X" variable. The solid dot would be an outlier in Y | X space because 2.8 is an unusual value of Y, given that X = 80. While Y = 2.8 is not an outlier in Y space, because it is only 1.41 standard deviations below the mean of the Y data, the more important point is that it is indeed an outlier in Y | X space. It is useful to investigate outliers in Y space just to get to know your data, but remember that regression is a model for $p(y \mid x)$, so for the purposes of regression, it is more important to diagnose outliers in Y | X space than it is to diagnose outliers in Y space.

Figure 16.3 shows four distinct cases where a point (the solid dot) is (a) neither an outlier in Y | X space nor an outlier in X space, (b) an outlier in Y | X space but not an outlier in X space, (c) not an outlier is Y | X space but an outlier in X space, and (d) an outlier in both Y | X space and an outlier in X space.

R code for Figure 16.3

```
set.seed(12345); X = runif(30, .5, 3.5)
beta0 = 1.0; beta1 = 1.5; sigma = 0.7
Y = beta0 + beta1*X + sigma*rnorm(30) # The regular process

# Suspicious data: Four cases
X.suspect1 = 1.5; Y.suspect1 = 3.3
```

(*Continued*)

```
X.suspect2 = 1.5; Y.suspect2 = 9.7
X.suspect3 = 9.0; Y.suspect3 = 14.5
X.suspect4 = 9.3; Y.suspect4 = 0.6

Y.all1 = c(Y, Y.suspect1); X.all1 = c(X, X.suspect1)
Y.all2 = c(Y, Y.suspect2); X.all2 = c(X, X.suspect2)
Y.all3 = c(Y, Y.suspect3); X.all3 = c(X, X.suspect3)
Y.all4 = c(Y, Y.suspect4); X.all4 = c(X, X.suspect4)

par(mfrow=c(2,2),mar = c(4,4,2,1))
plot(X.all1, Y.all1)
points(X.suspect1, Y.suspect1, pch=19)
plot(X.all2, Y.all2)
points(X.suspect2, Y.suspect2, pch=19)
plot(X.all3, Y.all3)
points(X.suspect3, Y.suspect3, pch=19)
plot(X.all4, Y.all4)
points(X.suspect4, Y.suspect4, pch=19)
```

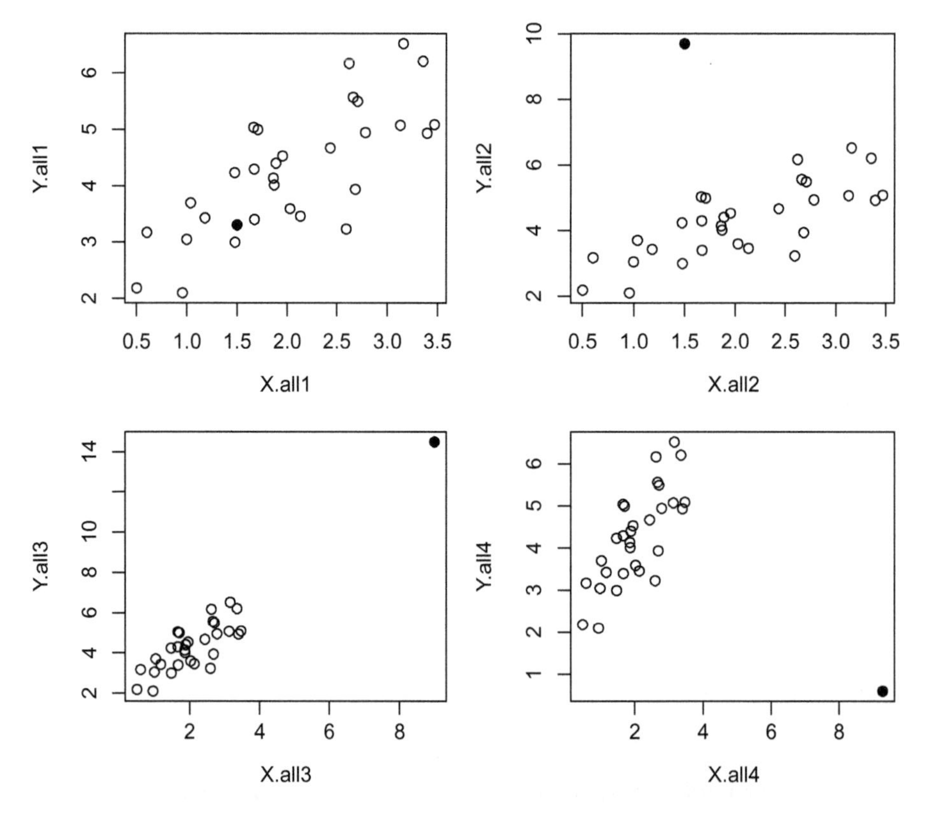

FIGURE 16.3
Upper left: The solid point is neither an outlier in $Y \mid X$ space nor an outlier in X space. Upper right: The solid point is an outlier in $Y \mid X$ space but not an outlier in X space. Lower left: The solid point is not an outlier in $Y \mid X$ space but is an outlier in X space. Lower right: The solid point is an outlier in $Y \mid X$ space and an outlier in X space.

In Figure 16.3, upper left panel, the solid point is not troubling at all. It is easily explained by the model where $p(y \mid x)$ are normal distributions strung together along a straight-line function of $X = x$. In the upper right panel, the observation is not unusual in its X value, but the Y value appears unusual if the distributions $p(y \mid x)$ are normal distributions. (On the other hand, this data point is explainable by a model where the $p(y \mid x)$ are distributions with heavier tails than the normal distribution.) In the lower left panel, the observation is unusual when the Y data are considered separate from the X data (i.e., it is an outlier in Y space), but since the point is near the trend predicted by X (assuming it is linear), it is not an outlier in $Y \mid X$ space.

The most troublesome case of all is the bottom right panel of Figure 16.3, where the data value is both an outlier in $Y \mid X$ space and an outlier in X space. This data point is not well explained by the model where $p(y \mid x)$ are normal distributions strung together along a straight-line function of $X = x$. However, it can be explained alternatively in at least two very distinct ways: (i) by a model where the $p(y \mid x)$ are distributions with heavier tails than the normal distribution, strung together as a straight-line function of $X = x$, and (ii) by a model where the $p(y \mid x)$ are normal distributions, strung together as a curved (concave in this case) function of $X = x$. There is only one data point, so it is difficult to tell which of these two models is best. You can fit those two models and compare penalized log-likelihoods, but since there is only one data point to distinguish the two, you should not put too much stock in the results. Instead, as with all cases involving sparse data, you need to use reasoned arguments involving subject matter considerations (informally, Bayesian arguments) to help decide whether heavy tails or curvature is the most plausible data-generating process.

16.4 Using the "Leverage" Statistic to Identify Outliers in X Space

When you have an outlier in X space, such a point is said to have high *leverage*. Like a person with a lever can move great weights, despite using little force, such an outlier has the capability to completely change the entire OLS-fitted regression line, despite being just one data point. See the lower right panel of Figure 16.3: The outlier value will affect the fitted regression line greatly. Without that data point, the OLS line will have a distinctly steep positive slope; including that data point "leverages" the OLS line so that it has a negative slope.

In regression analysis, the "hat matrix" is used to identify high-leverage outliers in X space. Recall from Chapter 7 that the OLS estimates are given by

$$\hat{\beta} = \left(\mathbf{X}^{\mathrm{T}}\mathbf{X}\right)^{-1}\mathbf{X}^{\mathrm{T}}Y$$

Thus, the predicted values are given by

$$\hat{Y} = \mathbf{X}\hat{\beta} = \mathbf{X}\left(\mathbf{X}^{\mathrm{T}}\mathbf{X}\right)^{-1}\mathbf{X}^{\mathrm{T}}Y = \mathbf{H}Y$$

where $\mathbf{H} = \mathbf{X}\left(\mathbf{X}^{\mathrm{T}}\mathbf{X}\right)^{-1}\mathbf{X}^{\mathrm{T}}$ is called the "hat" matrix. The meaning of "hat" is clear: The matrix $\mathbf{H}$ converts the $(n \times 1)$ data vector $\boldsymbol{Y}$ into the $(n \times 1)$ vector of predicted (hatted) values, $\hat{\boldsymbol{Y}}$.

Using the linearity property of covariance matrices presented in Chapter 7, the (conditional) covariance matrix of $\hat{\mathbf{Y}}$ is thus given by:

$$\mathbf{H}(\sigma^2\mathbf{I})\mathbf{H}^{\mathrm{T}} = \sigma^2\mathbf{H}\mathbf{H}^{\mathrm{T}} = \sigma^2\mathbf{H}\mathbf{H}$$

Now,

$$\mathbf{HH} = \mathbf{X}\left(\mathbf{X}^{\mathrm{T}}\mathbf{X}\right)^{-1}\mathbf{X}^{\mathrm{T}}\mathbf{X}\left(\mathbf{X}^{\mathrm{T}}\mathbf{X}\right)^{-1}\mathbf{X}^{\mathrm{T}} = \mathbf{X}\mathbf{I}\left(\mathbf{X}^{\mathrm{T}}\mathbf{X}\right)^{-1}\mathbf{X}^{\mathrm{T}} = \mathbf{X}\left(\mathbf{X}^{\mathrm{T}}\mathbf{X}\right)^{-1}\mathbf{X}^{\mathrm{T}} = \mathbf{H},$$

Hence we have the following result:

The conditional covariance matrix of $\hat{\mathbf{Y}}$ and the leverage statistic

The conditional covariance matrix of $\hat{\mathbf{Y}}$ is given by $\sigma^2\mathbf{H}$.

The *leverage* statistic is $h_{ii} = \left(i^{th} \text{ diagonal element of } \mathbf{H}\right)$.

It is interesting that the hat matrix, **H**, multiplied by itself gives you back that same matrix; i.e., **HH** = **H**. Matrices that have this property are called *idempotent* matrices, and are important in geometry.

Recall from Chapter 3 that, in simple regression, the conditional variance of the predicted values is given by:

$$\mathrm{Var}(\hat{Y}_i) = \mathrm{Var}(\hat{\beta}_0 + \hat{\beta}_1 x_i) = \sigma^2\left\{\frac{1}{n} + \frac{(x_i - \hat{\mu}_x)^2}{(n-1)\hat{\sigma}_x^2}\right\}.$$

Comparing terms, you can see that in the case of simple regression involving only one X variable, the ith diagonal element of $\mathbf{H}$, h_{ii}, is given by

$$h_{ii} = \frac{1}{n} + \frac{(x_i - \hat{\mu}_x)^2}{(n-1)\hat{\sigma}_x^2} = \frac{1}{n} + \frac{z_i^2}{(n-1)},$$

where z_i is the ith standardized X value. Thus, in the case of simple regression, the leverage statistic h_{ii} is a linear function the z_i^2 value for the ith observation on the X variable, implying that larger h_{ii} means that x_i is more of an outlier in X space.

The formula above for $\mathrm{Var}(\hat{Y}_i)$ shows you that a more remote (farther from the mean) X data value results in a less precise estimates of $\mathrm{E}(Y \mid X = x)$, because the variance of the estimate, $\mathrm{Var}(\hat{Y})$, is larger. This is intuitively sensible, right? If you try to estimate the mean of the distribution of Y for an extreme X value, you expect your estimate to be less accurate. After all, the fact that X is extreme means there is little data nearby with which to estimate the conditional mean of Y.

The statistic h_{ii} is called the *leverage* statistic because larger h_{ii} means that the X data for the single observation i is (are) potentially highly influential in determining the OLS fit. In the case of multiple regression, this h_{ii} statistic can identify vectors $(x_{i1}, x_{i2}, \ldots, x_{ik})$ that are far from the centroid $(\bar{x}_1, \bar{x}_2, \ldots, \bar{x}_k)$. With multiple X variables, the h_{ii} statistic can identify vectors such as the one shown in Figure 16.2, where the combination of values is unusual, but the individual components are not. (For reference, the leverage statistic is a function of a multivariate distance measure called *Mahalanobis distance*.)

To derive an ugly rule of thumb for how large h_{ii} has to be to be considered "large," we need to use matrix algebra. The *trace* of a square matrix, denoted $\text{tr}(\mathbf{A})$, is defined as the sum of its diagonal elements. With some attention to detail, you can prove that

$$\text{tr}(\mathbf{AB}) = \text{tr}(\mathbf{BA}),$$

provided that **A** and **B** can be multiplied in either order. Thus, the sum of the h_{ii} is given by

$$\sum h_{ii} = \text{tr}(\mathbf{H}) = \text{tr}\left\{\mathbf{X}\left(\mathbf{X}^{\mathrm{T}}\mathbf{X}\right)^{-1}\mathbf{X}^{\mathrm{T}}\right\} = \text{tr}\left\{\left(\mathbf{X}^{\mathrm{T}}\mathbf{X}\right)^{-1}\mathbf{X}^{\mathrm{T}}\mathbf{X}\right\} = \text{tr}\left(\mathbf{I}_{k+1}\right) = k+1$$

implying that the average value of the h_{ii} is $(k+1)/n$. Thus, an ugly rule of thumb is that an h_{ii} much larger than average, say, greater than twice the average, or greater than $2(k+1)/n$, is "large," and indicates that the vector $(x_{i1}, x_{i2}, \ldots, x_{ik})$ is far from the centroid.

But do not make too much of this "$> 2(k+1)/n$" ugly rule of thumb. There is no recipe for action here; most importantly, there is no suggestion that the observation should be deleted. The number $2(k+1)/n$ is just a useful reference point to gauge "outlier in X space" when using the leverage statistics h_{ii}; the main thing you need to know is that larger h_{ii} indicate greater remoteness of the point $(x_{i1}, x_{i2}, \ldots, x_{ik})$ in X-space. The logic for the ugly rule of thumb is simply the mathematical fact that the average of the h_{ii} values is $(k+1)/n$. (Mathematical facts are always preferable to ugly rules of thumb!)

The leverage statistics, h_{ii}, are always non-negative, and their average value is $(k+1)/n$, where k is the number of X variables.

You can annotate Figure 16.2 by labeling points according to their h_{ii} values as shown in the following code.

R code for Figure 16.4

```
# method 1: "By hand" calculation of hii.
set.seed(12345)
W = rnorm(100, 70,10)
U = .05*W + rnorm(100,0,.1)
W = c(W,80); U = c(U,2.8)
X = cbind(rep(1,101), W,U)
H = X %*% solve(t(X) %*% X) %*% t(X)
hii = diag(H)

# method 2: Requires a fitted lm object. So make up fake Y data.
Y = rnorm(101)
hii = hat(model.matrix(lm(Y~U+W)))
hii

plot(W,U, xlim=c(40,105), ylim=c(2.0,5.0))
identify(W,U, round(hii,3)) # now point and click on the graph
```

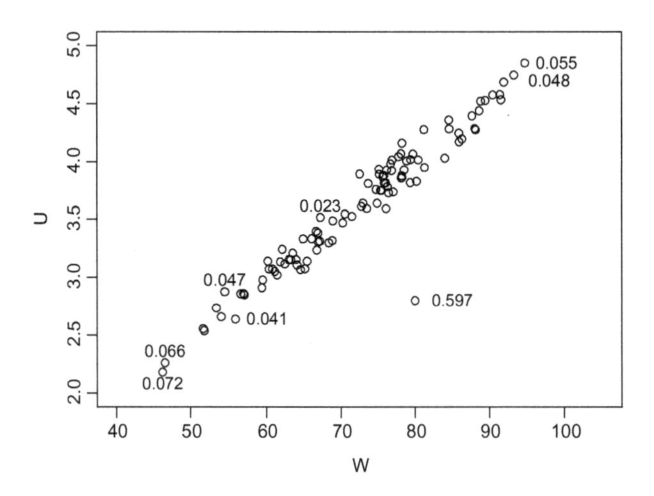

FIGURE 16.4
Scatterplot shown in Figure 16.2 with some points labeled by their h_{ii} statistic.

Notice in Figure 16.4 that the outlier indicated in Figure 16.2 has the highest h_{ii}, namely, 0.597. There are two X variables here, so the average of the h_{ii} values is $(2+1)/101 = 0.0297$; twice that value is 0.0594; the outlier exceeds that threshold by quite a bit, with $h_{ii} = 0.597$. But rather than worry about any ugly rule of thumb, just look at the data. The most extreme h_{ii}, 0.597, is much larger than the next largest value 0.072, which occurs for the data point in the lower left corner. Hence, the point in the lower right is much more remote than all other points.

If you only have two X variables, you can simply plot them to identify the outliers, as shown in Figure 16.4. The "`identify`" option shown in the `R` code above allows you to identify the particular observation i in the data set, simply by pointing and clicking on the suspicious data point.

With more than two X variables, you cannot draw the graphs as easily, because they are objects in higher-dimensional space. Still, the h_{ii} statistics tell you which observation vectors are outliers, even though you cannot draw the scatterplot to see them directly. The fact that h_{ii} can identify outliers in high-dimensional space, where you cannot directly see the data values as you can in a 2-D or 3-D scatterplot, is a wonderful feature of the h_{ii} (leverage) statistic.

The concept of "leverage" in multiple X dimensions is essentially the same as in one X dimension. Look at Figure 16.2 again. If the variables in Figure 16.2 are both X variables, then the Y variable axis will come off the screen (or page if you are reading this on paper) towards you. The fitted regression function will be a tilted plane, also in the space between you and the screen. And the outlier is a high-leverage point that will tip the entire plane towards the particular Y value that goes with that outlier X point. In summary:

A single outlier in X space has high leverage to move the entire fitted OLS regression function.

16.5 Using Standardized Residuals to Identify Outliers in $Y|X$ Space

An outlier in $Y \mid X$ space is simply a data point with a large deviation from the mean function, i.e., a large residual. But what is "large" for a residual? As with any univariate measure, we use distance in terms of number of standard deviations from the mean (also known as "standardized value" and "z score") to measure extremity.

With the homoscedastic model, you know that $\text{Var}(\varepsilon_i) = \sigma^2$, a constant. But you also know that $e_i = Y_i - (\hat{\beta}_0 + \hat{\beta}_1 x_i)$ differs from $\varepsilon_i = Y_i - (\beta_0 + \beta_1 x_i)$. As we hope is abundantly clear by now, there are major consequences that stem from the fact that estimates differ from their true values. One consequence is the fact that the variances of the e_i terms are *not constant*, even when the variance of the ε_i terms *is constant*.

The logic for this seeming contradiction involves leverage: When you have a high-leverage X data value (or vector of values), the OLS regression line is "leveraged" towards whatever the $Y_i = y_i$ value is that corresponds to the X_i data or vector. See the lower right panel of Figure 16.3 again: The outlier "leverages" the OLS line downward toward the Y value for the outlier.

So, if a high-leverage value pulls the regression line toward the Y value, then the residual e_i will tend to be closer to zero, i.e., have smaller variance, for such a value. The actual formula for the conditional variance of the residual is, in the homoscedastic model,

$$\text{Var}(e_i) = \sigma^2(1 - h_{ii}).$$

Thus, points with high-leverage h_{ii} give residuals e_i with smaller variance, even when the variance of the errors ε_i is constant.

Since we have already done most of the math, we can easily show you why it is true that the conditional variance is given by $\text{Var}(e_i) = \sigma^2(1 - h_{ii})$. Note that the vector of residuals is

$$e = \mathbf{Y} - \hat{\mathbf{Y}} = \mathbf{Y} - \mathbf{HY} = (\mathbf{I} - \mathbf{H})\mathbf{Y};$$

hence, by using the linearity property of covariance, you get

$$\text{Cov}(e) = (\mathbf{I} - \mathbf{H})(\sigma^2\mathbf{I})(\mathbf{I} - \mathbf{H})^{\text{T}} = \sigma^2(\mathbf{I} - \mathbf{H})(\mathbf{I} - \mathbf{H}) = \sigma^2(\mathbf{I} - \mathbf{H} - \mathbf{H} + \mathbf{HH}) = \sigma^2(\mathbf{I} - \mathbf{H}).$$

(As a side comment, note that both $\mathbf{H}$ and $(\mathbf{I} - \mathbf{H})$ are idempotent. These facts are important in n-dimensional geometry.)

The formula $\text{Var}(e_i) = \sigma^2(1 - h_{ii})$ is true because the diagonal elements of $\sigma^2(\mathbf{I} - \mathbf{H})$ are $\sigma^2(1 - h_{ii})$. Hence, the standardized sample residual for observation i is given as follows:

Standardized residual for observation *i*

$$r_i = \frac{e_i}{\hat{\sigma}\sqrt{1 - h_{ii}}}$$

In the formula for the standardized residual, the estimate $\hat{\sigma}$ is the usual OLS estimate, called "Residual standard error" in the `lm` summary. Since these r_i values are standardized, they are z-values, so you can use the ugly rule of thumb $|r_i| > 3$ to identify extreme points.

To understand these standardized residuals a little better, let's see how they look for our pathological example in the lower right corner of Figure 16.3. You can get these residuals by asking for `rstudent(fit)` in R, where "`fit`" is a fitted `lm` object.

R code for Figure 16.5

```
fit=lm(Y.all4~X.all4)
rstudent(fit)
plot(X.all4,Y.all4, xlim = c(0,10), ylim = c(0,8))

# now, point and click on the graph!
identify(X.all4,Y.all4, round(rstudent(fit),2))
abline(lsfit(X.all4,Y.all4), lty=2)
```

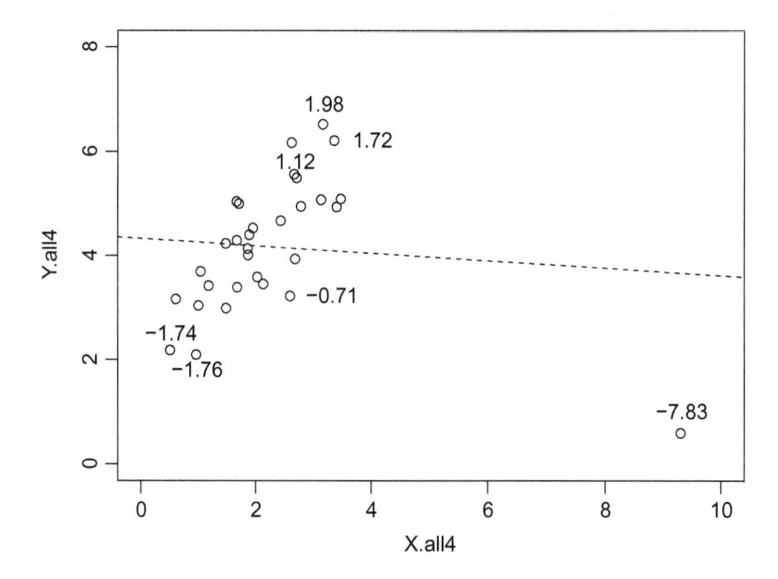

FIGURE 16.5
Data in lower right panel of Figure 16.3, labeled by standardized residual.

In Figure 16.5, you can see that the outlier point does not have an ordinary residual $|e_i|$ that is extremely large compared to the others, because the point is not particularly far from the line compared to the other points. However, the leverage for that data value is $h_{31,31} = 0.723$, (obtained using `hat(model.matrix(fit))[31]`), implying that the standard deviation of that particular residual is small, making its standardized residual (–7.83) much larger in magnitude than all the others seen in Figure 16.5.

16.6 Cook's Distance

The absolute worst kind of outlier for OLS estimates is one that has both large leverage (outlier in X space) *and* has a large (in magnitude) standardized residual (outlier in $Y \mid X$ space). Cook's distance is an outlier diagnostic that combines both. One (slightly mysteri ous-looking) formula for Cook's distance is given as follows:

Cook's distance for observation *i*

$$d_i = \frac{r_i^2 h_{ii}}{(k+1)(1-h_{ii})}$$

All else fixed, the formula for Cook's distance shows you that larger leverage h_{ii} will give you larger d_i. Also, larger squared standardized residual r_i^2 will give you larger Cook's distance. But large h_{ii}, coupled with large r_i^2, gives the most extreme Cook's distance.

The Cook's distance statistic d_i is a simple formula to compute by hand, but it is automatically available as `cooks.distance(fit)` for a fitted `lm` object. Let's see how this statistic looks for the pathological data set in the lower right panel of Figure 16.3.

R code for Figure 16.6

```
fit=lm(Y.all4~X.all4)
cooks.distance(fit)
plot(X.all4,Y.all4, xlim = c(0,10), ylim = c(0,8))
identify(X.all4,Y.all4, round(cooks.distance(fit),2))  # point and click!
abline(lsfit(X.all4,Y.all4), lty=2)
```

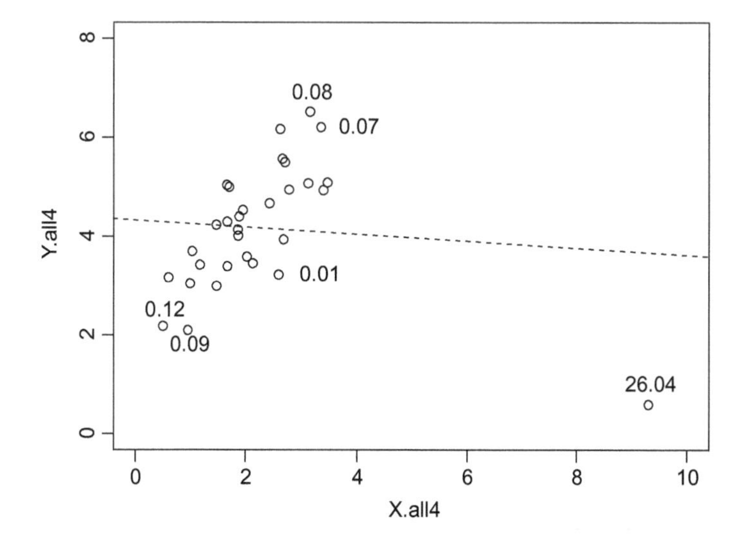

FIGURE 16.6
Data in lower right panel of Figure 16.3, labeled by Cook's distance.

The large value of Cook's distance shown in Figure 16.6 for the outlier means that the outlier has great influence on the fitted OLS function. Specifically, it means that the fitted OLS function will differ greatly, depending upon whether the outlier is in or out of the data set. In Figure 16.6, it is clear that if the outlier is removed, the OLS function will be dramatically different, having a steep positive slope rather than a negative slope.

In Figure 16.6, there is only one point that has a large Cook's distance; namely, the outlier in the lower right of the graph. You do not need an ugly rule of thumb here! But there are indeed several ugly rules of thumb that have been given for identifying "large" values of Cook's distance. We are loathe to give any of them, because then you might get the wrong message and think that there is some recipe for automatic exclusion of outliers based on the ugly rule of thumb. The only reason we are going to give any ugly rule of thumb at all is that this book uses `R`, and one of the "default" graphs for the `lm` function in `R` uses the following ugly rule of thumb:

A Cook's distance d_i greater than 0.5 or 1.0 indicates that observation *i* inordinately influences the OLS fit.

Notice that in Figure 16.6, we see a Cook's distance of 26.04.

Self-study question: Is $d_i = 26.04$ "large" according to the just-given ugly rule of thumb? Or is it "*larger* than *large*"? Or is it "*very much larger* than *very large*"? Or is it *very very much larger* than *very very large*?

A graph that uses the ugly rule(s) of thumb for Cook's distance is produced by the command `plot(fit)`, where "`fit`" is a fitted `lm` object. This command produces several graphs we have already discussed for assessing validity of the classical model assumptions. The last of these graphs is a plot of (h_{ii}, r_i), with contour lines demarking points having Cook's distance > 0.5 and > 1. Using the `lm` OLS fit to pathological data set in the lower right panel of Figure 16.3, if you enter `plot(fit)` and click through to the last graph, you get Figure 16.7.

Notice that the outer dashed curve envelope in Figure 16.7 is given by $d_i > 1$, or by

$$d_i = \frac{r_i^2 h_{ii}}{(k+1)(1-h_{ii})} > 1.$$

Equivalently,

$$r_i^2 > \frac{(k+1)(1-h_{ii})}{h_{ii}}.$$

Taking $\pm$ the square root of the right-hand side gives the upper and lower dashed curve envelopes of the standardized residual as a function of leverage, shown in Figure 16.7.

Notice also that `R` labels the points with observation numbers. This can be very handy if you have a large data set, so that you can easily go into your data set and investigate the outliers. In the pathological data set, the outlier is observation $i = 31$.

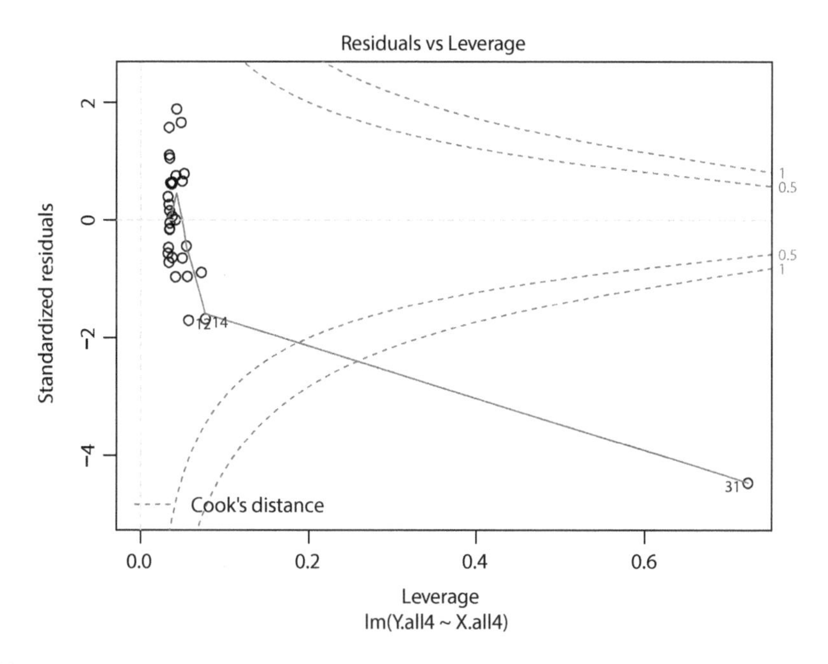

FIGURE 16.7
Plot of (h_{ii}, r_i) for the pathological data set. Points having Cook's distance greater than 1.0 are outside the outer dashed curve envelope.

16.6.1 Outlier Analysis Using the Data of the Crime Rate Prediction Model

In Chapter 11, we used regression to predict crime rate. The complete model that we considered is estimated using our familiar "`lm`" is given as follows:

```
library(MASS)
data("Boston"); attach(Boston)
fit = lm(crim ~ zn + indus + rm + age + dis + rad + tax + lstat + medv)
```

When you enter `plot(fit)`, you get four plots that you can use to assess the model assumptions and the outliers. The last of these four diagnostic plots is shown in Figure 16.8.

To understand Figure 16.8 more, let's do some investigation. First, the numbered data points refer to the values with largest Cook's distance. You can check this using

```
tail(order(cooks.distance(fit)))
```

which shows that the observations having largest Cook's distance are observations 405, 415, 411, 419, 406, and 381. Checking these values using `cooks.distance(fit)[c(405, 415, 411, 419, 406, 381)]`, we see that observation 381 has the largest Cook's distance, 0.322, and observations 406 and 419 have the next largest values of Cook's distance, 0.0965 and 0.0955. With the possible exception of observation 381, none of the observations are particularly influential to the OLS analysis, as their Cook's distance values are not particularly large.

On the other hand, Figure 16.8 shows some extreme outliers in X space, with leverage values h_{ii} far exceeding their average $(9+1)/506 = 0.0198$. Figure 16.8 also shows extreme residuals that are from 5 to 10 standard deviations from zero.

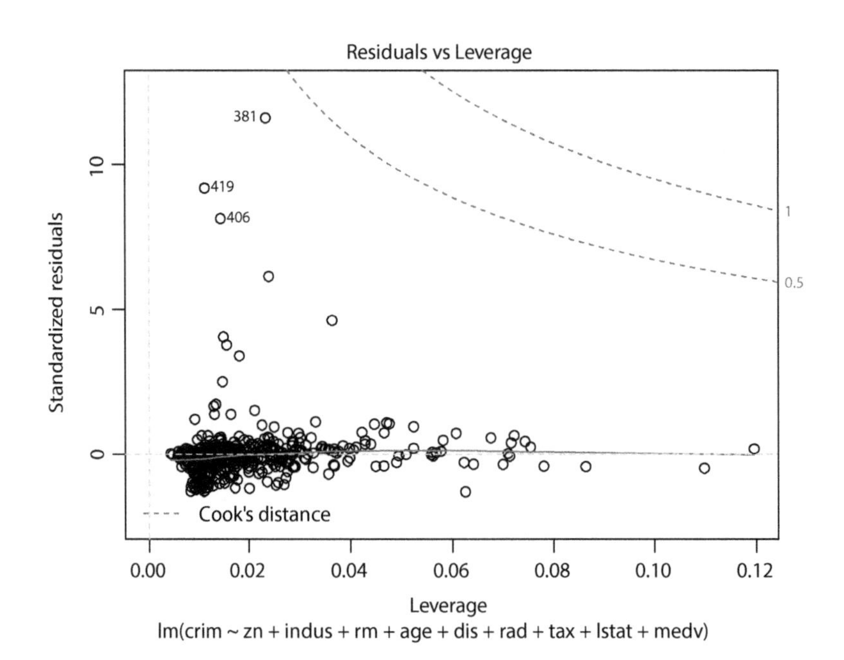

FIGURE 16.8
Plot of (leverage, standardized residual) (or (h_{ii}, r_i)) using the data for the crime rate prediction example.

Some detective work is needed to investigate these outliers. First, let's investigate the two towns having largest leverage, both shown in Figure 16.8 as having leverage (h_{ii}) greater than 0.10. By using `tail(order(hat(model.matrix(fit))))`, you can see that these two most extreme values belong to the cities that are observations 366 and 369 in the data set. Why are these observations "extreme" with regard to their X values? Perhaps they have outliers in one or another of their X variables. Checking, you can enter the command

```
round(scale(data.frame(zn,indus,rm,age,dis,rad,tax,lstat,medv))[369,],2)
```

which gives you the following z-values:

```
   zn indus    rm   age   dis   rad   tax lstat  medv
-0.49  1.01 -1.87  1.12 -1.17  1.66  1.53 -1.32  2.99
```

These values are for the town that is observation 369. Notice that this town has a "`medv`," or median home value, that is extremely high, nearly 3 standard deviations above the mean.

Replacing town "369" with town "366" in the analysis, you get the following:

```
   zn indus    rm   age   dis   rad   tax lstat  medv
-0.49  1.01 -3.88  0.69 -1.04  1.66  1.53 -0.77  0.54
```

This town's "`rm`," or average number of rooms per dwelling, is nearly four standard deviations below the mean.

Thus, the leverage values indicated towns with unusually high or low values of one of its X variables. But remember, as shown in Figures 16.2 and 16.4, a large leverage can also occur without a large or small z-value. In such a case you can examine the *combinations* of the z-values to identify what is unusual. For example, if X_1 and X_2 are highly positively correlated, then the combination $z_1 = 1.29$, $z_2 = -1.73$ would indicate a highly unusual observation, despite the fact that neither value is, by itself, unusual.

What can you do with the outlier information for these two cities? Mainly you can learn about your data! If you are studying different cities, then you should be curious about how they differ, which ones are unusual, and in what ways. This is the nature of *statistical analysis*, which is a way to learn about your subject matter by using data.

Despite the fact that these two cities are "high leverage" in their X variables, they are not particularly influential on the fitted OLS regression function because their residuals are so small. Their Cook's distance values are `cooks.distance(fit)[369] = 0.00047` and `cooks.distance(fit)[366] = 0.00296`. The reason that these points are not particularly influential is that, despite their high leverage, they have small residuals, as you can clearly see in Figure 16.8. On the other hand, the towns with high residuals require investigation. Observation 381 in particular is a town having crime rate that is more than 10 standard deviations from what is predicted by the regression model.

To investigate this town further, first note that the (marginal) crime rate values, obtained using `summary(crim)`, look as follows:

```
    Min.   1st Qu.    Median      Mean   3rd Qu.      Max.
 0.00632   0.08204   0.25651   3.61352   3.67708  88.97620
```

Notice that the range from the minimum and the median (0.00632 to 0.25651), as well as the range from the median to the maximum (0.25651 to 88.97620). Clearly, there is something quite unusual about the town having crime rate 88.97620. And sure enough, this is the town having the largest standard residual: `rstudent(fit)[381]` gives 13.58317, or 13.58 standard deviations above the regression prediction.

Clearly, there is more to explaining crime rates than what is in the regression model. If you were a law enforcement officer in town 381, or if you were considering moving to that town, you would certainly be interested in knowing the reasons for this extreme level of crime. Again, this is called *statistical analysis*, which means learning about your subject by using data.

Town #381 also has the largest Cook's distance, with $d_i = 0.323$. Thus, among all the towns in the study, the biggest change in the estimated regression model occurs when you remove town #381. While it is neither good statistical practice nor good scientific practice to delete of outliers, there is no harm is assessing the effect of an outlier via deletion. You can estimate the model without town #381 by using the following command:

```
fit1 = lm(crim ~ zn + indus + rm + age + dis + rad + tax + lstat + medv,
data=Boston[-381,])
```

The comparison of the fitted models, with and without the outlier, is shown in Table 16.1.

Notice in Table 16.1 that the results are not dramatically different when town #381 is removed. Because the Cook's distance measure for this town (0.323) is less than the smaller of the two ugly rule of thumb thresholds (0.5, 1.0), such small differences are expected.

While the outlier analysis turned up no major concerns in this example, it is a serious concern that the crime rate distribution is so obviously non-normal. But log transformation provides a simple and appropriate fix here. Figure 16.9 is a repeat of Figure 16.8, but using `Y=log(crim)`.

TABLE 16.1

Comparison of Coefficient Estimates and Standard Errors for the Crime Rate Model When Estimated Using the Full Data Set, Versus When Town #381 is Deleted

	Estimates Using All Data		Town #381 Deleted	
	Estimate	**Standard Error**	**Estimate**	**Standard Error**
(Intercept)	2.6525	4.6200	4.6483	3.9499
zn	0.0512	0.0181	0.0471	0.0155
indus	−0.1103	0.0801	−0.1108	0.0685
rm	0.6706	0.6077	−0.0669	0.5220
age	−0.0102	0.0173	−0.0064	0.0148
dis	−0.8797	0.2657	−0.7501	0.2272
rad	0.5656	0.0850	0.5399	0.0726
tax	−0.0039	0.0052	−0.0033	0.0044
lstat	0.1505	0.0755	0.1692	0.0645
medv	−0.1898	0.0561	−0.1217	0.0482

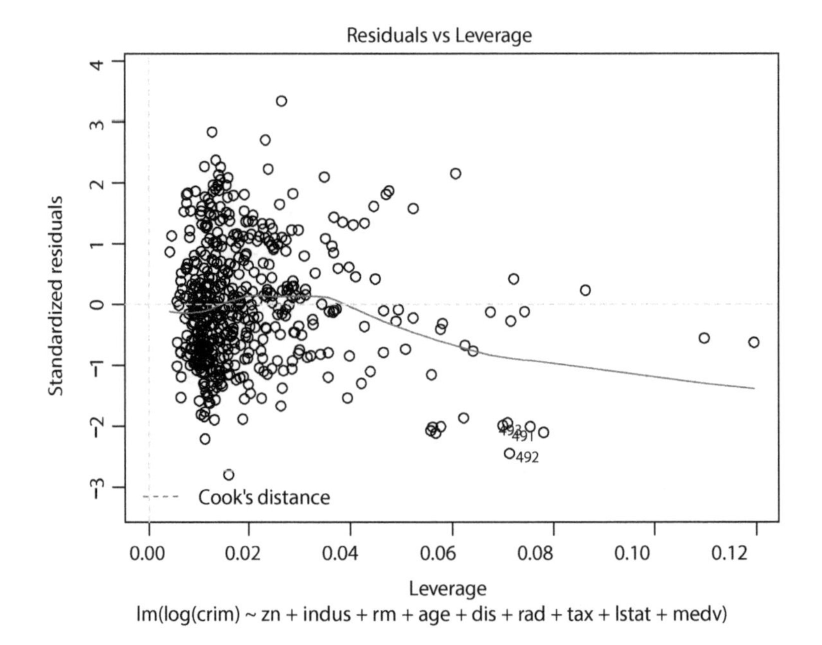

FIGURE 16.9
Plot of (leverage, standardized residual) (or (h_{ii}, r_i)) using the data for the crime rate prediction example, after log transforming crime rate.

Notice in Figure 16.9 that the Cook's distance boundaries are not even visible, hence no observations are close to influential. Further, all but one observation have a residual that is within three standard deviations of zero. The normal quantile-quantile plot (not shown here) looks like a straight-line, indicating that the distribution of the residuals is close to a normal distribution after log transforming crime rate.

This is not to say that town #381 is no longer interesting, though. Police and potential settlers still want to know why the town's crime rate is so high!

16.7 Strategies for Dealing with Outliers

The first thing to consider is, are the outlier simply mistakes, e.g., data entry errors? If so, then it makes sense to correct the data values. Deletion is a second-best option, but can cause bias for the same reasons as indicated in Chapter 15 on censored observations. Also, do not assume that data values that are outliers are the only possible mistakes. All else equal, it is indeed more likely that the grossly extreme values are mistakes than the typical values, so the outliers are a good place to start looking for mistakes. That having been said, it is likely that whatever mechanism caused the mistakes among the outlier data, also caused mistakes among the more typical data. So you should identify the mechanism, and track down all the mistakes, not just the ones that are outliers. Science is serious work, not a silly academic game. Just because an observation is not an outlier does not mean it is a good data value. As mentioned before: Garbage in, garbage out. If you estimate a model using data where many of the observations are simply wrong, how can you trust the results?

Outlier observations that are not mistakes are indications that your data-generating process produces occasional extreme values. There are several strategies for estimating such data-generating processes. On the design side, you might re-consider the nature of the variables you are collecting. Perhaps you are defining a ratio-type measure V/W. Such a ratio measure will automatically induce outliers when the variable W is close to zero. You might consider whether that ratio measure is needed, or whether you can just as well use the variables V and W in their raw forms.

Ratio measures or not, if some of your key variables are quite outlier-prone, then they are not as reliable as other measures that are not so outlier-prone. You might consider identifying alternative, more reliable measurements.

But if nothing can be done on the design and measurement side, and you are simply stuck with outlier measurements, there are a number of possibilities for data-centric analyses. Suggestions for dealing with outliers are as follows:

1. The simplest, and often best solution is the simple log transform, as discussed in Chapter 5, and as shown in the crime rate example above. This works very well for right-skewed (often, "size-related") data, where there are occasional large outliers. If there are 0's in the variable V, you can simply use $\log(V+1)$, and if there are negative values of V you can use $\log(V+m)$, where m is larger than the maximum absolute negative value. You have to be careful, though, that you do not induce negative skew by log transforming.
2. Recall that regression is a model for $p(y \mid x)$. All of the discussion in this chapter has centered on the effects of outliers on OLS estimates, which implicitly assume that the distributions $p(y \mid x)$ are normal distributions. With normal distributions, points that are far from the mean have extremely low likelihood. Recalling that OLS estimates are maximum likelihood estimates, the estimated OLS parameter estimates are thus greatly influenced by outliers, which pull the regression function closer to them so that their contributions to the likelihood do not make the overall likelihood so small (recall, you want to *maximize* the likelihood).

   A simple remedy is to assume non-normal distributions that are heavier-tailed than the normal distribution. Implicitly, log-transforming Y models the distributions $p(y \mid x)$ as lognormal, which is one example of a heavy-tailed distribution. If log-transforming does not work, perhaps because the distribution of Y

is symmetric to begin with, or perhaps because Y has many negative data values, then you might use maximum likelihood with the Laplace distribution, the logistic distribution, or the T distribution with small degrees of freedom. A benefit of this approach is that you have likelihood-based fit statistics to assess which distribution(s) are best supported by your data. A further example of this approach is given below in this section.

Sometimes it makes sense to keep the normal distributions, but fit a heteroscedastic model: The outliers may simply result from distributions $p(y \mid x)$ that have higher variance. The AIC statistic will tell you whether a heteroscedastic normal model fits your data better than a homoscedastic non-normal (e.g., T error distribution) model.

When you use likelihood-based methods, you have objective, likelihood-based methods for comparing the models to see which best fit the data. The next two recommendations are not likelihood-based, so you cannot use likelihood to see whether they fit your data better. Instead, you must use simulation to compare methods.

3. Quantile regression provides a model for relating the *median* of the distribution $p(y \mid x)$ as a function of $X = x$. Since the median is not affected by outliers, quantile regression estimates are similarly not affected. With quantile regression, you can also estimate quantiles other than the median (the 0.5 quantile). For example, you can model the 5th percentile (the 0.05 quantile) and the 95th percentile (the 0.95 quantile) of the distributions $p(y \mid x)$ as a function of $X = x$ to obtain a nonparametric 90% prediction interval for Y given $X = x$.
4. ***The Mother of All Blunt Axes***. This final option is one we do not recommend at all. We only discuss it because, strangely enough, it is nearly standard practice in some disciplines. The approach is to either delete outliers *en masse* (i.e., *truncate* the data) or Winsorize them. Outlier deletion *en masse* (*truncation*) means simply discarding outliers; Winsorization means replacing the outliers with a less extreme number, then performing the usual analysis. Winsorization is discussed in detail in the final section of this chapter.

If you think that "suggestion" 4. (truncation or Winsorization) sounds like scientific misconduct, you are thinking correctly. People have lost their jobs for selectively using data (e.g., deleting observations) or data fabrication (e.g., using data other than what was collected). A classical case involves the resignation of researcher Dirk Smeesters, PhD, a professor of Consumer Behavior and Society at Erasmus University in Rotterdam, who published some suspicious-looking results. The following is a quote from the news story on the incident:

> Smeesters denied having made up the suspect data, but was unable to provide the raw data behind the findings, asserting the files were lost in a computer crash…
>
> But he did admit to selective omission of data points that undercut the hypotheses he was promoting. However, he insisted that such omission was common practice in psychology and marketing research.

When this note was shared with a Marketing professor, the return comment was "No wonder I am having such a hard time getting published. Everyone else is cheating!" We include "suggestion" 4. (truncation or Winsorization), not as an actual suggestion for dealing with outliers, but instead to explain why you should *not* deal with outliers in this way.

We used suggestion 1. (log transformation) in the analysis of crime rate data shown above. Suggestion 2. (fitting a heavy-tailed distribution) is easily illustrated by using the pathological data set in the lower right panel of Figure 16.3. As discussed in Chapter 15, you can use the `survreg` function to estimate regression models $p(y \mid x)$ where the distributions have heavy tails. Just use the `survreg` function on data with no censored observations: Define a "censoring" variable that has all 1's to indicate that all observations are observed and none are censored.

In the example below, we use `dist = "t"` to model the distributions $p(y \mid x)$ as T distributions having different means, rather than as normal distributions having different means. The T distributions have heavier tails than the normal distribution, with smaller degrees of freedom indicating heavier tails. When the degrees of freedom parameter is $df = 1$, the T distribution is also known as the *Cauchy* distribution, which is famous for having heavy tails. The default of the `survreg` function is to use the T_4 distribution, which is the T distribution with four degrees of freedom.

16.7.1 Analysis of Data with an Extreme Outlier by Using Heavy-Tailed Distributions

```
library(survival)
one = rep(1,length(Y.all4))
fit.T = survreg(Surv(Y.all4, one) ~ X.all4, dist = "t")
summary(fit.T)
plot(X.all4, Y.all4)
abline(lsfit(X.all4, Y.all4),  lwd=2)
abline(fit.T$coefficients[1:2], lty=2, lwd=2)
logLik(lm(Y.all4~X.all4))
logLik(fit.T)
```

In Figure 16.10, the fitted line using the heavy-tailed distributions obviously looks better than the normality-assuming OLS fit. The log-likelihoods also show better fit, with −45.3 for the T_4 distributions vs. −51.3 for the normal distributions.

The next section provides a detailed discussion of our suggestion 3. for dealing with outliers, namely, quantile regression.

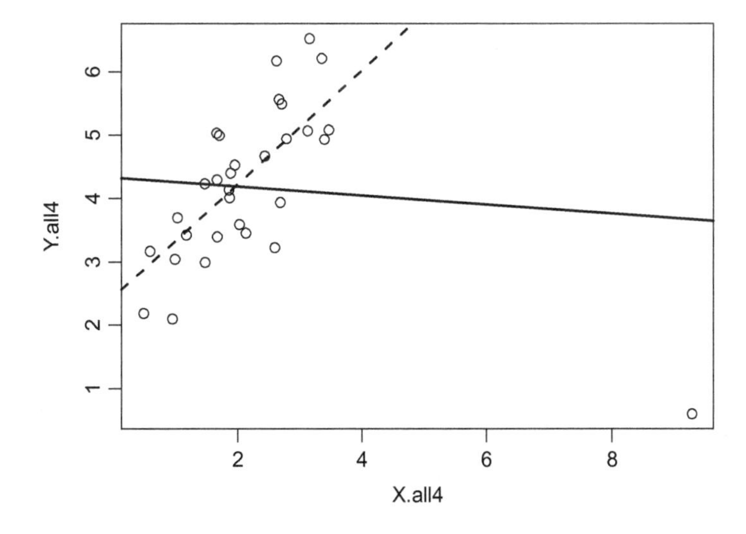

FIGURE 16.10
Estimated mean functions assuming $p(y \mid x)$ are normal distributions (the solid line, which is the OLS fit), and assuming the $p(y \mid x)$ are heavy-tailed distributions (specifically, T_4, dashed line).

16.8 Quantile Regression

Data from the United States Bureau of Labor Statistics (BLS) give us the 10th and 90th percentiles of the distribution of weekly U.S. salaries for the years 2002–2014. These data are entered and graphed as follows:

R code for Figure 16.11

```
Year = 2002:2014
Weekly.10 = c(295, 301, 305, 310, 319, 330, 346, 350, 353, 358, 359,
  372, 379)
Weekly.90 = c(1376, 1419, 1460, 1506, 1545, 1602, 1693, 1744, 1769,
  1802, 1875, 1891, 1898)
plot(Year, Weekly.10, ylim = c(0,2000), pch="+", ylab = "Weekly
  U.S. Salary")
points(Year, Weekly.10, type="l")
points(Year, Weekly.90, pch="+")
points(Year, Weekly.90, type="l", lty=2)
legend(2009, 1300, c("90th percentile", "10th percentile"), pch=c("+","+"),
  lty= c(2,1))
```

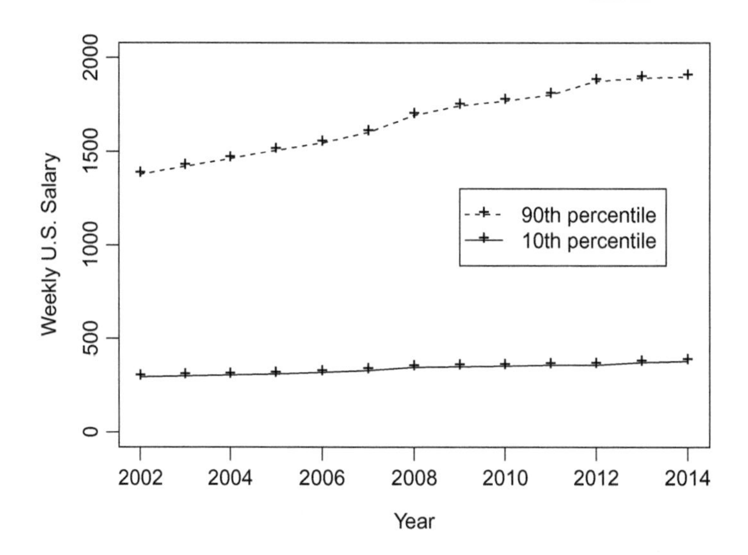

FIGURE 16.11
Quantiles (0.90 and 0.10) of the weekly U.S. salary distributions, 2002–2014.

Notice in Figure 16.11 that the quantile functions look reasonably well approximated by straight-lines, albeit with different slopes: The 90th percentile increases more rapidly than the 10th percentile over time. These relationships are *quantile regression* relationships. With quantile regression, you can model a quantile of the conditional distribution $p(y \mid x)$ as a function of $X = x$. In Figure 16.11, Y = weekly salary, X = year, and there are two conditional quantile functions represented:

$$y_{0.10} = \beta_{0(0.10)} + \beta_{1(0.10)}x,$$

which is a model for the 0.10 quantiles of the distributions $p(y \mid x)$, and

$$y_{0.90} = \beta_{0(0.90)} + \beta_{1(0.90)}x,$$

which is a model for the 0.90 quantiles of the distributions $p(y \mid x)$.

The β's of the two quantile lines are different because, as you can see in Figure 16.11, the lines themselves are different.

The fact that the 90th percentile line has a steeper slope than the 10th percentile line in the BLS example tells you that there is greater variation (i.e., heteroscedasticity) in the distribution of Salary for more recent years. You could make political arguments using these data. On the one hand, you might say it is troubling that the income gap is getting wider, that this is bad for society, and that a minimum wage hike or tax credits should be given to those who earn less. On the other hand, you could argue that income inequality is necessary for a free, efficient economy, and that we are only recently getting a more optimal disposition of income inequality, where goods and services will flow more freely for the benefit of everyone. We do not care which side of the political spectrum you are on, as long as you understand *quantile, regression,* and most importantly, that regression is a model for the *conditional distribution* of Y, given some predictor variables' values (i.e., regression is a model for $p(y \mid x)$).

The most famous quantile is the median (the 0.50 quantile). You can use quantile regression to estimate the *medians* of the conditional distributions $p(y \mid x)$ as a function of $X = x$. Such an estimate is a *robust* alternative to OLS, which estimates the *means* of the distributions $p(y \mid x)$ as a function of $X = x$.

A side comment about the word "robust": In English, "robust" means "strong." In statistics, a "robust method" is a method that "works well (is 'strong') despite violated assumptions." For example, the heteroscedasticity-consistent standard errors presented in Chapter 12 are sometimes called "robust standard errors," because they work well when homoscedasticity is violated. By the same token, the median works well as a measure of the center of a distribution when the distribution has extremely heavy tails, and is called a "robust measure of central tendency."

Recall that the OLS estimates are those values b_0, b_1 that minimize

$$\text{SSE} = \sum \{y_i - (b_0 + b_1 x_i)\}^2$$

If you do not have any X variable, and you choose the value b_0 to minimize

$$\text{SSE} = \sum (y_i - b_0)^2$$

then the minimizing value will be $b_0 = \bar{y}$, the average of the y_i values. This is a result you can prove by calculus, but it should be a fairly intuitive result, even without the proof.

On the other hand, if you choose b_0 to minimize the sum of *absolute* errors, rather than the sum of *squared* errors, i.e., if you choose b_0 to minimize

$$\text{SAE} = \sum |y_i - b_0|,$$

then the minimizing b_0 will be a *median* of the y_i values. This fact is not provable by calculus, because the absolute value function is non-differentiable. So to gain some insight into this minimization process, consider the following R code used to draw Figure 16.12.

R code for Figure 16.12

```
par(mfrow=c(1,2))
y1 = c(2.3, 4.5, 6.7, 13.4, 40.0)   # Median is 6.7
y2 = c(2.3, 4.5, 6.7, 13.4)         # For second graph
b0 = seq(2, 15,.001)
mat1 = outer(y1,b0,"-"); mat2 = outer(y2,b0,"-")
SAE1 = colSums(abs(mat1)); SAE2 = colSums(abs(mat2));
plot(b0, SAE1, type="l", ylab = "SAE"); abline(v = 6.7, lty=2)
plot(b0, SAE2, type="l", ylab = "SAE"); abline(v = (4.5+6.7)/2, lty=2)
```

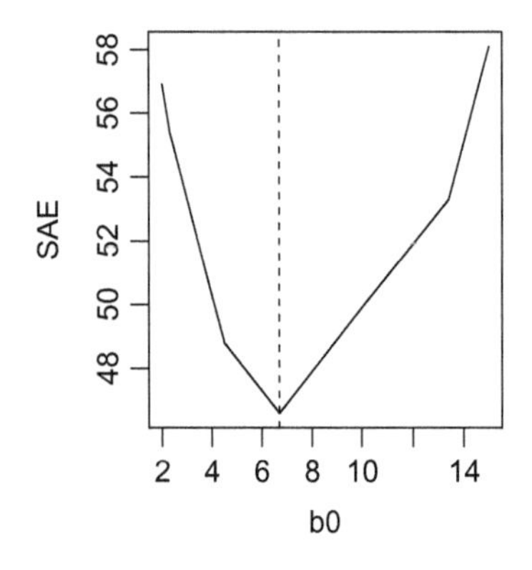

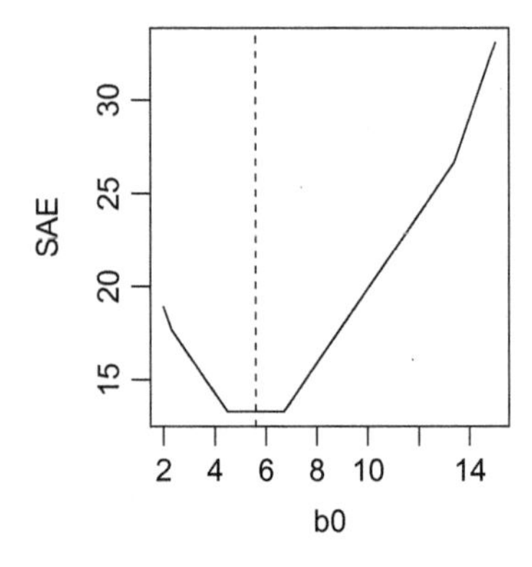

FIGURE 16.12
Graphs of sum of absolute errors as a function of b_0. In the left panel with $n = 5$ (an odd number of) data values 2.3, 4.5, 6.7, 13.4, and 40.0, the median 6.7 uniquely minimizes the sum of absolute errors. In the right panel with $n = 4$ (an even number of) data values 2.3, 4.5, 6.7, and 13.4, the usual median estimate $(4.5 + 6.7)/2$ does not uniquely minimize the sum of absolute errors. Rather, any number between 4.5 and 6.7 also gives the minimum.

Notice in Figure 16.12 that the usual median of a data set minimizes the SAE when n is odd, but in the case where n is even, there is no unique minimum. This explains why, when you perform a quantile regression analysis, you sometimes get a warning message saying that the "solution is non-unique."

The method for finding the median based on SAE minimization extends to a method for finding the quantile using minimization of a different function. Suppose you want to estimate the τ quantile of the data. Define the function

$$\rho_\tau(x) = \begin{cases} (1-\tau)\,|x|, & \text{if } x < 0 \\ \tau\,|x|, & \text{if } x \geq 0 \end{cases}$$

Then define

$$\text{SAE}(\tau) = \sum \rho_\tau(y_i - b_0)$$

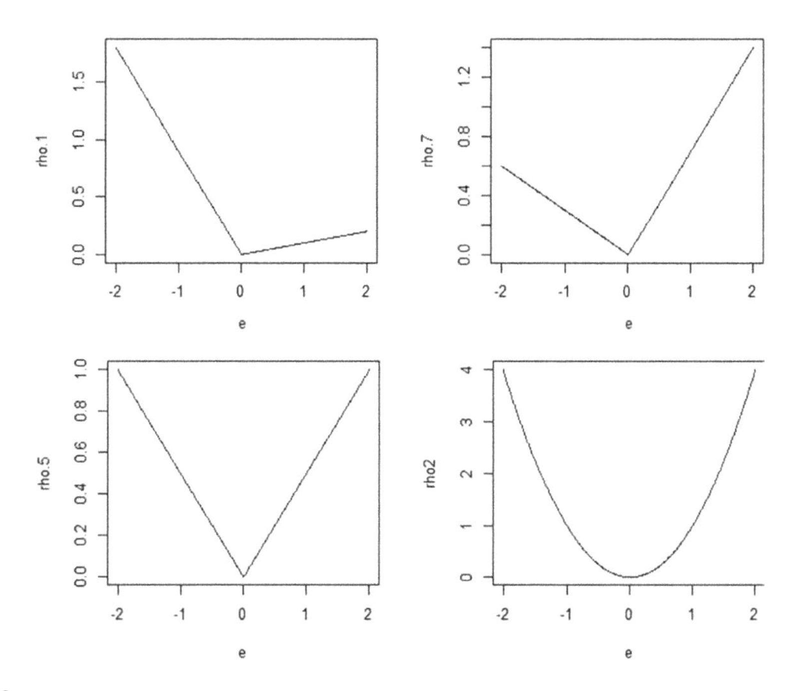

FIGURE 16.13
Functions ρ used in the minimization of $\sum \rho(e_i)$, where $e_i = y_i - (b_0 + b_1 x_i)$. Upper left: Minimization gives estimates of the 0.1 quantile. Upper right: Minimization gives estimates of the 0.7 quantile. Lower left: Minimization gives estimates of the 0.5 quantile or median. Lower right: Minimization of the quadratic function gives an estimate of the mean.

Then a (possibly non-unique) value b_0 that minimizes SAE(τ) can be called the τ quantile of the data set $\{y_1, y_2, \ldots, y_n\}$.

The OLS estimates arise from a similar definition: If you define $\rho(x) = x^2$, then $\sum \rho(y_i - b_0) = \sum (y_i - b_0)^2$ is just the SSE, and you know that minimizing SSE gives you the OLS estimates.

Thus, you get different estimates depending upon what function $\rho(\cdot)$ you use when you minimize $\sum \rho(e_i)$. Figure 16.13 shows graphs of four such functions.

To understand how the minimization works to obtain quantiles other than the median, consider a data set containing 100 y_i values, $\{1,2,3,\ldots,98,99,1000\}$. This is a simple data set in that the median should be around 50, the tenth percentile should be around 10, the 80th percentile should be around 80, etc. For good measure, and to show robustness to outliers, we have made the largest value in the data set 1,000 (an outlier) rather than 100.

Let's see which b_0 minimizes the function $\sum \rho_{0.1}(e_i)$.

R code for Figure 16.14

```
tau = .10  ## A quantile of interest, here .10
y = c(1:99, 1000)  ## The .10 quantile should be around 10.
b0 = seq(5, 15,.001)
mat = outer(y,b0,"-")
mat = ifelse(mat < 0, (1-tau)*abs(mat), tau*abs(mat))
SAE = colSums(mat)
plot(b0, SAE, type="l")
```

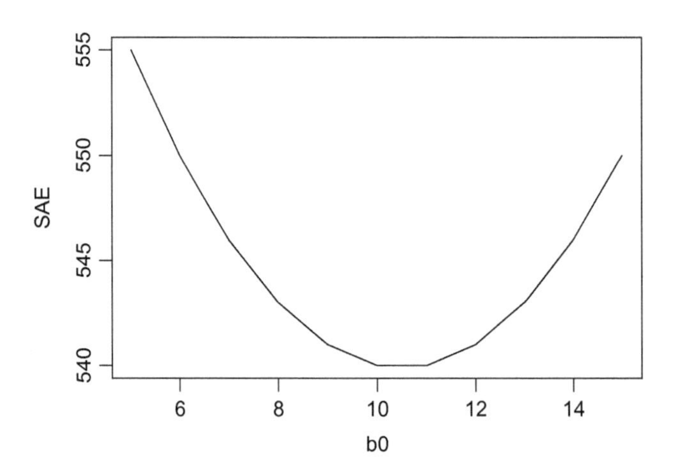

FIGURE 16.14
Plot of $(b_0, \sum \rho_{0.1}(y_i - b_0))$. The function is minimized by values between 10 and 11.

As shown in Figure 16.14, and as expected, values around 10 can be considered as estimates of the 10th percentile of these data. And again, there is no unique minimizing value: Any number between 10 and 11 will minimize the function, so all such numbers are legitimate estimates of the 10th percentile.

When you ask R for the 10th percentile of that data set, it picks a particular number in the minimizing range using an interpolation formula: For the data depicted in Figure 16.14, `quantile(y,.10)` gives 10.9; but again, any number in the range from 10 to 11 can be called the 10th percentile.

The real use of these $\rho_\tau(\cdot)$ functions is not to find quantiles of individual variables, but rather to estimate the quantiles of distributions $p(y \mid x)$. Recall that minimization of

$$\sum \rho\{y_i - (b_0 + b_1 x_i)\}\text{, where } \rho(x) = x^2$$

gives you the usual OLS values $\hat{\beta}_0$ and $\hat{\beta}_1$, and that $\hat{\beta}_0 + \hat{\beta}_1 x$ is an estimate of the *mean* of the distribution $p(y \mid x)$. By the same token, minimization of

$$\sum \rho_\tau\{y_i - (b_0 + b_1 x_i)\}$$

gives you values $\hat{\beta}_{0,\tau}$, $\hat{\beta}_{1,\tau}$ such that $\hat{\beta}_{0,\tau} + \hat{\beta}_{1,\tau} x$ is an estimate of the τ *quantile* of the distribution $p(y \mid x)$.

Don't believe this can possibly work? Maybe a simulation study will convince you!

Suppose

$$Y = 2.1 + 0.7X + \varepsilon,$$

where

$$\text{Var}(\varepsilon \mid X = x) = (0.4x)^2$$

This is a heteroscedastic model. Further, suppose the X data are uniformly distributed on (10, 20), and that the ε values are normally distributed.

Then the distribution of $Y \mid X = x$ is normal, with mean $2.1 + 0.7x$ and standard deviation $0.4x$. Thus, the 0.90 quantile of the distribution $p(y \mid x)$ is

$$2.1 + 0.7x + z_{0.90}(0.4x),$$

where $z_{0.90} = 1.28155$ is the 0.90 quantile of the standard normal distribution. Collecting terms algebraically, the true 0.90 quantile function is thus

$$y_{0.90} = 2.1 + 1.212621x$$

Notice in particular that the heteroscedasticity in this model causes the slope of the 0.90 quantile function (1.212621) to be larger than the slope of the median function (0.7). See Figure 16.11 again to understand.

If the quantile regression method works, then minimization of $\sum \rho_{0.90}\{y_i - (b_0 + b_1 x_i)\}$ should give estimates that are close to 2.1 and 1.212621 in this case, and those estimates should get closer to those targets as the sample size gets larger. Further, the 95% confidence intervals for the true parameters (often obtained by using the bootstrap) should trap the true parameters in approximately 95% of the intervals.

The "rq" function in the "quantreg" library performs quantile regression. Rather than using calculus, the software uses linear programming algorithms to find the estimates. We used this software in Chapter 2 to find estimates that minimize the sum of absolute errors, thus finding maximum likelihood estimates when the error terms have the Laplace distribution.

16.8.1 Simulation Study to Validate the Quantile Regression Estimates

The simulation code is given as follows:

```
n = 1000
beta0 = 2.1; beta1 = 0.7
set.seed(12345)
X = runif(n, 10, 20)
Y = beta0 + beta1*X + .4*X*rnorm(n)
# True slope of the .90 quantile function
beta1.90 = beta1 + qnorm(.90)*.4
beta1.90
library(quantreg)
fit.90 = rq(Y ~ X, tau = .90)
summary(fit.90)
library(quantreg)
fit.90 = rq(Y ~ X, tau = .90)
summary(fit.90)
```

The resulting output is given as follows:

```
Call: rq(formula = Y ~ X, tau = 0.9)

tau: [1] 0.9

Coefficients:
            coefficients lower bd upper bd
(Intercept)  1.65748     -0.33148  4.30665
X            1.21470      1.04248  1.36427
```

```
Warning message:
In rq.fit.br(x, y, tau = tau, ci = TRUE, ...) : Solution may be nonunique
```

Notice the following about the output.

- The "solution may be non-unique" comment is explained by the function to be minimized: Sometimes it is flat at the minimum as shown in the figures above.
- While this default can be over-ridden, with smaller sample sizes the `rq` function gives intervals for the coefficients rather than standard errors or p-values. This is because the distribution of the estimates is non-symmetric with smaller sample sizes. While you may be thinking that $n = 1{,}000$ in this study is actually "large," remember that we are estimating the 0.90 quantile. Thus, you need data on either side of the quantile to get estimates. Here, there are only 100 observations above the 0.90 quantile, so 100 is a more relevant sample size than 1,000.
- The intervals nicely contain the true parameters, 2.1, and 1.212621. In particular, the slope of the mean (and median, because of normality) function, $\beta_1 = 0.7$, is well *outside* the interval `1.04248 1.36427`, so it is clear that the quantile regression method is not estimating the same thing that is estimated by OLS.

If you change the sample size to $n = 100{,}000$, you get the following:

```
Call: rq(formula = Y ~ X, tau = 0.9)

tau: [1] 0.9

Coefficients:
            Value    Std. Error t value   Pr(>|t|)
(Intercept)  1.90385  0.16094    11.82965  0.00000
X            1.22385  0.01135   107.80454  0.00000
```

Notice now that the standard errors are given. With larger sample sizes, the distribution of the estimates is closer to symmetric, so it makes sense to take the usual estimates plus or minus two standard errors as intervals. Notice also that the estimates have tightened considerably around the theoretical targets 2.1 and 1.212621, and that these targets are still well within plus or minus two standard errors of the estimates.

The simulation above is not a proof that the method works in general, just an illustration that it works in our simulation model. But hopefully, this analysis makes you feel more comfortable that the method is doing what it is supposed to. There are mathematical proofs that quantile regression produces consistent estimates, though, so the simulation results were not just coincidences.

Notice that the simulation model of the example above is a heteroscedastic normal model. In Chapter 12, you learned how to estimate such models using maximum likelihood. This brings up a good question: Which is better, quantile regression or maximum likelihood? The answer is, hands down, maximum likelihood—if you have specified the distributions $p(y \mid x)$ correctly in your likelihood function. On the other hand, if you specify these distributions incorrectly, quantile regression may be better. You can use simulation to give a more precise answer.

As always, the answer to the question "Is maximum likelihood good" is not as simple as "correct distribution, good; incorrect distribution, bad." Rather an answer that involves of *degree* of misspecification of the distribution. You can (and should) use maximum likelihood, even when you have not specified the distributions correctly. Just realize that the closer your specified distributions are to the true distributions of the data-generating process (DGP), the better your results will be. You can guess which distributions to use by using your subject matter knowledge about your DGP, and by looking at scatterplots. You can also check which distributions fit the data better using penalized likelihood statistics such as AIC, so you do not have to make a wild guess as to which distributions you should use.

16.8.2 Quantile Regression Models for Personal Assets

In addition to being robust to outliers, quantile regression also handles heteroscedasticity very nicely, as is shown in Figure 16.11 and in the simulation study above. The following R code estimates the 25th, 50th, and 75th percentiles of the $p(y \mid x)$ distributions in the personal assets example (Y = Personal Assets, X = Age) and displays them, along with the data scatter, in Figure 16.15.

```
Worth = read.table("https://raw.githubusercontent.com/andrea2719/
URA-DataSets/master/Pass.txt")
attach(Worth)
library(quantreg)
fit.25 = rq(P.assets ~ Age, tau = .25)    # .25 quantile
b = coefficients(fit.25); b0 = b[1]; b1 = b[2]
x = seq(20, 70)
y.25 = b0 + b1*x
fit.50 = rq(P.assets ~ Age, tau = .50)   # .5 quantile (median)
b = coefficients(fit.50); b0 = b[1]; b1 = b[2]
y.50 = b0 + b1*x
fit.75 = rq(P.assets ~ Age, tau = .75)    # .75 quantile
b = coefficients(fit.75); b0 = b[1]; b1 = b[2]
y.75 = b0 + b1*x

plot(Age, P.assets, xlim = c(20,70))
points(x, y.25, type="l"); points(x, y.50, type="l", lty=2); points(x, y.75,
type="l", lty=3)
abline(lsfit(Age, P.assets), col = "gray", lwd=2)
legend("topleft", c("tau=.25", "tau=.50", "tau=.75", "OLS"),
lty=c(1,2,3,1), lwd=c(1,1,1,2), col=c(1,1,1,"gray"))
```

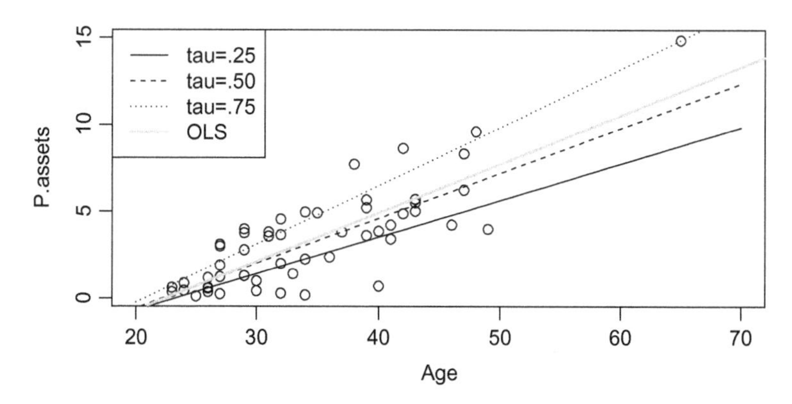

FIGURE 16.15
The personal assets data set, with fitted 0.25, 0.50, and 0.75 quantile functions and the OLS-fitted mean function.

Note the following in Figure 16.15:

- Heteroscedasticity is apparent in that the difference between the 0.25 and 0.75 quantiles of the $p(y \mid x)$ distributions increases with larger values of the Age variable. This increase is due to different slopes of the quantile functions.
- The median and mean estimates ($\tau = 0.50$ and OLS) are similar, with the OLS estimate being pulled towards the outlier point in the upper right. The median estimate is not so affected by this outlier.
- Even though it is an extrapolation, the quantile functions cross near 0, suggesting that, e.g., 0.25 quantile can be higher than the 0.50 quantile, which is of course impossible. Crossings are due to randomness of the estimates in this example.
- There are only $n = 53$ observations in this data set, so estimates of quantiles closer to 0.00 or 1.00 will be unreliable due to extremely small sample sizes on the rare side of the quantile.

Self-study question: What happens to the quantile regression estimates when you have an extreme outlier, such as in the data set shown in the lower right panel of Figure 16.3?

16.9 Outlier Deletion *en masse* and Winsorization

As noted above, the outlier detection rules are not meant to be recipes for deleting them. On the other hand, there is no harm in *investigating the effect* of outliers by using deletion. You can (and in many cases, should) delete an outlier, or outliers, and compare the results with and without the outlier(s). If your conclusions are essentially the same either way, then you have confidence that your results are not inordinately influenced by an aberrant observation or two. If you do this, all analyses should be disclosed—there is no harm in doing any number of analyses of a given data set as long as the nature of the data exploration is fully disclosed. The problems noted earlier of academic misconduct arise when much data-tinkering is done behind the scenes, then a particular chosen analysis is put forward without disclosure.

It is possible that much of the current "replication crisis" in science is caused by such behind-the-scenes data-tinkering. If an analyst manipulates a data set in unreasonable ways to "prove" his or her preferred theory, is it any surprise that the results do not replicate when carried out in a fresh context with new data?

There are two main types of outlier deletion in regression: (1) Finding outliers that are influential, e.g., via Cook's distance, and deleting them to assess their effects, and (2) Deleting all extremes beyond a certain quantile for a given variable (e.g., less than the 0.05 quantile or greater than the 0.95 quantile) and deleting them. The first method is called *sensitivity analysis*, and the second is called *truncation*, which is an example of deleting outlier *en masse*.

Researchers are generally aware that outlier deletion is a slippery slope on the way to academic misconduct. As well, outlier deletion reduces the sample size, and in studies with small to medium sample sizes, researchers simply cannot afford to throw away data. An alternative to truncation is a closely related method called *Winsorization*. Rather than delete the extreme observations *en masse*, Winsorization means simply replacing those that are less than the 0.05 quantile with the 0.05 quantile; those that are greater than the 0.95 quantile are replaced with the 0.95 quantile. Other quantiles can be used as well; quantiles 0.05 and 0.95 are defaults of the `Winsorize` function in the `DescTools` library. The typical logic used by a Winsorizer to motivate this practice goes something like this:

> "Well, I know I have outliers, and I know that they cause problem with OLS. But I am reluctant to throw away the data completely because that seems wrong. So instead, I will keep the data value in my data set, just replace its value with something less extreme. That way, I still acknowledge that 'there is something out there,' but now it is not so far 'out there' to cause problems."

Truncation and Winsorization violate the entire notion of modeling the distribution of Y as a function of $X = x$. If the real process you are studying produces occasional outliers in Y, then a model that is faithful to your reality will similarly produce occasional outliers of the same type and magnitude that are typically seen in reality.

As originally developed for regression applications, Winsorized regression used the Winsorized *residuals* from OLS regressions (Yale & Forsythe, 1976). While this approach has its problems, it is much more sensible than the common practice of Winsorizing the Y and X variables. However, the latter practice is what is usually used, so we will discuss that one.

To introduce Winsorization, we show via an `R` example how to do it, and we also show that it can give you quite different results than OLS regression. The following code reads a data set and performs two different analyses, one based on OLS with the raw X and Y data, and one with the Winsorized X and Y data. The `R` code for the OLS analysis is as follows:

```
WD = read.table("https://raw.githubusercontent.com/andrea2719/
URA-DataSets/master/Testdata.txt")
summary(lm(Y ~ X, data=WD))
```

The OLS analysis gives the following:

```
Coefficients:
            Estimate Std. Error t value Pr(>|t|)
(Intercept)  209.176      6.598  31.702   <2e-16 ***
X              2.557      1.089   2.349    0.019 *
```

The analysis of the same data, following Winsorization, is as follows:

```
library(DescTools)
Yw = Winsorize(WD$Y); Xw = Winsorize(WD$X)
summary(lm(Yw ~ Xw))
```

This gives the following output:

```
Coefficients:
             Estimate Std. Error t value Pr(>|t|)
(Intercept) 220.3851     2.5009  88.122<   2e-16 ***
Xw           -1.3767     0.4137  -3.328 0.000906 ***
```

In the OLS analysis, the effect of the X variable on Y is deemed *positive* ($\hat{\beta}_1 = 2.557$), differing from zero by more than a chance amount ($p = 0.019$); while in the Winsorized analysis, the effect of the same X variable on Y is deemed *negative* ($\hat{\beta}_1 = -1.3767$), also differing from zero by more than a chance amount ($p = 0.0009$).

In truth, the data set in the above analysis was simulated from a data-generating process where $\beta_1 = 2.5$ for the conditional mean function, a value which is nicely consistent with the OLS estimate $\hat{\beta}_1 = 2.557$. Clearly, Winsorization does not correctly estimate the conditional mean function. It does something quite different.

To understand what Winsorization (and any statistical procedure, for that matter) does, it is a very good idea to use simulation, where you know the true data-generating process, so you can then compare the outputs of the procedure to the known truth. The example above, where the OLS and Winsorized estimates gave opposite results, was such an example, but the simulation model was somewhat complicated. To start with a simpler simulation model, let us suppose that the true data-generating process is

$$Y = 2.0 + 0.1X + 1.5\varepsilon$$

where $\varepsilon \sim T_3$; as noted above, the T distribution with 3 degrees of freedom is a heavy-tailed distribution. The true parameters are $\beta_0 = 2.0$, $\beta_1 = 0.1$, and $\gamma = 1.5$. The parameter γ is called the "scale" parameter (a "scale parameter" is a parameter that is similar, but not necessarily identical to a standard deviation). We will first generate X data from the lognormal$(1,1)$ distribution, and then generate the Y data from the model above, and then Winsorize both X and Y.

R code to see what Winsorization does to regression data

```
set.seed(123)
X = exp(rnorm(1000, 1, 1))
Y = 2 + .10*X + 1.5*rt(1000,3)

library(DescTools)
X.w = Winsorize(X)
Y.w = Winsorize(Y)
par(mfrow=c(2,1), mar=c(4,4,1,1))
plot(X, Y, cex=.5, xlim = c(0,80), ylim = c(-10,20)); abline(v=10, lwd=2,
col="gray"); plot(X.w, Y.w, cex=.5, xlim = c(0,80), ylim = c(-10,20));
abline(v=10, lwd=2, col="gray")
```

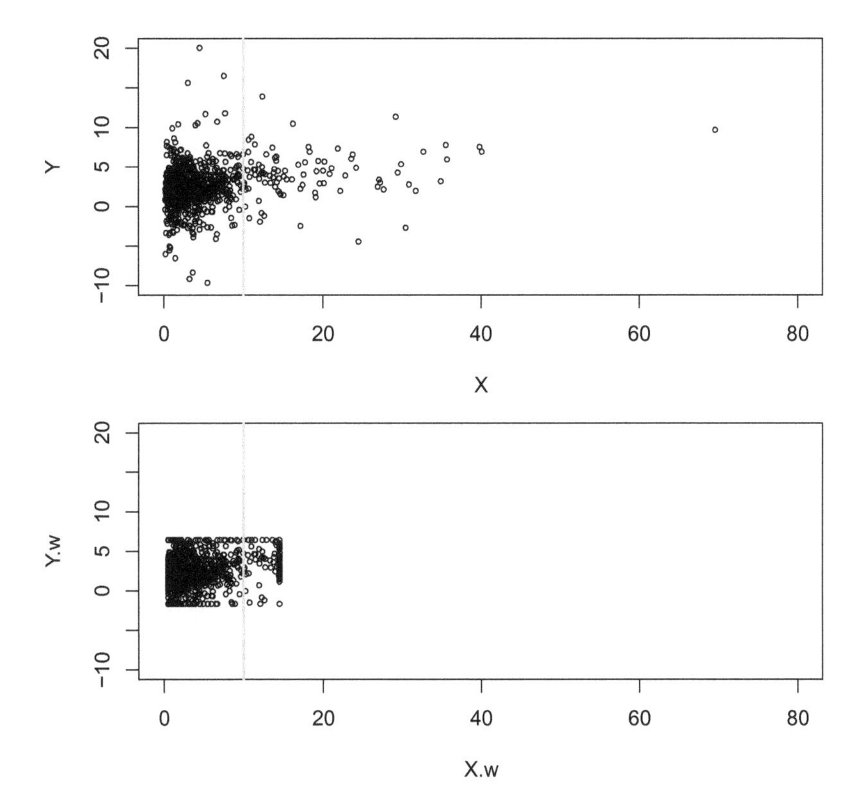

FIGURE 16.16
Top panel: Typical regression (X,Y) data. Bottom panel: Same data as in the top panel, but using the Winsorized X and Y data. Vertical lines at $X = 10$ are shown in both panels to visualize the distributions $p(y \mid X = 10)$ that are suggested by the data.

From Figure 16.16, it is clear that even the most useful and basic regression tool, and the one tool that you should use before doing anything else with your data, namely, the scatterplot, becomes useless once you Winsorize. Instead of showing you the actual data, the Winsorized scatterplot shows you a strange-looking box, with data values piled up on the boundaries. Winsorized regression uses the "box data" of the bottom panel of Figure 16.16, and applies least squares regression to it. Let's explore the consequences of this practice.

First, the regression model is a model for the real conditional distributions $p(y \mid x)$. Looking at the bottom panel of Figure 16.16, see the vertical strip above $X = 10$, and consider the distribution suggested by the Winsorized Y data in this strip. Clearly, it is not the real distribution of the Y data—there are observations "piled up" at the Winsorized values in the left and right tails, falsely indicating discreteness at those two points, and the actual tail data themselves are absent. On the other hand, the same vertical line above $X = 10$ in the top panel, with the raw data, provides correct information about the true distribution $p(y \mid X = 10)$; which, in this case, is a (shifted) T_3 distribution. Thus, Winsorization gives you a greatly distorted picture of the real data-generating processes that are at work to produce your data.

Second, recall the discussion of the "Call Center" censored data in Chapter 15: Winsorized values are actually censored data values. Recall also that treating the censored data values "as is" (a) causes bias, and (b) cannot be used in the likelihood function in the ordinary way. So, Winsorization causes bias, and you cannot use likelihood to see whether the model "Winsorize, then assume normal" is better than any other model. By the way, truncation, in which the extremes are simply deleted, suffers the same problems of bias, also as noted in the discussion of the "Call Center" data in Chapter 15.

Third, almost everything we have discussed in this book is no longer relevant with Winsorized data:

- Graphical analysis is moot, as shown in Figure 16.16.
- Parameter interpretation is also moot: While in regressions we have discussed so far, the parameters have clear and specific interpretations in terms of the distributions $p(y \mid x)$, (e.g., in terms of means, quantiles, etc.); after Winsorizing, the parameters have no clear meaning whatsoever. See the data analysis example above, for example: The estimated OLS slope was significantly positive and the estimated slope after Winsorizing was significantly negative.
- The bottom panel of Figure 16.16 shows that assessments of linearity, heteroscedasticity, and normality are also moot after Winsorizing.
- Likelihood-based methods, the related Bayesian methods, and many other statistical methods are also unsupportable after you Winsorize; the remainder of this section provides further examples.

So what good *can* be stated for Winsorizing? If you look around, you will find some research touting its benefits in terms of (a) more accurate estimates of β_1 than OLS despite its bias (the variance-bias trade-off again), and (b) more powerful tests of the hypothesis $H_0 : \beta_1 = 0$ than OLS. To see if these benefits hold using the model that gives Figure 16.16, we will simulate many samples, fit the OLS model to the raw data and also to the Winsorized data, store the estimates of β_1 and the p-values for testing $H_0 : \beta_1 = 0$ using both fits, and summarize the results. The `R` code for this study is based upon the more elaborate code given in Appendix A at the end of this chapter, and the results are shown in Table 16.2.

Unlike the data analysis above which indicated gross bias in the Winsorized estimates (recall that the Winsorized and OLS estimates had significantly opposite signs in that example), the results of this particular simulation study show that the Winsorized estimates are only slightly biased, with an average 0.115, only 0.015 larger than the target $\beta_1 = 0.10$. The OLS estimates are, as theory predicts, unbiased.

TABLE 16.2

Simulation Results Comparing Various Estimation/Testing Methods for the Data-Generating Process that Produced Figure 16.16. All Estimates are of the True Slope $\beta_1 = 0.1$, and All Results are Based on 20,000 Simulated Data Sets of Size $n = 100$ Each

Method	Average	S. Dev.	RMSE	Power
OLS	0.100	0.0523	0.0523	0.586
Winsorized	0.115	0.0538	0.0560	0.615

One might have hoped that the bias in the Winsorized estimates was offset by a decrease in variance, thus achieving more accurate estimates than OLS via the variance-bias trade-off, but sadly this is not the case here: The standard deviations are 0.0538 for the Winsorized estimates and 0.0523 for the OLS estimates. The reason for this is that when you Winsorize the X data, you reduce the standard deviation of the X data. And recall that a larger standard deviation of X decreases the standard deviation of the OLS estimate of β_1.

On the other hand, the Winsorized estimates in this simulation give rise to a more powerful test of $H_0: \beta_1 = 0$, with $p < 0.05$ observed in 61.5% of the samples versus 58.6% in the case of OLS. So, while the Winsorized estimates are not preferred on the basis of accuracy in this example, Winsorization does provide a more powerful test of the hypothesis $H_0: \beta_1 = 0$ in this simulation setting.

Self-study question: Why is the percentage of samples where $p < 0.05$ called the "power" of the test in this simulation study?

Let's change the simulation study slightly in order to greatly favor Winsorization. First, we will generate X data from a uniform distribution over the range from 0 to 20 instead of a lognormal distribution, so that Winsorization will only barely affect the range of the X data. Second, we will make the error distribution the T_1 (Cauchy) distribution instead of the T_3 distribution, giving a much more outlier-prone process in $Y \mid X$ space. See Figure 16.17 for a typical data set from this data-generating process.

Using the Cauchy error model, which makes the conditional distributions $p(y \mid x)$ shifted Cauchy distributions (shifted to the median value $\beta_0 + \beta_1 x$), Table 16.3 reports the results of a more elaborate simulation study than shown in Table 16.2.

Those who promote Winsorization are extremely happy with the first two lines in Table 16.3: The OLS results are horrible—the RMSE of the OLS estimates is 14.79 compared to 0.080 for the Winsorized estimates, indicating that the OLS estimates are 185 (= 14.79/0.08) times worse. They are even happier that OLS has essentially no power, with 6.6% correct rejections of the null hypothesis $H_0: \beta_1 = 0$. The power of the Winsorized test is better, although not stellar: 26.3% of the simulations correctly rejected the null hypothesis $H_0: \beta_1 = 0$.

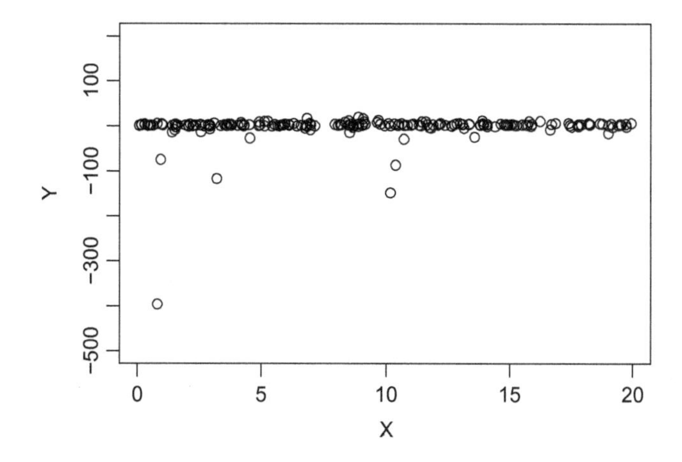

FIGURE 16.17
A typical sample where X is uniformly distributed between 0 and 20, and $p(y \mid x)$ is a shifted Cauchy distribution.

TABLE 16.3

Simulation Results Comparing Various Estimation/Testing Methods for the Data-Generating Process that Produced Figure 16.17. All Estimates are of the True Slope $\beta_1 = 0.1$, and All Results are Based on 20,000 Simulated Data Sets of Size $n = 100$ Each. Full R Code to Perform the Simulation is Given in Appendix A at the End of This Chapter

Method	Average	S. Dev.	RMSE	Power	Type I Error Rate
OLS	0.004	14.79	14.79	0.066	0.035
Winsorized	0.093	0.080	0.080	0.263	0.048
Truncated	0.090	0.058	0.059	0.376	0.048
Quantile Regression	0.101	0.043	0.043	0.642	0.060
ML (Cauchy errors)	0.101	0.038	0.038	0.763	0.058

Self-study question: Why were these rejections "correct"?

The truncated results shown in Table 16.3 delete all observations where the Y data is outside the 5th and 95th percentiles, followed by OLS analysis, and are slightly better in terms of accuracy and power than the Winsorized results, at least in this simulation study. (But truncation sometimes gives worse results than Winsorization as shown below.)

So there you have it. With a horribly outlier-prone data-generating process, Winsorization and truncation greatly outperform OLS.

But there is much more to this story! Who ever said you should use OLS with outlier-prone data-generating processes anyway? We have repeated throughout this book that OLS estimates can only be considered best when the distributions $p(y \mid x)$ are normal distributions because OLS estimates are maximum likelihood estimates in that case. But outlier-prone distributions, like the Cauchy, are quite different from normal distributions.

Thus, better comparators to Winsorization/truncation would be maximum likelihood regression (using a heavy-tailed distribution) and quantile regression. As shown in Table 16.3, the RMSE of the quantile regression estimates is estimated to be 0.043, approximately *half* that of the Winsorized estimates' RMSE of 0.080. Thus, the Winsorized estimates tend to be *twice as far* from the target $\beta_1 = 0.10$ as are the quantile regression estimates. Further, the quantile regressions correctly rejected the null hypothesis $(H_0 : \beta_1 = 0)$ in 64.2% of the studies. The power of the Winsorized test, while much better than OLS, was a comparatively paltry 26.3%.

Further, quantile regression is not even the best method. Using maximum likelihood with Cauchy error distributions provides still more accurate estimates and more powerful tests than the quantile regression method. See the last line of Table 16.3.

So far, the story looks pretty bad for Winsorizing and truncation. But it gets even worse.

The reason that the emphasis of this book is on the conditional distributions $p(y \mid x)$ is that it is a realistic and correct model. The relationship between Y and $X = x$ is not deterministic; rather, for any given $X = x$, there is a distribution $p(y \mid x)$ of possible Y values. From this realistic and correct standpoint, it is clear that any single number that you might give for Y when $X = x$ must be the wrong number. Instead, there is a *range of possible values of* Y for that $X = x$.

In Chapter 3, you saw how to estimate this range using a *prediction interval* for Y. This interval strongly depends on the normality assumption. The hope of Winsorizing is that, once the data are Winsorized, the usual OLS-based methods will work reasonably well.

Consider the same data-generating process analyzed above in Table 16.3. Let's see how well the 90% prediction intervals look with the Winsorized and truncated data, as compared to the corresponding intervals obtained by using quantile regression with $\tau = 0.05$ and $\tau = 0.95$. Let's choose a very large sample size, n = 200,000, to see whether the procedures are *consistent*.

A *consistent* procedure is one that works better and better with larger sample sizes. Intuitively, consistency is one of the most important properties that you would like an estimation procedure to have because it just makes sense that you should have better and better information with larger and larger data sets. Thus, you should avoid using inconsistent methods. With the current "data science revolution," where large data sets are commonplace, consistency is a very practical concern, rather than a theoretical concept.

The true model for the data-generating process studied above is

$$Y = 2.0 + 0.10X + 1.5\varepsilon,$$

where

$$\varepsilon \sim \text{Cauchy} \left(\text{or } T_1\right)$$

Thus, the true 90% central range of the $p(y \mid x)$ distribution is the range from

$$2.0 + 0.10x + 1.5\varepsilon_{0.05}$$

to

$$2.0 + 0.10x + 1.5\varepsilon_{0.95};$$

using `qt(0.05,1)= -6.313752` and `qt(0.95,1) = 6.313752`, the true 90% range is from $2.0 + 0.10x + 1.5(-6.313752)$ to $2.0 + 0.10x + 1.5(6.313752)$, or from

$$y_{0.05} = -7.47 + 0.10x$$

to

$$y_{0.95} = 11.47 + 0.10x.$$

These are the *true* quantile regression functions for quantiles $\tau = 0.05$ and $\tau = 0.95$, respectively.

Figure 16.18 shows the results of estimating this prediction interval using n = 200,000 observations, using standard prediction intervals obtained after Winsorization, as well as by using quantile regression.

R code for Figure 16.18

```
n = 200000; beta0 = 2; beta1 = .10; scale = 1.5;
set.seed(12345)
library(DescTools); library(quantreg)

X = runif(n, 0, 20); epsilon = scale*rt(n,1)
Y = beta0 + beta1*X + epsilon

X.w = Winsorize(X); Y.w = Winsorize(Y)
fit.wins = lm(Y.w~X.w)

L = quantile(Y, .05); U = quantile(Y, .95)
Y.t = Y[Y>L & Y<U]; X.t = X[Y>L & Y<U]
fit.trunc = lm(Y.t~X.t)

X.new = seq(0, 20, .1)
X.new = data.frame(X.new); names(X.new) = c("X")
Xw.new = data.frame(X.new); names(Xw.new) = c("X.w")
Xt.new = data.frame(X.new); names(Xt.new) = c("X.t")

pred.w = predict(fit.wins, Xw.new, interval="prediction", level=.90)
pred.t = predict(fit.trunc, Xt.new, interval="prediction", level=.90)

fit.quant.05 = rq(Y~X,tau=.05); fit.quant.95 = rq(Y~X,tau=.95)
lower.W = predict(fit.quant.05, X.new)
upper.W = predict(fit.quant.95, X.new)

x.plot = X.new[,1]
true.quant.05 = beta0 + beta1*x.plot + scale*qt(.05,1)
true.quant.95 = beta0 + beta1*x.plot + scale*qt(.95,1)

plot(x.plot, lower.W, type="l", lty=2, ylim = c(-10,15), xlab="X",
ylab="Prediction Limits")
points(x.plot, upper.W, type="l", lty=2)
points(x.plot, true.quant.05, type="l")
points(x.plot, true.quant.95, type="l")
points(x.plot,  beta0 + beta1*x.plot, type="l", lty=2, lwd=3)
points(x.plot, pred.w[,2], type="l", lty=3)
points(x.plot, pred.w[,3], type="l", lty=3)
points(x.plot, pred.t[,2], type="l", lty=4)
points(x.plot, pred.t[,3], type="l", lty=4)
```

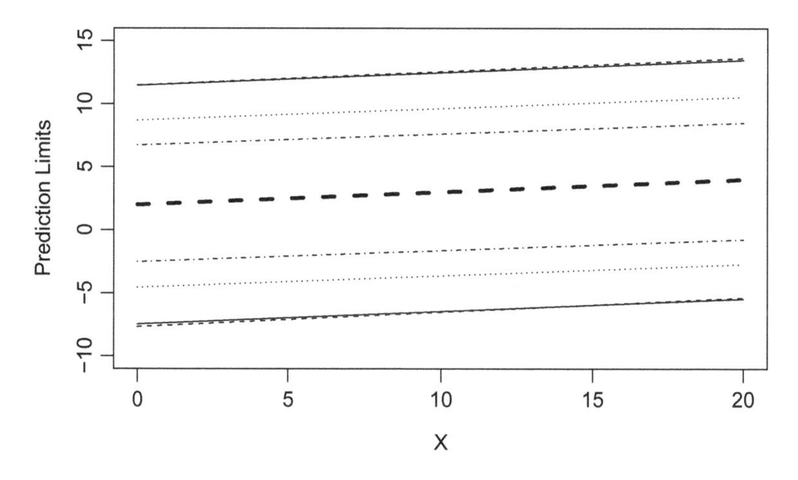

FIGURE 16.18
Graph showing inconsistency of 90% prediction limits using Winsorized and truncated data, and consistency of the quantile regression limits. Solid lines are true 90% limits. Thin dashed lines are estimated 90% prediction limits using quantile regression on $n = 200{,}000$ observations with $\tau = 0.05$ (lower limit) and $\tau = 0.95$ (upper limit), and are quite close to the true 90% limits, indicating consistency of the quantile regression-based estimates. Dotted lines are estimated 90% prediction limits using Winsorized data with the same $n = 200{,}000$ observations, and dash-dot lines are 90% prediction limits using the truncated data, both showing inconsistency. Thick dashed line is the true median function.

Figure 16.18 illustrates the inconsistency of the Winsorized and truncated prediction limits. The logic is clear: For the Winsorized data, if you replace the extreme Y data with less extreme values, you cannot expect to estimate the true distribution of the Y values correctly, even with extremely large sample sizes such as the $n = 200{,}000$ used in Figure 16.18. For the truncated data, the problem is even worse. On the other hand, the quantile regression estimates show consistent behavior, with estimated prediction limits nearly equal to the true limits, despite the horribly outlier-prone (Cauchy) error distribution. And as suggested by Table 16.3, Maximum Likelihood estimates would be even more accurate.

To summarize, Winsorizing (and/or truncation) is a questionable scientific practice since (i) it does not model the real distributions $p(y \mid x)$ correctly, (ii) it often performs worse than other methods as shown in Tables 16.2, 16.3 and Figure 16.18, and (iii) it opens the door to selective data usage, which is a slippery slope on the way to academic misconduct. Therefore, any researcher who plans to use Winsorizing or truncation should first justify their choice by using simulations such as displayed in Tables 16.2, 16.3, and Figure 16.18, but modified to mimic both the kind of data they intend to analyze and the kind of analysis they wish to perform.

Appendix A: R Code to Perform the Simulation Study Given Table 16.3

```
n = 100  # Sample size per data set
beta0 = 2; beta1 = .10; scale = 1.5  # true parameter values
set.seed(12345)
nsim = 20000  ## Simulations. WARNING!! 20000 takes about an hour.
              ## Change nsim to 500 first.

# Initializations
b1.ols = numeric(nsim); pv.ols = numeric(nsim); pv0.ols = numeric(nsim)
b1.wins = numeric(nsim); pv.wins = numeric(nsim); pv0.wins = numeric(nsim)
b1.trunc = numeric(nsim); pv.trunc = numeric(nsim); pv0.trunc = numeric(nsim)
b1.quant = numeric(nsim); pv.quant = numeric(nsim); pv0.quant = numeric(nsim)
b1.ml = numeric(nsim); pv.ml = numeric(nsim); pv0.ml = numeric(nsim)

# libraries
library(DescTools); library(maxLik); library(quantreg)

# ML estimation using Cauchy errors
loglik <- function(param) {
 b0 = param[1]; b1 = param[2]; g0 = param[3]
 ln.scale = g0
 z = (Y - b0 - b1*X)/exp(ln.scale)
 ll = sum(dt(z,log=T,1) - ln.scale)
 ll
}

# Main simulation loop
for (i in 1:nsim) {
 X = runif(n, 0, 20)
 epsilon = scale*rt(n,1)
 Y = beta0 + beta1*X + epsilon
 Y0 = beta0 + (0)*X + epsilon
 fit.ols = summary(lm(Y~X))$coefficients
 b1.ols[i] = fit.ols[2,1]
 pv.ols[i] = fit.ols[2,4]
 pv0.ols[i] = summary(lm(Y0~X))$coefficients[2,4]

 X.w = Winsorize(X); Y.w = Winsorize(Y)
 Y0.w = Winsorize(Y0)
 fit.wins = summary(lm(Y.w~X.w))$coefficients
 b1.wins[i] = fit.wins[2,1]
 pv.wins[i] = fit.wins[2,4]
 pv0.wins[i] = summary(lm(Y0.w~X.w))$coefficients[2,4]

 L = quantile(Y, .05); U = quantile(Y, 1-.05)
 Y.t = Y[Y>L & Y<U]; X.t = X[Y>L & Y<U]
 L0 = quantile(Y0, .05); U0 = quantile(Y0, 1-.05)
 Y0.t = Y0[Y0>L0 & Y0<U0]; X0.t = X[Y0>L0 & Y0<U0]
 fit.trunc = summary(lm(Y.t~X.t))$coefficients
 b1.trunc[i] = fit.trunc[2,1]
```

```
 pv.trunc[i] = fit.trunc[2,4]
 pv0.trunc[i] = summary(lm(Y0.t~X0.t))$coefficients[2,4]

 fit.quant = summary(rq(Y~X), se="iid")$coefficients
 b1.quant[i] = fit.quant[2,1]
 pv.quant[i] = fit.quant[2,4]
 fit0.quant = summary(rq(Y0~X), se="iid")$coefficients
 pv0.quant[i] = fit0.quant[2,4]

 fit.ml = summary(maxLik(loglik, start=c(0,0,0)))$estimate
 b1.ml[i] = fit.ml[2,1]; pv.ml[i] = fit.ml[2,4]
 Y = Y0
 pv0.ml[i] = summary(maxLik(loglik, start=c(0,0,0)))$estimate[2,4]
}


## OLS results: mean, sdev, rmse, and power
mean(b1.ols); sd(b1.ols); sqrt(mean ((b1.ols - .10)^2) );
mean(pv.ols <.05)

## Winsorized results: mean, sdev, rmse, and power
mean(b1.wins); sd(b1.wins); sqrt(mean ((b1.wins - .10)^2) );
mean(pv.wins <.05)

## Truncation results: mean, sdev, rmse, and power
mean(b1.trunc); sd(b1.trunc); sqrt(mean ((b1.trunc - .10)^2) );
mean(pv.trunc <.05)

## Quantile regression results: mean, sdev, rmse, and power
mean(b1.quant); sd(b1.quant); sqrt(mean ((b1.quant - .10)^2) );
mean(pv.quant < 0.05)

## ML results: mean, sdev, rmse, and power
mean(b1.ml); sd(b1.ml); sqrt(mean ((b1.ml - .10)^2) ); mean(pv.ml <.05)

## Type I error rates:
mean(pv0.ols <.05)  #OLS
mean(pv0.wins <.05) # Winsorized
mean(pv0.trunc <.05) # Truncated
mean(pv0.quant <.05) # Quantile
mean(pv0.ml <.05) # ML
```

References

Taleb, N. N. (2007). *The black swan: The impact of the highly improbable* (Vol. 2). New York, NY: Random house.

Yale, C., & Forsythe, A. B. (1976). Winsorized regression. *Technometrics, 18*(3), 291–300.

Exercises

1. Consider the following code: There is a normal process and a "suspect" data value.

```
set.seed(12345)
X.0 = runif(30, .5, 3.5)
beta0 = 1.0
beta1 = 1.5
sigma = 0.7
Y.0 = beta0 + beta1*X.0 + sigma*rnorm(30)    # The regular process

# Suspicious data value
X.suspect = 9.3
Y.suspect = 0.6

# The complete data
Y = c(Y.0, Y.suspect)
X = c(X.0, X.suspect)
```

   a. Graph the X, Y scatterplot. Identify the suspicious point by making the point solid on the plot.
   b. Estimate the OLS line and overlay it on the graph.
   c. Estimate the line using `survreg`, with no censored observations, and `dist= "t"`. Overlay that line as well on the graph, using a different line type.
   d. Compare models 1.b and 1.c using the AIC statistic and comment.
   e. Inserting `parms = 3` into the survreg fit along with `dist = "t"` fits the conditional T distribution with three degrees of freedom (`df`). Create a new graph of the data, with overlayed straight-line fits using t error distributions with df = 3, 4, 5, 10, 15, 20, 50, and ∞. Overlay the least-squares fit as well. Briefly explain the lesson of this graph.

2. The data set `Firms = read.csv("https://raw.githubusercontent.com/andrea2719/URA-DataSets/master/Firms.csv")` contains measurements of Income and Assets of various firms. Consider the prediction of Assets (Y) using Income (X). Perform an analysis of the outliers in the data set, following Sections 16.2 through 16.4 of the book. (Be thoughtful in your analysis, as in the book. This will take some work.)

3. The data set `Firms = read.csv("https://raw.githubusercontent.com/andrea2719/URA-DataSets/master/Firms.csv")` contains measurements of Income and Assets of various firms. Consider the prediction of Assets (Y) using Income (X).
   a. Fit the median regression, the 0.95 quantile regression, and the 0.05 quantile regression functions. Report the estimated regression functions and interpret them in terms of conditional distributions.
   b. Graph the scatterplot and overlay the three regression lines in 3.a.
   c. Use the results of 3.a to construct a 90% prediction interval for Y when $X = 30$. Interpret the result in terms of Firms, Assets and Income.

d. Graph the scatterplot again, and overlay the OLS regression line and corresponding 90% prediction intervals.

e. Construct a 90% prediction interval for Y when X = 30 using the classical regression model. Interpret the result in terms of Firms, Assets and Income.

f. Comparing your interpretations of 3.c and 3.e, which approach seems more reasonable, classical regression or quantile regression?

4. Regression is a model for the distribution of potentially observable Y values, given values of the particular X variables. If you knew this distribution, you could answer questions like, "What is the range of potentially observable Y data when X = 10? How does this range differ from the range when X = 20? Of course, we don't use the entire range because that could be infinite. Instead, we settle for the "essential range," usually defined as the central 90% or 95% of the distribution $p(y \mid x)$. This range defines the *prediction limits*.
Use the following auditing data set
`fees=read.csv("https://raw.githubusercontent.com/andrea2719/URA-DataSets/master/auditfees.csv", header=T)` and define:

Y = Audit.fees/1000000 (Audit fees of a firm in millions of $)

X = Matchfy.balsh.total.assets/1000000000 (firm assets in billions of $)

(Changing the units so that the numbers are "nice" in this way is a "best practice." Not only are the outputs easier to read, the computer algorithms also work better because there are fewer "overflow" and "underflow" errors, which are cases where the limits of the computer round-off error comes into play. And, unlike nonlinear transformations such as the log transform, transforming the variables like this to change the units of measurement has no effect whatsoever on the validity (or lack of validity) of the model.)

a. Plot X, Y. Label the axes appropriately (not as "X" or "Y")

b. Overlay the (i) 0.50 quantile regression line, the (ii) OLS line, and (iii) the line obtained by OLS regression on the Winsorized X and Y variables (X_w and Y_w).

c. Draw a new X, Y scatterplot, and overlay the 90% prediction limits obtained by quantile regression.

d. Repeat 4.c but using OLS.

e. Repeat 4.c Again, use (X, Y) for the scatterplot, but overlay the OLS line from the Winsorized variables (X_w, Y_w).

f. How many points are below the lower line in 4.c? What % is this? What is the target %?

g. How many points are above the upper line in 4.c? What % is this? What is the target %?

h. How many points are below the lower line in 4.d? What % is this? What is the target %?

i. How many points are above the upper line in 4.d? What % is this? What is the target %?

j. How many points are below the lower line in 4.e? What % is this? What is the target %?

k. How many points are above the upper line in 4.e? What % is this? What is the target %?
l. Summarizing these results, which of the three methods is preferred for constructing prediction limits?
m. Why might a firm be interested in the prediction limits for Audit Fees? Suppose the firm has 10 billion in assets, to be specific.

17

Neural Network Regression

As we have repeated often throughout this book, the classical regression model is clearly wrong in many ways. One way that the model is wrong is in the assumption that the conditional distributions are normal distributions. Instead, the distribution of $Y \mid \mathbf{X} = \mathbf{x}$ is correctly given as $p(y \mid \mathbf{x})$, where $p(y \mid \mathbf{x})$ can be any distribution (recall that $\mathbf{x} = (x_1, x_2, \ldots, x_k)$, a particular set of values that the X variables may take.)

Another way that the classical model is wrong is in the assumption that the conditional mean function is given by $\mathrm{E}(Y \mid X_1 = x_1, X_2 = x_2, \ldots, X_k = x_k) = \beta_0 + \beta_1 x_1 + \beta_2 x_2 + \cdots + \beta_k x_k$. Instead, the conditional mean function (assuming the mean is finite) is correctly given as

$$\mathrm{E}(Y \mid X_1 = x_1, X_2 = x_2, \ldots, X_k = x_k) = f(x_1, x_2, \ldots, x_k)$$

where $f(x_1, x_2, \ldots, x_k)$ can have any function form whatsoever in (k + 1)—dimensional space, from curved hyperplanes, to functions with "bumps," to discontinuous functions, depending upon the Nature of the process that you are studying. Just because you like planar functions does not mean that Nature likes them! Your personal preferences, likes and dislikes, do not change Nature's workings. (Recall that "Nature" refers to the data-generating process; it can contain elements of biology, and/or physics, and/or economics, and/or psychology, etc.)

As we noted in Chapter 1, Section 1.7, even the simple linear model $f(x_1) = \beta_0 + \beta_1 x_1$ is usually wrong: We gave a logical argument that there must be (with 100% probability) some curvature whenever the X_1 variable has at least three levels and there is dependence between Y and X_1. This argument becomes more relevant in multidimensional space, where the curvature can take so many forms, including twists, turns and bumps.

The goal of neural network (NN) regression is to approximate Nature's true function $f(x_1, x_2, \ldots, x_k)$ by using a non-linear (or non-planar) function $g(x_1, x_2, \ldots, x_k)$. In particular, these NN approximating functions $g(x_1, x_2, \ldots, x_k)$ allow complex curvatures and interactions between various combinations of X variables, which can be very tricky to tease out using the classical regression model with polynomial and product interaction terms. But they also allow smoother, nearly linear (or planar) functions as well. By allowing greater flexibility, and therefore greater realism, than the rigid classical (planar) function $\beta_0 + \beta_1 x_1 + \beta_2 x_2 + \cdots + \beta_k x_k$, NN regression models can give you predicted values $\hat{Y} = \hat{g}(x_1, x_2, \ldots, x_k)$ that are sometimes much more accurate than those provided by the classical model.

A disadvantage of NN models is that the estimated NN model $\hat{Y} = \hat{g}(x_1, x_2, \ldots, x_k)$ is not simple to interpret. (See below for explicit function form representations of the NN model $g(x_1, x_2, \ldots, x_k)$.) Thus, instead of trying to interpret the function form, users of NN models usually just treat them as a "black box" to produce a predicted value $\hat{Y} = \hat{g}(x_1, x_2, \ldots, x_k)$. In this regard, NN regression is similar to LOESS, which also provides predicted values $\hat{Y}$ using a function form that is usually too complicated to examine.

In many practical cases, you simply do not care that the predicted values $\hat{Y} = \hat{g}(x_1, x_2, \ldots, x_k)$ have a complicated function form. If you only care about getting a $\hat{Y}$ that is close to Y, then NN regression can be ideal. Examples include:

- $\hat{Y}$ = predicted distance your car can travel without refueling, given current amount of fuel in the tank and current fuel economy.
- $\hat{Y}$ = predicted EEG reading of a heart patient, given current measurements of pulse, sweat, and blood pressure.
- $\hat{Y}$ = credit worthiness of a customer who is applying for a loan ("credit score"), given current financial data and recent credit history.

These examples are all cases where you want a good prediction $\left(\hat{Y}\right)$ of Y, but you do not necessarily care what the function $\hat{Y} = \hat{g}(x_1, x_2, \ldots, x_k)$ looks like.

17.1 Universal Approximators

Since a main goal of regression analysis (not just NN regression) is to approximate Nature's unknown function $f\left(x_1, x_2, \ldots, x_k\right) = \mathrm{E}\left(Y \mid X_1 = x_1, X_2 = x_2, \ldots, X_k = x_k\right)$ by using a known function $g\left(x_1, x_2, \ldots, x_k\right)$, we first discuss function approximation. As noted repeatedly throughout this book, functions $g\left(x_1, x_2, \ldots, x_k\right)$ that are linear, quadratic, interaction, exponential, and others, are nearly always different from Nature's function $f\left(x_1, x_2, \ldots, x_k\right)$. But some classes of functions $g(\cdot)$ can get arbitrarily close to $f(\cdot)$; such a class of functions is called a "universal approximator."

The class of polynomial functions is an example of a universal approximator. That is, given any continuous function $f\left(x_1, x_2, \ldots, x_k\right)$ defined over a bounded set of values for your X variables, you can approximate $f\left(x_1, x_2, \ldots, x_k\right)$ arbitrarily well by some polynomial function $g\left(x_1, x_2, \ldots, x_k\right)$. You might need an extremely high order of polynomial (recall that such functions involve all possible interaction terms up to the given order), but still, there *exists* a polynomial $g(\cdot)$ such that the maximum difference between the functions $f(\cdot)$ and $g(\cdot)$ is as small as you would like. This result is mathematically proven, and is called the Stone-Weierstrass Theorem.

The Stone-Weierstrass Theorem seems to suggest that you should use high order polynomial models to estimate regression functions, because, as the Stone-Weierstrass Theorem dictates, such models are less biased. However, you also know from the variance-bias trade-off discussed in Chapter 11 that, while such higher-order models will be less biased, such models also require many more estimates of parameters. Hence, high order polynomial models will suffer from extremely high variance when estimated using data, and therefore provide inaccurate predictions, despite their low bias. (Note that "high variance" translates to "inordinately influenced by the idiosyncratic data," since the results will *vary greatly* from one idiosyncratic data set to another idiosyncratic data set, even when both data sets come from the same data-generating process.)

In addition, polynomial models can have extremely poor behavior at the extremes of the data, or where data are sparse. At the extremes, the predictions shoot off quickly to positive or negative infinity, because that's how high-order polynomials terms like x^3, x^4, x^5, etc., behave. Where data are sparse, the behavior can also be erratic: See Figure 11.5, for example.

There exist universal approximators other than polynomials that do not have such wild behavior. Fourier analysis gives you functions $g(\cdot)$ involving sines and cosines that can approximate general functions $f(\cdot)$; this class functions is a universal approximator that does not have as wild extrapolation properties as does the polynomial class. Being based on sines and cosines, Fourier functions $g(\cdot)$ tend to be better for approximating functions $f(\cdot)$ that are cyclical, such as time-series data with an annual seasonal component.

There are infinitely many universal approximators, depending on the particular set of "basis functions" that you use. In polynomial approximators, the basis functions are polynomials; in Fourier analysis, the basis functions involve sines and cosines. Pick an appropriate set of basis functions, and voilà! You can construct a universal approximator.

17.2 Neural Network and Polynomial Approximations of a Known Noiseless Function

"Neural network" is just a fancy-sounding name (meant to suggest that it somehow operates like the brain!) for a universal approximator that involves a particular type of basis function (called an "activation function" in the NN jargon), which is typically the same function used in logistic regression, $1/(1+e^{-x})$. Like polynomial functions are constructed of functions involving various orders of polynomial terms, their interactions, and constant multipliers, NN functions are constructed of logistic functions.

The Neural Network regression model is one of many models that you can use to fit non-linear (or non-planar) processes.

The following example gives the NN regression function explicitly. First, consider the following true, "noiseless" (meaning no random error) function,

$$Y = f(x_1, x_2) = -1 + \sin(x_1) + 3\sin(x_1) \times \cos(x_2),$$

which is *not* a neural network function. Suppose further that the range of possible X values is $0 < x_1,\ x_2 < 3.14159265\ldots$. The goal is to find a function $g(x_1, x_2)$ that is close to $f(x_1, x_2)$ over the possible range; you can do this by using a variety of methods, including by using polynomial approximation and by using neural network approximation.

This function is graphed in Figure 17.1.

R code for Figure 17.1

```
x = seq(0,pi,pi/40)
x1 = rep(x, each = 41); x2 = rep(x,41)
Y  = -1 + sin(x1) + 3*sin(x1) * cos(x2)

library(lattice)
wireframe(Y ~ x1*x2,
  xlab = "x1", ylab = "x2",
  main = "True Function", xlim = c(0,pi), ylim = c(0,pi),
  drape = TRUE,
  colorkey = FALSE,
  scales = list(arrows=FALSE,cex=.8,tick.number = 5),
  screen = list(z = -70, x = -70))
```

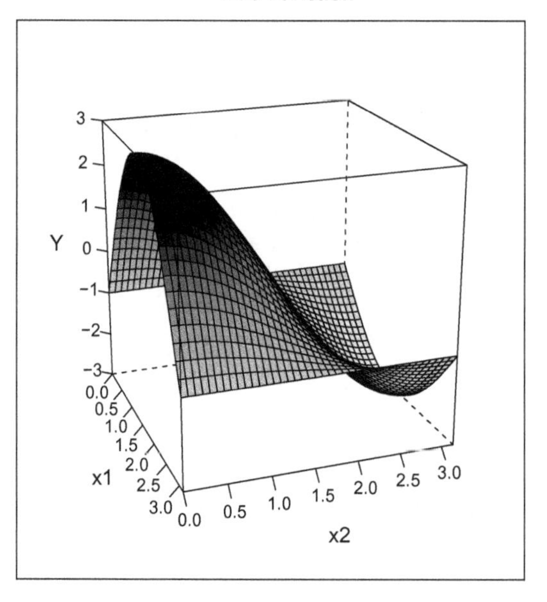

FIGURE 17.1
A true function $Y = -1 + \sin(x_1) + 3\sin(x_1) \times \cos(x_2)$, to be approximated by using neural network and polynomial approximating functions.

Now, suppose you did not know the function form of the true regression function $f(x_1, x_2)$ that is graphed in Figure 17.1, but you wanted to approximate it so that you can get an accurate value of Y for given values (x_1, x_2). A universal approximator is a class of functions that provide better and better approximations to any function as the complexity (number of parameters) of the approximating function increases. For example, you can approximate the function $f(x_1, x_2)$ shown in Figure 17.1 using polynomial functions $g(x_1, x_2)$ of degree 1, 2, and 3 as shown in the display that follows.

Polynomial approximations

Polynomial approximation of degree 1:

$$g(x_1, x_2) = \beta_0 + \beta_1 x_1 + \beta_2 x_2$$

Polynomial approximation of degree 2:

$$g(x_1, x_2) = \beta_0 + \beta_1 x_1 + \beta_2 x_2 + \beta_3 x_1^2 + \beta_4 x_2^2 + \beta_5 x_1 x_2$$

Polynomial approximation of degree 3:

$$g(x_1, x_2) = \beta_0 + \beta_1 x_1 + \beta_2 x_2 + \beta_3 x_1^2 + \beta_4 x_2^2 + \beta_5 x_1 x_2 + \beta_6 x_1^3 + \beta_7 x_2^3 + \beta_8 x_1^2 x_2 + \beta_9 x_1 x_2^2$$

As discussed in Chapter 11, the more complex models have less bias, because they have greater flexibility. Because of this, and because of the Stone-Weierstrass Theorem, you can expect that the degree 3 polynomial approximation $g(x_1, x_2)$ will provide a more accurate approximation to the true regression function $f(x_1, x_2)$, for some specific values of the β parameters. You can demonstrate that this is true by fitting the three models, graphing the fitted functions, and comparing them with the true function in Figure 17.1. The following code makes the assumption that the data come from the function of Figure 17.1 without error (i.e., "noiseless"); hence, all deviations between the fitted polynomial functions and the true function are indications of bias, not idiosyncratic variance. The actual data are $n =$ 1,681 ($= 41^2$) evaluations of the function $-1 + \sin(x_1) + 3\sin(x_1) \times \cos(x_2)$, for x_1 and x_2 in the grid of values where $(x_1, x_2) \in \{0, \pi/40, 2\pi/40, \ldots, 40\pi/40\} \times \{0, \pi/40, 2\pi/40, \ldots, 40\pi/40\}$.

R code for Figure 17.2

```
x = seq(0,pi,pi/40)
x1 = rep(x, each = 41); x2 = rep(x,41)
Y  = -1 + sin(x1) + 3*sin(x1) * cos(x2)

library(lattice); library(grid)
par(mfrow=c(2,2))
poly1 = lm(Y ~ x1 + x2)
poly2 = lm(Y ~ x1 + x2 + I(x1^2) + I(x2^2) + I(x1*x2))
poly3 = lm(Y ~ x1 + x2 + I(x1^2) + I(x2^2) + I(x1*x2) + I(x1^3)
         + I(x2^3) + I(x1^2*x2) + I(x1*x2^2) )
xpred = data.frame(x1,x2)
yhat1 = predict(poly1,xpred)
yhat2 = predict(poly2,xpred)
yhat3 = predict(poly3,xpred)
wf <- function(y) {
```

(Continued)

```
  wireframe(y ~ x1*x2,
  xlab = "x1", ylab = "x2", zlab= " ",
  main = "", xlim = c(0,pi), ylim = c(0,pi),
  drape = TRUE,
  colorkey = FALSE,
  scales = list(arrows=FALSE,cex=.8,tick.number = 5),
  screen = list(z = -70, x = -70))
 }
g1 = wf(Y); g2 = wf(yhat1); g3 = wf(yhat2); g4 = wf(yhat3)
print(g1, split = c(1, 1, 2, 2), more = TRUE)
print(g2, split = c(2, 1, 2, 2), more = TRUE)
print(g3, split = c(1, 2, 2, 2), more = TRUE)
print(g4, split = c(2, 2, 2, 2))
```

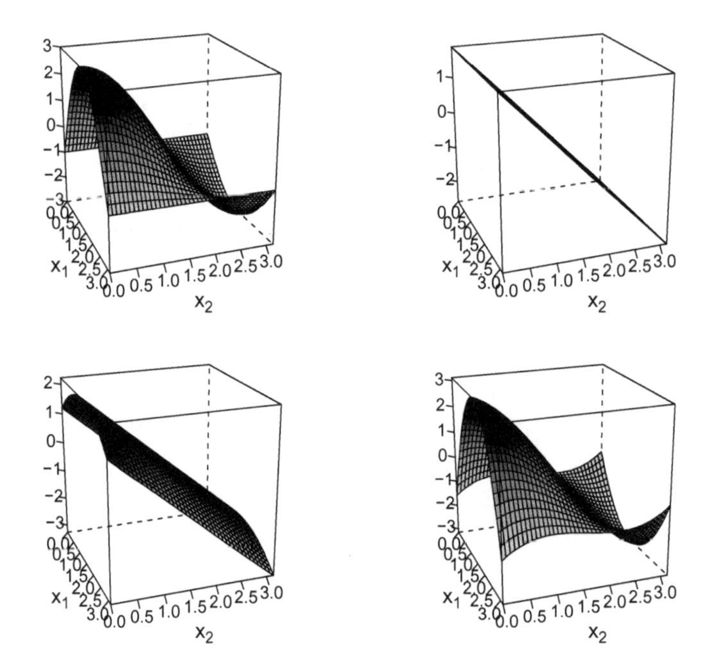

FIGURE 17.2
Upper left panel: True function $Y = -1 + \sin(x_1) + 3\sin(x_1) \times \cos(x_2)$. Upper right: True function approximated by a first-degree polynomial (a plane). Lower left: True function approximated by a second-degree polynomial. Lower right: True function approximated by a third-degree polynomial.

Figure 17.2 illustrates the idea that the class of polynomial functions is a universal approximator: As the approximating function becomes more complex, it can approximate the given function with less error. However, the problem with polynomial regression models that was noted in Chapter 11 is that they can give quite wild (extremely large or extremely small) results in regions where there is no X data available to estimate the function. In other words, if the best approximation shown in the lower right of Figure 17.2 were extended beyond the ranges from 0 to 3.14 for the X variables, that function would be very far from the true Y. (The higher the order of the polynomial term, the faster it tends toward plus or minus infinity.)

Similar to polynomial models, NN regression models form a class of functions that is a universal approximator. However, NN regression models are constructed from functions (activation functions) that do not have such wild extrapolation properties as do polynomials. The S-shaped logistic regression function is typically used as an activation function, and this function is bounded as x tends toward infinity, unlike polynomial functions that shoot off toward negative and positive infinity.

Also, unlike polynomial models, NN regression functions are not linear in the parameters (the β's) and thus cannot be estimated using the `lm` function of `R`. Alternative, iterative approaches (similar to those used to obtain maximum likelihood estimates) are used instead to find estimates that minimize the sum of squared errors.

The display shows NN functions $g(x_1, x_2)$ with 1, 2 and 3 "nodes."

Neural Network approximations

Neural network approximation with 1 node in a single hidden layer (NN(1)):

$$g(x_1,x_2)=\gamma_0+\frac{\gamma_1}{1+e^{-(\beta_{10}+\beta_{11}x_1+\beta_{12}x_2)}}$$

Neural network approximation with 2 nodes in a single hidden layer (NN(2)):

$$g(x_1,x_2)=\gamma_0+\frac{\gamma_1}{1+e^{-(\beta_{10}+\beta_{11}x_1+\beta_{12}x_2)}}+\frac{\gamma_2}{1+e^{-(\beta_{20}+\beta_{21}x_1+\beta_{22}x_2)}}$$

Neural network approximation with 3 nodes in a single hidden layer (NN(3)):

$$g(x_1,x_2)=\gamma_0+\frac{\gamma_1}{1+e^{-(\beta_{10}+\beta_{11}x_1+\beta_{12}x_2)}}+\frac{\gamma_2}{1+e^{-(\beta_{20}+\beta_{21}x_1+\beta_{22}x_2)}}+\frac{\gamma_3}{1+e^{-(\beta_{30}+\beta_{31}x_1+\beta_{32}x_2)}}$$

There are 5 parameters to estimate (the γ's and the β's) in NN(1), and 9 and 13 in NN(2) and NN(3), respectively. Recall that with more parameters, the model becomes more flexible. Figure 17.3 shows how the fitted functions NN(1), NN(2) and NN(3) look compared to the true function.

`R` code for Figure 17.3

```
library(neuralnet)
df = data.frame(Y, x1,x2); set.seed(123)
nn.fit1 = neuralnet(Y ~ x1+x2, data=df, hidden = 1)
nn.fit2 = neuralnet(Y ~ x1+x2, data=df, hidden = 2)
nn.fit3 = neuralnet(Y ~ x1+x2, data=df, hidden = 3)
yhat.1 = compute(nn.fit1, xpred)$net.result
yhat.2 = compute(nn.fit2, xpred)$net.result
yhat.3 = compute(nn.fit3, xpred)$net.result
g1 = wf(Y); g2 = wf(yhat.1); g3 = wf(yhat.2); g4 = wf(yhat.3)
print(g1, split = c(1, 1, 2, 2), more = TRUE)
print(g2, split = c(2, 1, 2, 2), more = TRUE)
print(g3, split = c(1, 2, 2, 2), more = TRUE)
print(g4, split = c(2, 2, 2, 2))
```

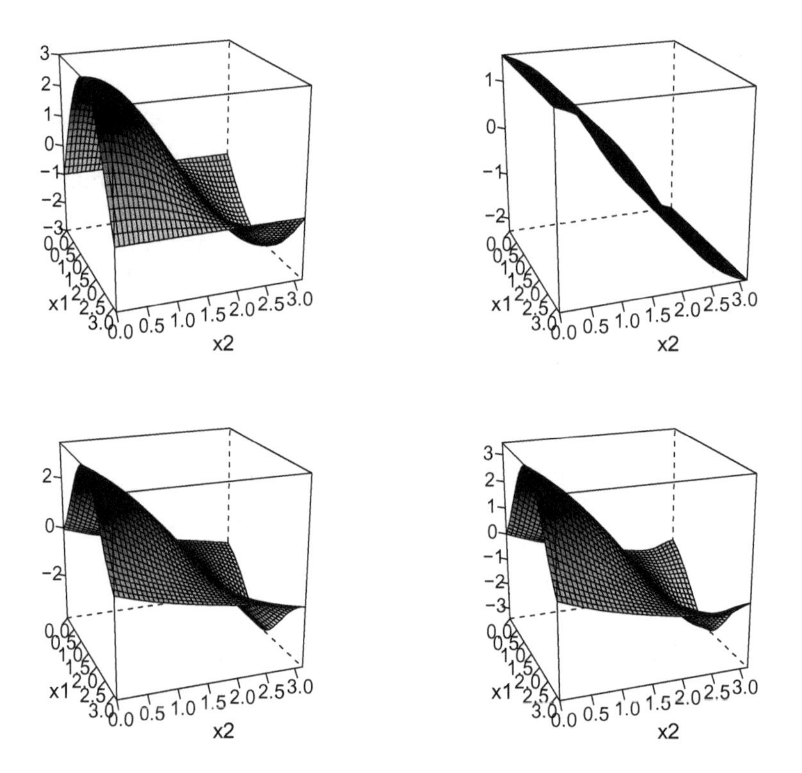

FIGURE 17.3
Upper left panel: True function $Y = -1 + \sin(x_1) + 3\sin(x_1) \times \cos(x_2)$. Upper right: True function approximated by a NN with one node. Lower left: True function approximated by a NN with two nodes. Lower right: True function approximated by a NN with three nodes.

Figure 17.3 illustrates the idea that the class of NN functions, like the class of polynomial functions, is a universal approximator: As the approximating function becomes more complex, it can approximate the true function with less error.

Note also that the NN model with two nodes seems to be better than the 2nd degree polynomial model (compare the lower left panels of Figures 17.2 and 17.3). However, the comparison is not entirely fair since the NN model has more parameters (9 versus 6), and hence is expected to fit better for that reason alone.

The reason for the "neural" and "node" terms is that the approximating functions can be represented in a way that resembles the brains neurons connected by synapses. Figure 17.4 shows this graph for the model with three nodes; you can obtain this graph simply by entering `plot(nn.fit3)`.

Notes on Figure 17.4:

1. While the schematic of Figure 17.4 resembles neurons and synapses, the relationship to "neural brain structure" is superficial. In reality, all that Figure 17.4 gives you is the fitted mathematical model whose equation is given above (the NN with three nodes). The neural network regression model is simply a non-linear regression model. It is not an imitation of the brain.
2. The coefficients on the left are the β's and those on the right are the γ's.
3. The circled "1" terms indicate intercepts: Left, β's, right γ.

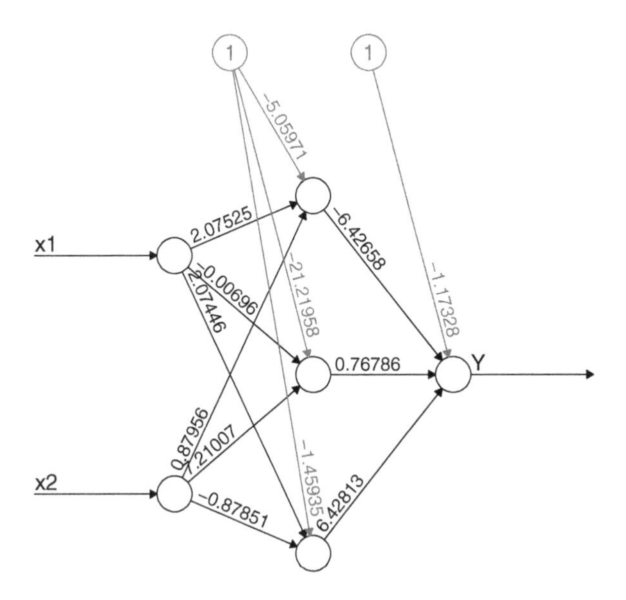

FIGURE 17.4
Schematic plot of the NN model having three nodes.

4. Putting it all together, the schematic gives a fitted function as follows:

$$\hat{g}(x_1,x_2) = -1.17328 + \frac{-6.42658}{1+e^{-(-5.05971+2.07525x_1+0.87956x_2)}} + \frac{0.76786}{1+e^{-(-21.21958-0.00696x_1+7.21007x_2)}} + \frac{6.42813}{1+e^{-(-1.45935+2.07446x_1-0.87851x_2)}}$$

The reason for saying that this is "*a* fitted function" instead of "*the* fitted function" is that the parameters of NNs are not uniquely identifiable, in the same way that the parameters of the ANOVA indicator variable model are not uniquely identifiable unless you drop an indicator variable (Chapter 10). Thus, if you fit a NN to the same data using a different value of `set.seed`, you will likely get different parameter estimates, but the sum of squared errors should be nearly the same. This is another reason why you ordinarily do not care about the specific function form of the NN regression: You do not even get the same parameter estimates when you repeat the analysis on the same data! However, if you just want the predicted values to be close to the actual values, then you do not care about the parameter estimates; you only care that the model gives predicted values that are close to the actual values.

5. The mathematical function form of $\hat{g}(x_1,x_2)$ shown above for the three-node NN model does not at all resemble the true function form $f(x_1,x_2)=-1+\sin(x_1)+3\sin(x_1)\times\cos(x_2)$, but as shown in the bottom right panel of Figure 17.3, it nonetheless provides an excellent approximation.

6. The note "Error: 37.10478" is the sum of squared errors between the "Y" data values and the fitted function, divided by 2. The division by 2 is just a NN-specific oddity; it makes no difference whether you minimize the sum of squared errors or the sum

of squared errors divided by 2. You can calculate this value "manually" by entering the command `sum((yhat.3 - Y)^2)/2`, which returns the value 37.10477983.

7. The note "Steps: 30463" refers to the number of steps taken by the algorithm to estimate the coefficients to minimize the Error, within the convergence range. A larger number of steps takes more time. Typically, more highly parameterized models require more steps.
8. You can verify that the function form indicated in 4. above by calculating the predicted values "by hand" as follows:

```
b10 = nn.fit3$weights[[1]][[1]][1,1]; b10
b11 = nn.fit3$weights[[1]][[1]][2,1]; b11
b12 = nn.fit3$weights[[1]][[1]][3,1]; b12

b20 = nn.fit3$weights[[1]][[1]][1,2]; b20
b21 = nn.fit3$weights[[1]][[1]][2,2]; b21
b22 = nn.fit3$weights[[1]][[1]][3,2]; b22

b30 = nn.fit3$weights[[1]][[1]][1,3]; b30
b31 = nn.fit3$weights[[1]][[1]][2,3]; b31
b32 = nn.fit3$weights[[1]][[1]][3,3]; b32

# Linear functions
L1 = b10 + b11*x1 + b12*x2
L2 = b20 + b21*x1 + b22*x2
L3 = b30 + b31*x1 + b32*x2

#Nodes
N1 = 1/(1 + exp(-L1))
N2 = 1/(1 + exp(-L2))
N3 = 1/(1 + exp(-L3))

# Node weights
g0 = nn.fit3$weights[[1]][[2]][1,1]; g0
g1 = nn.fit3$weights[[1]][[2]][2,1]; g1
g2 = nn.fit3$weights[[1]][[2]][3,1]; g2
g3 = nn.fit3$weights[[1]][[2]][4,1]; g3

#Final prediction
yhat.3.check = g0 + g1*N1 + g2*N2 + g3*N3

#The hand calculation agrees with the software within machine error:
sum ( (yhat.3 - yhat.3.check)^2 )
```

9. The parameters (the γ's and the β's) of the NN model with a given number of nodes are chosen to minimize the sum of squared errors. Thus, the estimated NN model is a maximum likelihood-estimated model assuming

$$Y \mid \mathbf{X} = x \sim \mathrm{N}(g(x), \sigma^2),$$

where $g(x)$ is the specific NN function that is assumed. Better NN model estimates are available by using maximum likelihood with appropriately chosen non-normal distributions.

10. This model is a NN with a single hidden layer having three nodes. Multiple hidden layers with varying numbers of nodes per layer can be estimated, leading to a very rich class of highly parameterized models.
11. As discussed in Chapter 11, the more highly parameterized models have less bias (since the true function is closer to a function in the given class), but more variance (since more parameters must be estimated, and each estimate has random error). To select a particular NN model among the myriads of possibilities involving nodes and layers, you can use cross-validation, as discussed in Chapter 11, and as demonstrated in the next section of this chapter.

17.3 Neural Network and Polynomial Approximations in a Real Example: Predicting Charitable Contributions

Let's see how the neural network works in a real example that we have used before; namely, the predictions of charitable contributions (in log scale) as a function of income (in log scale) and number of dependents.

First, a data pre-processing note: When using the `neuralnet` function in `R`, you will receive the following warning message when the algorithm used to select the parameters failed to converge:

```
Warning message:
algorithm did not converge in 1 of 1 repetition(s) within the stepmax
```

Similar to maximum likelihood analyses, numerical analysis problems such as overflow, underflow, and non-convergence can be lessened by re-scaling the input variables (e.g., 3.421 million of dollars is better for input than 3,421,000 dollars). Standardization via z-scores is one such re-scaling.

Thus, in the code below, we transform the X variables to each have means 0.0 and standard deviations of 1.0 in the analysis. We have also increased the "`stepmax`" specification from the default `1e5` to `1e6` to allow the algorithm to run longer. The code is as follows:

```
char = read.csv("https://raw.githubusercontent.com/andrea2719/
URA-DataSets/master/charitytax.csv")
attach(char)
INC.S = scale(INCOME); D.S = scale(DEPS)
f = as.formula(paste("CHARITY ~ INC.S + D.S"))
dat = data.frame(CHARITY, INC.S, D.S)
library(neuralnet)

##Fit various NN models
# Single hidden layer models
fit1 = neuralnet(f, data=dat, hidden=c(1), stepmax=1e6)
fit2 = neuralnet(f, data=dat, hidden=c(2), stepmax=1e6)
fit3 = neuralnet(f, data=dat, hidden=c(3), stepmax=1e6)
 # ... continue through fit12...
```

TABLE 17.1

RMSEs for the Various Fits for the Models That Are Fit Using the R Code Above

Model	RMSE	RMSPE
NN(1)	1.213	1.224
NN(2)	1.206	1.226
NN(3)	1.189	1.211
NN(4)	1.182	1.268
NN(5)	1.120	1.222
NN(6)	1.106	1.219
NN(7)	1.109	1.222
NN(8)	1.111	1.241
NN(9)	1.094	1.247
NN(10)	1.082	1.231
NN(11)	1.068	1.260
NN(12)	1.066	1.255

Note: Root mean squared out-of-sample prediction errors (RMSPEs) from 10-fold cross-validation are also included.

You can check the conformance of the fitted NN values (the $\hat{Y}$ values) with the actual Y values by finding the root mean squared errors (RMSE). For example, the NN model with four nodes in a single hidden layer has RMSE equal to

```
rmse = sqrt(mean( (CHARITY - fit4$net.result[[1]])^2 ))
```

Table 17.1 displays root mean square errors, both in-sample and out-of-sample (code is given below for the out-of-sample calculations).

Notice that the models have better in-sample prediction accuracy (smaller RMSE) as the complexity increases, no matter whether polynomial or NN. Among the models estimated here, the NN(12) comes closest to the data values, having the lowest RMSE (1.066).

Self-study question: Can you find a NN model with RMSE smaller than 1.066?

To understand the NN(12) model, let's graph it as a 3-D function, as shown above. Fortunately, like the examples above, this model is also a function of two X variables, and hence can be displayed using a 3-D plot. The code that follows shows how to draw this graph, and also shows how to back-transform the standardized X variables.

R code for Figure 17.5

```
inc.pred = seq(min(INCOME), max(INCOME), (max(INCOME)-min(INCOME))/20 )
inc.pred.s = (inc.pred - mean(INCOME))/sd(INCOME)
inc.pred.s = rep(inc.pred.s,7)
deps.pred = seq(0,6,1)
deps.pred.s = (deps.pred - mean(DEPS))/sd(DEPS)
deps.pred.s = rep(deps.pred.s, each=21)
```

(*Continued*)

```
all.pred = data.frame(inc.pred.s, deps.pred.s)
names(all.pred) = c("INC.S", "D.S")
yhat.plot = compute(fit12, all.pred)$net.result

INC.plot = all.pred[,1]*sd(INCOME) + mean(INCOME)
DEPS.plot = all.pred[,2]*sd(DEPS) + mean(DEPS)

library(lattice)
wireframe(yhat.plot ~ INC.plot*DEPS.plot,
  xlab = "ln(INC)", ylab = "DEPS", zlab = "ln(CHAR)",
  main = "NN(12) Regression Function",
  drape = TRUE,
  colorkey = FALSE,
  scales = list(arrows=FALSE,cex=.8,tick.number = 5),
  screen = list(z = -70, x = -70))
```

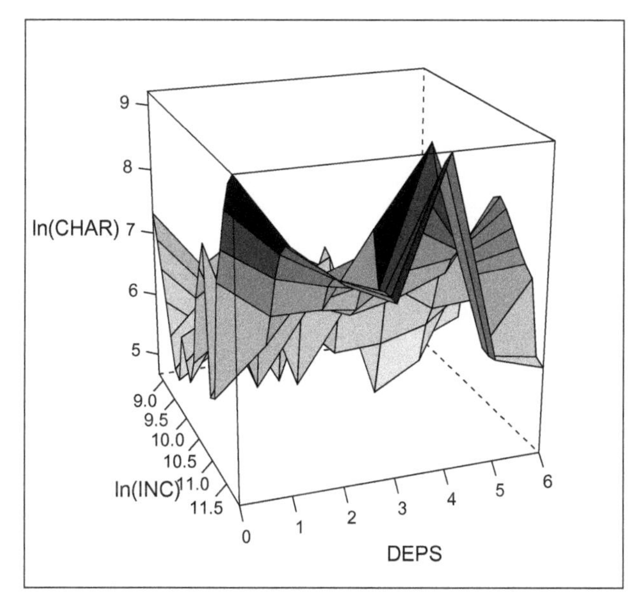

FIGURE 17.5
Fitted NN function with 12 nodes in a single hidden layer to predict ln(Charitable Contributions) as a function of ln(Income) and number of dependents.

Notice in Figure 17.5 that the function looks unusually "wiggly." This "wiggly" appearance seems illogical: Logic suggests that the true charitable contribution mean function should be a fairly smooth function of income and number of dependents.

To understand the fitted NN(12) function better, let us use the same trick that we used in Chapter 9 to view a graph in multiple dimensions suing a simple 2-D graph. This trick is very useful because you can use it with any number of X variables, whereas you can only use the 3-D plot such as shown in Figure 17.6 when you have two X variables.

The following code graphs 2-D plots of the NN(12) predicted log charitable contributions as a function of DEPS, when Income is held fixed at its 10th, 50th, and 90th percentiles.

R code for Figure 17.6

```
dep.p = (0:6 - mean(DEPS))/sd(DEPS)
p10 = cbind(quantile(INC.S, .10), dep.p); p10 = data.frame(p10)
colnames(p10) = c("INC.S", "D.S")
yhat.10 = compute(fit12, p10)$net.result

p50 = cbind(quantile(INC.S, .50), dep.p); p50 = data.frame(p50)
colnames(p50) = c("INC.S", "D.S")
yhat.50 = compute(fit12, p50)$net.result

p90 = cbind(quantile(INC.S, .90), dep.p); p90 = data.frame(p90)
colnames(p90) = c("INC.S", "D.S")
yhat.90 = compute(fit12, p90)$net.result

plot(0:6,yhat.10, ylim = c(5,8.5), xlab="Number of Dependents",
ylab="Predicted ln(Charity)")
points(0:6,yhat.10, type="l",lty=1)
points(0:6,yhat.50, pch=2); points(0:6,yhat.50, type="l",lty=2)
points(0:6,yhat.90, pch=3); points(0:6,yhat.90, type="l",lty=3)
legend("topright", c(".9 Income Quantile", ".5 Income Quantile",
    ".1 Income Quantile"), lty=c(3,2,1), pch=c(3,2,1))
```

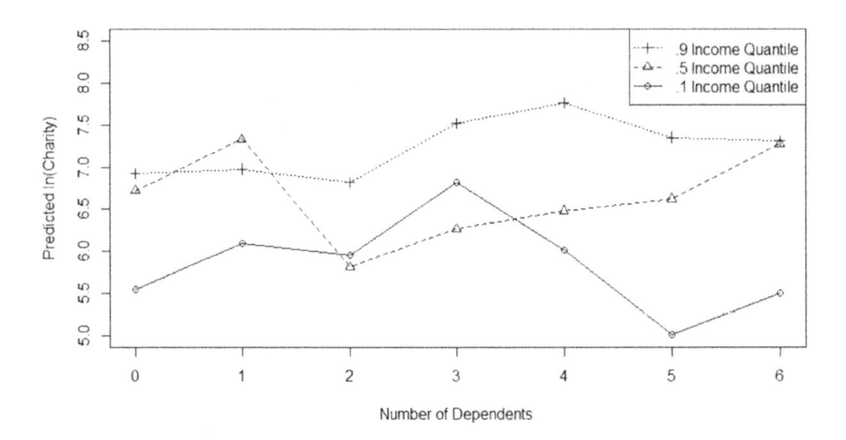

FIGURE 17.6
Predictions of the effect of number of dependents on log charitable contributions using the fitted NN(12) function when Income is held fixed at its 10th, 50th, and 90th percentiles.

Notes on Figure 17.6:

- It is clear that NN models allow interaction because the effects of dependents on the response are different for different values of income.
- It is also clear that NN models allow bumps and twists in the regression functions, unlike classical regression models.
- Unlike the artificial noiseless example of Section 17.2 above, the data in this example are real and therefore have random noise—that is, families with the same dependents and income give different (random) amounts to charity. Because of

> this reason, the NN fitted model will chase the random noise in addition to the true mean function. Hence, NN models suffer from the same "overfitting" problem noted in Chapter 11 that can occur highly parameterized polynomial models; see Figure 11.5 in particular. Here, the overfitting is manifested by illogical conclusions, such as that people with more income and the same number of dependents give less to charity, and in the extreme non-monotonic character of the relationship between charitable contributions and number of dependents.

To solve the overfitting problem, you can use the same out-of-sample RMSPE statistic based on cross-validation that we discussed in Chapter 11. The following code finds RMSPE for the models described above using ten-fold cross-validation, with the resulting RMSPE values reported above in Table 17.1.

```
# R Code to Find Ten-Fold Cross-Validated RMSPE for a NN model
dat.r = dat[sample(nrow(dat)),]
folds = cut(seq(1,nrow(dat.r)),breaks=10,labels=FALSE)

# Change neural net model and then run the following code block
sspe = 0
for(i in 1:10){
    testIndexes = which(folds==i,arr.ind=TRUE)
    testData = dat.r[testIndexes, ]
    trainData = dat.r[-testIndexes, ]
    fit.tr = neuralnet(f, data=trainData, hidden=c(1), stepmax=1e6)
    pred.te = compute(fit.tr, testData[,2:3])$net.result
    sspe = sspe + sum( (testData[,1] - pred.te)^2 )
}

rmspe = sqrt(sspe/nrow(dat))
```

Among the NN models, the NN(3) model has the best out-of-sample prediction error (RMSPE = 1.211). Figure 17.7 is a repeat of Figure 17.6, but using the NN(3) model.

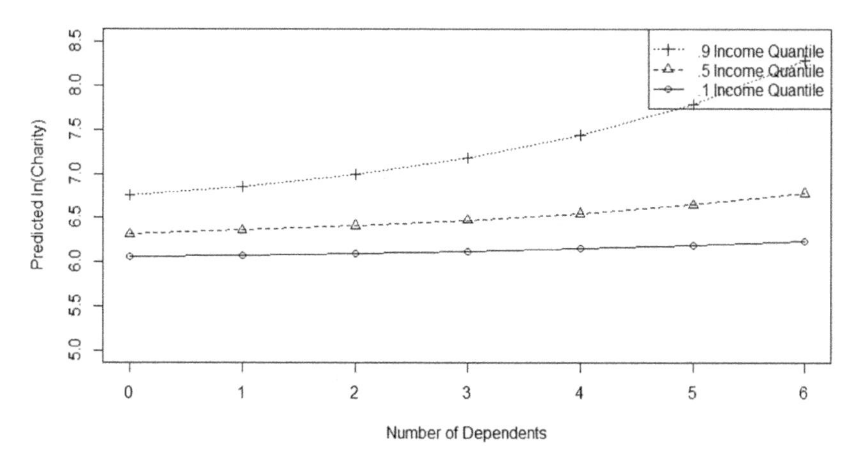

FIGURE 17.7
Predictions of the effect of number of dependents on log charitable contributions using the fitted NN(3) function when Income is held fixed at its 10th, 50th, and 90th percentiles.

The NN(3) model looks much smoother than the NN(12), and the smoothness makes much more sense, considering the subject matter of charitable contributions. This is another indication that the NN(12) model was overfit.

Polynomial regression models are valid comparators to NN regression models in that both are universal approximators. So for comparison, let us also consider some polynomial models to predict $Y = \ln(\text{CHARITY})$ as a function of $I = \ln(\text{INCOME})$ and $D = \text{DEPS}$. We will start by considering the complete third-degree polynomial model:

$$\ln(\text{CHARITY}) = \beta_0 + \beta_1 I + \beta_2 D + \beta_3 I^2 + \beta_4 D^2 + \beta_5 ID + \beta_6 I^3 + \beta_7 D^3 + \beta_8 I^2 D + \beta_9 ID^2 + \varepsilon$$

Recall that the variable inclusion principle that was discussed in Section 9.4, where lower-order terms are always kept in the model whenever higher-order terms are included, is needed to ensure that the parameters have their intended interpretations. But here, following the "black box" view of the neural network, we do not care about parameters, all we care about is getting a model that provides predicted values that are close to what might be observed. Here, the variable inclusion principle is not relevant.

Hence, we will select a subset of polynomial terms via *backward elimination* (see Section 11.4), not worrying about whether the dropped terms obey the inclusion principle. The following code shows what happens when you start with all terms, and drop them sequentially, using the BIC statistic to guide the selections.

```
C = CHARITY; I = INCOME; D = DEPS
I2 = I^2; D2 = D^2; ID = I*D
I3 = I^3; D3 = D^3; ID2 = I *D^2; I2D = I^2 *D
library(leaps)
dat = data.frame(C,I,D,I2,D2,ID,I3,D3,ID2,I2D)
back = regsubsets(C~I+D+I2+D2+ID+I3+D3+ID2+
    I2D, data = dat, method = "backward")
cbind(summary(back)$bic, summary(back)$which)
```

The output is as follows:

```
                  (Intercept) I D I2 D2 ID I3 D3 ID2 I2D
1 -45.84623412              1 0 0  0  0  0  1  0   0   0
2 -47.78705419              1 0 0  0  0  0  1  0   1   0
3 -46.01795114              1 0 0  0  1  0  1  0   1   0
4 -41.26716188              1 0 1  0  1  0  1  0   1   0
5 -45.86159306              1 0 1  0  1  1  1  0   1   0
6 -43.99301493              1 0 1  1  1  1  1  0   1   0
7 -37.96285628              1 0 1  1  1  1  1  0   1   1
8 -31.83926677              1 0 1  1  1  1  1  1   1   1
```

To understand, start at the bottom. At first, all 9 polynomial terms are in the model. Then (see the bottom line), the term "`I`" (INCOME) is removed. Then at the second from bottom step, the "`D3`" $\left(\text{DEPS}^3\right)$ term was removed, and so on. So we will evaluate the 8 models shown above, as well as the complete model, using the RMSPE statistic.

Polynomial models are fit much more quickly than neural network models because they do not require multiple iterations. It is better to use the n-fold cross-validation RMSPE rather than 10-fold when computationally feasible; Section 11.4 discusses this issue and shows how to find the statistics by using the `CVlm` function in the `DAAG` library. As an alternative, you can modify the 10-fold cross-validation code above to do n-fold cross-validation; the following code shows how to do this for the complete third-degree polynomial model.

```
# R Code to find n-Fold Cross-Validated RMSPE for an lm model
f = as.formula(paste("C~I+D+I2+D2+ID+I3+D3+ID2+I2D")) # The specific model
n = nrow(dat)
 folds = cut(seq(1,n),breaks=n,labels=FALSE)
 # Change lm model and then run the following code block
 sspe = 0
 for(i in 1:n){
     testIndexes = which(folds==i,arr.ind=TRUE)
     testData = dat[testIndexes, ]
     trainData = dat[-testIndexes, ]
     fit.tr = lm(f, data=trainData)
     pred.te = predict(fit.tr, testData)
     sspe = sspe + sum( (testData[,1] - pred.te)^2 )
 }
 rmspe = sqrt(sspe/nrow(dat)); rmspe
 rmse = mean(lm(f)$residuals^2); rmse
```

The results give the in-sample RMSE as 1.196 and the out-of-sample RMSPE as 1.218. Table 17.2 shows the RMSEs and the n-fold cross-validated RMSPEs for the various polynomial models chosen through backward elimination as shown above.

Results in Table 17.2 are also displayed graphically in Figure 17.8.

According to the RMSPE criterion, the model using 6 polynomial terms (terms involving INCOME and DEPS) predicts best. Using the entire sample, this fitted model is:

$$\text{Predicted CHARITY} = 12.15 + 8.95D - 0.30I^2 - 1.92D^2 - 0.85ID + 0.02I^3 + 0.18ID^2$$

Our point in showing this function is to underscore that there is really no point in trying to interpret the model in any way other than graphically, just like the NN functions. Figure 17.9 repeats Figures 17.6 and 17.7 using this selected polynomial model.

TABLE 17.2

RMSEs and RMSPEs for Polynomial Models Selected Via Backwards Elimination to Predict ln(CHARITY)

Terms in Model	RMSE	RMSPE
None (intercept only)	1.312	1.315
I^3	1.234	1.239
I^3, ID^2	1.223	1.230
D^2, I^3, ID^2	1.217	1.226
D, D^2, I^3, ID^2	1.215	1.227
D, D^2, ID, I^3, ID^2	1.202	1.216
D, I^2, D^2, ID, I^3, ID^2	1.196	1.212
D, I^2, D^2, ID, I^3, I^2D, ID^2	1.196	1.214
D, I^2, D^2, ID, I^3, D^3, I^2D, ID^2	1.196	1.216
I, D, I^2, D^2, ID, I^3, D^3, I^2D, ID^2	1.196	1.218

Note: RMSPE calculations use n-fold cross-validation.

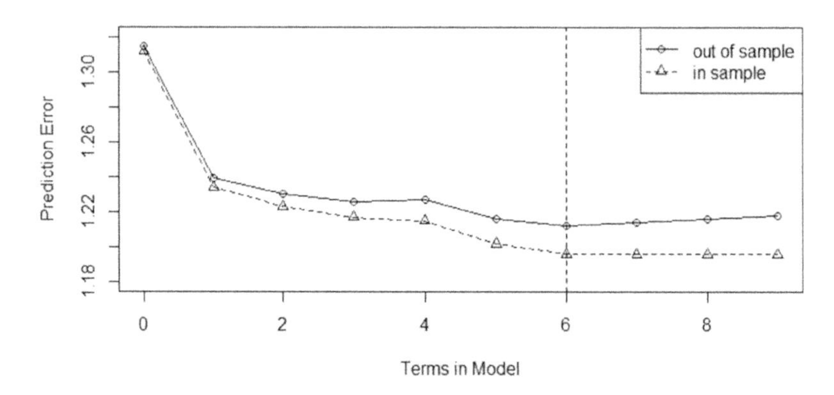

FIGURE 17.8
RMSEs and RMSPEs of polynomial models to predict ln(CHARITY) selected via backward elimination. The model with 6 terms has the lowest RMSPE.

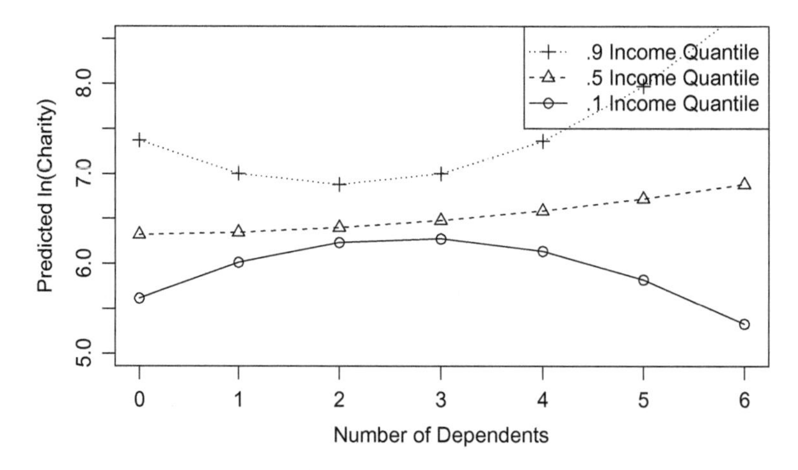

FIGURE 17.9
Predictions of the effect of number of dependents on log charitable contributions using the fitted 3rd-degree polynomial function when Income is held fixed at its 10th, 50th, and 90th percentiles.

Which model is right, the polynomial model shown in Figure 17.9 or the Neural Network model shown in Figure 17.7? Both have nearly identical RMSPE, so either can be used. Both models are wrong (all models are wrong!); hence, subject matter (here, regarding charity) is important, in addition to statistical measures, to help decide which form to use.

Self-study question: Based on the graphs (not the fit statistics) and on the subject matter concerning charitable contributions, which graph better represents reality in your opinion, Figure 17.7 or Figure 17.9?

Like classical regression models using polynomial terms, neural network regression is just another model you can use. In some cases, NN regression can offer great improvements over classical regression, in other cases you can use either. Cases that favor NN are the more "wild" cases, with twists and turns that can be difficult to model with polynomials; in addition, the extrapolation problems with polynomials are problematic. But when the functions are relatively smooth and gradual, simple linear models, perhaps including an interaction or quadratic term, are equally as viable as NN models. In such cases, the classical models have the advantage that their function forms are interpretable, and hence they can more easily facilitate insights concerning the subject matter.

Finally, note that, like most regression models, the NN regression model is a predictive model only, and thus cannot allow causal inferences. In particular, as discussed in Chapter 10, the increase in charitable contributions as a function of dependents is clearly not a causal effect; instead, it is explained by unobserved confounding variables.

Exercises

1. Change the Y value in the code for the "noiseless" example to `Y = Y + rnorm(length(Y))`. Thus there is random error. Compare the RMPSE values using ten-fold cross-validation for the single hidden layer NN models having one, two and three nodes, and pick the best model.
2. Show that the "by hand" predicted values and the fitted values obtained using the `compute` function are also the same for the one-node and two-node models.
3. Treat DEPS as a categorical variable. Experiment with various models using INCOME, INCOME^2, and interactions involving such terms and the categorical version of DEPS to find a model with smaller RMSPE than any of the polynomials or NN models considered. Graph the resulting estimated model as shown in Figures 17.7 and 17.9, and compare. Which model is best in terms of out-of-sample prediction accuracy? Which do you think would be best for obtaining a prediction of CHARITY when DEPS=7 (i.e., an extrapolation), for any INCOME?
4. Use the auditing data set

   `fees=read.csv("https://raw.githubusercontent.com/andrea2719/URA-DataSets/master/auditfees.csv", header=T)` and define:

   Y = Audit.fees/1000000 (Audit fees of a firm in millions of $)

   X = Matchfy.balsh.total.assets/1000000000 (firm assets in billions of $)

   Fit a neural network regression model to predict Y as a function of X. Use ten-fold cross-validation to select the particular neural network. Graph the fitted selected neural network function atop the (X, Y) scatterplot, along with the usual OLS fitted line, and comment.

18

Regression Trees

You should by now have an overwhelmingly clear understanding that the regression model is a model for $p(y \mid x)$, which is the conditional distribution of Y for a given $X = x$. You should also have an understanding that, when the $p(y \mid x)$ model continuously "morphs" as a function of x, such as most models we have presented (including the neural network models of the preceding chapter), then the estimated model "borrows strength" from X values nearby x to allow estimation of $p(y \mid x)$, even when there are no observations where $x_i = x$ in the data set. For instance, when predicting CHARITY (= ln(Charitable Contributions)) in terms of INCOME (=ln(Adjusted Gross Income)), there are no observations i where $\text{INCOME}_i = 10.0$. Nevertheless, by specifying and estimating the model

$$\text{CHARITY}_i \mid \text{INCOME}_i = \text{income}_i \sim_{\text{independent}} \text{N}(\beta_0 + \beta_1(\text{income}_i), \sigma^2)$$

you can estimate this distribution as

$$\text{CHARITY} \mid \text{INCOME} = 10 \sim \text{N}(6.09, 1.24^2)$$

Self-study questions: Where did the numbers 6.09 and 1.24 come from? How did you use the normality, linearity, and constant variance assumptions to get this distribution? And why would the distribution be called "approximate" even if all those assumptions were true?

However, there is no free lunch! Since there are no data values where INCOME = 10, you had to make various assumptions concerning the data-generating process in order to allow such an estimation. The normality, linearity and constant variance assumptions are used in obvious ways to estimate the distribution. In addition, the normality, linearity, constant variance, and independence assumptions ensure that the OLS estimates are maximum likelihood estimates, and are therefore better than other estimates such as WLS, quantile regression, ML using non-normal distributions, or estimates based on Winsorized or truncated data.

When you have a large sample size, you might not need so many assumptions. For example, in the same data set where CHARITY is predicted in terms of DEPS (number of dependents), you can estimate the distribution of

$$\text{CHARITY} \mid \text{DEPS} = 2$$

without using any assumption of the classical regression model. You can do this by finding the *subset* of the data set having DEPS = 2, and then by constructing the histogram of the CHARITY variable using only the data in that subset. There are 120 such observations where DEPS = 2, enough to obtain a reasonable estimate of $p(y \mid X = 2)$ by using the histogram. The following code and Figure 18.1 illustrate how to estimate this conditional distribution without making any of the (usually highly questionable) regression assumptions.

R Code for Figure 18.1

```
charity = read.csv("https://raw.githubusercontent.com/andrea2719/
URA-DataSets/master/charitytax.csv")
attach(charity)
Y2 = subset(CHARITY, DEPS ==2)
hist(Y2, breaks=10, freq=F, main="", xlab="Log Charitable
Contributions When DEPS = 2")
rug(Y2)
legend("topleft", c(paste("mean =", round(mean(Y2),2)),
paste("sd =", round(sd(Y2),2))))
```

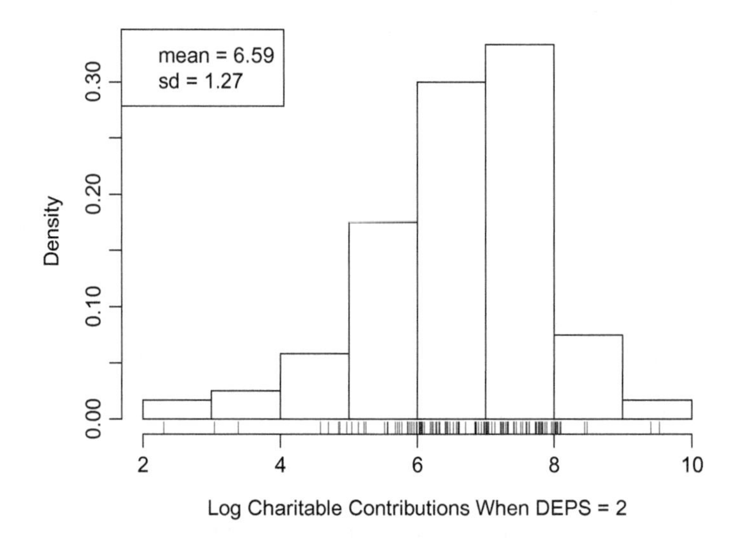

FIGURE 18.1
Estimated conditional distribution (histogram) of charitable contributions (in log scale), given that DEPS = 2. Actual data values are shown under the histogram by using the "rug" command.

Notice that the distribution in Figure 18.1 appears non-normal with left skew, whereas the classical regression model assumes a symmetric bell curve distribution. Thus, the "subsetting" method is clearly preferable in the sense that the estimated distribution follows the actual data, rather than making the (always incorrect) normality assumption.

Self-study question: Is the estimated normal distribution that is predicted from OLS analysis similar to the histogram in Figure 18.1?

The main reason that the method of subsetting the data shown in Figure 18.1 is not used more often is simply that, as in the case of estimating $p(\text{CHARITY} \mid \text{INCOME} = 10)$, there is not enough data in the subset to obtain an adequate estimate of the distribution.

Tree regression is a technique that uses the subsetting approach to estimate the conditional distributions, as shown in Figure 18.1. When there are no repeats of Y values for given X values, the tree regression algorithm divides the X data into *intervals*, where there are sufficient data values within the interval ranges to estimate the distributions.

For example, the INCOME variable in the Charitable Contributions data set is nearly continuous, so you cannot use the subset approach as you can with the discrete DEPS variable. However, you can easily estimate the distribution of CHARITY when using a *range* of INCOME values. For example, half of the $n = 470$ observations have INCOME less than 10.533 (the median). Using INCOME either above or below the median, you can estimate conditional CHARITY distributions as shown in Figure 18.2.

R Code for Figure 18.2

```
Y.LO.I = subset(CHARITY, INCOME <= 10.533)
Y.HI.I = subset(CHARITY, INCOME > 10.533)
hist(Y.LO.I, breaks=10, freq=F, main="", lty="blank",
 col="gray", xlab="Log Charitable Contributions", xlim = c(0,12))
hist(Y.HI.I, breaks=seq(1.9, 10.9,.5), freq=F, add=TRUE)
```

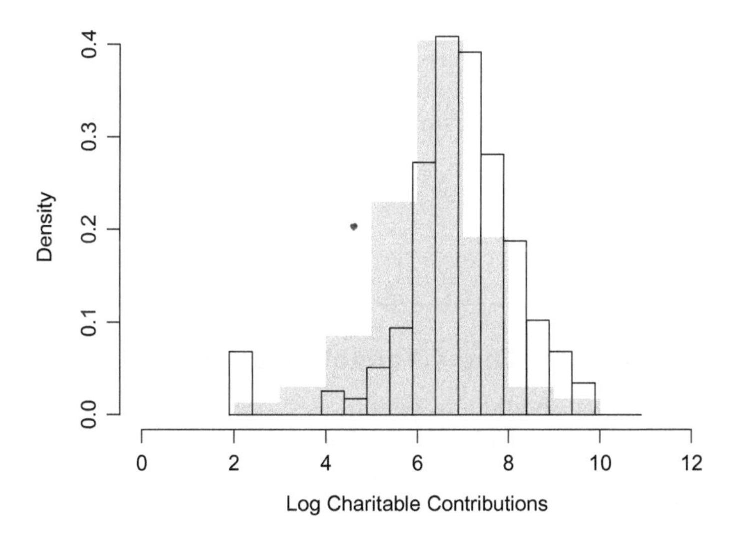

FIGURE 18.2
Estimated conditional distributions of charitable contributions (in log scale), given low INCOME (solid gray) and high INCOME (solid lines). Low INCOME refers to ln(AGI) $\le$ 10.533; high INCOME refers to ln(AGI) > 10.533.

In Figure 18.2, the two histograms are estimates of the distributions $p(y \mid \text{INCOME} \le 10.533)$ and $p(y \mid \text{INCOME} > 10.533)$. Such conditional distributions are different from all others discussed in this book: Usually, the condition is given as "$X = x$," which states that the X variable is *exactly equal* to some particular value x. But here, the condition is given either as "$X \le x$" or as "$X > x$," either of which refers to a *range* of possible values of X. Still, you can use the range-based conditional distribution to *approximate* the conditional distribution for a particular X. For example, if you are interested in the conditional distribution $p(\text{CHARITY} \mid \text{INCOME} = 10)$, you could use the solid histogram in Figure 18.2 as an approximation, since 10 is less than 10.533. But here is the key point: You can also use the same gray distribution in Figure 18.2 to approximate $p(\text{CHARITY} \mid \text{INCOME} = 9)$, since 9 is also less than 10.533. In fact, for *every* INCOME value x, where $x \le 10.533$, you can use the same distribution (the gray one) to approximate $p(y \mid x)$.

You should see immediately that there are some serious biases when you use the same distribution for different values of x. After all, you know that people at the low end of the ln(INCOME) range have less money to give to charity than people who have income near, but still slightly below, the median. Thus, the lower income people obviously have a different distribution of charitable contributions. Nevertheless, the approach that is taken by tree regression is exactly as suggested: The conditional distribution $p(y \mid X = x)$ for every single x in a common range is estimated to be exactly the same distribution.

The major limitation of Tree Regression

With tree regression, the estimated conditional distributions of Y are identical for all X values within a given subset range.

The limitation that regression trees assume identical distributions within X ranges is balanced by the "common sense" appeal of estimating distributions by using actual data values, rather than by imposing some special form (normal, lognormal, T, Poisson, etc.).

18.1 Tree Regression with One X Variable and One Split

To learn anything new, you should always start as simply as possible to make everything easier to understand. Consider again the charitable contributions data and the prediction of CHARITY in terms of INCOME. Tree regression first divides the data into two groups, depending on whether INCOME is above or below a particular threshold, as can be seen in the following code and its output.

```
library(rpart)
fit1 = rpart(CHARITY ~ INCOME, maxdepth=1)
summary(fit1)
```

The output is as follows:

```
Call:
rpart(formula = CHARITY ~ INCOME, maxdepth = 1)
  n= 470
```

```
        CP nsplit rel error    xerror       xstd
1 0.107652      0  1.000000 1.0024235 0.08856043
2 0.010000      1  0.892348 0.9300641 0.08511286

Variable importance
INCOME
   100

Node number 1: 470 observations,    complexity param=0.107652
  mean=6.57715, MSE=1.722029
  left son=2 (359 obs) right son=3 (111 obs)
  Primary splits:
      INCOME < 10.90845 to the left,  improve=0.107652, (0 missing)

Node number 2: 359 observations
  mean=6.337738, MSE=1.64523

Node number 3: 111 observations
  mean=7.351464, MSE=1.185475
```

We do not intend to explain every bit of the `rpart` output, for the simple reason that there are many different tree regression tools, even inside the `R` package, that use different statistics and methods. Instead, we will highlight some of the main points that are applicable to most implementations of tree regression.

The output shows that the data are split into two groups, one where INCOME < 10.90845, and the other where INCOME ≥ 10.90845. Within the group of 359 tax returns where INCOME < 10.90845, the estimated mean and variance (the "n" variance formula, not "$n - 1$"), of the conditional distribution of CHARITY are respectively 6.337738 and 1.64523. Within the group of 111 tax returns where INCOME ≥ 10.90845, the estimated mean and variance of the conditional distribution of CHARITY are respectively 7.351464 and 1.185475.

Self-study question: How can you verify ("by hand") the mean and variance numbers given in the `rpart` output?

Figure 18.3 shows the two estimated conditional distributions suggested by the tree regression.

`R` code for Figure 18.3

```
Y.Lower = subset(CHARITY, INCOME < 10.90845)
Y.Upper = subset(CHARITY, INCOME >= 10.90845)
hist(Y.Lower, freq=F, lty="blank", xlim = c(2, 10.2), breaks=10,
ylim = c(0, .5), xlab = "Logged Charitable Contributions",
main = "Estimated Conditional Distributions", col="gray")
hist(Y.Upper, freq=F, lty=1, add=T, breaks=seq(1.9, 10.9,.5))
```

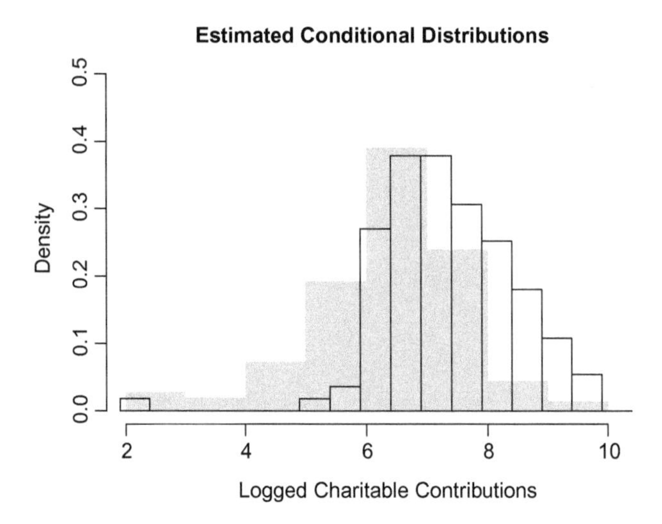

FIGURE 18.3
Estimated conditional distributions $p(\text{CHARITY} \mid \text{INCOME} < 10.908)$ (solid gray) and $p(\text{CHARITY} \mid \text{INCOME} \geq 10.908)$ (solid lines).

Figure 18.3 differs from Figure 18.2 in one very important way: The cut point 10.908 in Figure 18.3 is chosen by the tree regression algorithm to maximally separate the two groups based on their charitable contributions. The median cut point 10.533 (used in Figure 18.2) does not separate the groups maximally: The average of CHARITY in the lower split is 6.24, and 6.91 in the upper split, a 0.77 difference. Using the cut point 10.908 (used in Figure 18.3), the average of CHARITY in the lower split is 6.34, and 7.35 in the upper split, a 1.01 difference. Thus, the optimal split given by tree regression finds a bigger difference in charitable contributions between the lower and upper-income groups.

Another view of the results of the analysis, and the reason the method is called "tree regression," is shown in the "tree diagram" of Figure 18.4, which is obtained using the following code:

R code for Figure 18.4

```
library(rpart.plot)
rpart.plot(fit1, extra=1, digits=4)
```

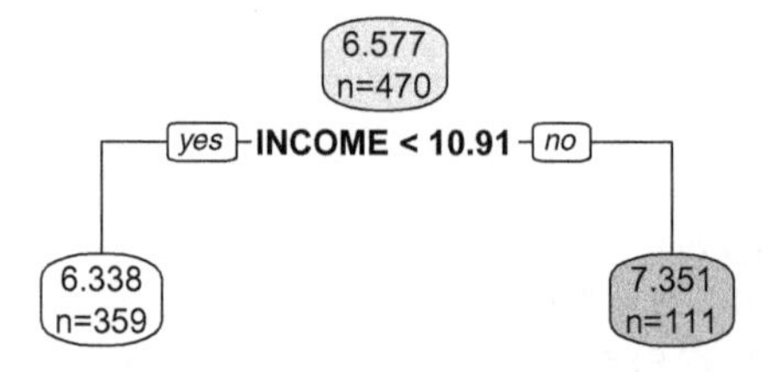

FIGURE 18.4
Tree representation of the results of the tree regression analysis of the charitable contributions data. Mean (logged) charitable contributions are shown in the "leafs" (also called "nodes"), as well as the sample sizes within the subgroups.

Figure 18.4 shows an upside-down tree with only two branches. With more branches, as is typically the case, the graph looks more like a real tree.

In the case of the Charity data, the only predicted values are 6.34 (for any $X < 10.9$) and 7.35 (for any $X \geq 10.9$). Figure 18.5 displays the fitted function, with the OLS and LOESS fits included for the sake of comparison.

R code for Figure 18.5

```
plot(INCOME, CHARITY, col="gray", pch="+")
points(c(8.5,10.9), c(6.34, 6.34), type="l", lwd=2)
points(c(10.9,12.0), c(7.35, 7.35), type="l", lwd=2)
points(c(10.9, 10.9), c(6.34,7.35), type = "l", col="gray", lwd=2)
abline(lsfit(INCOME, CHARITY),  lty=2, lwd=2)
add.loess(INCOME, CHARITY, lty=3, lwd=2)
legend("topleft", c("Tree", "OLS", "LOESS"), lty=1:3, lwd=c(2,2,2))
```

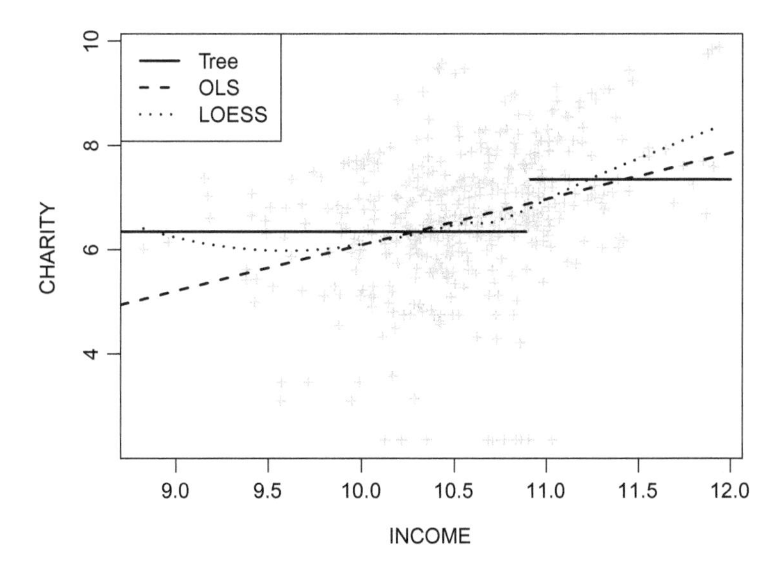

FIGURE 18.5
Comparing predicted values using tree regression (solid) with OLS (dashed) and LOESS (dotted) fits in the charitable contributions data set.

The tree regression is clearly the least realistic of the three fits for these data for two reasons: (i) there is a discontinuity at INCOME = 10.9, and (ii) the function is flat within the two regions. However, what the tree regression loses in terms of realism, it makes up for in simplicity and interpretability. The model simply states that the predicted CHARITY is 6.34 when INCOME is less than 10.9; otherwise the predicted CHARITY is 7.35. With linear regression, the interpretation requires an understanding of slope parameters, and with LOESS, the interpretation is even more difficult.

18.2 Choosing the Split Value

The split value (or split point or cut point) of INCOME = 10.9 is the value that "best" separates the charitable contributions distributions shown in Figure 18.3. As discussed in Chapter 8, see Figure 8.1 in particular, the R^2 statistic is a useful measure of separation of conditional distributions: Larger R^2 implies better separation.

Thus, a common method for choosing the split value is to pick the value giving the highest R^2 from an indicator variable model, where the indicator variable is either 0 or 1, for the left and right intervals that are determined by a given split value. The following code finds the R^2 values for all split values "by hand" and graphs them. Note that the maximum $R^2 = 0.1077$ occurs at INCOME = 10.9, as in the `rpart` output.

R code for Figure 18.6

```
unique = sort(unique(INCOME))
n.unique = length(unique)
midpoints = (unique[1:(n.unique-1)] + unique[2:n.unique])/2
n.midpoints = n.unique - 1
rsq = numeric(n.midpoints)

for (i in 1:n.midpoints) {
  Indicator = INCOME > midpoints[i]
  rsq[i] =  summary(lm(CHARITY~Indicator))$r.squared
}
plot(midpoints, rsq, xlab="INCOME Split Point", ylab="R Squared")
abline(v=10.90845)
```

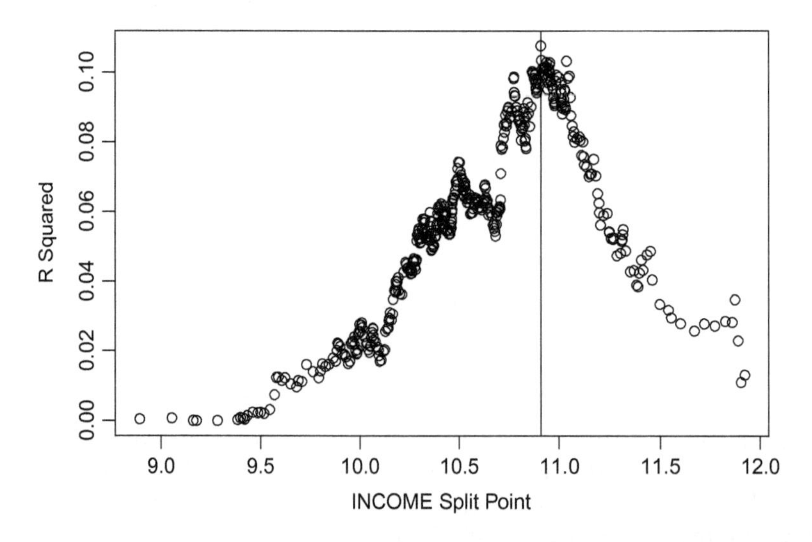

FIGURE 18.6
R-squared statistics for indicator variable models to predict CHARITY that use different split points for INCOME. The maximum $R^2 = 0.1077$ occurs for splitting at INCOME = 10.90845. By comparison, the R^2 for splitting at the median INCOME (10.533), as shown in Figure 18.2, is $R^2 = 0.0661$.

While the "by hand" analysis above picked the same value chosen by the `rpart` software, this does not always happen. Typically, the software restricts the allowable split points so that there are sufficient numbers of observations within each node; hence it is possible that the "by hand" analysis can find a split point with even higher R^2 than the one chosen by the software.

18.3 Multiple Splits on a Single *X* Variable

Tree regression is also called "recursive partitioning," which is the reason that the `R` function is named "`rpart`." The "recursive" aspect refers to the fact that the same partitioning algorithm that determines the first split also determines subsequent splits; i.e., the algorithm is used "recursively." The following code uses the same example as above, but without using "`maxdepth=1`," which forced a single split.

`R` code for Figure 18.7

```
charity = read.csv("https://raw.githubusercontent.com/andrea2719/
URA-DataSets/master/charitytax.csv")
attach(charity)
fit2 = rpart(CHARITY ~ INCOME)
summary(fit2)
rpart.plot(fit2, extra=1, digits=3)
```

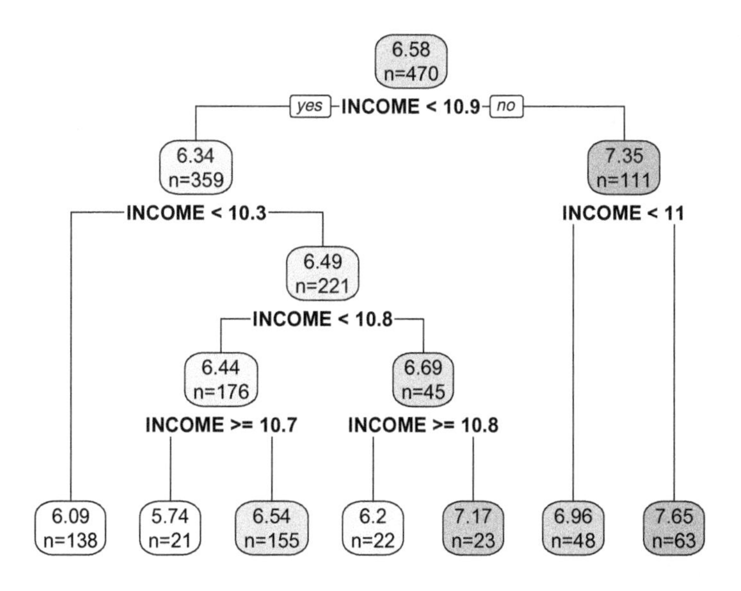

FIGURE 18.7
Tree representation of the results of the tree regression analysis of the charitable contributions data, using only INCOME as a predictor. Means of ln(charitable contributions) are shown in the "leafs" (also called "nodes"), as well as the sample sizes within the subgroups.

Notes on Figure 18.7:

- To understand the "recursive" aspect of tree regression, look at the top node of the (upside-down) tree in Figure 18.7. The right branch refers to the subset of $n = 111$ tax returns having INCOME $\geq$ 10.91. The "recursion" refers to the fact that the software analyzes this data set in exactly the same way as the original ($n = 470$) data set. Namely, the split point for INCOME in this subset is determined by checking all the possible splits, and choosing the one with maximum R^2. Within this subset, splitting at INCOME = 11.04 maximizes the R^2 for predicting CHARITY.
- Notice, again on the right branches, that the algorithm stopped after the split at 10.91, while in the left branches there are further splits. To keep the tree from splitting forever, tree regression software employs various stopping criteria. In `rpart`, the splitting stops by default when the number of observations in a subset is too small, or when the increase in R^2 is too small. Sometimes analysts override the default stopping criteria to allow the tree to grow very large, and then "prune" it back by choosing a tree to minimize the out-of-sample prediction error; see Chapter 11.
- The bottom nodes in Figure 18.7 display the final model. When you follow the branches, you can determine particular ranges. For example, the $n = 22$ observations yielding 6.202 for predicted CHARITY are in the group where (starting from the top), INCOME < 10.91, *and* INCOME $\geq$ 10.32, *and* INCOME $\geq$ 10.78, *and* INCOME $\geq$ 10.84. Putting it all together, this group refers to the subset of tax returns where 10.84 $\leq$ INCOME < 10.91. Using the actual (not rounded off) cut points, you can verify the numbers in Figure 18.7 using `mean(CHARITY[INCOME >= 10.83595 & INCOME < 10.90845])` and `length(CHARITY[INCOME >= 10.83595 & INCOME < 10.90845])`, giving 6.201693 and 22, respectively.

It is easier to visualize the final model by using a graph of its estimated mean function.

R code for Figure 18.8

```
Income.plot = seq(8.5, 12, .001)
Income.pred = data.frame(Income.plot)
names(Income.pred) = c("INCOME")
Yhat2 = predict(fit2, Income.pred)
plot(INCOME, CHARITY, pch = "+", col="gray")
abline(lsfit(INCOME, CHARITY), lwd=2, lty=2)
points(Income.plot, Yhat2, type="l", lwd=2)
add.loess(INCOME, CHARITY,  lwd=2, lty=3)
legend("topleft", c("Tree", "OLS", "LOESS"), lty=1:3, lwd=c(2,2,2))
```

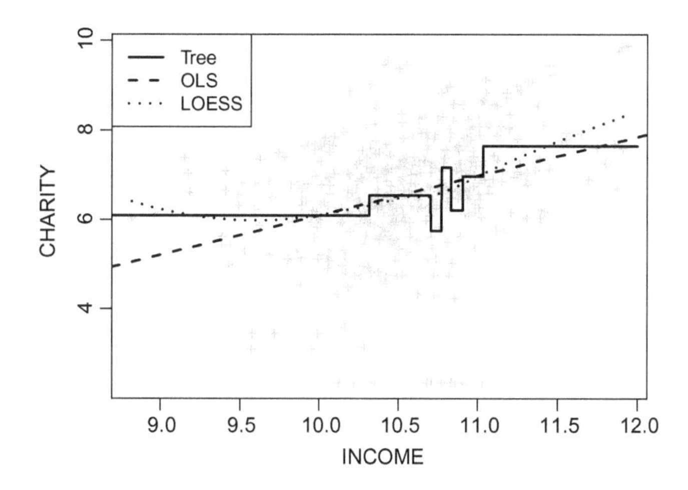

FIGURE 18.8
Comparing predicted values using tree regression (solid) with OLS (dashed) and LOESS (dotted) fits in the charitable contributions data set. Here, the tree is allowed to split as far as the `rpart` function will allow.

Notice in Figure 18.8 that the predicted CHARITY is constant within interval ranges defined by INCOME. Notice also that it seems that the model is "overfit"; that is, it is determined too much by idiosyncrasies of the given data set. There is no logical reason that charitable contributions should have such an up-then-down-then-up-then-down-then-up relation to INCOME. Calculation of cross-validated RMSPE is left as an exercise.

18.4 Tree Regression with Multiple *X* Variables

While the example just given shows how tree regression works, it also emphasizes the discontinuous aspect of the fitted model, which is the main limitation of tree regression. With multiple variables, this limitation remains, but the benefit of easy interpretability becomes more apparent.

Multiple regression can be much harder to interpret than simple regression. The "held fixed" concept can be mysterious. The nature of the fitted function is hard to understand, since it exists in $(k + 1)$ – dimensional space and cannot be easily visualized. Tree regression overcomes these obstacles by giving the "held fixed" results in a very intuitive, easy-to-understand form, with all "held fixed" interpretation referring to subsets of the actual data. Consider the taxpayer data again, but using the variables DEPS and INCOME as in the neural net and polynomial predictions of Chapter 17. To start simply, we allow a tree with maximum depth = 2.

R code for Figure 18.9

```
fit3 = rpart(CHARITY ~ INCOME + DEPS, maxdepth=2)
rpart.plot(fit3, extra=1, digits=3)
```

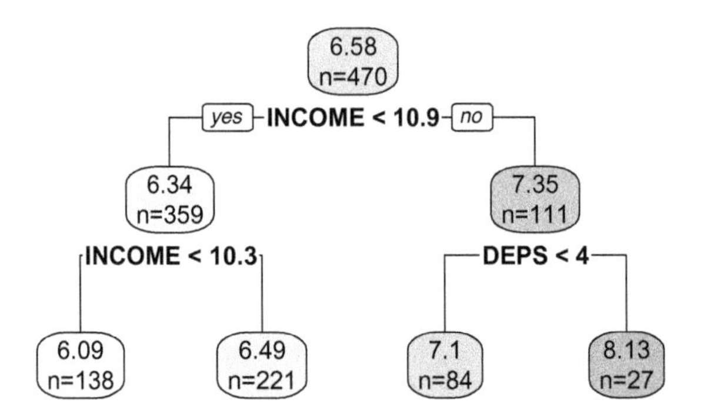

FIGURE 18.9
Tree representation of the results of the tree regression analysis of the charitable contributions data, using INCOME and DEPS as predictors. Means of ln(charitable contributions) are shown in the "leafs" (also called "nodes"), as well as the sample sizes within the subgroups.

The tree diagram facilitates understanding of the "held fixed" concept in regression, as well as the interaction concept. When INCOME is held fixed at some "large" value (defined in the top node as any value ≥ 10.9), then there is an effect of DEPS, in that DEPS = 3 or less gives an estimated mean(CHARITY) of 7.1, while DEPS = 4 or more gives an estimated mean(CHARITY) of 8.13. Thus, there is an effect of DEPS on CHARITY when INCOME is "held fixed" at high values. On the other hand, the model estimates no effect of DEPS when CHARITY is "held fixed" at low values (values < 10.9 as shown in the top split), since there are no splits involving DEPS in that arm. The fact that there is no split for DEPS in the lower arm also illustrates interaction, in that the effect of DEPS on CHARITY depends on the value of INCOME.

The splits above correspond to the three-dimensional prediction surface shown in Figure 18.10.

R code for Figure 18.10

```
inc.plot = seq(min(INCOME), max(INCOME), (max(INCOME)-min(INCOME))/20 )
inc.plot = rep(inc.plot,7)
deps.plot = seq(0,6,1)
deps.plot = rep(deps.plot, each=21)
all.pred = data.frame(inc.plot, deps.plot)
names(all.pred) = c("INCOME", "DEPS")

fit2 = rpart(CHARITY ~ INCOME + DEPS, maxdepth=2)
yhat.plot = predict(fit2, all.pred)

library(lattice)
wireframe(yhat.plot ~ inc.plot*deps.plot,
  xlab = "INC", ylab = "DEPS", zlab = "CHAR",
  main = "Tree Regression Function",
  drape = TRUE,
  colorkey = FALSE,
  scales = list(arrows=FALSE,cex=.8,tick.number = 5),
  screen = list(z = -20, x = -60))
```

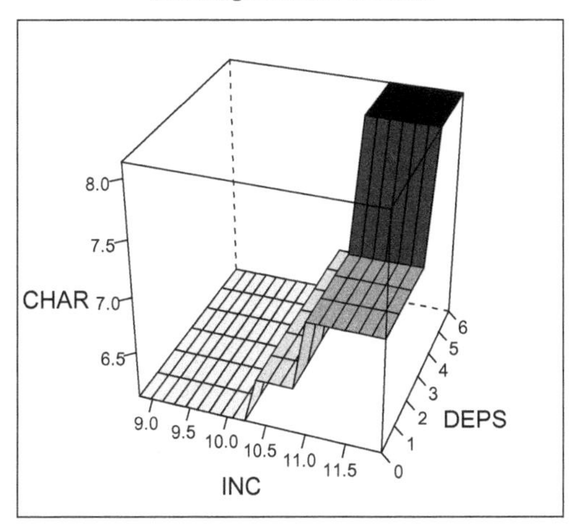

FIGURE 18.10
Three-dimensional representation of the tree model displayed in Figure 18.9. The effect of DEPS is apparent where INCOME ≥ 10.9, and no effect of DEPS is apparent where INCOME < 10.9. Within the INCOME < 10.9 data, there is a further split at INCOME < 10.3, indicating lower mean charitable contributions in that group.

The limitations of the tree regression model are clearly seen in Figure 18.10. No one really believes that the real charitable giving process has a mean function that is a discontinuous function of INCOME and DEPS as shown in the graph. But again, what tree regression loses in realism, it gains in simplicity and ease of use.

Now consider a more elaborate model using the variables DEPS, INCOME, AGE (65 or older) and MS (marital status). The following code shows how to do fit the model and to graph the tree diagram.

R code for Figure 18.11

```
fit3 = rpart(CHARITY ~ INCOME + DEPS + AGE + MS)
rpart.plot(fit3, extra=1, digits=3)
```

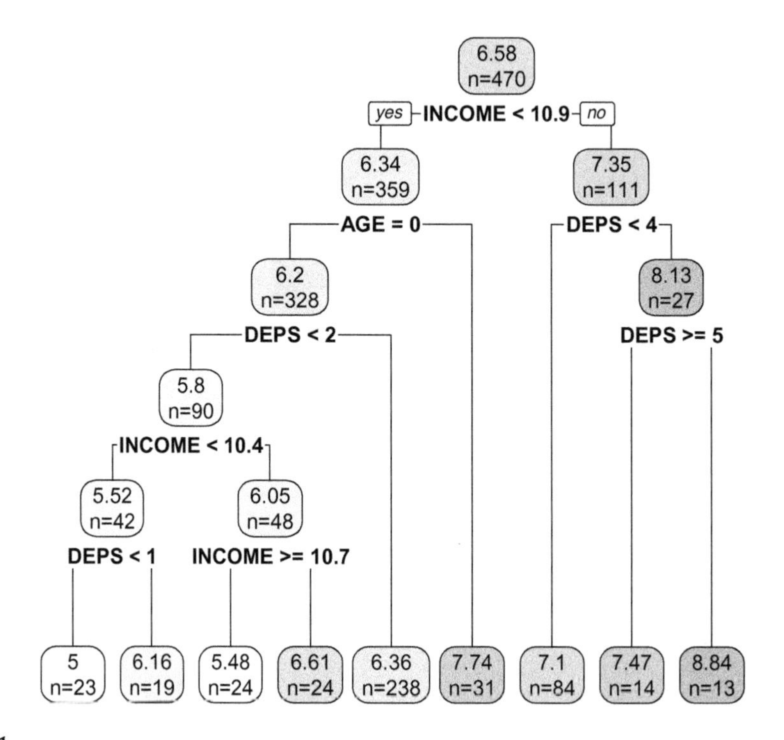

FIGURE 18.11
Tree representation of the results of the tree regression analysis of the charitable contributions data, using INCOME, DEPS, MS, and AGE to predict CHARITY.

Notes on Figure 18.11:

- While the tree starts the same as with a single X variable, the algorithm is different. First, it tries each variable, INCOME, DEPS, MS and AGE, one at a time, to see which predicts CHARITY best, in terms of R^2 statistic at maximizing split point. In this case, INCOME was best among the four predictors.
- The second step involves recursion, but again it involves all variables. The variable with the highest R^2 is chosen, using the split point that maximizes the R^2 statistic.
- The splitting continues until the stopping criteria are satisfied.

For example, one interesting terminal node is the group with estimated second-largest mean charitable contributions, which is the node having $n = 31$ tax returns, and mean CHARITY = 7.74. This group actually has lower INCOME (<10.91), but are older taxpayers. Thus, the model estimates that older taxpayers tend to be highly charitable, even when they have low income. This conclusion is quite reasonable and believable but would have been difficult to find using standard regression models. A main benefit of tree regression is its ability to find and display such interesting "pockets" of predictive activity.

Exercises

1. Find the cross-validated RMSPE for the model shown in Figure 18.8 and compare it to the RMSPE of the linear and LOESS fitted models.
2. Suppose Y = Production Cost and X = Widgets. Tree regression gives the intervals Widgets < 1050 and Widgets ≥ 1050. How do you estimate $p(y \mid X = 1000)$ when using tree regression?

19

Bookend

This book has demonstrated that the regression model is a model for the *conditional distributions* of Y, given specific values of your predictor variable(s). It is the model that you specify for the conditional distributions $p(y \mid X = x)$ when you analyze your data. To summarize, it is best to view the regression model in terms of conditional distributions (rather than in terms of linear or other functions) for the following reasons:

- This model is correct in any situation where regression is an appropriate method to analyze your data. For any specific value of X, there is variation in your Y variable, and the model $p(y \mid x)$ correctly states that there is variation in Y given any value of X. In other words, unlike most models, this one is *right*.
- The usual linear/planar functions of regression analysis, e.g., $\beta_0 + \beta_1 x_1 + ... + \beta_k x_k$, fit naturally within the conditional distribution framework as the means (or in some cases, medians) of these conditional distributions. But the conditional distribution framework allows you to study variances and quantiles with equal ease. In other words, focus on the means (or medians) is needlessly restrictive.
- To construct valid prediction ranges for Y given your X values, you have to model the true distributions $p(y \mid x)$ of possible Y values for given $X = x$ as accurately as possible. You cannot assume normality, and the Central Limit Theorem offers no help here.
- Specifying models $p(y \mid x)$ that continuously "morph" as a function of x allows you to estimate conditional distributions $p(y \mid x)$ for X values where you have no data at all.
- All regression models fall under this umbrella: They are fundamentally all the same, whether normal regression, ANOVA, Poisson regression, non-linear regression, heteroscedastic regression, survival analysis regression, quantile regression, neural network regression, tree regression, and even regression for non-numeric (nominal) data, when viewed from the $p(y \mid x)$ perspective.
- Ordinary least squares is best only when the distributions $p(y \mid x)$ are normal, homoscedastic, and have mean functions that are identical to what you specify (e.g., linear or planar). Otherwise, a different estimation procedure (e.g., maximum likelihood) will give better estimates and predictions.
- To estimate the models via maximum likelihood, you use the conditional probability distributions $p(y \mid x)$ directly to create the likelihood function that is maximized.
- You can determine which model $p(y \mid x)$ best fits your data by using likelihood-based statistics, which are also determined by the specific distributions $p(y \mid x)$ that you specify.
- When you specify a grossly incorrect model for the $p(y \mid x)$ conditional distributions, your estimates and inferences will likewise be grossly incorrect. The conditional distribution framework focuses your attention squarely on the assumptions that you make when you perform regression, which is good scientific practice.

Continuing on the last bullet point, good science means modeling your processes faithfully, so that you can understand Nature as well as possible. What we have seen instead, too often, is grossly flawed scientific practices based largely on overuse of p-values. One problem is that researchers routinely "prove" null models based on large (e.g., $p > 0.05$) p-values. They "prove" that certain β coefficients are zero, and they "prove" that certain model assumptions are true. You should by now have a clear understanding that such conclusions are simply gross scientific malpractice: No p-value, no matter how large, allows you to state that a β is zero, or that an assumption is true, or that a particular model is the correct model.

Another problem with p-values concerns the opposite "$p < 0.05$" determination. What we have also seen, too often, is that researchers engage in "p-hacking with OLS," where they try a bunch of OLS regressions this way and that way, until they get a $p < 0.05$ result. Then they make up a story about why the result is important, pretending to ignore all the p-hacking that they did originally, and try to publish the results. This practice is also known as "p-hacking, followed by HARKing." (To "HARK" is to "hypothesize after results are known.") Such practices are bad science at best, and scientific misconduct at worse. The practice of "p-hacking" is substantively no different than "selective omission of data points that undercut the [proposed] hypotheses" discussed in Chapter 16, because either way, you are just messing around with the data to get it to tell the story *you* want to tell, rather than objectively letting the data tell you the story that *Nature* wants to tell.

Good science involves using the data to tell the true story of Nature, not twisting the data to match your desires. The latter is what sleazy (i.e., most!) politicians do. It should not be your goal. Current research shows that much of the "replication crisis" in the sciences is caused by p-hacking and selective use of data (e.g., Harris 2017; Head et al., 2015). It is easy to understand the problem: If you force your data to tell an incorrect story, is it any surprise that others who try to replicate your findings will not be able to do so?

Thus, our emphasis is on correctly telling the story of Nature, whether that Nature be psychological, economic, biological, medical, physical, or combinations thereof. And variability is real in Nature, and Nature's variability is precisely what the model $p(y \mid x)$ describes. That is why you should model $p(y \mid x)$: Because it is Nature's model.

Once you understand that you are after Nature, you realize that anything you might assume about $p(y \mid x)$ will be wrong in various ways, including whatever mean function you assume, and whatever specific distributions you assume. You then realize, correctly, that all research that has ever been done is similarly wrong. So it is not so bad! You will be wrong like everyone else! But you can be better than others by being less wrong. Aim for specifying regression models $p(y \mid x)$ that better represent Nature. And do not misinterpret the story that $p > 0.05$ tells you about Nature, and do not engage in p-hacking to tell a story that you wish to tell. Aim to tell the story that Nature wants to tell, even if that story is not the one that you wish to tell.

This book has a strong likelihood-based flavor, but we are not dogmatic "likelihoodists." Our reason for choosing the likelihood point of view is simply that it focuses your attention squarely on Nature's data-generating processes. Other techniques (such as sloppy analysis that prove assumptions are true, that some β's are 0, that other β's are in the wrong directions) allow you to be unethical with seeming technical justification, and ignore the fact that the scientific goal is to model Nature's processes.

Likelihood-based methods lead you directly and naturally toward Bayesian methods. While Bayesian methods are not covered in this book, this book is an ideal precursor to the study of Bayesian regression. Once you understand likelihood functions, it is just one small step to Bayesian posterior distribution functions. Both methods start with your specification of the conditional distributions $p(y \mid x, \theta)$ for your potentially observable $Y \mid X = x$ data. Your model $p(y \mid x, \theta)$ involves (i) the conditional probability distributions, whether normal, multinomial, Poisson, etc., and (ii) the linkage of these distributions to $X = x$, where (typically) they continuously "morph" as a function of the X variables. Likelihood-based methods and Bayesian methods use exactly the same likelihood function: Under the conditional independence assumption, it is

$$L(\theta \mid \text{data}) = p(y_1 \mid x_1, \theta) \times \ldots \times p(y_n \mid x_n, \theta),$$

as shown numerous times in the book. We used $L(\theta \mid \text{data})$ to get maximum likelihood estimates, standard errors, likelihood ratio tests, and likelihood-based measures of fit. Bayesian analysis also uses this likelihood function to construct the posterior distribution as

$$p(\theta \mid \text{data}) \propto L(\theta \mid \text{data}) \times p(\theta),$$

where $p(\theta)$ is your prior distribution.

We must reiterate an important point about likelihood-based methods, and that is that likelihood-based methods can be called a "gold standard." They point you to the best possible analysis of your data. For example, the classical OLS regression methods are really only best with normally distributed, homoscedastic, independent data whose conditional mean function is linear; in this case OLS coincides with maximum likelihood. In other cases, there will be different maximum likelihood estimates, and they will be better than OLS. Of course, you never know the true distributions, you never know the true conditional mean functions, you never know the true conditional variance functions, and you usually never know the true dependence structure of Nature's data-generating process. But at least with likelihood-based methods, you have penalized likelihood-based statistics (such as AIC and BIC) to help identify which models are more consistent with your data.

You may have come away from this book wanting more "recipes" for "how to do regression." Let us reiterate that your goal is to model $p(y \mid x)$ well. Depending on the particulars of your study, there will be different logical steps you should take in order to do that. Since you presumably know something about your subject area before even looking at your data, you can start this modeling process from pure *a priori* considerations. For example, if you know, *a priori* (before collecting your data) that your dependent variable Y is a count (0, 1, 2, 3, …), then you should consider Poisson and negative binomial regression *a priori*. If your dependent variable is likely outlier-prone, then you should consider quantile regression and maximum likelihood estimation using heavy-tailed distributions *a priori*. If your dependent variable is a time-to-event measurement, then you know *a priori* that you should use survival analysis, and so on.

A reasonable first step *after* collecting data is to draw scatterplots, look at summary statistics, and generally get to know your data. Then the specific steps you should take will involve the particular models that you use. But for all models, we have emphasized a holistic approach to the analysis of regression data, including subject matter considerations, graphical analysis, model and variable selection, and statistical measures of fit, including penalized likelihood and out-of-sample prediction accuracy. We have also emphasized that, consistent with good science and directives from the scientific community, model assessments and conclusions based on tests of hypotheses (p-values) should be giving low priority or abandoned altogether. We have consistently and correctly described p-value based methods as a way to determine what is capable of being explained by a data-generating process where the questioned effect is non-existent, and nothing more. If you do use p-value based methods, at least you will interpret them correctly if you follow the examples given in this book.

A final note. We mentioned early on in this book that simulation is a key to understanding regression. As demonstrated throughout this book, you can construct simulation models to resemble your own data-generating process and potentially observable data sets. With such a simulation model, you can try out different methods for yourself. Since you know the true parameter values of your simulation model, you can determine which method tends to be best based on repeated simulations. Then you can pick the best method for your real data based on that simulation study. To ensure that your simulation is not too narrow, you should also perform sensitivity analysis, where you consider alternative simulation models that are also plausible models for your data-generating process. By following this advice, you can understand the regression model, and you can pick the appropriate regression analysis logically, rather than by resorting to illogical testing/p-value based approaches, or other similar approaches based on unsupported rules of thumb.

References

Harris, R. (2017). *Rigor Mortis: How sloppy science creates worthless cures, crushes hope and wastes billions.* New York: Basic Books.

Head, M. L., Holman, L., Lanfear, R., et al. (2015). The extent and consequences of P-hacking in science. *PLoS Biology, 13*(3), e1002106. https://doi.org/10.1371/journal.pbio.1002106

Index

Note: Page numbers in italic refer to figures and those in bold to tables.

A

additivity property, covariance, 162
`add.loess` function, 34
Akaike's Information Criterion (AIC), 285
American Statistical Association (ASA), 90, 94
ANCOVA, 240
  indicator variables and ordinary *X* variables, 229–232, *233*
  with interaction, 232–236
approximate prediction interval, 69
ASA (American Statistical Association), 90, 94
`as.factor` function, 249–250
autocorrelation coefficient, 320
autoregressive model, 68

B

Bayesian Information Criterion (BIC) statistics, 285–287, *286*
Bayesian parlance, 204
Bernoulli distribution, 331
best linear unbiased estimators (BLUEs), 53, 303
bias-variance tradeoff, xiv, 278–279, **284**
  demonstrate, 279–283
BIC (Bayesian Information Criterion) statistics, 285–287, *286*
bivariate normal distribution, 36–40, *37*
block diagonal covariance structure, 320
BLUEs (best linear unbiased estimators), 53, 303
Blunt Axe approach, 311–316
Box-Cox transformation, 136–138
Breusch-Pagan test, 107

C

Calisota Tech University (CTU), 275
Car Sales data, 110, *110*
  curvature testing with, *102*, 102–103
  with LOESS smooth, 99, *100*
  normality assumption using, 113
  transformations of *X* with, 121–124
Cauchy distribution, 427, *441*
causal interpretation, 155
causal model, 165
causation, prediction, 153–156, *154*
censored data models, 379–384, **380**, **384**
  regression analysis with, 384–391
chance-only (null) model, 86, 241, 243–246
charitable contributions study, *265*, 322–324
Charity data set, 127, 132, 477
Chow test, 261
classical interaction model, 202
classical multiple regression model, 156–161, *157*, **159**, 207
classical regression model, 21–22, *22*, 57, **60**, 67, 120, 173, 186, 295
commodity price *versus* commodity stockpile, 259–261
computer time data *versus* GB RAM, *135*
conditional covariance matrix, OLS estimates, 179–180
conditional distribution, **349**, **355**
  approach, 392
  charitable contributions, *472*, *473*
conditional distributions, regression models in, 2–3
  area of circle, 3–4, *4*
  person's financial assets, 4–6, *5*
conditional independence assumption, 21
conditional mean function, 19–20, *206*
conditional mean model, 164
conditional variance, classical regression model, 66
confidence interval, practical interpretation, 76–78
confounding variable, 155
consistency, 161
  IV estimator, 166
  variance and covariance estimates, 163
consistent, 314
constant variance assumptions, 20, 96
  hypothesis testing methods, 107–109
Cook's distance, 419–421, *419*, *421*
correct functional specification assumption, 19–20
correlation parameter, 36–37
covariance matrices, linearity property, 179
covariance *versus* regression coefficient, 162
covariate, 240

Cox proportional hazards regression (CPHR) model, 392–394, **394**, *395*
credible interval, 204
crime rate prediction model, 421–424, *422*, *424*, **424**
cross-validation (CV), 288
CTU (Calisota Tech University), 275
cubic term, 213
cumulative distribution function, 402
CV (cross-validation), 288

D

data analysis tools, 94
data-generating process (DGP), 6–8, 93, 368, 383, **440**, **441**, 490
Data Science, 146
data set indicating success, 334, **334**
degree of polynomial, 201
degrees of freedom, 65, 122
deterministic relationship, 1–2
Deviance residuals, 367
DGP, *see* data-generating process (DGP)
discrete distribution, generic, 329–330
dispersion, 368
divorce data set, **386**
Durbin-Watson test, 111

E

elasticity, *140*, 140–143, *142*
electroencephalograph (EEG) data, 410
estimated standard deviation, 73
estimator, 53
Euclidean distance, 406
`EuStockMarkets` data set, 11
exact inferences, 73–75
  confidence interval, practical interpretation, 76–78
  $E(Y|X = x)$, confidence intervals for, 78–80
  p-value, 88–91
  $Y|X = x$, prediction intervals for, 80–82
exactness and non-exactness via simulation, 75–76
"explained by chance alone" phenomenon, 83–86
exponential distribution, *381*, 382
extrapolation concepts, 7, *7*

F

firm-level data analysis, 219
first-degree polynomial, *456*
`fit.main`, 242
`fitted.values`, 342
fixed cost, 14
fixed effects, 266–269
forecasting, 1
foundations, instrumental variables, 161–162
four-way ANOVA, 257
F statistic, 242–243
F test, 189
  simulation study, 190–191
full model *versus* restricted model F tests, 240–242
full *versus* restricted (null) model, 241, 325

G

garbage in, garbage out (GIGO), 410
Gauss-Markov (G-M) model, 303
  OLS estimates, unbiasedness of, 174–176
  and theorem, 52–55, 59, 70
  WLS, 303–304
generalization, 6–7
generalized least squares (GLS), 321
general multinomial regression model, 344
general standard deviation functions, 305–310, *308*
GIGO (garbage in, garbage out), 410
Glejser test, 107
global variables, 273
GLS (generalized least squares), 321
G-M model, *see* Gauss-Markov (G-M) model
gold standard method, 489
grade point average (GPA), 274, *276*
  data analysis, 16–17, *17*, 181–183, *182*
  data set, **247**
  predicting Hans' graduate, 275–278, *282*
graphical/descriptive methods, 94
  versus testing methods, 95–96
grossly incorrect interpretation, p-value, 90–91

H

hat matrix, 413
Hausman test, 269
HC, *see* heteroscedasticity-consistent (HC)
heavy-tailed distributions, 409
"held fixed" concept, 155–156, 481–482
heteroscedasticity, 16
heteroscedasticity-consistent (HC), 312
  standard errors, 70
heteroscedastic model, 426
heteroscedastic processes, 132–133
homoscedasticity, 20, 104–105

personal assets data, 108–109
production cost data, 108
hyperplane, 171
hypothesis testing and p-values, 82–83
null distribution, T statistic, 86–88
person's identification number, 83–86
hypothetical mean charitable contributions, 257, *257*

I

idealism, moderating effect, 209–213
idempotent matrices, 414
independence assumption and repeated measurements, 268–270
indicator variable, 222–223
additional applications of, 257–270
insignificance, 100
instrumental variables to estimate causal effect, 161–166
interaction model, 217
interaction/moderator analysis, 206–207, *210*
parameter interpretation, interaction models, 208–209
path diagrams, 207–208
interaction term, 213
intercept of line, 28
intercept term, 214–215, *216*
interpolation concepts, 7, *7*
interval censored data, 402, **402**
inverse transformation, 134–136

J

jittered scatterplot, *338*

L

LAD (Least Absolute Deviations) estimates, 51
lagged error term, 109
Laplace distribution, 49, *50*
latent variable, 352
Law of Total Expectation, *147*, 147–153
Law of Total Variance, 151–153
Least Absolute Deviations (LAD) estimates, 51
least squares estimates, 47–48
matrix form, 170–172
least squares estimation method, 43
least squares phrase, 46
leverage statistic, 413–416, *415*
likelihood-based methods, 43, 488
likelihood function, 381, **381**
likelihood ratio tests (LRT), 324–325
linear and quadratic fits, *101*, 341, *341*
linearity assumptions, 28–29, *29*, 93, 96
hypothesis testing methods, 100–103
using graphical methods, 97–99
linear models, 142
linear regression function, 28–32, *31*
linear term, 213–214
ln(Y) transformation, 132–133, *134*
LOESS method, 32–36
fit, problems, 121
smooth, plot, *98*, 98–100
logarithmic transformation, 125–126, *126*
charity data set, log likelihoods with, 132
log transforming income, 127–129
maximized log likelihood, 129–132, *130*, *131*
log charitable contributions, *465*, *468*
logistic regression function, 332
logistic regression model, 331–334, *332*
trash can, wadded-up paper into, 334–342
log-likelihood function, 46
lognormal distributions, 5, 130–131, *131*
lognormal regression model, 133, *133*, *134*
log odds function, 333
log transforming income, 127–129
longitudinal data, 267
LRT (likelihood ratio tests), 324–325

M

Mahalanobis distance, 415
mathematical theorem, 38, 58, 73, 174
matrix algebra, 169, 194
maximum likelihood estimates (MLEs), 43–45, **44**, 131
in classical regression model, 45–48
with non-normal distributions, 49–52
MC, *see* multicollinearity (MC)
measurable variables, 273
measurement error, 176–178, *178*
Mega-Hans model, 282, *283*
misanthropy, 209, *212*
MLEs, *see* maximum likelihood estimates (MLEs)
MLR (multinomial logistic regression) model, 344, 357
multicollinearity (MC), 192–195, *192*, *193*
on T statistics, 195–198
X variables, 198
multinomial distribution, 343
multinomial logistic regression (MLR) model, 344, 357

multinomial regression, 343–345
  laundry, 346–351, **347**, *350*
  model, **346**
multiple indicator variables, 223–229
multiple regression model, 9, 481
  data set, **170**
  and interpretation, 156–160
  overview, 145–146
  prediction, 146–153
multiple splits on single *X* variable, 479–481

N

naïve guess line, 39, *39*
negative binomial (NB) regression, 370–372, *372*
  financial planners, predicting number, 372–376, *375*, *376*
  on replicability and preregistration, 376–377
nested model sequence, version 1 and 2, 255–256
neural network (NN) regression, 451–452
  polynomial approximations, noiseless function, 453–461, *454*, *458*
  predicting charitable contributions, 461–469, **462**, *463*
n-fold CV, 288
NN regression, *see* neural network (NN) regression
no-interaction model, 235–236, *237*, *251*, 254
non-independent observations, GLS, 317–322
non-jittered data, 335, *335*
non-linear conditional mean function, 64, *64*
normality assumption, 21
  car sales data, 113, *113*
  graphical methods, 111–113
  using testing methods, 114
normally distributed models, 358
null distribution, 325
  T statistic, 86–88
null (chance-only) model, *191*, 243–246, *244*

O

`observed` variable, 385
OLS estimates, *see* ordinary least squares (OLS) estimates
ordinal probit model, 361
ordinal regression models, 354, 358
ordinal variable, indicator variables to, 261–266
ordinal *Y*, models for, 351–358
ordinary least squares (OLS) estimates, 46, 48, *51*, 296, *302*, *304*
  estimator, 164
  -fitted mean function, *435*
  unbiasedness, 61–63, *63*
ordinary least-squares fit, *260*
ordinary linear regression, *248*
outliers, 49, 408–410
  en masse/Winsorization, 436–445
  in regression data, 411–413
  strategies for dealing, 425–427
  using heavy-tailed distributions, 427, *427*
  in *X* space, 413–416
  in *Y*|*X* space, 417–418
out-of-sample prediction accuracy, 287–288
  in SSE, *289*, 289–294

P

parameter interpretation, interaction models, 208–209
parameters, variance functions in, 310
penalized fit, variable selection, 284–285
  predicting crime rate, 285–287
Personal Assets data, 106, *107*, *297*
  and analysis, 15–16, *16*
p-hacking, 83, 488
piecewise linear regression, 257–258
Poisson probability distribution, 363, *363*
Poisson regression model, 361–365, *362*, *365*
  financial planners, predicting number, 365–370
polynomial approximations, noiseless function, 453–461
polynomial models, 202, *468*
"population" terminology, xiii, 8–10, **9**
practical significance, 95
practical *versus* statistical significance, 103–104
predicting charitable contributions, 461–469
predicting Hans' graduate GPA, 275–278, *282*
predicting loan repayment, 146–147
predictive association interpretation, 154
predictive MC, 198, 246
predictive modeling, 146
probit function, 357
probit regression model, 357–358
process distribution, 10
product complexity data, *204*
Production Cost data, 22–28, 100–101, *138*
  and analysis, *14*, 14–15
  set, **59**, *69*

proportional hazards regression model, 392–395
p-value, 88–90
  based testing methods, 95
  exact inferences, 88–91
  grossly incorrect interpretation of, 90–91

Q

quadratic functions, *202*
quadratic models, 216–217
  in one *X* variable, 202–204
  problems, 121
  in two/more *X* variables, 205
quadratic response surface, *205*
quadratic term, 213
quantile regression, *428*, 428–433, *430*, *431*, *432*
  estimates, 433–435
  for personal assets, 435–436

R

random effects, 269–270
randomness assumption, 18–19
random-*X* simulation, 27–28
random-*X* *versus* fixed-*X*, 12–14, **13**
recursive partitioning, 479
regime analysis, 258
regression analysis, censored data, 384–385
regression coefficient, 163–164
regression data
  analysis, 97
  set, **10**, 10–12, **12**, **59**
regression models, 1–2, 207, 329, 487
  assumptions of, 17–22
  in conditional distributions, 2–6
  for European stock market data, 12, **12**
  and generalization, 6–8
  matrix form, 172–173
  using `R` analyses, 14–17
  using simulation, 22–28
"regression to the mean" concept, 36, 39–40
regression trees, 471–474
  with multiple *X* variables, *479*, 481–485
  one *X* variable and one split, 474–477, *476*
replicability and preregistration, 376–377
residuals, evaluate normality, 114–115, *115*
Residual standard error, 67, 72, 418
right censored data, 379
robust measure of central tendency, 429
root mean square prediction errors (RMSEs), 283, **467**
R-squared statistic, 185–187, *187*
  adjusted, 188

S

SAE (sum of absolute errors), 51
sandwich estimate, 314
scale parameter, 438
second-degree polynomial, *456*
semi-parametric model, 392
sensitivity analysis method, 436
Shapiro-Wilk test, 114
shrinkage estimates, 269–270
σ^2, practical use, 65–69, *67*
signed distance, 406
significance level, 89
simulation models, 22, 57
single nominal variable, 219–222, *224*
social security number (SSN), 83–85
solid point, *406*, *412*
split value, *478*, 478–479
SSE (sum of squared errors), 46
`sseplot` function, 47, *47*
SSN (social security number), 83–85
standard errors, 70–73
  OLS estimates, 178–181
standardized value, 417
statistical analysis, 423
statistically insignificant, 78
statistical model, 17–18
statistics, MC, 195–198
sum of absolute errors (SAE), 51
sum of squared errors (SSE), 46
survivor function, 382, 389, *390*

T

Taylor's Theorem, 100, 288
ten-fold CV, 287
10th percentile, 432
third-degree polynomial, *456*
three-dimensional representation, tree model, *483*
three-D scatterplot, *194*
three nodes, NN model, *459*
Tobit model, 395–398, *397*, *398*
  days lost, predicting number, 398–401, *399*, *400*
transformation of *X* data, *118*, 118–124, *119*
trashcan experiment, 13–14
tree regression, 476, *476*, *477*, *481*, *482*, *484*
truncated data, *445*

truncation, 409–410, 436
Tukey's Method, 227
two nominal variables (two-way ANOVA), 246–255, *250*, *253*

U

ugly rule of thumb, 406–407, 415
unbiasedness, 59–61
unconditional distribution, *112*
unconditional unbiasedness, 175
uncorrelated errors assumption, 21
  using graphical methods, 109–110
  using testing methods, 111
universal approximators, 452–453
unobserved confounding variables, 266–268

V

variable cost, 14
variable inclusion principles, 198, 213–217
various data sets, Box-Cox analysis, 137

W

Wald standard errors, 326–327
weighted least squares (WLS) estimates, 296–302, *301*
  G-M, 303–304
White's test, 107
Widgets data, 14, 27
Winsorization, 426, 437–440, *439*
WLS estimates, *see* weighted least squares (WLS) estimates

X

X matrix, 169, 174
X variables, MC, 198

Z

zeta (ξ) parameter, 353
z-values, 406–407

Lightning Source UK Ltd.
Milton Keynes UK
UKHW050205181022
410458UK00030B/238